D0080994

INTRODUCTION *to* Food Science

INTRODUCTION *to* Food Science

Rick Parker

DELMAR
CENGAGE Learning

Australia • Brazil • Japan • Korea • Mexico • Singapore • Spain • United Kingdom • United States

Introduction to Food Science
Rick Parker

Business Unit Director: Susan L. Simpfenderfer

Executive Editor: Marlene McHugh Pratt

Acquisitions Editor: Zina M. Lawrence

Developmental Editors: Andrea Edwards Myers and Judith Boyd Nelson

Editorial Assistant: Elizabeth Gallagher

Executive Production Manager: Wendy A. Troeger

Production Manager: Carolyn Miller

Production Editor: Kathryn B. Kucharek

Executive Marketing Manager: Donna J. Lewis

Channel Manager: Nigar Hale

Cover Images: PhotoDisc

Cover Design: Dutton & Sherman Design

Composition: Larry O'Brien

Library of Congress Cataloging-in-Publication Data

Parker, R. O.
Introduction to food science / Rick Parker.
p. cm.
Includes bibliographical references and index.
ISBN 0-7668-1314-2
1. Food. 2. Food industry and trade. I. Title.
TP370.P33 2002
664—dc21
00-052395

ISBN-13: 978-0-7668-1314-4

ISBN-10: 0-7668-1314-2

Delmar Cengage Learning
5 Maxwell Drive
Clifton Park, NY 12065-2919
USA

Printed in the United States of America
11 12 16 15

Contents

Preface

As the title of the book suggests, science is an important component of the book. Food science as understood by humans represents a specific body of knowledge that approaches and solves problems by the scientific method–a continuous cycle of observation, hypothesis, predictions, experiments, and results. The *science* of food science is emphasized throughout the book.

Introduction to Food Science makes teaching easy. The chapters are based on a thorough, easy-to-follow outline. Developing a lesson plan is simple. Each chapter in this book starts with a list of learning objectives. These help the student identify what concepts are really important from all the information in the chapter. The beginning of each chapter also features a list of key terms. Knowing the meaning of these key terms is essential to reading and understanding the chapter. Many of the words are defined within the text, and all are defined in the extensive Glossary.

Throughout the book, numerous tables, charts, graphs, photographs, and illustrations provide quick and understandable access to information without wading through excess words. Students will quickly learn how to read these and grasp the information they contain. To help maintain interest, each chapter contains short, informative sidebars, and the book contains a series of photographs that feature the many aspects of food science.

Knowledge and information alone are useless unless they can be applied. In the Student Activities section at the end of each chapter, students and instructors will find opportunities for learning by doing. For more information the student can go to the list of Resources, including URLs for Internet sites. Also, at the end of each chapter students can test their understanding by answering the Review Questions.

The 28 chapters of *Introduction to Food Science* are divided into four sections. Section I, Introduction and Background, provides the necessary introductory and background information for understanding the science of foods. This includes chapters on a review of basic chemistry, nutrition and digestion, food composition and quality, unit operations, and food deterioration. These chapters are the foundation.

Section II, Preservation, groups the chapters that relate to the methods of food preservation including heat, cold, drying, radiant and electrical energy, fermentation, microorganisms, biotechnology, chemicals, and packaging. These chapters are the basics of food science.

Section III, Foods and Food Products, includes chapters on milk, meat, poultry and eggs, fish and shellfish, cereal grains, legumes and oilseeds, fruits and vegetables, fats and oils, candies and sweets, and beverages. These chapters are the application of food science.

Finally, Section IV, Related Issues, includes five chapters covering environmental concerns, food safety, regulations and labeling, world food needs, and career opportunities. These chapters represent the challenges of food science.

The Appendix contains helpful tables with information for converting units of measure, and for making contact with the industry and agencies. Most important, the Appendix contains a food composition table that students can learn to read and use. Also, the Appendix lists the Web addresses (URLs) for agencies and other food science-related Internet sites.

ACKNOWLEDGMENTS

Without the support of my wife Marilyn, this book would still be a dream or idea. As I have discovered, writing each textbook required the goodwill of 10 years of marriage. Since we have been married a little over 30 years and this is the fourth book for Delmar, this book was completed on "borrowed time!" She is a friend who critiques ideas, types parts of the manuscript, writes questions and answers, organizes artwork, takes photographs, and checks format. She is a partner in the production of a textbook and in all other aspects of my life.

Four young sons at home were patient and helpful during the time required for all the steps in the production of this book. My two teenage sons, Cole and Morgan, even proved helpful with some of the checking, searching, and revising. All four tolerated my distractions as long as we found frequent time together.

Unless otherwise noted, Marilyn or I took the photographs in the book.

Finally, I appreciate the support, understanding, help, and encouragement of Judith Boyd Nelson and Andrea Edwards Myers, and the rest of the Delmar team.

The author and Delmar wish to express their since appreciation to the following reviewers:

Daniel Andrews
Wauneta-Palisade High School
Wauneta, Nebraska

Roy Crawford
Lancaster High School
Lancaster, Texas

Diane Ryberg
Eau Claire North High School
Eau Claire, Wisconsin

Dr. Janelle Walter
Baylor University
Waco, Texas

ABOUT THE AUTHOR

R. O. (Rick) Parker grew up on an irrigated farm in southern Idaho. His love of agriculture guided his education. Starting at Brigham Young University, he received his bachelor's degree and then moved to Ames, Iowa, where he finished his Ph.D. in animal physiology at Iowa State University. After completing his Ph.D., he and his wife Marilyn and their children moved to Edmonton, Alberta, Canada, where he completed a postdoctorate at the University of Alberta. The next move was to Laramie, Wyoming, where he was a research and teaching associate at the University of Wyoming.

The author currently works as a division director at the College of Southern Idaho in Twin Falls. As director, he works with faculty in agriculture, information technology, and business management. He serves on the governor's Idaho Food Quality Assurance Commission, and he serves as Chair of the College-wide Curriculum Committee. Occasionally, he teaches a computer class, agriculture class, or a writing class.

He is also the author of three other Delmar texts–*Aquaculture Science, 2nd Edition, Introduction to Plant Science,* and *Equine Science.*

To Marilyn, wife, mother, partner, and one true love for over 30 years, through good times and bad, helping me enjoy the journey.

SECTION *One*

Introduction and Background

Chapter 1

Overview of Food Science

Objectives

After reading this chapter, you should be able to:

- Name the four parts of the food industry
- Describe consumer food buying trends
- Divide the food industry by major product lines
- Compare spending for food in the United States to that in other countries
- List four consumption trends
- Discuss trends in meal purchases
- Identify allied industries
- Explain how the food industry is international

Key Terms

allied industry
consumer
distribution
expenditures
manufacturing
marketing
per capita
production
tariffs
trends

Consumers vote every day in the marketplace with their dollars, and the market listens carefully to their votes. A continuous feedback exists from consumers, who respond to the offerings of marketers trying to meet the perceived wants of consumers. Changes in the makeup of the population, lifestyles, incomes, and attitudes on food safety, health, and convenience have drastically altered the conditions facing producers and marketers of food products. Food manufacturers and distributors maintain vigorous efforts to meet changing consumer demands.

PARTS OF THE FOOD INDUSTRY

The food industry is divided into four major segments:

- Production
- Manufacturing/processing
- Distribution
- Marketing

Production includes such industries as farming, ranching, orchard management, fishing, and aquaculture. Technologies involved in production of the raw materials include the selection of plant and animal varieties, cultivation, growth, harvest, slaughter, and the storage and handling of the raw materials. **Manufacturing** converts raw agricultural products to more refined or finished products. Manufacturing requires many unit operations and processes that are at the core of food technology. **Distribution** deals with those aspects conducive to product sales, including: product form, weight and bulk, transportation, storage requirements, and storage stability. **Marketing** is the selling of foods and involves wholesale, retail, institutions, and restaurants (Figure 1-1).

FIGURE 1-1
Food science makes possible healthy nutritious meals for everyone. *(Courtesy USDA Photography Library)*

These four divisions are rather artificial and the divisions actually overlap one another. Nevertheless the food industry requires planning and synchronization in all its divisions to be successful.

Another way of dividing the food industry is along major product lines:

- Cereals and bakery products
- Meats, fish, and poultry
- Dairy products
- Fruits and vegetables
- Sugars and other sweets
- Fats and oils
- Nonalcoholic beverages/alcoholic beverages

These divisions are typically where **consumer** consumption is measured and reported.

TRENDS

Although food spending has increased considerably over the years, the increase has not matched the gain in disposable income. As a result, the percentage of income spent for food has declined. The decline is the direct result of the income-inelastic nature of the aggregate demand for food: As income rises, the proportion spent for food declines (Figure 1-2). The **expenditures** for food require a large share of income when income is relatively low–in any country.

Americans spent only about 8 percent of their personal consumption expenditures for food to be eaten at home (Figure 1-3). This compares with 10 percent for Canada and 11 percent for the United Kingdom. In less developed countries, such as India and the Philippines, at-home food expenditures often account for more than 50 percent of a household's budget (Table 1-1).

Americans do not have the highest **per capita** income (the average Swiss income is higher). Yet, in relation to total per capita personal consumption expenditures, Americans spend the lowest percentage on food. Factors other than income alone influence food expenditures in developed nations. Thanks to abundant arable land and a varied climate, Americans do not have to rely as heavily on imported foods as do some other nations. The American farm-to-consumer distribution system is highly successful at moving large amounts of perishable food over long distances with a minimum of spoilage or delay. Finally, American farmers use a

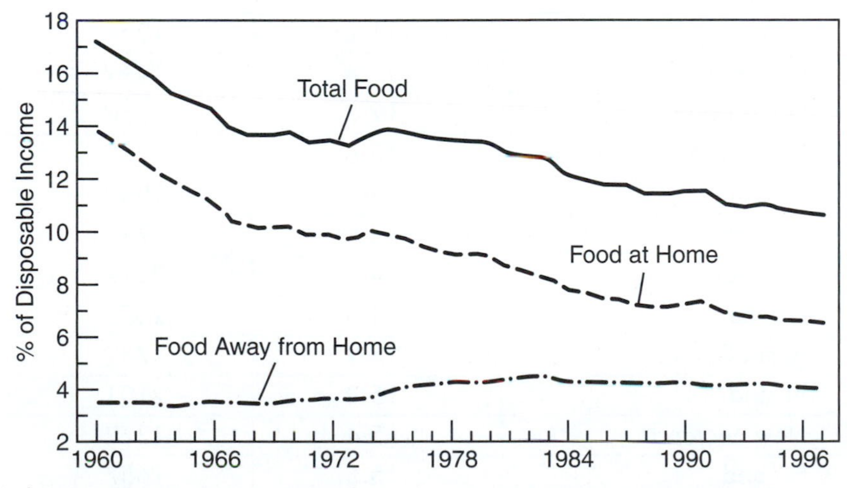

FIGURE 1-2
Over the years Americans have spent less of their disposable income for food. *(Source: USDA Economic Research Service)*

FIGURE 1-3

Total food expenditures increased over the years, but percent of income spent for food decreased. *(Source: U.S. Department of Labor, Bureau of Labor Statistics)*

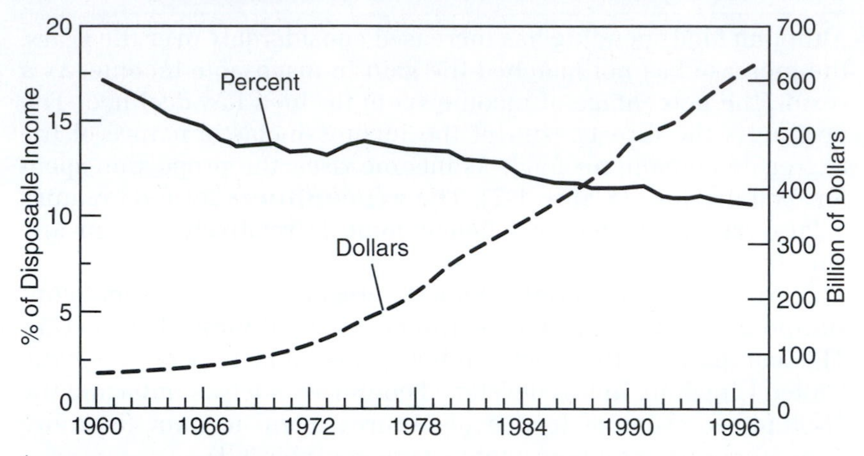

[1]Total food expenditures have been increasing, yet the percent of income spent for food has been decreasing.

TABLE 1-1 Percent of Total Personal Consumption Expenditures Spent on Food Consumed at Home by Selected Countries

Country	Percent of Total Personal Consumption Expenditures on Food	Personal Expenditures on Food (dollars)
United States	7.9	1381
Canada	10.3	1193
United Kingdom	11.2	1254
Netherlands	11.4	1499
Hong Kong	12.3	1550
Singapore	13.8	1279
Belgium	13.9	1949
Sweden	14.6	1784
Denmark	14.7	2212
France	14.8	2053
Australia	14.9	1732
Austria	15.3	2101
New Zealand	15.4	1372
Finland	15.5	1657
Puerto Rico	16.8	1141

TABLE 1-1 Percent of Total Personal Consumption Expenditures Spent on Food Consumed at Home by Selected Countries *(concluded)*

Country	Percent of Total Personal Consumption Expenditures on Food	Personal Expenditures on Food (dollars)
Italy	17.2	1890
Germany	17.3	2133
Japan	17.6	3842
Spain	18.2	1411
Ireland	19.0	1550
Iceland	19.0	2629
Norway	19.8	2449
Israel	20.5	1869
Portugal	23.2	3557
Thailand	23.3	317
Switzerland	24.4	3886
Fiji	24.4	918
Mexico	24.5	1101
South Africa	27.5	508
Hungary	27.5	653
Cyprus	28.3	1688
Korea, Republic of	29.1	1544
Colombia	29.6	263
Peru	31.0	788
Greece	31.7	1709
Malta	32.3	1496
Ecuador	32.8	307
Bolivia	34.8	240
Venezuela	38.2	737
Sri Lanka	49.3	108
India	51.3	100
Philippines	55.6	364

Source: Economic Research Service (ERS), United States Department of Agriculture (USDA).

tremendous wealth of agricultural information and state-of-the-art farming equipment. This allows them to produce food efficiently.

Consumption **trends** change over the years, and this influences what the food industry does. For example, since 1970 each American consumed, on average, 81 pounds more of commercially grown vegetables; 65 pounds more of grain products; 57 pounds more of fruit; 32 pounds more of caloric sweeteners (sugars and syrups); 13 pounds more of total red meat, poultry, and fish (boneless, trimmed equivalent); 17 pounds more of cheese; 13 pounds more of added fats and oils; 3 gallons more of beer; 70 fewer eggs; 10 gallons less of coffee; and 7 gallons less of milk (Figure 1-4 and Figure 1-5).

However, demand for individual foods is more responsive to prices as consumers substitute among alternative food commodities. Rising incomes increase expenditures on more expensive foods as consumers demand more convenience and quality. Short-period changes in consumption reflect mostly changes in supply rather than changes in consumer tastes. Demographic factors, such as changes in household size and in the age distribution of the population, can bring about changes in consumption.

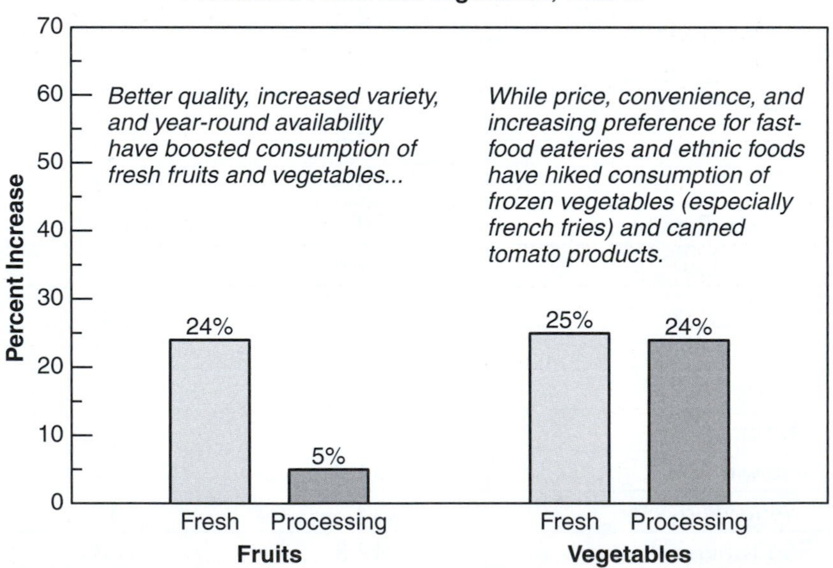

FIGURE 1-4

Consumption of fruits and vegetables has increased. *(Source:* USDA Economic Research Service)

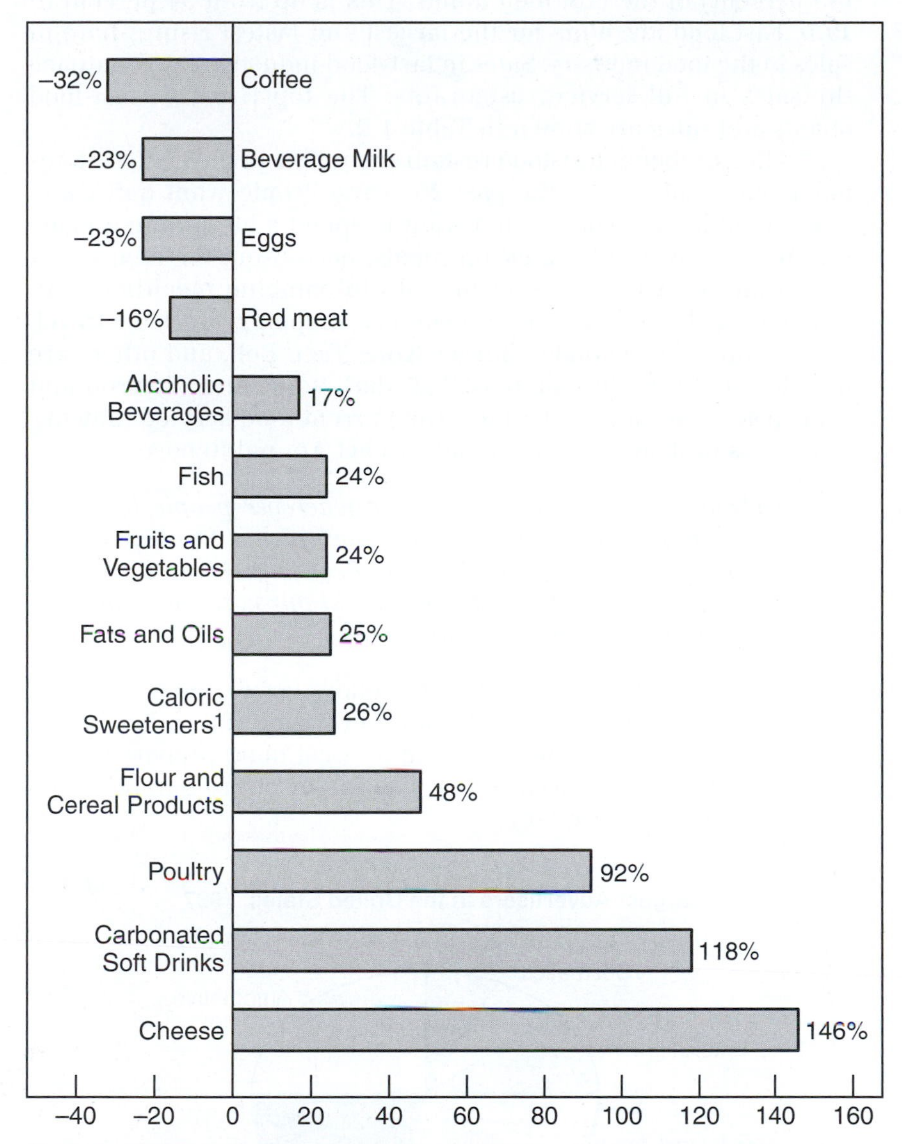

FIGURE 1-5

Per capita consumption of some foods has decreased, whereas others have dramatically increased. (*Source:* USDA Economic Research Service)

TABLE 1-2 Top 5 Restaurant Chains in the United States, 1997

Chain	Units
McDonald's	14,204
Subway	11,462
Pizza Hut	8,998
Burger King	7,584
KFC	5,120

Source: Food Review, September-December, 1998.

Away-from-home meals and snacks now capture almost half (45 percent) of the U.S. food dollar. This is up from 34 percent in 1970. Fast food accounts for the largest and fastest rising share of sales in the food industry. Sales in fast-food industries now outpace the sales in full-service restaurants. The top five U.S. fast-food chains and sales are shown in Table 1-2.

The number of fast-food restaurant outlets in the United States has risen steadily over the past 25 years. People want quick and convenient meals. They do not want to spend a lot of time preparing meals, traveling to pick up meals, or waiting for meals in a restaurant. Consumers seem to want to combine mealtime with time engaged in other activities such as shopping, work, or travel. For example, McDonald's, Burger King, Taco Bell, and others are now located in outlets such as Wal-Mart stores and Chevron and Amoco service stations. Perhaps the current food service industry strategy is best stated in McDonald's 1994 Annual Report:

> *McDonald's wants to have a site wherever people live, work, shop, play, or gather. Our Convenience Strategy is to monitor the changing lifestyles of consumers and intercept them at every turn. As we expand our customer convenience, we gain market share.*

The food industry is big and it employs large numbers of people in a variety of occupations because everyone eats, more people eat away from home (Table 1-3), and they eat more prepared products. Advertising (media) also plays an important role in influencing food trends (Figure 1-6).

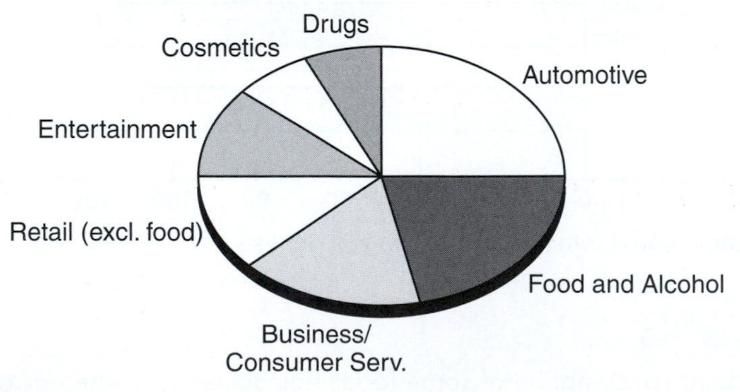

FIGURE 1-6

The food industry represents some of the largest advertisers. *(Source: USDA Economic Research Service)*

TABLE 1-3 Per Capita Food Expenditures

		U.S. Per Capita Food Expenditures					
		Current prices			1988 prices		
	U.S. Resident Population, July 1	At home	Away from home	Total	At home	Away from home	Total
Year	(millions)	(dollars)					
1953	158.242	298	96	394	1,042	508	1,550
1960	179.979	306	109	415	1,044	487	1,531
1965	193.526	318	135	453	1,025	545	1,570
1970	203.984	387	194	581	1,065	603	1,668
1975	215.465	567	316	883	1,011	676	1,687
1980	227.225	828	529	1,357	1,067	772	1,839
1985	237.924	1,009	710	1,719	1,120	797	1,917
1990	249.439	1,270	997	2,267	1,120	910	2,030
1991	252.127	1,309	1,016	2,325	1,123	897	2,020
1992	254.995	1,300	1,040	2,340	1,097	900	1,997
1993	257.746	1,303	1,090	2,393	1,046	927	1,973
1994	260.289	1,350	1,133	2,483	1,082	948	2,030
1995	262.765	1,387	1,167	2,554	1,079	954	2,033
1996	265.190	1,423	1,207	2,630	1,066	963	2,029
1997	267.744	1,463	1,266	2,729	1,054	982	2,036
1998	270.299	1,487	1,311	2,798	1,051	991	2,042

Source: Economic Research Service (ERS), United States Department of Agriculture (USDA), 1999.

ALLIED INDUSTRIES

Many companies do not sell food directly but they are deeply involved in the food industry. These are called **allied industries**. Allied industries produce nonfood items that are necessary for marketing food. The packaging industry is a good example. Some specific examples include cans, food color and flavor, paper products, and plastic products (see Figure 1-7). Chemical manufacturers represent another group of allied industries. They supply the acidulants, preservatives, enzymes, stabilizers, and other chemicals used in foods.

FIGURE 1-7
A part of the food industry includes companies that manufacture packaging materials.

New food products and safe foods require new food processing methods and systems. The food machinery and equipment manufacturers are another aspect of the allied industries. They develop pasteurizers, evaporators, microwave ovens, infrared cookers, freeze-drying systems, liquid nitrogen freezers, instrumentation, and computer controls (see Chapter 7).

Finally, keeping the food supply safe and healthy and keeping consumers informed require monitoring and regulatory agencies such as the Food & Drug Administration (FDA), lawyers, consumer action and information agencies, and other regulatory agencies.

If recent trends in the U.S. food industry continue, food production may be increasingly dominated by firms exercising control over several stages of food production. Vertical coordination seems to be a way of the future. This refers to the way products are acquired or traded in a market. Food industry firms form three basic types of vertical coordination:

- Open production. A firm purchases a commodity from a producer at a market price determined at the time of purchase.
- Contract production. A firm commits to purchase a commodity from a producer at a price formula established in advance of the purchase.
- Vertical integration. A single firm controls the flow of a commodity across two or more stages of food production.

The food industry has traditionally operated in an open production system. However, more discriminating consumers, plus new technological developments that allow farm product differentiation, are contributing to a decrease in open production and an increase in contract production and vertical integration. Also fueling this trend are changing demographics and the increasing value of a person's time, both of which have contributed to consumer preferences for a wide variety of safe, nutritious, and convenient food products.

Providing food products with specific characteristics preferred by more discriminating consumers will likely involve increasingly more detailed raw commodity products, such as a frying chicken of a specific weight and size, or a corn kernel with a specific protein content. This effort to carefully tailor raw commodities with processing in mind is already underway in some food industries, accompanied by changes in vertical coordination.

INTERNATIONAL ACTIVITIES

Food is a global commodity. Foods are traded and shipped around the world. For example, the modern grocery store sells food from all over the world. These foods might include cheeses from Europe, beef from Australia, strawberries from Mexico, and apples from Argentina. Also, many U.S. companies have established subsidiaries in other countries, and fast-food companies such as McDonald's and Pizza Hut are opening outlets all over the world. Major food companies such as Kraft-General Foods, CPC International, H.J. Heinz, Borden, Campbell Soup, Nabisco Brands, Coca-Cola, Pepsico, Beatrice Companies, Ralston Purina, and General Mills all have extensive overseas operations. Table 1-4 lists the top 50 international food processing firms, their headquarters, and annual sales.

The processed food sector is a major participant in the global economy. The United States accounts for about one fourth of the industrialized world's total production of processed foods. Six of the largest 10, and 21 of the largest 50 food processing firms in the world are headquartered in the United States. Through a combination of imports and exports of foods and food ingredients, foreign production by U.S. food firms, host production by foreign food firms, and other international commercial strategies, the U.S. processed foods market is truly global in scope. Easily recognized U.S. food brands are so well received internationally that many consumers in other countries accept them as leading local brands. In terms of international trade, the processed foods sector surpasses agricultural commodities by a considerable margin.

TABLE 1-4 Top 50 of the World's Largest Food Processing Firms

Rank	Company	Headquarters	Processed Food Sales ($ billions)	Total Company Sales ($ billions)
1	Nestle S.A.	Switzerland	38.8	40.4
2	Philip Morris/Kraft Foods	USA	33.4	64.1
3	Unilever	UK/Netherlands	26.7	54.1
4	ConAgra	USA	24.8	24.8
5	Pepsico	USA	19.1	30.4
6	Cargill	USA	18.7	51.0
7	Coca-Cola	USA	18.0	18.0
8	Danone S.A.	France	14.2	14.2
9	Archer Daniels Midland	USA	13.3	13.3
10	Mars, Inc.	USA	13.0	13.0
11	Grand Metropolitan Plc.	UK	12.7	14.5
12	IBP, Inc.	USA	12.5	12.5
13	Kirin Brewery Co. Ltd.	Japan	11.6	11.6
14	CPC International Inc.	USA	9.8	9.8
15	Anheuser-Busch Co. Inc.	USA	9.6	10.3
16	Sara Lee	USA	9.4	18.6
17	Associated British Food	UK	9.2	9.2
18	H.J. Heinz	USA	9.1	9.1
19	Asahi Breweries	Japan	9.1	9.1
20	Eridania Beghin-Say	France	9.1	9.1
21	R.J.R. Nabisco Inc.	USA	8.3	16.0
22	Norvartis	Switzerland	8.1	25.9
23	Cadbury Schweppes Plc.	UK	7.7	7.7
24	Campbell Soup	USA	7.7	7.7
25	Guinness Plc.	UK	7.6	7.6

TABLE 1-4 Top 50 of the World's Largest Food Processing Firms *(concluded)*

Rank	Company	Headquarters	Processed Food Sales ($ billions)	Total Company Sales ($ billions)
26	Kellogg Company	USA	7.0	7.0
27	Tate & Lyle Plc.	UK	6.8	8.3
28	Nippon Meat Packers Inc.	Japan	6.8	6.8
29	Tyson Foods Inc.	USA	6.4	6.4
30	Heineken	Netherlands	6.4	6.4
31	Quaker Oats Company	USA	6.4	6.4
32	Allied-Domercq Plc.	UK	6.2	8.7
33	Suntory Ltd.	Japan	6.1	6.1
34	South Africa Breweries	S. Africa	5.7	5.7
35	Sapporo Breweries Ltd.	Japan	5.6	5.6
36	United Biscuits	UK	5.5	5.5
37	General Mills Inc.	USA	5.4	5.4
38	Ajinomoto Co., Inc.	Japan	5.3	6.3
39	Hillsdown Holdings Plc.	UK	5.2	5.2
40	Yamazaki Baking Co. Ltd.	Japan	4.7	4.7
41	Joseph E. Seagram & Sons Inc.	Canada	4.7	7.2
42	Snow Brand Milk Prod. Co. Ltd.	Japan	4.5	4.5
43	Besnier SA	France	4.5	4.5
44	Sudzucker Group	Germany	4.4	NA
45	Brahma	Brazil	4.4	4.4
46	Dalgety Plc.	UK	4.1	7.0
47	Mid America Dairymen Inc.	USA	4.1	4.1
48	The Proctor & Gamble Co.	USA	4.0	33.4
49	Ralston Purina Co.	USA	3.9	6.1
50	Dole Food Co.	USA	3.8	3.8

Source: Economic Research Service (ERS), United States Department of Agriculture (USDA), 1997.

World trade imports are also represented by products not grown in the United States such as coffee, tea, cocoa, and spices. Worldwide the demand for cereal grains and soybeans increased, so the United States is the largest exporter of these foods.

Aside from the worldwide demand for food and food products, the recent trends to decrease trade **tariffs** has stimulated the international activities in the food industry.

Improvements in transportation and communication have also increased the international activities of food industries. Products move around the world by air freight in hours or days and communications take place around the world in seconds.

National infrastructure policies affect the ability of a nation's firms to pursue global marketing strategies. For processed foods, particularly important linkages exist with the communications and transportation sectors. Technical innovations in both communications and transportation enhance efficiency in the production and distribution of processed foods, improve managerial control and responsiveness, and help identify and fulfill new commercial opportunities. In the United States, policies that have reduced direct government control in these sectors and that have fostered evolution of competitive communications and transportation industries are tied directly to international commercial gains in processed foods.

NEW FRUIT COMING TO A SUPERMARKET NEAR YOU

All across Southeast Asia, people eat the extremely popular durian or stinkfruit *(Durio zibethinus)*. In Thailand, it is called the King of Fruits. Though the flesh of the durian is sweet and mild, the aroma is so strong that many westerners cannot eat it without gagging. Westerners who try them are often flabbergasted at the smell, which has been described as a blend of decayed onion, turpentine, garlic, Limburger cheese, and resin. Eating it on commercial flights has been banned by several Asian airlines.

This spiky-skinned, brownish green fruit can grow as big as cantaloupe. It will soon be available in the United States and Europe.

For more information about the durian, visit these Web sites:

<www.sunfood.net/fruits/durian.html>
<www.durian.net/durian_daniel_eat.htm>
<www.proscitech.com.au/trop/d.htm>

RESPONSIVENESS TO CHANGE

Total food consumed by each individual (per capita food consumption) changes little from year to year. The kinds of foods consumed change continually, contributing to competition and making the industry change frequently. Over 10,000 new food products are introduced each year.

The kinds of foods people eat change in response to many influences, such as demographic shifts; supply of ingredients; availability and costs of energy; politics; scientific advances in nutrition, health, and food safety; and changes in lifestyle.

Attitudes toward foods change consumption patterns. This means that the industry must respond too. For example, fresh and frozen fruit consumption increased about 24 percent the last 27 years, whereas red meat consumption declined about 16 percent. Per capita consumption of poultry, carbonated soft drinks, and cheese increased markedly. During this time, consumer attitudes about fat, cholesterol, and fiber changed.

Changes in government regulation of food additives, food composition standards, and labeling also require the food industry to be responsive to change. Finally, technical innovations such as ingredient modifications, new processing methods, new packaging methods, and cooking advances also force the industry to respond.

INTERRELATED OPERATIONS

Food production relies on a highly advanced and organized industry. Decisions to produce a product are not random. The industry is a systematic and rhythmic process. Throughout production, manufacturing/processing, distribution, and marketing, the costs and availability are carefully monitored and controlled. Further, because the industry is high-volume, low-markup, small losses anywhere along the chain can mean large losses to the food producer.

Any trend toward contract production and vertical integration, as opposed to open production, implies that firms at one stage of production exert more control over the quality of output at other stages. For example, pasta processors who prefer a specific type of wheat for a specific type of pasta gain control over planting decisions or seed selection that were previously made by farmers who sold their wheat on the spot market. Farmers are compensated for relinquishing control through bonuses for quality and through reduced uncertainty.

Recent changes in vertical coordination have been accompanied by an increase in concentration in the food sector. These developments have raised two primary policy concerns: market power in the processing sector and environmental protection.

Summary

The food industry is divided into production, manufacturing/processing, distribution, and marketing. The industry is highly responsive to change, and interrelated with others. Consumers drive the food industry, and to some extent the food industry drives the consumer, making changes in food consumption, food types, and meals purchased. Food is now a global commodity due to changes in export/import laws, transportation, and processing and communication.

Review Questions

Success in any career requires knowledge. Test your knowledge of this chapter by answering these questions or solving these problems.

1. Away-from-home meals captures ______ percent of the U.S. food dollar.
2. Why have the international activities of food industries increased?
3. Name all seven product lines along which the food industry is divided.
4. List the four artificial divisions of the food industry.
5. Consumption of cheese has __________, whereas consumption of red meat has __________ over the last 27 years.
6. List four reasons that influence people and the kind of food they eat.
7. About __________ new food products are introduced each year.
8. Explain how the consumer votes in the marketplace.
9. Define an allied industry.
10. Compare the spending on food in the United States to that of Spain and Greece.

Student Activities

1. Make a list of foreign foods sold in a supermarket. Bring this list to class for discussion.
2. Pick one of the food companies listed in Table 1-4 and visit their Web site. Report on the type of consumer information available.
3. Search the Internet or some other resource and look for trade tariffs that affect food. Report on your findings.
4. Conduct a contest to see who can bring the most creative and novel food packaging to class.

Resources

Potter, N. N., and J. H. Hotchkiss. 1995. *Food science,* 5th ed. New York: Chapman and Hall.

Vaclavik, V. A., and E. W. Christina. 1999. *Essentials of food science.* Gaithersburg, MD: Aspen Publishers, Inc.

Vieira, E. R. 1996. *Elementary food science,* 4th ed. New York: Chapman and Hall.

Internet

Internet sites represent a vast resource of information. The URLs (uniform resource locator) for the World Wide Web sites can change. Using one of the search engines on the Internet such as Yahoo!, HotBot, AltaVista, Excite, Dogpile, About, or Google, find more information by searching for these words or phrases: food industry, food technology, imported/exported food, food tariff, food expenditures, international food companies, Food & Drug Administration, names of particular food companies or brands. Also, Table A-7 provides a listing of some useful Internet sites that can be used as a starting point.

Chapter 2

Review of Chemistry

Objectives

After reading this chapter, you should be able to:

- Describe the chemical properties of an element
- Name the three elements most important to life
- Explain how covalent, hydrogen, and ionic bonds are formed
- Define a molecule
- Identify symbols for hydroxyl, amino, ammonia, methyl, and carboxyl
- Discuss oxidation-reduction reactions
- Describe the two divisions of metabolism

Key Terms

anabolism
atomic number
catabolism
covalent bond
electrons
electron-transfer
electronegativity
element
hydrogen bonds
ionic bonds
metabolism
molecule
neutrons
oxidation-reduction reactions
protons
Van der Waals bond

The study of food science requires an understanding of simple chemistry and organic chemistry. Carbohydrates, lipids, and proteins are all important to food science, and they are composed of smaller building blocks. This chapter contains a review of important chemical interactions and concepts encountered in food science.

ELEMENTS

The atom is the smallest unit of an **element** that still exhibits the properties of that element. Atomic structure is shown in Figure 2-1. Atoms consist of a nucleus containing **protons** (1 mass unit, positive charge) and **neutrons** (1 mass unit, no charge). Surrounding the nucleus are a number of **electrons**. In its elemental state, the number of electrons of an atom equals the number of protons, and the atom is electrically neutral. Electrons travel around the nucleus at very high speed, each one traveling in one of several possible energy levels (also called shells, orbits, or orbitals). Each level has a maximum number of electrons–two in the first, eight in the second, and so on. The energy level nearest the nucleus is the one with the lowest energy electrons. Orbitals further from the nucleus contain higher energy electrons.

The **atomic number** of an atom is the total number of protons. The atomic weight of an atom is the total number of protons plus neutrons. The periodic table shows that columns in the table contain elements with the same number of electrons in their outermost energy levels or have full energy levels. On a periodic table the elements important to life include carbon, hydrogen, nitrogen, and oxygen (Figure 2-2).

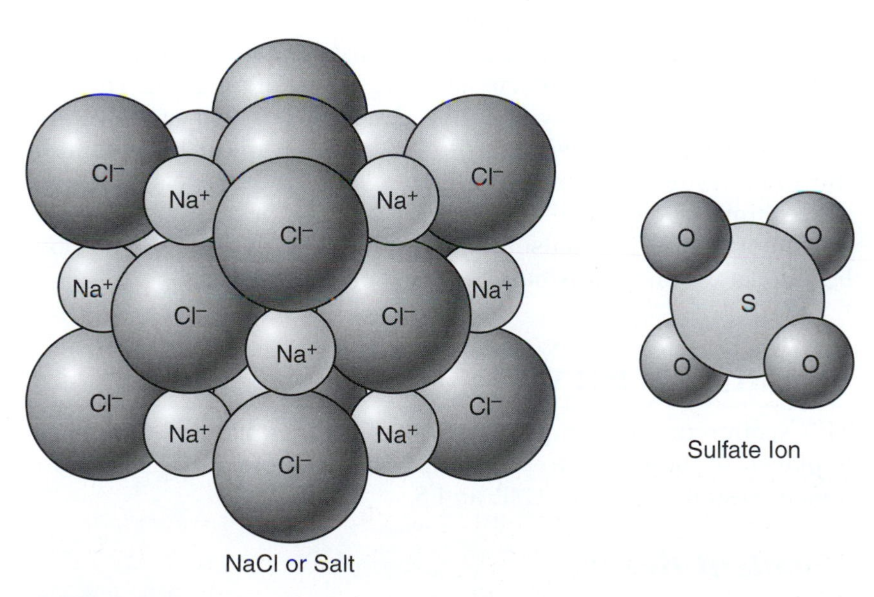

FIGURE 2-1

Atoms combined to form molecules represented by nucleus and electron cloud.

Periodic Table of the Elements

1 H																	2 He
3 Li	4 Be											5 B	6 C	7 N	8 O	9 F	10 Ne
11 Na	12 Mg											13 Al	14 Si	15 P	16 S	17 Cl	18 Ar
19 K	20 Ca	21 Sc	22 Ti	23 V	24 Cr	25 Mn	26 Fe	27 Co	28 Ni	29 Cu	30 Zn	31 Ga	32 Ge	33 As	34 Se	35 Br	36 Kr
37 Rb	38 Sr	39 Y	40 Zr	41 Nb	42 Mo	43 Tc	44 Ru	45 Rh	46 Pd	47 Ag	48 Cd	49 In	50 Sn	51 Sb	52 Te	53 I	54 Xe
55 Cs	56 Ba	57 La	72 Hf	73 Ta	74 W	75 Re	76 Os	77 Ir	78 Pt	79 Au	80 Hg	81 Tl	82 Pb	83 Bi	84 Po	85 At	86 Rn
87 Fr	88 Ra	89 Ac	104 Rf	105 Db	106 Sg	107 Bh	108 Hs	109 Mt	110 Uun								

58 Ce	59 Pr	60 Nd	61 Pm	62 Sm	63 Eu	64 Gd	65 Tb	66 Dy	67 Ho	68 Er	69 Tm	70 Yb	71 Lu
90 Th	91 Pa	92 U	93 Np	94 Pu	95 Am	96 Cm	97 Bk	98 Cf	99 Es	100 Fm	101 Md	102 No	103 Lr

FIGURE 2-2

Periodic Table of the elements provides information about the atoms.

The element is determined by the number of protons in the atom. Isotopes (different forms) of elements are determined by the number of neutrons in the atom.

Chemical properties of an element are determined by the number of electrons in the outermost energy level of an atom. The outermost level interacts with other atoms when two atoms come together. Elements in vertical columns of the periodic table contain the same number of electrons in their outermost levels, and they share similar chemical properties. Arsenic occurs right below phosphorus and is a poison. Silicon appears right below carbon, but life as we know it is based on carbon, not silicon.

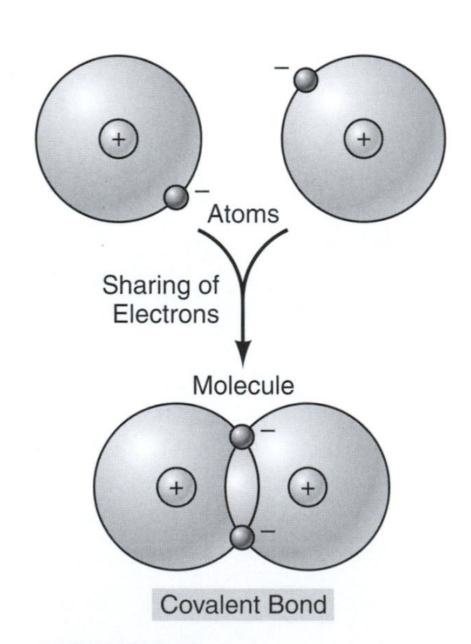

FIGURE 2-3

Sharing of electrons in a covalent bond.

CHEMICAL BONDS

This section provides a quick review of chemical bonds. Emphasis is placed on bonds between the six major elements found in biological systems: H, C, N, O, P, and S.

Covalent Bonds

Covalent bonds are the strongest chemical bonds, and are formed by the sharing of a pair of electrons (Figure 2-3). Once formed, covalent bonds rarely break spontaneously, because of simple energetic considerations–the thermal energy of a **molecule** at room temperature is much lower than the energy required to break a covalent bond. Covalent bonds can be single, double, and triple.

Carbon-carbon bonds are unusually strong and stable covalent bonds.

The major organic elements have standard bonding capabilities, as shown in Figure 2-4.

Covalent bonds can also have partial charges when the atoms involved have a different **electronegativity**. Water is perhaps the most obvious example of a molecule with partial charges. The symbols delta+ and delta– are used to indicate partial charges.

Oxygen, because of its high electronegativity, attracts the electrons away from the hydrogen atoms, resulting in a partial negative charge on the oxygen and a partial positive charge on each of the hydrogens.

The possibility of **hydrogen bonds** (H-bonds) is a consequence of partial charges.

Hydrogen Bonds

Hydrogen bonds are formed when a hydrogen atom is shared between two molecules (Figure 2-5).

Hydrogen bonds have polarity. A hydrogen atom covalently attached to a very electronegative atom (N, O, or P) shares its partial positive charge with a second electronegative atom (N, O, or P). One common example involves the hydrogen bonding between water molecules.

Hydrogen bonds are frequently found in proteins and nucleic acids (as in DNA), and by reinforcing each other serve to keep the protein (or nucleic acid) structure secure. Because the hydrogen atoms in the protein could also hydrogen bond to the surrounding water, the relative strength of protein-protein hydrogen bonds versus protein-water bonds is smaller.

Ionic Bonds

Ionic bonds are formed when there is a complete transfer of electrons from one atom to another, resulting in two ions, one positively charged and the other negatively charged (Figure 2-6, page 26). For example, when a sodium atom (Na) donates the one electron in its outer valence shell to a chlorine (Cl) atom, which needs one electron to fill its outer valence shell, NaCl (table salt) results. The symbol for sodium chloride is Na+Cl–. Ionic bonds are often 4 to 7 kcal/mol in strength.

Functional Groups	Class of Molecules	Formula	Example
Hydroxyl — OH	Alcohols	R — OH	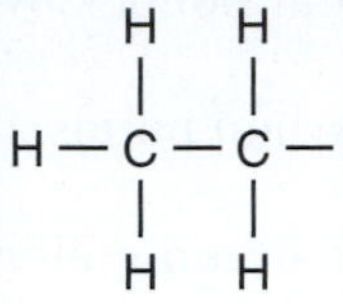 Ethanol
Carboxyl — CHO	Aldehydes	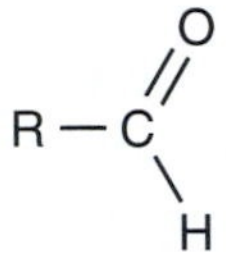 R — C(=O) — H	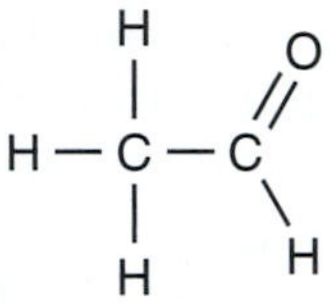 Acetaldehyde
\CO/	Ketones	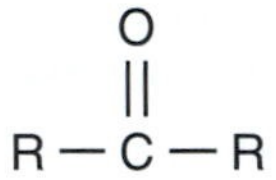 R — C(=O) — R	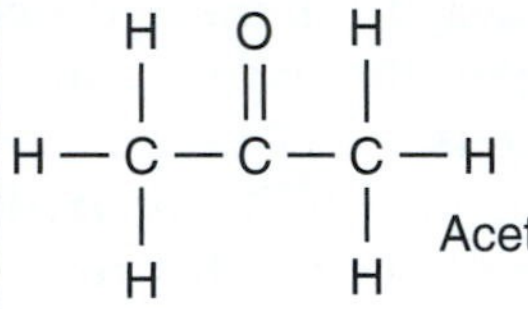 Acetone
Carboxyl — COOH	Carboxylic Acids	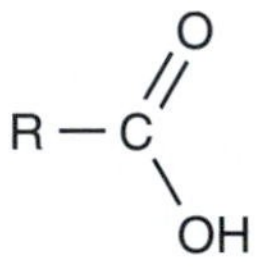 R — C(=O) — OH	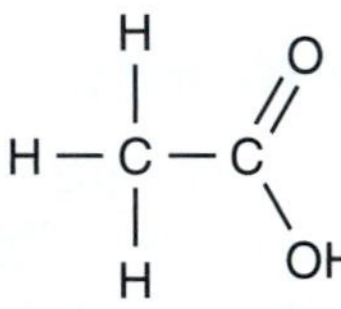 Acetic Acid
Amino — NH_z	Amines	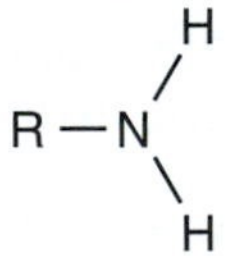 R — N(H) — H	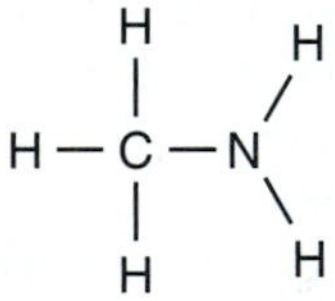 Methylamine
Phosphate — OPO_3^{-2}	Organic Phosphates	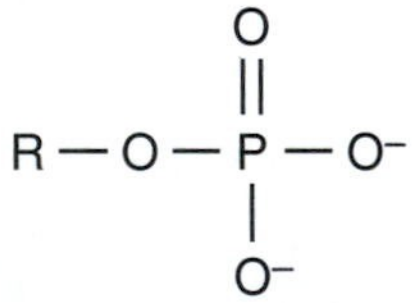 R — O — P(=O)(— O^-) — O^-	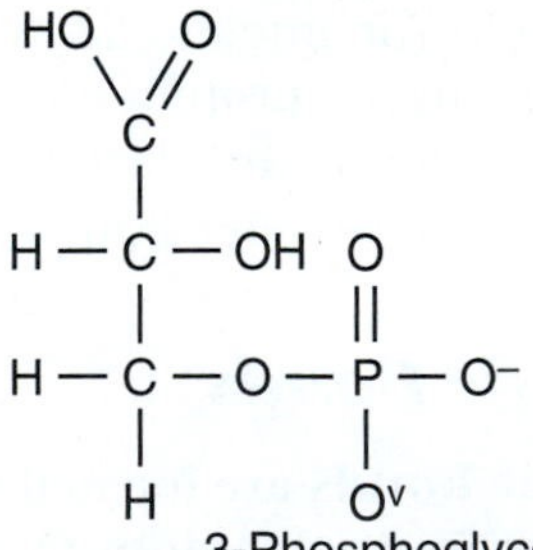 3-Phosphoglyceric acid
Sulfhydryl — SH	Thiols	R — SH	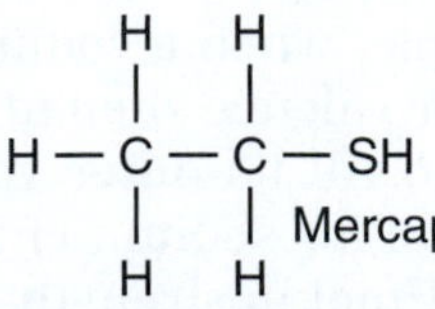 Mercaptoethanol

FIGURE 2-4

Representation of functional groups found in organic molecules.

HYDROGEN BONDS

Hydrogen bonds form when a hydrogen atom is "sandwiched" between two electron-attracting atoms (usually oxygen or nitrogen).

Hydrogen bonds are strongest when the three atoms are in a straight line:

O—H|||||||||O   N—H|||||||||O

Examples in macromolecules:

Amino acids in polypeptide chain hydrogen-bonded together.

Two bases, G and C, hydrogen-bonded in DNA or RNA.

HYDROGEN BONDS IN WATER

Any molecules that can form hydrogen bonds to each other can alternatively form hydrogen bonds to water molecules. Because of this competition with water molecules, the hydrogen bonds formed between two molecules dissolved in water are relatively weak.

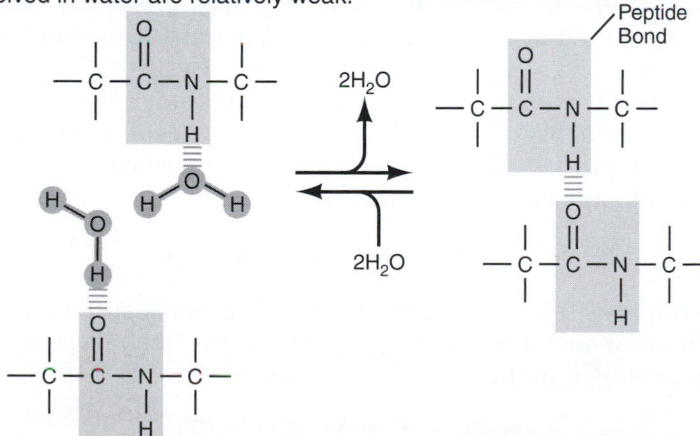

FIGURE 2-5

Description of a hydrogen bond.

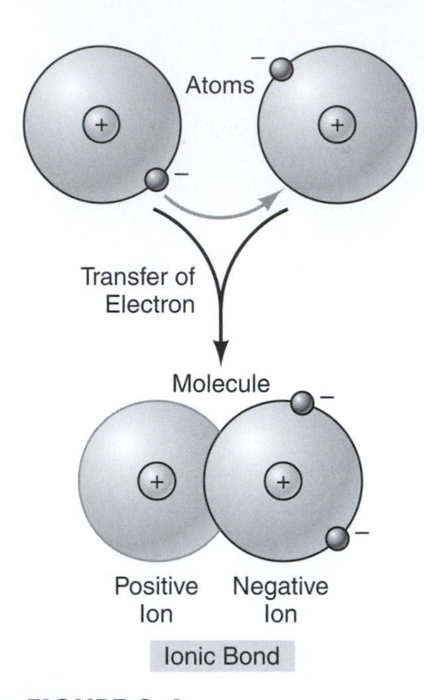

FIGURE 2-6

Representation of an ionic bond.

Van der Waals Bonds

Van der Waals bonds or interactions are very weak bonds (generally no greater than 1 kcal/mol) formed between nonpolar molecules or nonpolar parts of a molecule (Figure 2-7). The weak bond is created because a C-H bond can have a transient dipole and induce a transient dipole in another C-H bond.

MOLECULES

Molecules are the smallest identifiable unit into which a pure substance can be divided and still retain the composition and chemical properties of that substance. The division of a sample of a substance into progressively smaller parts produces no change in either its composition or its chemical properties.

Molecules are held together by shared electron pairs, or covalent bonds. Such bonds are directional, meaning that the atoms adopt specific positions relative to one another so as to maximize the bond strengths.

The molecular weight of a molecule is the sum of the atomic weights of its component atoms. If a substance has a molecular weight 32, then 32 grams of the substance is termed one mole. For example, the molecular weight of NaCl is 58.5 (23 + 35.5) and one mole of NaCl weighs 58.5 grams.

REACTIONS

The rusting of metals, the process involved in photography, the way living systems produce and use energy, and the operation of a car battery are but a few examples of a very common and important type of chemical reaction. These chemical changes are all classified as **electron-transfer** or **oxidation-reduction reactions**.

The term oxidation was derived from the observation that almost all elements reacted with oxygen to form compounds called oxides. A typical example is the corrosion or rusting of iron, as described by the chemical equation:

$$4Fe + 3O_2 \longrightarrow 2Fe_2O_3$$

Reduction was the term originally used to describe the removal of oxygen from metal ores, which reduced the metal ore to pure metal, as shown in this chemical reaction:

$$2Fe_2O_3 + 3C \longrightarrow 3CO_2 + 4Fe$$

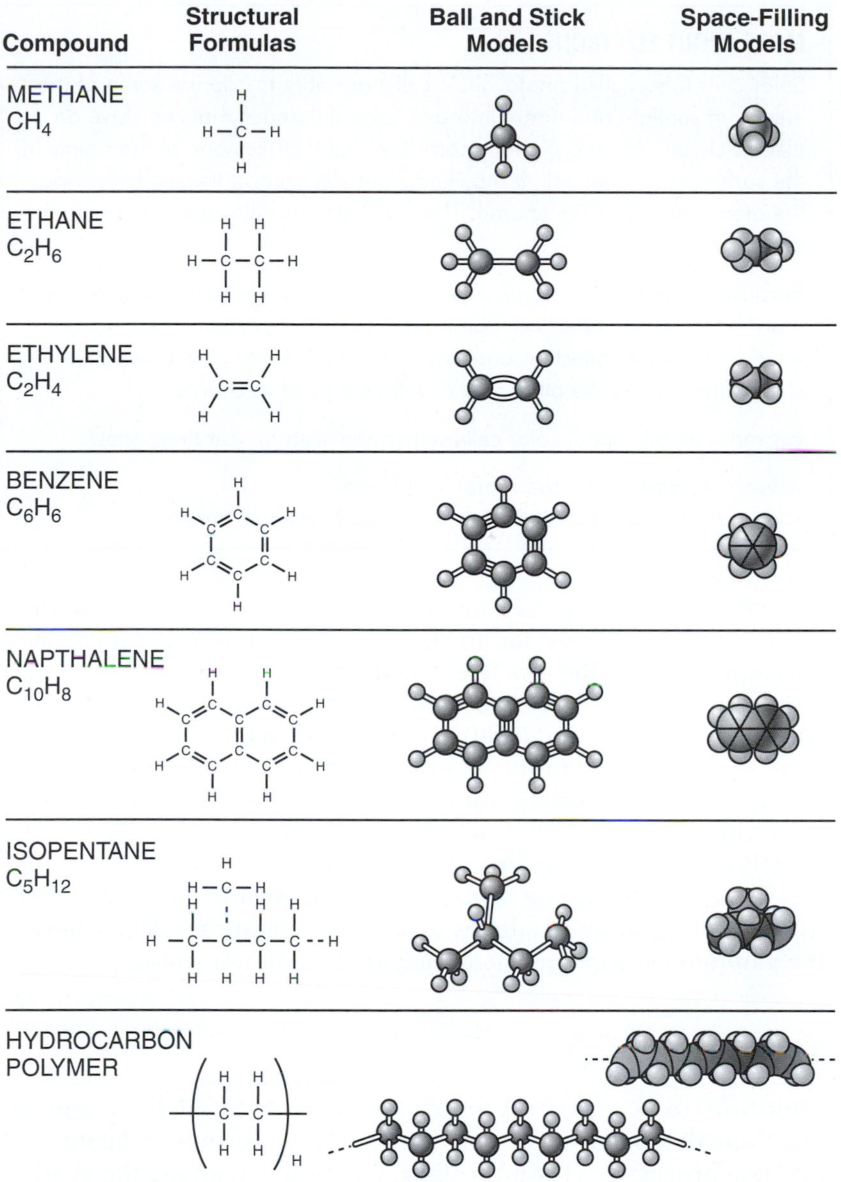

Compound	Structural Formulas	Ball and Stick Models	Space-Filling Models
METHANE CH_4			
ETHANE C_2H_6			
ETHYLENE C_2H_4			
BENZENE C_6H_6			
NAPTHALENE $C_{10}H_8$			
ISOPENTANE C_5H_{12}			
HYDROCARBON POLYMER			

FIGURE 2-7
Four different methods of describing organic molecules.

Based on these two examples, oxidation can be defined very simply as the addition of oxygen. Reduction can be defined as the removal of oxygen.

The logical starting point in the discussion of oxidation-reduction reactions is the atom, and the terms and conventions used by chemists in describing this phenomenon. All atoms are electrically

MORE ABOUT ELECTRONS

Solar cells (also called photovoltaic cells) are able to capture some of the energy in sunlight and turn it into a voltage difference that can drive an electric circuit. When a photon (particle of light) strikes one of the atoms in the surface of a solar cell, it may knock an electron off the atom, leaving the atom with a positive charge. The freed electron flies away, carrying the photon's energy.

Because of the pattern of impurities in the solar cell, electrons move much more easily in one direction than in the opposite direction. The electrons freed by the light collect on one side of the cell, developing a negative charge there while the other side develops a positive charge.

For more details about solar cells search the Web or visit these sites:

<www.howstuffworks.com/solar-cell.htm>
<www3.umassd.edu/Public/Exhibit/enl600/assign6.html>

neutral even though they are comprised of charged, subatomic particles. The terms oxidation state or oxidation number have been developed to describe this "electrical state" of the atom. The oxidation state or oxidation number of an atom is simply defined as the sum of the negative and positive charges in an atom. Because every atom contains equal numbers of positive and negative charges, the oxidation state or oxidation number of any atom is always zero.

Oxidation-reduction reactions always involve a change in the oxidation state of the atoms or ions involved. This change in oxidation state is due to the loss or gain of electrons. The loss of electrons from an atom produces a positive oxidation state, whereas the gain of electrons results in negative oxidation states.

METABOLISM

Metabolism is a general term used to refer to all the chemical reactions that occur in a living system. Metabolism can be divided into two processes: (1) **anabolism**, reactions involving the synthesis of compounds, and (2) **catabolism**, reactions involving the breakdown of compounds. In terms of oxidation-reduction principles, anabolic reactions are primarily characterized by reduction reactions, such as the dark reaction in photosynthesis where carbon dioxide is reduced to form glucose. Catabolic reactions are primarily oxidation reactions. Although catabolism involves many separate reactions, an example of such a process can be described by the oxidation of glucose, as shown here:

$$C_6H_{12}O_6 + 6\ O_2 \longrightarrow 6\ CO_2 + 6\ H_2O + \text{Energy}$$

In this reaction, the carbon atoms in glucose are oxidized, undergoing an increase in oxidation state (each carbon loses two electrons) as they are converted to carbon dioxide. At the same time, each oxygen atom is reduced by gaining two electrons when it is converted to water. Part of the energy is released as heat, and the remainder is stored in the chemical bonds of energetic compounds such as adenosine triphosphate (ATP) and nicotinamide adenine dinucleotide (NADH).

Catabolic reactions can be divided into many different groups of reactions called catabolic pathways. In these pathways, glycolysis, the citric acid cycle, and electron transport (Figure 2-8), the carbon atoms are slowly oxidized by a series of reactions that gradually modify the carbon skeleton of the compound as well as the oxidation state of carbon. Coupled to these reactions are other reversible oxidation-reduction reactions designed to capture the energy released and temporarily store it within the chemical bonds of compounds called adenosine triphosphate (ATP) and nicotamide dinucleotide (NADH). These compounds are then use to provide energy for driving the cellular machinery.

ORGANIC CHEMISTRY

Nutrients are chemicals and require an understanding of organic chemistry. Table 2-1 indicates some of the basics required for an understanding.

Organic chemistry involves carbon-containing molecules, and all carbon atoms have four bonds to account for. In carbohydrates, fats, and proteins, each carbon can connect to:

- Another carbon
- A hydroxyl
- A hydrogen
- An amino group
- An oxygen (double bond)

Table 2-1 Chemical Symbols and Representations

Symbol	Represents
C	Carbon atom
H	Hydrogen atom
O	Oxygen atom
N	Nitrogen atom
OH	Hydroxyl (alcohol)
NH_3	Ammonia
NH_2	Amino group
CH_3	Methyl group
COOH	Carboxyl (acid)

Summary

Atoms are the smallest unit of an element that still shows properties of the element. Atoms bond with other atoms in chemical bonds such as covalent bonds, hydrogen bonds, ionic bonds, and Van der Waals bonds. Molecules are the smallest identifiable units

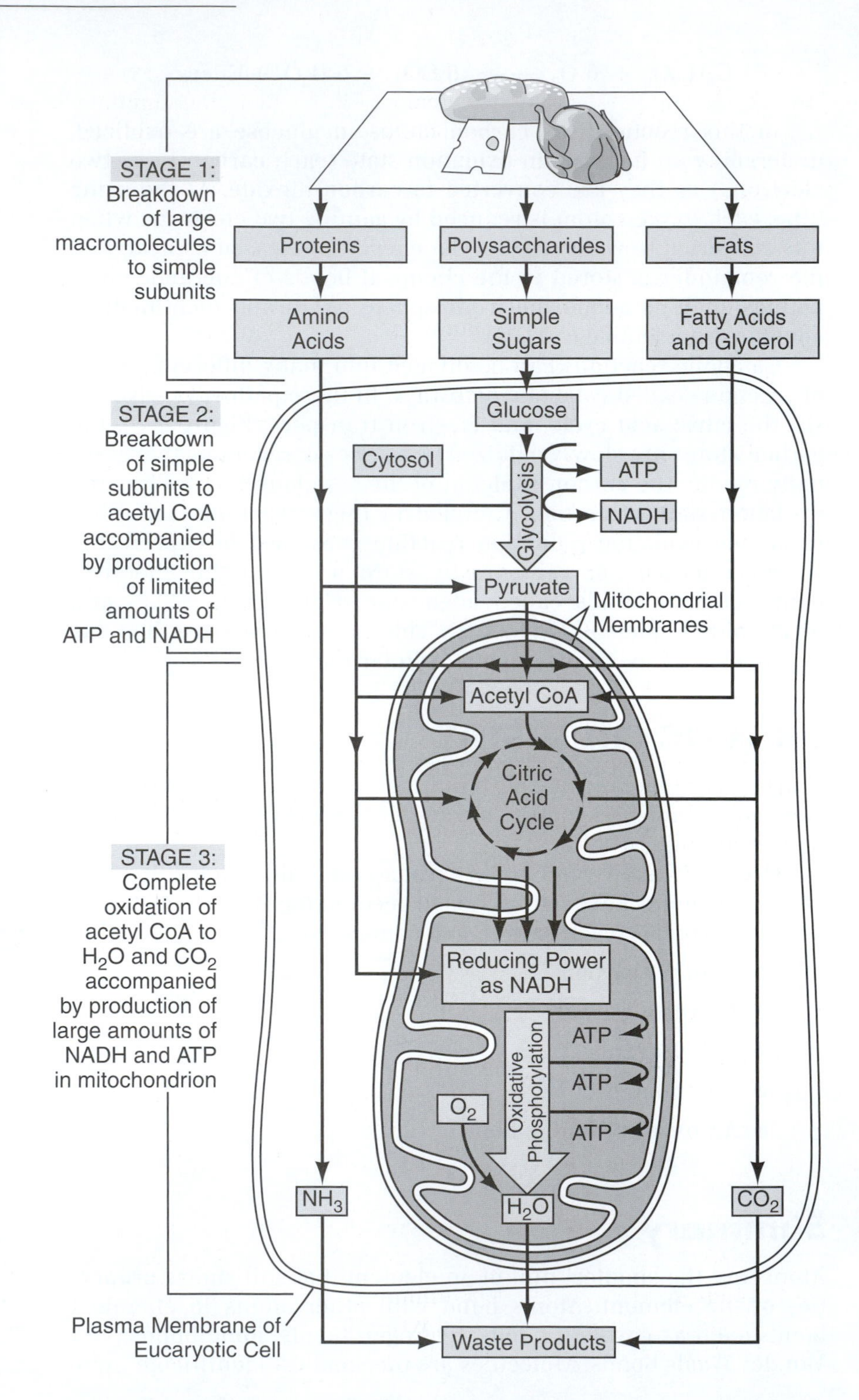

FIGURE 2-8

Overview of the chemical processes of life.

of pure substances. These molecules are formed by chemical reactions or become involved in chemical reactions. Reactions including those of metabolism can be classified as oxidation-reduction. Organic chemistry involves carbon-containing molecules; such groups are hydroxyl, amino, ammonia, methyl, and carboxyl.

Review Questions

Success in any career requires knowledge. Test your knowledge of this chapter by answering these questions or solving these problems.

1. The atom is the smallest unit of an ___________ that still exhibits the properties of that ___________.
2. Define a molecule.
3. Name and describe the two divisions of metabolism.
4. List the elements most important to life.
5. How are covalent bonds formed?
6. The atomic number of an atom is the total number of ___________. The atomic weight of an atom is the total number of ___________ plus ___________.
7. Salt is an example of a/an ___________ bond.
8. Explain the oxidation-reduction reaction.
9. Chemical properties of an element are determined by the number of ___________ in the outermost energy level of an atom.
10. All carbon atoms have four bonds to account for. What are the names of the bonds?

Student Activities

1. Using one of the minerals, develop a report or presentation of its properties as an element.
2. Develop a presentation or a report that describes the oxidation-reduction reactions in glycolysis, electron-transport, or the citric acid cycle.
3. Make a set of flash cards to learn the chemical symbols of all the elements and molecules important to nutrition.
4. Identify an everyday or common chemical reaction. Demonstrate it in class and write the chemical formula for the reaction.

Resources

Corriher, S. O. 1997. *Cookwise: The hows and whys of successful cooking.* New York: William Morrow and Company, Inc.

Cremer, M. L. 1998. *Quality food in quantity. Management and science.* Berkeley, CA: McCutchan Publishing Corporation.

Potter, N. N., and J. H. Hotchkiss. 1995. *Food science,* 5th ed. New York: Chapman and Hall.

Smoot, R. C., R. G. Smith, and J. Price. 1990. *Chemistry–A modern course.* Columbus, OH: Merrill Publishing Company.

Vaclavik, V. A., and E. W. Christina. 1999. *Essentials of food science.* Gaithersburg, MD: Aspen Publishers, Inc.

Vieira, E. R. 1996. *Elementary food science,* 4th ed. New York: Chapman and Hall.

Internet

Internet sites represent a vast resource of information. The URLs (uniform resource locator) for the World Wide Web sites can change. Using one of the search engines on the Internet such as Yahoo!, HotBot, AltaVista, Excite, Dogpile, About, or Google, find more information by searching for these words or phrases: metabolism or metabolic reactions, catabolism or catabolic reactions, anabolism or anabolic reactions. Also, Table A-7 provides a listing of some useful Internet sites that can be used as a starting point.

Chapter 3

Chemistry of Foods

Objectives

After reading this chapter, you should be able to:

- Name four carbohydrates and describe their chemical makeup
- Classify carbohydrates
- Compare the sweetness of various sugars
- Name three uses of carbohydrates in foods
- Describe the chemical makeup of proteins
- Discuss the use of proteins in foods
- List six functions of protein in the body
- Name three functions of protein in food
- Classify lipids
- Discuss the use of lipids or fats in foods
- Identify saturated and unsaturated fats
- List the fat- and water-soluble vitamins
- Name ten minerals important in nutrition
- List two functions of water in the body
- Identify biotin, choline, and phytochemicals

Key Terms

amino acids
biotin
birefringence
cellulose
choline
disaccharides
fatty acids
gelatinization
gum

homeostasis
hydrolysis
inversion
kilocalories
lipid
macrominerals
microminerals
monosaccharides
oil
oligosaccharides
osmotic pressure
oxidation
peptide bond
phospholipid
phytochemical
polymer
polysaccharide
rancidity
saturated
triglycerides
unsaturated

Nutrition is the process by which the foods people eat provide the nutrients they need to grow and stay healthy. Nutrients are naturally occurring chemical substances found in food. There are six categories of nutrients: proteins, lipids, carbohydrates, vitamins, minerals, and water. The chemistry of these nutrients influences the characteristics of our food.

Proteins, fats, and carbohydrates in food provide the energy, or **kilocalories** (kcal), our bodies need to function. Each gram of protein and carbohydrate has 4 kilocalories; each gram of fat has 9 kilocalories. Food science uses the metric system–grams, milligrams, and micrograms–to measure the amounts of nutrients in foods.

CARBOHYDRATES

The carbohydrates in our diet come from plant foods. Simple carbohydrates include the different forms of sugar (**monosaccharides** and **disaccharides**); complex carbohydrates (**polysaccharides**) include starches and dietary fiber.

Carbohydrates are called carbohydrates because they are essentially hydrates of carbon. Specifically they are composed of carbon and water and have a composition of $C_n(H_2O)_n$. The major nutritional role of carbohydrates is to provide energy; digestible carbohydrates provide 4 kilocalories per gram. No single carbohydrate is essential, but carbohydrates do participate in many required functions in the body.

Function in Food

Carbohydrates perform these functions in food:

- Flavor enhancing and sweetening due to caramelization
- Water binding

- Contributing to texture (starch, gluten)
- Hygroscopic nature/water absorption
- Providing source of yeast food
- Regulating gelation of pectin dispersing molecules of protein or starch
- Acting to subdivide shortening for creaming control crystallization
- Preventing spoilage
- Delaying coagulation protein
- Giving structure due to crystals
- Affecting osmosis
- Affecting color of fruits
- Affecting texture (viscosity, structure)
- Contributing flavor other than sweetness

Depending on the food, carbohydrates play many roles (Table 3-1). For example, in lollipops the sugars, glucose and/or glucose and fructose, will control crystallization, give structures due to sucrose, and serve as a flavor enhancer and sweetener due to all three sugars. In a more complex system such as a pineapple upside down cake, carbohydrates play many roles as it consists of all categories of carbohydrates–monosaccharides, disaccharides, and polysaccharides.

Monosaccharide

Monosaccharides may have 6 carbons and are called hexoses, or they may have 5 carbons and are called pentoses (Table 3-2). Glucose (sometimes called dextrose), fructose, and galactose are three common hexoses. Ribose and deoxyribose are two common pentoses. Figure 3-1 (page 40) shows some of the structure of some of the monosaccharides.

Disaccharides

Two monosaccharides may be linked together to form a disaccharide. Sucrose is the most common disaccharide and is made of one molecule each of glucose and fructose. Sucrose is commonly referred to as sugar. Lactose is the major sugar in milk and is made up of one molecule of glucose and one of galactose. Maltose is a disaccharide made from two molecules of glucose (Table 3-2). This linkage is formed by the removal of water (dehydration) and is broken by adding water back (**hydrolysis**).

TABLE 3-1 Carbohydrate Applications in Food

Industry/Product	Properties and Benefits
Baking and Snack Foods	
Cream-type Fillings	Film-forming; smooth texture
Fruit Leather	Film-forming; crystallization inhibitor; good humectant; sweetness moderator
Glazes/Frostings	Crystallization inhibitor; film-forming; improved adherence; sweetness control
Beverages	
Flavored Drinks	Low sweetness; improved body
Infant Formulas	Complete solubility; low sweetness; easy digestibility; bland flavor
Sports/Special Diets	Low osmolality; complete solubility; bland flavor; readily digestible
Binders	
Cereals/Snacks	Low hygroscopicity; good adhesion
Nut/Snack Coatings	Flavor carrier; film-forming
Carriers	
Artificial Sweeteners	Low hygroscopicity; neutral flavor; quick dispersion
Gums and Hydrocolloids	Good solubility; standardizes viscosity
Confectionery	
Candies	Good solubility; humectant properties; inhibits bloom
Pan Coatings	Binding; film-forming; drying agent
Dairy	
Coffee Whiteners	Fat dispersant; improved mouthfeel
Imitation Cheeses	Processing aid; contributes texture

TABLE 3-1 Carbohydrate Applications in Food *(concluded)*

Industry/Product	Properties and Benefits
Dry Mixes	
Powdered Drinks	Bulking agent; rapid dispersiblity; nonsweet carrier
Soup and Sauce Mixes	Low hydroscopicity; good solubility; provides body; protects flavor
Spice Blends	Low hydgroscopicity; bulking agent; nonsweet carrier
Fat Reduction	
Baking	Film-forming; humectant
Frozen Desserts	Minimal freezing point depression; lactose/ice crystal inhibitor; provides body; improves melt
Meats	
Processed Meats	Low sweetness; nonmeat solids; moisture retention; browning control
Spices and Seasonings	Carrier; moisture management
Pharmaceutical	
Tableting	Good binding; low hygroscopicity; directly compressible
Sauces and Salad Dressings	
Cheese/White Sauce	Low sweetness; smooth mouthfeel; does not mask flavors
Salad Dressings	Provides body and cling; decreased gum "stringiness"
Tomato Sauces	Provides body; brilliant sheen, intensifies color
Spray-Drying Aid	
Cheese/Fats	Encapsulates fats and flavors; protects proteins; improves flowability
Flavors	Encapsulates; low hygroscopicity; bland flavor
Fruit Juices and Syrups	Low hygroscopicity; high solubility

TABLE 3-2 Carbohydrates and Characteristics

Name/ Classification	End Products (Hydrolysis)	Source, Function, or Characteristics
Glucose	Glucose	Fruits, honey, corn syrup
Fructose	Fructose	Fruits, honey, corn syrup
Galactose	Galactose	Does not occur in free form in foods
Mannose	Mannose	Does not occur in free form in foods
Ribose	Ribose	Derived from pentoses of fruits and nucleic acids of meat products and seafood, does not occur in free forms in foods; an aldose
Xylose	Xylose	An aldose
Arabinose	Arabinose	An aldose
Sucrose	Glucose fructose	Beet and cane sugars, molasses, maple syrup, comes in many crystal sizes and grades
Lactose	Glucose galactose	Milk and milk products
Maltose	Glucose	Malt products, low concentrations in plants and processed foods
Starch	Glucose	Branches [amylopectin] contributed viscosity; linear [amylose] contributes gelling when gelatinized; granule is important to viscosity and gel formation
Dextrins	Glucose	Usually considered to be hydrolysis products of incompletely broken down starch fractions
Glycogen	Glucose	Meat products and seafood
Cellulose	Glucose	Comprises skeletal structure of plant cell. Indigestible stable cell structural framework of stalks and leaves of vegetables, fruits, and coverings of seeds.
Hemicellulose	Glucose	Comprises some of the plant skeletal structure; amorphous heterogeneous substance; pentose and uronic acid predominant
Pectic substances	Galactose	Cell cementing compound; fruits and vegetables; pectin will form gel with appropriate concentration, amount of sugar, and pH. Amorphous substances in the matrix of plant skeletal structure; contains minor amounts of neutral monomers such as arabinose, amylose, galactose, mannose

TABLE 3-2 Carbohydrates and Characteristics *(concluded)*

Name/ Classification	End Products (Hydrolysis)	Source, Function, or Characteristics
Malin	Fructose	Matrix
Galactogens	Galactose	Monomers such as arabinose, xylose, mannose, raffinose
Mannosans	Mannose	Polymers of mannose found in plants
Raffinose	Glucose fructose galactose	Cottonseed meal, sugar beets, and molasses
Pentosans	Pentoses	Found with cellulose in woody plants

Not all sugars have the same sweetness. A cola-type soft drink has between 10 and 12 percent sugars. Depending on the formulation, the sugar might be all sucrose or a blend of sucrose, glucose, and fructose. Milk, on the other hand, contains a little less than half this much sugar (lactose) and is not sweet. Table 3-3 lists the relative sweetness of some common sugars.

Sugars in Food

Color, texture, and flavor are all sensory characteristics that sugar plays in most foods. The study of sugars can be approached from their chemical structure, their properties, their characteristics, or their variety or source. Of the three sensory characteristics, sugars generally play a major role as a sweetener or in texture development. As a contributor to color, sugar participates in two phenomena: the Maillard reaction and caramelization.

Honey, sorghum/molasses, maple syrup, and selected fruit juice and pulps serve as a sweetener substitute for cane sugar and sugar beet sugar. Sugar-based sweeteners are those developed from corn starch. Processing the cane and beet sugars in the United States produces a granulated sugar, a brown sugar, and liquid sugars. The sugar's source and type impact sweetness and their interactive functioning.

Acid will hydrolyze and invert the disaccharide sugars into their component monosaccharides. Any product with an acid compound can bring about the hydrolysis of sucrose into fructose and glucose. Glucose and fructose are more soluble and more hygroscopic than sucrose, and they enhance browning.

Inversion of sugars refers to the hydrolysis of sucrose into fructose and glucose to form these sugars, which are sometimes

TABLE 3-3 Relative Sweetness of Some Sugars

Sugar	Relative Sweetness
Fructose	174
Invert Sugar	126
Sucrose	100
Glucose	74
Maltose	32
Galactose	32
Lactose	16

Monosaccharides

Monosaccharides usually have the general formula $(CH_2O)_n$, where *n* can be 3, 4, 5, 6, 7, or 8, and have two or more hydroxyl groups. They either contain an aldehyde group (–CHO) and are called aldoses or a ketone group (>C=O) and are called ketoses.

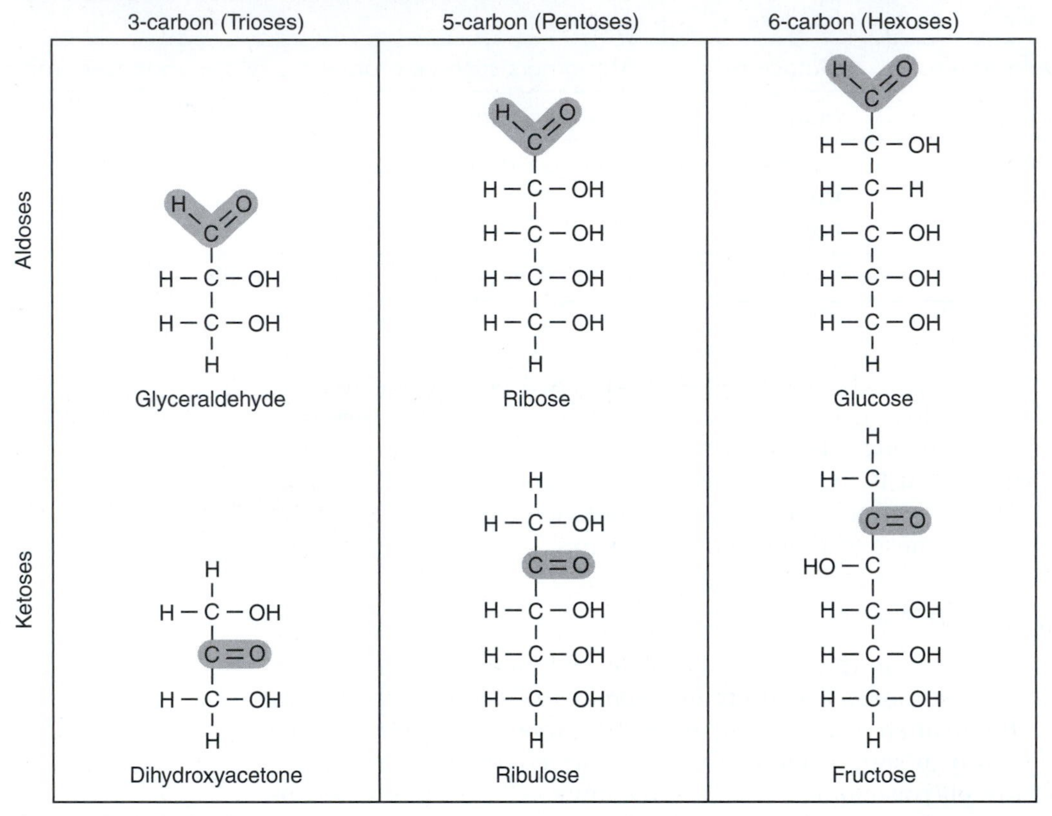

Ring Formation

In aqueous solution, the aldehyde or ketone group of a sugar molecule tends to react with a hydroxyl group or the same molecule, thereby closing the molecule into a ring.

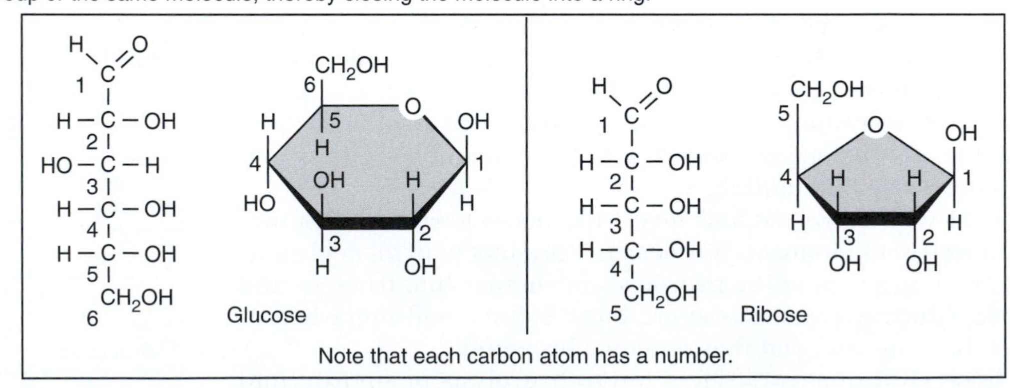

FIGURE 3-1 Representation of monosaccharides.

referred to as invert sugars. This inversion is thought to take place due to the presence of either enzyme or acid.

Caramelization. Caramelization is the application of heat to the point that sugars dehydrate and breakdown and polymerize (Figure 3-2).

Although a relatively complex reaction, it can be done simply as in the making of peanut brittle. The brown, flavorful peanut brittle is a result of this caramelization process. Researchers attribute the caramelization reaction to a range of browning reactions and flavor development. Once the melting point has been reached, sugars will caramelize. Each sugar has its own caramelization temperature.

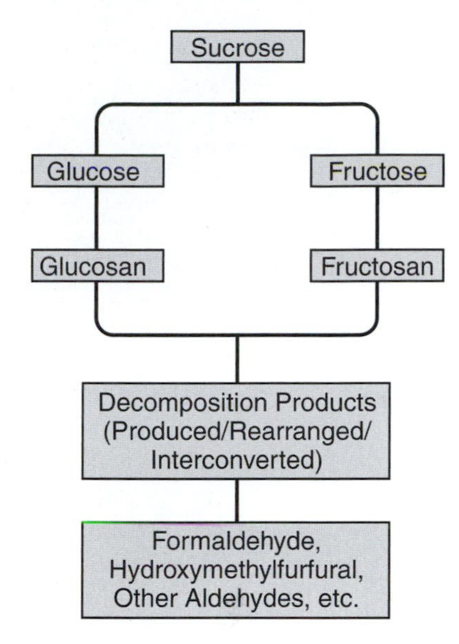

FIGURE 3-2

Representation of caramelization of sugars.

Crystallization. Crystallization of sugar can be a problem in a variety of products. For example, the crystallization of lactose in a glassy state will make nonfat milk difficult to disperse. If too great an amount of milk solids are added to a frozen dessert, a gritty texture results due to the lactose crystals. Crystallization of sugar is a major factor in the candy manufacturing industry (see Chapter 22).

Candies can be divided into two groups, crystalline and noncrystalline. Crystalline candies include fudge, fondant, and any other candies that have crystals as an important structural component. Divinity is a crystalline candy but is a special case as the crystals are dispersed in a foam. Noncrystalline candies include caramels, brittles, taffies, marshmallows, and **gum** drops. Marshmallows and gum drops are also special classes of candies as they contain a gelling substance.

Crystallization is a complex process with many interrelated factors. The nature of the crystallizing substance is important for crystallization. The rate of crystallization is the speed at which nuclei grow into crystals. This rate is dependent upon the concentration of the solute in the solution, as a more concentrated (more supersaturated) syrup will crystallize more rapidly than a less concentrated syrup. At a higher temperature the rate of crystallization is slow and becomes more rapid at a lower temperature. Agitation distributes the crystal forming nuclei and hastens crystallization. Impurities in the solution usually delay crystallization and in some cases such as caramels may prevent crystal formation. Fat and protein decrease the number and size of crystals.

Polysaccharides

Combinations of more than two sugars are often referred to as **oligosaccharides**, unless they are very large and then they are called polysaccharides. Raffinose and stachyose are two oligosaccharides of interest because they are hard to digest. Raffinose contains one molecule each of glucose, fructose, and galactose. Stachyose is very similar to raffinose except it contains two molecules of galactose.

Polysaccharides may be added to foods for a variety of reasons. Nutritionally, they are generally added to increase the dietary fiber content. Functionally, polysaccharides are added to thicken, form gels, bind water, and stabilize proteins. Starch is the most common polysaccharide added to food products. For some uses, starch may be chemically modified to improve stability or to alter its functional properties. **Cellulose** and cellulose derivatives are also added to a number of food products. The term "gum" is used to describe some of the naturally occurring polysaccharides added to food.

Some naturally occurring polysaccharides added to foods include: agar, gum tragacanth, algin, locust bean (carob) gum, carrageenan, starch, cellulose, pectin, guar gum, xanthan gum, and gum arabic.

Starch. Starch is a polysaccharide made up of glucose units linked together to form long chains. The number of glucose molecules joined in a single starch molecule varies from five hundred to several hundred thousand, depending on the type of starch. Starch is the storage form of energy for plants. Glycogen is the storage form of energy for animals. The plant directs the starch molecules to the amyloplasts, where they are deposited to form granules. In plants and in the extracted concentrate, starch exists as granules and varies in diameter from 2 to 130 microns. The size and shape of the granule is characteristic of the plant from which it came and serves as a way of identifying the source of a particular starch. The structure of the granule of grain is crystalline with the starch molecules orienting in such a way as to form radially oriented crystals (Figure 3-3).

Two types of starch molecules exist–amylose and amylopectin. Amylose averages 20 to 30 percent of the total amount of starch in most native starches. Some starches, such as waxy cornstarch, contain only amylopectin. Others may contain only amylose.

Amylose molecules contribute to gel formation. This is because the linear chains can orient parallel to each other, moving close enough together to bond. Probably due to the ease with which they can slip past each other in the cooked paste, they do not contribute significantly to viscosity. The branched amylopectin molecules give viscosity to the cooked paste. This is partially due

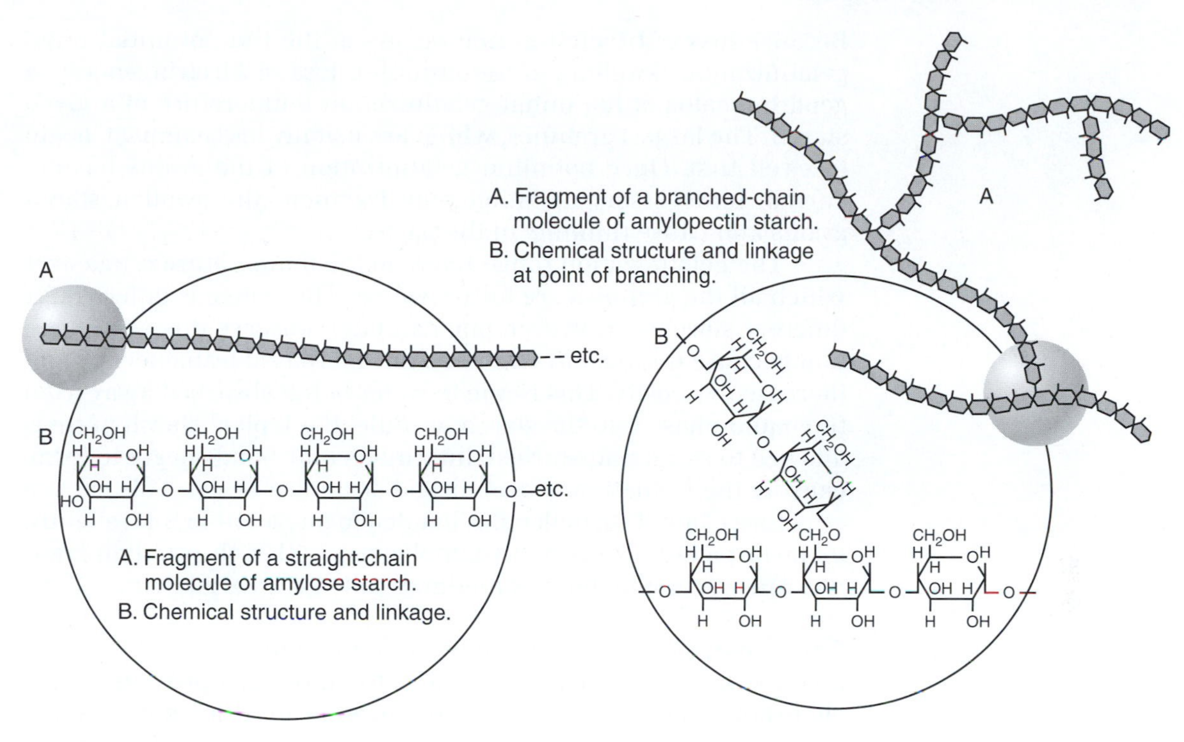

FIGURE 3-3

Straight chain of the starch amylose and the branched chain of the starch amylopectin.

to the role it serves in maintaining the swollen granule. Their side chains and bulky shape keep them from orienting closely enough to bond together, so they do not usually contribute to gel formation.

Different plants have different relative amounts of amylose and amylopectin. These different proportions of the two types of starch within the starch grains of the plant give each starch its characteristic properties in cooking and gel formation. Starch in its processed, commercial form is composed of starch grains or granules with most of the moisture removed. It is insoluble in water. When put in cold water, the grains may absorb a small amount of the liquid. Up to 140° to 158°F (60° to 70°C) the swelling is reversible, the degree of reversibility being dependent upon the particular starch. With higher temperatures an irreversible swelling called **gelatinization** begins.

Starch begins to gelatinize between 140° and 158°F (60° and 70°C), the exact temperature dependent is the specific starch. For example, different starches exhibit different granular densities, which affect the ease with which these granules can absorb water.

Because loss of **birefringence** occurs at the time of initial rapid gelatinization (swelling of the granule), loss of birefringence is a good indicator of the initial gelatinization temperature of a given starch. The largest granules, which are usually less compact, begin to swell first. Once optimum gelatinization of the grains has occurred, unnecessary agitation may fragment the swollen starch grains and cause thinning of the paste.

The gelatinization range refers to the temperature range over which all the granules are fully swollen. This range is different for different starches. However, one can often observe this gelatinization because it is usually evidenced by increased translucency and increased viscosity. This is due to water being absorbed away from the liquid phase into the starch granule. If a typical starch paste is allowed to stand undisturbed, intermolecular bonds begin to form, causing the formation of a semirigid structure or gel. This gel is a structure of amylose molecules bonded to one another and, slightly, to the branches of amylopectin molecules within the swollen granule. This phenomenon is sometimes called retrogradation.

Cellulose. Cellulose is the most common polysaccharide and the major component of plant cell walls. Cellulose is a **polymer** (long chain) of glucose molecules linked together by 1 to 4 linkages and cannot by digested by humans. Thus, cellulose is a major component of dietary fiber. Pectin is a polymer of galacturonic acid and is not digested. In plants, pectin "cements" cells together.

Complex carbohydrates that cannot be digested are generally called fiber. In the past, fiber was considered to be a non-nutritive substance. The FDA recognizes label claims that a diet high in fiber may offer protection against some forms of cancer, especially of the large intestine. Some forms of fiber may aid in reduction of serum cholesterol. These types of fiber appear to bind the cholesterol and make it less available for absorption. People who consume large quantities of fiber often complain of problems with gas and diarrhea. These probably result from fermentation of the undigested material in the large intestine.

PROTEINS

Proteins contain **amino acids**, sometimes referred to as the building blocks of protein. Dietary protein is supplied from plant and animal sources. Proteins are needed to build and repair body tissue and for the metabolic functions of our bodies. Figure 3-4 shows some diagrams representing different proteins.

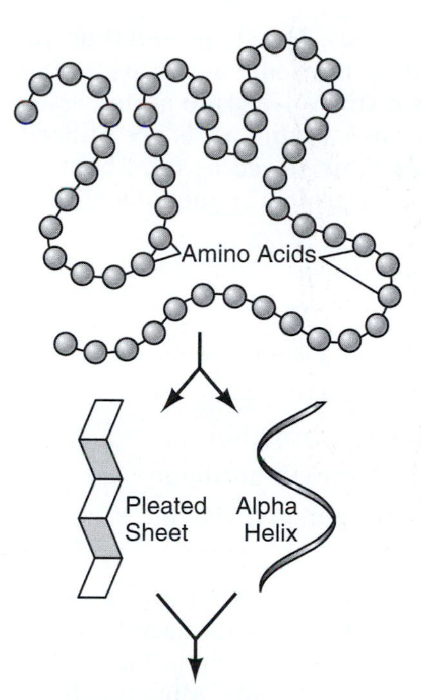

Primary Protein Structure
Is sequence of a chain of amino acids

Secondary Protein Structure
Occurs when the sequence of amino acids are linked by hydrogen bonds

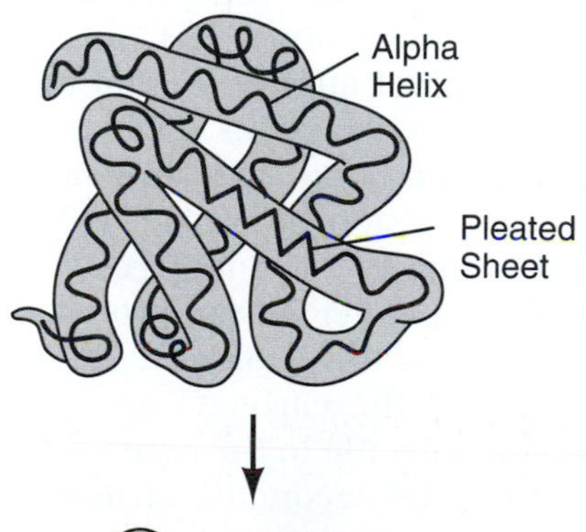

Tertiary Protein Structure
Occurs when certain attractions are present between alpha helices and pleated sheets

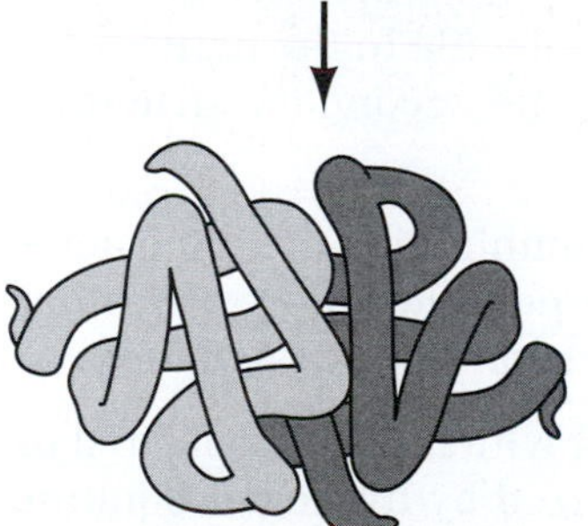

Quaternary Protein Structure
Is a protein consisting of more than one amino acid chain

FIGURE 3-4
Four elements of protein structure.

Proteins are polymers of amino acids. The shape and thus the function of a protein is determined by the sequence of its amino acids. Proteins must be broken down (hydrolyzed) to amino acids before they can be used. Once absorbed, amino acids are utilized to make proteins, converted to energy, or stored as fat. About 20 percent of the human body is made of protein. Functions of proteins include:

- Enzymes such as trypsin and pepsin
- Storage such as ovalbumin and ferritin
- Transport such as hemoglobin and lipoproteins
- Contractile such as actin and myosin
- Protective such as antibodies and thrombin
- Hormones such as insulin and growth hormone
- Structural such as keratin, collagen, and elastin
- Membranes

Amino acids contain an amino group ($-NH_2$) and an acid group ($-COOH$). There are twenty amino acids that are found in proteins. Some of the twenty amino acids are shown in Figure 3-5.

Amino acids join by forming **peptide bonds**. A peptide bond is formed by the condensation of the amino group ($-NH_2$) of one amino acid with the acid group ($-COOH$) of another amino acid resulting in the loss of water. Condensation reactions involve the removal of water (H_2O) and formation of a bond. The reversal of this is hydrolysis, which involves the addition of water. Peptide bonds are not easily broken. Cooking would not normally result in the breaking of peptide bonds to yield amino acids from proteins.

The amino acids form proteins due to the reaction between the amino group of one amino acid and the carboxyl group of another. The conformation of a protein molecule in the native state is determined by the primary structure, the secondary structure, and a tertiary structure.

Primary. The primary structure is the combination of amino acids in a proper sequence by means of the peptide bonds. No other forces or bonds are implied by this structural level designation.

Secondary. Secondary structure is that which forms a pleated or helix structure. The alpha-helix is stabilized by hydrogen bonding between carboxyl and the amide groups of the peptide bonds that generally appear in a regular sequence along the chain of amino acids.

Tertiary. A tertiary structure is the folding of the coiled chain or chains. Covalent, hydrogen, and Van der Waals forces may be involved in the structural organization of protein molecules.

Nonpolar Side Chains

Alanine (Ala, or A)

Valine (Val, or V)

Leucine (Leu, or L)

Isoleucine (Ile, or I)

Proline (Pro, or P)

Phenylalanine (Phe, or F)

Methionine (Met, or M)

Tryptophan (Trp, or W)

Glycine (Gly, or G)

Cysteine (Cys, or C)

Disulfide bonds can form between two cysteine side chains in proteins.

$-CH_2-S-S-CH_2-$

Uncharged Polar Side Chains

Asparagine (Asn, or N)

Glutamine (Gln, or Q)

Although the amide N is not charged at neutral pH, it is polar.

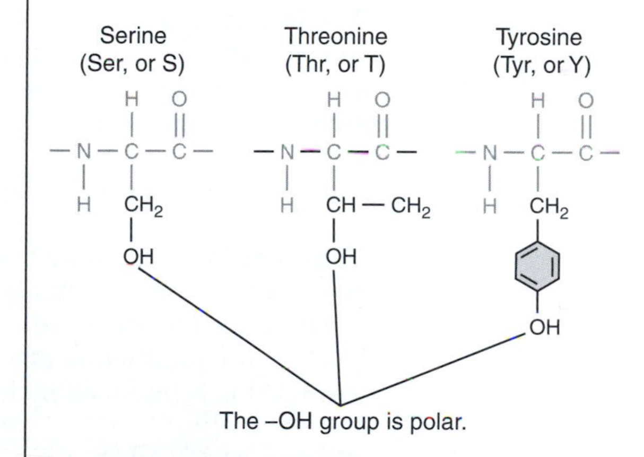

Acidic Side Chains

Aspartic Acid (Asp, or D)

Glutamic Acid (Glu, or E)

FIGURE 3-5 Representation and classification of essential amino acids.

Functions of Proteins in Foods

Proteins are amphophilic because they have polar (hydrophilic) and nonpolar (hydrophobic) side chains in one molecule, forming interfacial films similar to small molecular synthetic emulsifiers.

Proteins are fundamental food components, both functionally and nutritionally. A basic understanding of protein structure and characteristics is critical for the total understanding of how these function in foods. The actual characteristics of the proteins will be influenced if one can whip, beat, add ingredients, or heat to form the type of food. The characteristics of the proteins will influence how they will behave in colloidal systems, how they contribute to the color, texture, and flavor of foods.

Color. The role of protein in color of foods is not clear-cut. In most instances it may either play a role through its interaction or as part of a complex molecule. One of the biggest roles is through the Maillard Reaction, a browning reaction between an amino group and a reducing group of a carbohydrate. The reactions between proteins and sugars is the most common of these browning reactions that occur in baked products and a number of other foods. Selected color pigments, such as chlorophyll, are bound in the chloroplasts in a protein-**lipid** matrix.

Texture. Texture contributions of protein foods are not uncommon. For example, custards are protein gels where the gel strength is influenced by the ovalbumin denaturation. Another clear example is in the production of yogurt. In this particular case the texture of yogurt is influenced by the gelation of casein.

Flavor. The contribution of flavor is not as clear-cut. Some proteins and the amino acids may add flavor. Amino acids may contribute bitterness, sweetness, and other flavors.

LIPIDS

Lipids include fats and **oils** from plants and animals. Cholesterol is a fat found only in animal products. Lipids are of special interest because they are linked to the development of heart disease, the leading cause of death among Americans.

Lipids are the substances in foods that are soluble in organic solvents. This category includes **triglycerides**, **fatty acids**, **phospholipids**, some pigments, some vitamins, and cholesterol. Figure 3-6 illustrates the structure of some lipids found in food.

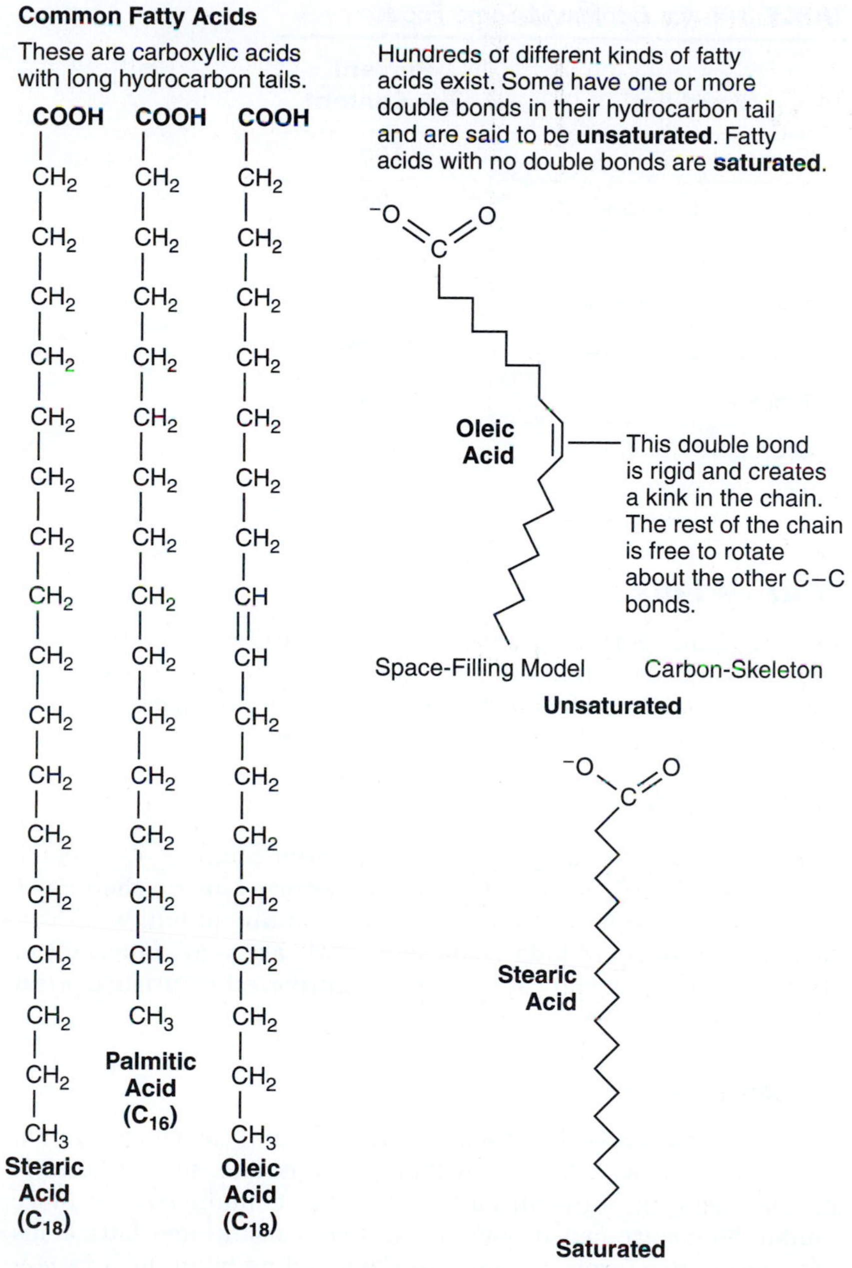

Common Fatty Acids

These are carboxylic acids with long hydrocarbon tails.

Stearic Acid (C_{18})	Palmitic Acid (C_{16})	Oleic Acid (C_{18})
COOH	COOH	COOH
CH_2	CH_2	CH_2
CH_2	CH_2	CH_2
CH_2	CH_2	CH_2
CH_2	CH_2	CH_2
CH_2	CH_2	CH_2
CH_2	CH_2	CH_2
CH_2	CH_2	CH_2
CH_2	CH_2	CH
CH_2	CH_2	CH
CH_2	CH_2	CH_2
CH_2	CH_2	CH_2
CH_2	CH_2	CH_2
CH_2	CH_2	CH_2
CH_2	CH_2	CH_2
CH_2	CH_3	CH_2
CH_2		CH_2
CH_3		CH_3

Hundreds of different kinds of fatty acids exist. Some have one or more double bonds in their hydrocarbon tail and are said to be **unsaturated**. Fatty acids with no double bonds are **saturated**.

FIGURE 3-6

Representation of fatty acids.

TABLE 3-4 Fat Content of Some Foods

Food Fat	Percent Fat Content
Oils and shortening	100
Butter and margarine	80
Most nuts	60
Peanut butter, bacon	50
Cheese, beef roasts	30–35
Franks	25–30
Lean pork, ice cream	12–14
Milk, shellfish	2–4

Role of Fats in Food

In food (Table 3-4), fats provide a source of essential fatty acids, add caloric density (energy), act as carriers for flavors, carry fat-soluble vitamins, contribute to texture and mouthfeel, become precursors of flavor, and provide heat transfer medium (in frying).

Fatty Acids

Naturally occurring fatty acids have an even number of carbons. Short-chain fatty acids are important as odors. Longer-chain fatty acids are not volatile and do not contribute much to flavor. Reaction products of long-chain fatty acids are very important to the flavor of foods. Fatty acids may be **saturated** or **unsaturated** (Table 3-5).

Double Bonds

Fatty acid molecules that are unsaturated contain what are known as double bonds. A fatty acid that contains one double bond is called mono-unsaturated. Fatty acids that contain two or more double bonds are called polyunsaturated. Unsaturated fatty acids can exist in two forms, cis and trans, depending upon the arrangement of the portions of the fatty acid molecules around the double bonds. Naturally occurring fatty acids are in the cis conformation. The double bonds in lipid molecules are very reactive toward oxygen. The products of lipid **oxidation** have undesirable flavors, and lipid oxidation leads to what is termed as **rancidity**.

TABLE 3-5 Characteristics of Fatty Acids

Fatty Acid	Number of Carbon Atoms	Melting Point (degrees C)
Saturated Fatty Acids		
Butyric	4	–7.9
Caproic	4	–3.9
Caprylic	8	16.3
Capric	10	31.3
Lauric	12	44.0
Myristic	14	54.4
Palmitic	16	62.8
Stearic	18	69.6
Arachidic	20	75.4
Behenic	22	80.0
Lignoceric	24	84.2
Unsaturated Fatty Acids		
Palmitoleic	16	–0.5 to 0.5
Oleic	18	13
Linoleic	18	–5 to –12
Linolenic	18	–14.5
Arachidonic	20	–49.5

Some food additives function to inhibit the oxidation of food lipids. These molecules are called antioxidants. The antioxidants most commonly added to foods are:

- Butylated hydroxytoluene (BHT)
- Butylated hydroxy anisole (BHA)
- Vitamin C
- Vitamin E

Triglycerides

Food fats made up of three molecules of fatty acids connected to a molecule of glycerol are known as triglycerides. The vast majority of foods that we consume contain fat in the form of triglycerides. Triglycerides can be broken apart by enzymes called lipases. The products of lipolysis often have soapy flavors. The food industry uses these products as emulsifiers. A triglyceride molecule that has had one fatty acid removed is called a diglyceride and one that has had two fatty acids removed is called a monoglyceride. Both mono- and diglycerides are used as emulsifiers.

Phospholipids

Some fatty acids are connected to glycerol molecules that contain a molecule of phosphorus. These special lipids are known as phospholipids. They play important roles in the body but are not essential nutrients because the body can synthesize them in adequate quantities. Probably the best known of the phospholipids is lecithin.

Cholesterol

Cholesterol is a compound produced by the body that has received considerable attention due to its reported link to heart disease. The reason that cholesterol is not considered an essential nutrient is because the body can produce all the cholesterol that it needs. The average American consumes from 400 to 800 mg/day of cholesterol and synthesizes from 1,000 to 2,000 mg/day. The more cholesterol that is consumed, the less the body produces and vice versa. This explains why it is so difficult to decrease serum cholesterol by dietary means alone. If you consume less cholesterol, your body produces more to keep the supply constant.

Cholesterol is used by the body for:

- Bile salts
- Membrane structure
- Myelin synthesis
- Vitamin D synthesis
- Steroid hormone synthesis

Some people have a genetic problem with the system that regulates cholesterol synthesis, and they produce excessive amounts. These people generally have greatly elevated serum cholesterol levels. This is of concern because high serum cholesterol is a risk factor for coronary heart disease.

VITAMINS

Vitamins are chemical compounds in our food that are needed in very small amounts (in milligrams and micrograms) to regulate the chemical reactions in our bodies. The vitamins are divided into fat-soluble and water-soluble vitamins. Table A-8 lists the vitamin content of common foods.

Fat-Soluble Vitamins

Fat soluble vitamins include vitamins A, D, E, and K.

Vitamin A. Vitamin A occurs in preformed state and as a precursor. Three active forms are retinol, retinal, and retinoic acid.

In food, most preformed vitamin A is found in the form of retinol. All three forms of vitamin A can be formed from the plant pigments carotenes. The most common form is beta-carotene. Vitamin A is susceptible to oxidation, but is relatively heat stable.

Vitamin D. The active form of vitamin D is cholecalciferol (vitamin D_3). It can be produced from cholesterol by the action of ultraviolet light. It can also be formed from a protovitamin. It is stored in the liver and functions in the absorption of the minerals calcium and phosphorus. Vitamin D also acts directly on bone, and affects reabsorption of calcium and phosphorus by the kidney.

Vitamin E. Vitamin E or alpha tocopherol is widely available in a normal diet. It functions to detoxify oxidizing radicals that arise in metabolism, to stabilize cell membranes, to regulate oxidation reactions, and to protect vitamin A and polyunsaturated fatty acids from oxidation.

Vitamin K. Dietary and intestinal bacterial sources contribute to the supply of vitamin K. Storage in the body is minimal, and vitamin K functions in normal blood clotting.

Water-Soluble Vitamins

The water-soluble vitamins include the B vitamins and vitamin C. B vitamins include: thiamin, riboflavin, niacin, vitamin B_6, pantothenic acid, folic acid, **biotin**, and cobalamin (vitamin B_{12}).

Thiamin. Thiamin functions in carbohydrate metabolism. It makes ribose to form RNA, and it maintains the normal appetite and normal muscle tone in the digestive tract.

Riboflavin. Riboflavin functions as part of a coenzyme involved in oxidation-reduction reactions in energy production.

Niacin. Niacin functions as a component of two coenzymes involved in oxidation-reduction reactions releasing energy from food.

Vitamin B_6. The functions of vitamin B_6 include the metabolism of amino acids and the conversion of glycogen to glucose.

Pantothenic Acid. Pantothenic acid is a part of coenzyme A, which is involved in synthesis and breakdown of fats, carbohydrates, and proteins. It is also part of the enzyme, fatty acid synthetase.

Folic Acid. The coenzyme form of folic acid is tetrahydrofolic acid. It functions in the transfer of formyl and hydroxymethyl groups. Folic acid is required for synthesis of purines and pyrimidines and for efficient use of the amino acid histidine.

Biotin. Biotin functions in fatty acid synthesis.

Cobalamin. Cobalamin or vitamin B_{12} is required for nucleic acid synthesis, amino acid synthesis, blood cell formation, neural function, and growth. Cobalamin is found only in animal products.

Vitamin C. Ascorbic acid or vitamin C functions in wound healing, collagen synthesis, iron absorption, and as an antioxidant. Vitamin C is necessary for conversion of proline to hydroxyproline and lysine to hydroxylysine. It is involved in iron absorption and the conversion of amino acids to neurotransmitters. It is the least stable of all vitamins. It oxidizes readily in light or air, when heated, or in alkaline solutions. Degradation is enhanced by presence of iron and copper.

MINERALS

Minerals, also needed only in small amounts, have many different functions. Some minerals assist in the body's chemical reactions and others help form body structures. Minerals are important for energy transfer and as an integral part of vitamins, hormones, and amino acids. Depending on the amount in the body, minerals in the diet are classified as **macrominerals** or **microminerals** (sometimes called trace minerals). Table A-8 provides the mineral content of some common foods. The seven macrominerals are:

- Calcium (Ca)
- Phosphorus (P)
- Potassium (K)
- Sodium (Na)
- Chloride (Cl)
- Magnesium (Mg)
- Sulfur (S)

Calcium

Calcium is involved in **homeostasis**–the functions that maintain life–blood clotting mechanisms, and muscle contractions. Calcium also makes up 35 percent of the bone structure.

Phosphorus

This mineral makes up 14 to 17 percent of the skeleton. Phosphorus is required for many energy-transfer reactions, and for the synthesis of some lipids and proteins.

Potassium

Potassium maintains the acid-base balance and **osmotic pressure** inside the cells.

Sodium

Sodium maintains the acid-base balance outside the cells and regulates the osmosis of body fluids. Sodium is also involved in nerve and muscle function.

Chloride

In the diet, chloride normally accompanies sodium as NaCl or salt. This is an important extracellular anion (negative charge) involved in acid-base balance and osmotic regulation. Chloride is an essential component of bile, hydrochloric acid, and gastric secretions.

Magnesium

More than half of the magnesium found in the body exists in the skeleton. Magnesium is an activator of many enzymes.

Sulfur

Sulfur is a component of many biochemicals in the body, including amino acids, biotin, thiamin, insulin, and chondroitin sulfate.

Microminerals important in nutrition include:

- Chromium (Cr)
- Cobalt (Co)
- Copper (Cu)
- Fluorine (F)
- Iodine (I)
- Iron (Fe)
- Manganese (Mn)
- Molybdenum (Mo)
- Nickel (Ni)
- Selenium (Se)
- Silicon (Si)
- Tin (Sn)
- Vanadium (V)
- Zinc (Zn)

Chromium

In 1959, chromium was shown to be the factor responsible for improved glucose tolerance in rats. Now evidence exists to support the contention that chromium is essential for humans. It is involved in glucose tolerance, stimulation of fatty acid synthesis, insulin metabolism, and protein digestion.

Cobalt

Cobalt is a part of vitamin B_{12}. Microflora in the cecum and colon use dietary cobalt to make vitamin B_{12}.

Copper

Copper is essential for several copper-dependent enzymes.

Fluorine

Fluorine is involved in bone and teeth development.

Iodine

Iodine is essential for the production of the thyroid hormones. These hormones regulate basal metabolism.

Iron

In the body, about 60 percent of the iron is in the red blood cells and 20 percent is in the muscles.

Manganese

Manganese is necessary for carbohydrate and fat metabolism and for the synthesis of cartilage.

Molybdenum

Molybdenum is part of the enzyme xanthine oxidase.

Nickel

Nickel was found associated with the protein nickeloplasmin. In 1970, it was shown to be essential for chickens.

Selenium

Selenium is essential for detoxification of certain peroxides that are toxic to cell membranes. Selenium is closely connected with vitamin E in that they work together to scavenge free radicals.

Silicon

In 1972, silicon was shown to be essential in young chickens. May be important for bone development and required in very small amounts for humans.

Tin

Possibly tin is required by humans. Tin has been shown to have a growth promoting effect in rats. No requirement is known for humans.

Vanadium

Vanadium was shown to be essential in rats in 1971. In chickens, it increased growth rate and increased hematocrit. Probably a small amount is required for humans.

Zinc

Zinc is a component of many enzymes.

Although not specifically discussed, cadmium (Cd), boron (B), and aluminum (Al) are also considered microminerals.

STUDYING THE LARGEST KNOWN PROTEIN

The largest known single-chain protein is found in muscle cells, and is referred to by two names: titin and connectin. This huge molecule helps to maintain resting tension in muscle tissue and takes part in the contraction of muscle fibers.

A molecule of titin can be nearly 1 micron long (0.000001 meter or 0.00004 inch); which is bigger than some cells. Each molecule consists of about 30,000 amino acids (the basic building blocks of proteins).

Scientists have recently used "optical tweezers" to study titin by carefully stretching individual molecules. Optical tweezers move and trap very tiny objects using light. They can hold objects as small as single cells (or even viruses) by shining a focused laser beam onto them. With optical tweezers, it is possible to trap and move living cells (or even internal parts of cells) without damaging them. The technique has even been used to insert new genes into cells.

Using optical tweezers, scientists found that a molecule of titin is something like a series of springs connected by looser chains, allowing it to stretch and return to its original shape easily.

To find out about how titin's structure was revealed, visit this Web site:

<www.embl-heidelberg.de/ExternalInfo/>

WATER

Fifty to 60 percent of human body weight consists of water. In the body, water performs these important functions:

- Carries nutrients and wastes
- Maintains structure of molecules
- Participates in chemical reactions
- Acts as a solvent for nutrients
- Lubricates and cushions joints, spinal cord, and fetus (during pregnancy)
- Helps regulate body temperature
- Maintains blood volume

Dehydration occurs when water output exceeds intake. Signs of dehydration include dry skin, dry mucous membranes, rapid heartbeat, low blood pressure, and weakness. Humans require 7 to 11 cups (56 to 88 ounces) per day.

Water sources for the body include:

- Water (100%)
- Fruits and vegetables (90% to 99%)
- Fruit juices (80% to 89%)
- Pasta, legumes, beef, and dairy (10% to 60%)
- Crackers and cereals (1% to 9%)

The exact water content of specific foods can be determined by using Table A-8.

BIOTIN

Biotin is also known as vitamin H and coenzyme R. It is found primarily in the liver, kidney, and muscle. Biotin functions as an essential cofactor for four carboxylases that catalyze the incorporation of cellular bicarbonate into the carbon backbone of organic compounds. Biotin is routinely provided to individuals receiving total intravenous feeding and is incorporated into almost all nutritionally complete dietary supplements and infant formulas. In larger doses, biotin is also used to treat inborn errors of metabolism.

Biotin is widely distributed in food stuffs, but the amounts are small relative to other vitamins. Biotin deficiency is rare in the absence of total intravenous feedings without added biotin or the chronic ingestion of raw egg white.

CHOLINE

Choline, a dietary component of many foods, is part of several major phospholipids (including phosphatidylcholine–also called lecithin) that are critical for normal membrane structure and function. The major precursor of betaine, it is used by the kidney to maintain water balance and by the liver as a source of methyl-groups for methionine formation. Also, choline is used to produce the important neurotransmitter acetylcholine. In the body choline is mainly found in phospholipids, such as lecithin (phosphatidylcholine) and sphingomyelin.

A choline deficiency in healthy humans is difficult to demonstrate. Choline and choline esters can be found in significant amounts in many foods consumed by humans. Some of the choline is added during processing (especially in the preparation of infant formula).

PHYTOCHEMICALS

Plants manufacture chemicals, known as **phytochemicals**, that have multiple functions. Some attract insects to encourage fertilization;

others provide defenses against predators such as viruses and animals. Phytochemicals exhibit diversified physiologic and pharmacologic effects. Active derivatives extracted from leaves, stems, roots, flowers, and fruits of plants may be classified into three main categories:

1. Toxic and no discernible therapeutic use; compounds such as pyrrolizidine alkaloids, nicotine, and hydrazine derivatives

2. Toxic but useful for treatment of disease when used in controlled amounts or for defined clinical conditions; compounds such as morphine, digitalis, and vinca alkaloids

3. Chemopreventative activity; compounds useful against diseases such as atherosclerosis, cancer, and diverticular disease

Most active chemopreventative phytochemicals are high-molecular-weight fibers such as celluloids, pectins, lignins, and low-molecular-weight compounds such as carotenoids, dithiolthiones, flavanoids, indole carbinols, isothiocyanates, mono- and triterpenoids, and thioallyl derivatives.

The majority of phytochemicals that have chemopreventative activity have no clearly defined role as essential nutrients except for the vitamins (ascorbate, tocopherols). Although phytochemical deficiencies have not been identified, their low concentrations in the diet have been associated with increased risks for cancer, cardiovascular diseases, and diabetes. A variety of data suggest that the best way to obtain chemopreventative phytochemicals is to include increased quantities of fruits and vegetables in the diet. All plants are sources of high-molecular-weight fibers. Specific low-molecular-weight phytochemicals with chemopreventative activity are contained within a variety of plants.

Plants such as cabbage and broccoli are excellent sources of indoles, dithiolthiones, isothiocyanates, and chlorophyllins. Legumes (soybeans, peanuts, beans, and peas) contain flavanoids, isoflavanoids, and other polyphenols that act as antioxidants and estrogenic agonists/antagonists. Citrus fruits and licorice root contain mono- and triterpenes that act as antioxidants, cholesterol synthesis inhibitors, and stifle growth of rapidly dividing cells. Thioallyl derivatives are found in garlic, leeks, and onion, and prevent thrombi formation, decrease cholesterol synthesis, and prevent DNA damage.

Summary

Besides contributing to nutrition, carbohydrates, proteins, and lipids function in food. For example, carbohydrates enhance flavor, contribute to texture, prevent spoilage, and influence color. The function of carbohydrates in foods to some extent depends on their type –monosaccharides, disaccharides, or polysaccharides. Starch is a polysaccharide whose characteristics depend on the type of plant producing the starch. Cellulose is a nondigestible polysaccharide that contributes to the characteristics of food, and demonstrates some health benefits. Proteins in food can act as emulsifiers and also influence the color, flavor, and texture of food. Lipids contribute to the texture, flavor, and heat transfer of foods. Lipids also carry the flavors and the fat-soluble vitamins. Food provides the vitamins and minerals necessary for normal growth and health. Although not a nutrient, water is necessary for a solvent of all nutrients. Biotin, choline, and phytochemicals are nutrients that seem to have some health benefits but do not have clearly defined requirements.

Review Questions

Success in any career requires knowledge. Test your knowledge of this chapter by answering these questions or solving these problems.

1. What is the chemical composition of a carbohydrate?
2. List the three functions of proteins in food.
3. What is the difference between a monosaccharide and a disaccharide?
4. Name five functions carbohydrates play in foods.
5. Explain two functions of water in the body.
6. Triglycerides, fatty acids, phospholipids, some pigments, some vitamins, and cholesterol are classed as __________.
7. Fatty acid molecules that are __________ contain what are known as double bonds. A fatty acid that contains one double bond is called mono __________. Fatty acids that contain two or more double bonds are called __________.
8. List the fat- and water-soluble vitamins.
9. __________ is part of several major phospholipids critical for normal membrane structure and function, is used by the kidney to maintain water balance, and is used to produce the important neurotransmitter acetylcholine.
10. Name ten minerals important in nutrition.

Student Activities

1. Develop a report or presentation on one of the beneficial phytochemicals.
2. Taste samples of food that are pure or almost pure protein, starch, and lipid.
3. Using food composition tables (see Table A-8), analyze the approximate amounts of energy, fat, protein, vitamins, and minerals in a recent meal.
4. Calculate how much you could reduce your monthly fat intake by switching from whole milk (3.25% fat) to 1% fat milk.
5. Use a match to burn a potato chip. Explain why it burns so readily.

Resources

Brody, J. E. 1981. *Jane Brody's nutrition book.* New York: Bantam Books.

Corriher, S. O. 1997. *Cookwise: The hows and whys of successful cooking.* New York: William Morrow and Company, Inc.

Ensminger, A. H., M. E. Ensminger, J. E. Konlande, and J. R. Robson. 1994. *Foods and nutrition encyclopedia.* 2 Vols. Boca Raton, FL: CRC Press.

Potter, N. N., and J. H. Hotchkiss. 1995. *Food science,* 5th ed. New York: Chapman and Hall.

Vaclavik, V. A., and E. W. Christina. 1999. *Essentials of food science.* Gaithersburg, MD: Aspen Publishers, Inc.

Vieira, E. R. 1996. *Elementary food science,* 4th ed. New York: Chapman and Hall.

Internet

Internet sites represent a vast resource of information. The URLs (uniform resource locator) for the World Wide Web sites can change. Using one of the search engines on the Internet such as Yahoo!, HotBot, AltaVista, Excite, Dogpile, About, or Google, find more information by searching for these words or phrases: carbohydrates, fats, proteins, crystallization, gelatinization, caramelization, cellulose, fiber, vitamins, minerals, triglycerides, monosaccharides, disaccharides. Also, Table A-7 provides a listing of some useful Internet sites that can be used as a starting point.

Chapter 4

Nutrition and Digestion

Objectives

After reading this chapter, you should be able to:

- Identify nutritional needs using RDA or DRI
- Discuss the functions of energy, carbohydrates, fats, and proteins in the body
- Provide the caloric content of proteins, carbohydrates, fats, and alcohol
- List the essential amino acids
- Name two protein-deficiency diseases
- Describe protein quality
- Name an essential fatty acid
- List the water- and fat-soluble vitamins and their functions
- List six minerals required by the body
- Describe the process of digestion
- Identify the organs involved in digestion
- Discuss the relationship of diet to health

Key Terms

absorption
bioavailability
BV
coenzyme
digestion
DRI
enzyme
essential amino acid
fiber
limiting amino acid
NPU
PER
protein quality
RDA
stability
vegan

People require energy and certain other essential nutrients. These nutrients are essential because the body cannot make them and must obtain them from food. Essential nutrients include vitamins, minerals, certain amino acids, and certain fatty acids. Foods also contain other components such as fiber that are important for health. Although each of these food components has a specific function in the body, all of them together are required for overall health. The digestive system breaks down food into nutrients for absorption.

NUTRIENT NEEDS

In the United States the nutritional needs of the public are estimated and expressed in the Recommended Dietary Allowances (**RDA**). These were initially established during World War II to determine in a time of possible shortage, what levels of nutrients were required to ensure that the nutrition of the people would be safeguarded. The RDA are established by the Food and Nutrition Board of the National Research Council, whose members come from the National Academy of Sciences, the National Academy of Engineering, and the Institute of Medicine.

The first RDA were published in 1943 by a group known as the National Nutrition Program, a forerunner of the Food and Nutrition Board. Initially, the RDA were intended as a guide for planning and procuring food supplies for national defense. Now RDA are considered to be goals for the average daily amounts of nutrients that population groups should consume over a period of time.

The RDA are the levels of intake of essential nutrients considered, in the judgment of the Food and Nutrition Board on the basis of available scientific knowledge, to meet the known nutrition needs of practically all healthy persons. The NAS-NRC recognizes that diets are more than combinations of nutrients and should satisfy social and psychological needs as well.

As the needs for nutrients have been clearly defined, the RDA have been revised at roughly five-year intervals. The Ninth Edition of the RDA was published in 1980. The Tenth Edition was due to be released in 1986, but controversy regarding some of its recommendations delayed its publication until 1989.

The requirement for a nutrient is the minimum intake that will maintain normal functions and health. In practice, estimates of nutrient requirements are determined by a number of techniques, including

- Collection of data on nutrient intake from apparently normal, healthy people.
- Determinations of the amount of nutrient required to prevent disease states (generally epidemiological data).
- Biochemical assessments of tissue saturation or adequacy of molecular function.

A major revision is in process to replace the RDA. The revised recommendations are called Dietary Reference Intakes (**DRI**). Until 1997, the RDA were the only standards available. They will continue to be useful until DRI can be established for all nutrients. Figure 4-1 presents the RDA and the DRI.

WATER

Water is essential. About 65 percent of the adult body is made up of water. Lack of water can cause death more quickly than lack of any other nutrient. All the chemical reactions that occur in the body take place in water. Water also reacts during the chemical processes, regulates body temperature, transports nutrients and wastes, and dissolves nutrients. An adult should drink six to eight glasses of water each day.

Energy

The carbohydrates, fats, and proteins in food supply energy, which is measured in calories. Carbohydrates and proteins provide about 4 calories per gram. Fat contributes more than twice as much–about 9 calories per gram. Alcohol, although not a nutrient, also supplies energy–about 7 calories per gram. Foods that are high in fat are also high in calories. However, many lowfat or nonfat foods can also be high in calories.

Calorie needs vary by age and level of activity (Figure 4-2). Many older adults need less food, in part due to decreased activity, relative to younger, more active individuals. People who are trying to lose weight and eating little food may need to select more nutrient-dense foods in order to meet their nutrient needs in a satisfying diet.

Carbohydrates

Sugars are carbohydrates. Dietary carbohydrates also include the complex carbohydrates starch and **fiber**. During **digestion** all carbohydrates except fiber break down into sugars. Sugars and

1989 Recommended Dietary Allowances (RDA)

AGE (YR)	(kcal) ENERGY	(g) PROTEIN	(µg RE) VITAMIN A	(mg α-TE) VITAMIN E	(µg) VITAMIN K	(mg) VITAMIN C	(mg) THIAMIN	(mg) RIBOFLAVIN	(mg NE) NIACIN	(mg) VITAMIN B_6	(µg) FOLATE	(µg) VITAMIN B_{12}	(mg) IRON	(mg) ZINC	(µg) IODINE	(µg) SELENIUM
Infants																
0.0–0.5	650	13	375	3	5	30	0.3	0.4	5	0.3	25	0.3	6	5	40	10
0.5–1.0	850	14	375	4	10	35	0.4	0.5	6	0.6	35	0.5	10	5	50	15
Children																
1–3	1300	16	400	6	15	40	0.7	0.8	9	1.0	50	0.7	10	10	70	20
4–6	1800	24	500	7	20	45	0.9	1.1	12	1.1	75	1.0	10	10	90	20
7–10	2000	28	700	7	30	45	1.0	1.2	13	1.4	100	1.4	10	10	120	30
Males																
11–14	2500	45	1000	10	45	50	1.3	1.5	17	1.7	150	2.0	12	15	150	40
15–18	3000	59	1000	10	65	60	1.5	1.8	20	2.0	200	2.0	12	15	150	50
19–24	2900	58	1000	10	70	60	1.5	1.7	19	2.0	200	2.0	10	15	150	70
25–50	2900	63	1000	10	80	60	1.5	1.7	19	2.0	200	2.0	10	15	150	70
51+	2300	63	1000	10	80	60	1.2	1.4	15	2.0	200	2.0	10	15	150	70
Females																
11–14	2200	46	800	8	45	50	1.1	1.3	15	1.4	150	2.0	15	12	150	45
15–18	2200	44	800	8	55	60	1.1	1.3	15	1.5	180	2.0	15	12	150	50
19–24	2200	46	800	8	60	60	1.1	1.3	15	1.6	180	2.0	15	12	150	55
25–50	2200	50	800	8	65	60	1.1	1.3	15	1.6	180	2.0	15	12	150	55
51+	1900	50	800	8	65	60	1.0	1.2	13	1.6	180	2.0	10	12	150	55
Pregnant	+300	60	800	10	65	70	1.5	1.6	17	2.2	400	2.2	30	15	175	65
Lactating																
1st 6 mo.	+500	65	1300	12	65	95	1.6	1.8	20	2.1	280	2,6	15	19	200	75
2nd 6 mo.	+500	62	1200	11	65	90	1.6	1.7	20	2.1	260	2.6	15	16	200	75

1997 Dietary Reference Intakes (DRI)

AGE (YR)	(µg) VITAMIN D	(mg) CALCIUM	(mg) PHOSPHORUS	(mg) MAGNESIUM	(mg) FLUORIDE
Infants					
0.0–0.5	5	210	100	30	0.01
0.5–1.0	5	270	275	75	0.5
Children					
1–3	5	500	460	80	0.7
4–8	5	800	500	130	1.1
Males					
9–13	5	1300	1250	240	2.0
14–18	5	1300	1250	410	3.2
19–30	5	1000	700	400	3.8
31–50	5	1000	700	420	3.8
51–70	10	1200	700	420	3.8
71+	10	1200	700	420	3.8
Females					
9–13	5	1300	1250	240	2.0
14–18	5	1300	1250	360	2.9
19–30	5	1000	700	310	3.1
31–50	5	1000	700	320	3.1
51–70	10	1200	700	320	3.1
71+	10	1200	700	320	3.1
Pregnant			*	+40	
Lactating	*	*	*	*	

* *Values are the same as for other women of comparable age.*

FIGURE 4-1

Recommended Dietary Allowances (RDA) and Dietary Reference Intakes (DRI). (*Source:* National Academy of Science)

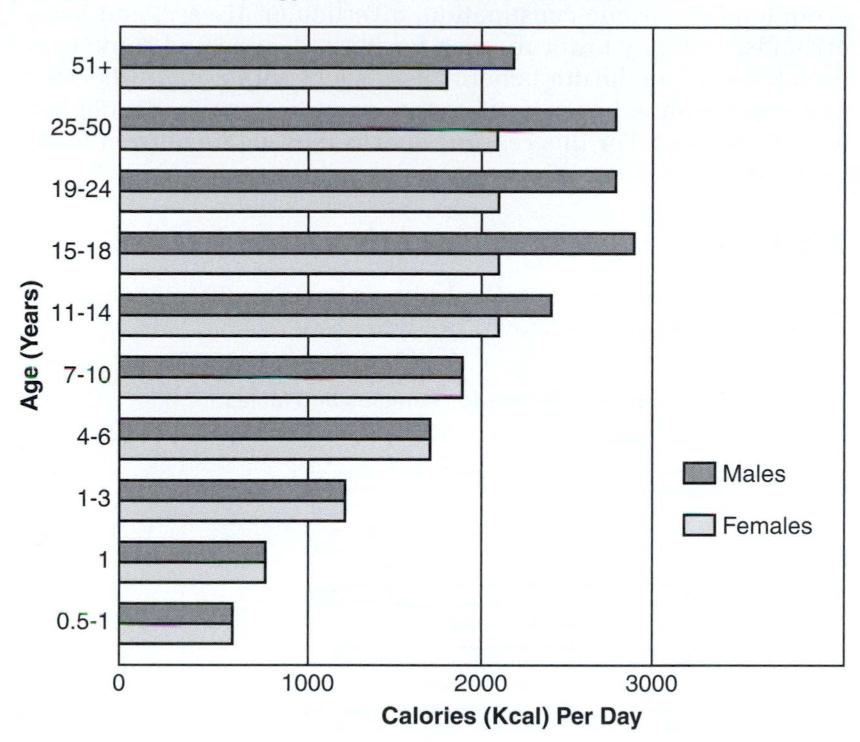

FIGURE 4-2

Energy needs vary for males and females and change as people age. (*Source:* National Academy of Science)

starches occur naturally in many foods that also supply other nutrients. Examples of these foods include milk, fruits, some vegetables, breads, cereals, and grains. Americans eat sugars in many forms, and most people like their taste.

Some sugars are used as natural preservatives, thickeners, and baking aids in foods; they are often added to foods during processing and preparation or when they are eaten. The body cannot tell the difference between naturally occurring and added sugars because they are identical chemically.

Fiber

Fiber is found only in plant foods like whole-grain breads and cereals, beans and peas, and other vegetables and fruits. Individuals should choose a variety of foods daily because the types of fiber in food vary. Eating a variety of fiber-containing plant foods is important for proper bowel function. Fiber can also reduce symptoms of chronic constipation, diverticular disease, and hemorrhoids, and may lower the risk for heart disease and some cancers. Some of the health benefits associated with a high-fiber diet may come from other components present in these foods, not just from fiber itself. For this reason, fiber is best obtained from foods rather than supplements.

Protein

Depending on age and gender, humans require different levels of protein in their diet, as shown in Figure 4-3.

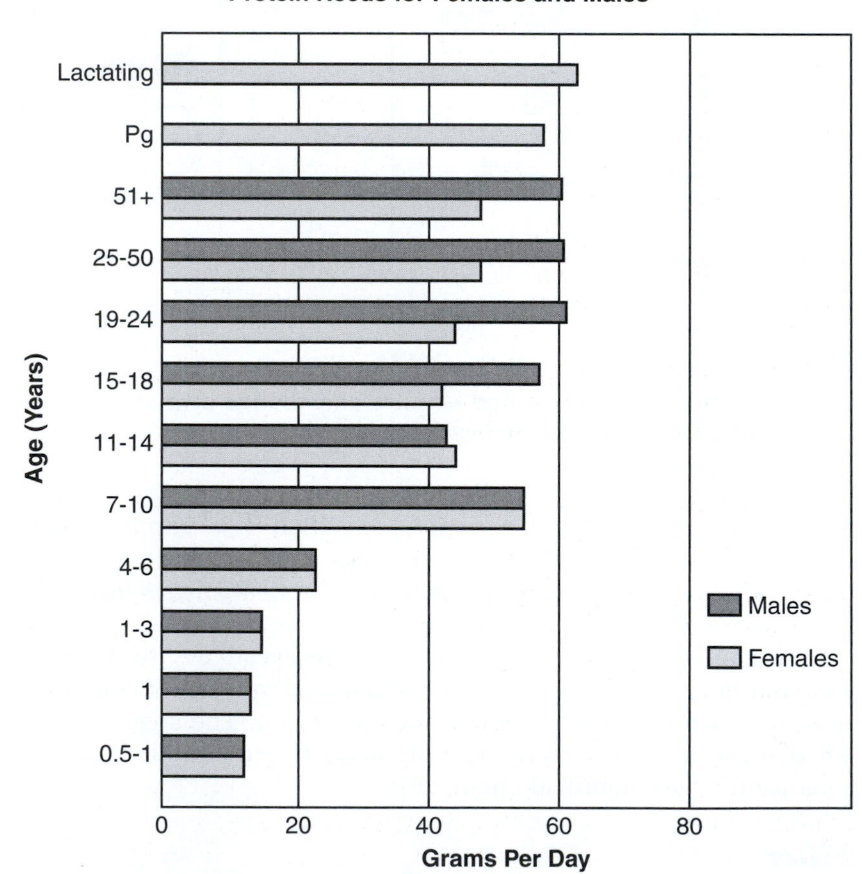

FIGURE 4-3

Protein needs vary for males and females and change as people age. Lactation and pregnancy increase protein needs for females. (*Source:* National Academy of Science)

Requirement for Protein. Humans need the amino acids that the body cannot synthesize in large enough amounts. These are known as **essential amino acids**. They include:

- Phenylalanine
- Tryptophan
- Histidine
- Valine
- Leucine
- Isoleucine
- Lysine
- Methionine
- Threonine
- Arginine

Additional Needs for Nitrogen in Protein. The nitrogen in protein is also used for the synthesis of purines, pyrimidines, porphyrin in nucleic acids, adenosine triphosphate (ATP), hemoglobin, and cytochromes.

Protein Deficiencies. Protein deficiencies in the diet of people from developing countries can lead to the dietary diseases kwashiorkor or marasmus.

- Kwashiorkor results from a protein deficient diet that contains sufficient calories. Common symptoms include a bloated belly and extreme apathy.
- Marasmus results from a deficiency of both protein and calories. Symptoms include a very low body weight and muscle wasting.

Protein Quality. Protein quality describes the nutritive value of a protein. It is ultimately related to providing the amino acids needed for protein synthesis. The body cannot make part of a protein and then wait for an amino acid that is lacking to be supplied to finish the protein. If all the necessary amino acids are present, protein synthesis can occur. If all but one are present, no synthesis can occur. The amino acid that is present in the lowest quantity compared to need is called the **limiting amino acid**.

For example, assume that a protein provides all the essential amino acids in optimal proportions except one. If 90 percent of the required amount of the limiting amino acid is present, all amino acids will be used to the point that 90 percent of the required protein is synthesized. If only 50 percent, then one half of all the essential amino acids will be used for energy.

GOOD FATS

Fish contain high concentrations of a unique type of fat, omega-3 polyunsaturated fatty acids (PUFAs), specifically decosahexaenoic acid (DHA) and ecospentanoic acid (EPA). Due to the health benefits associated with omega-3 fatty acids, health professionals encourage people to eat foods that contain high concentrations. Results of studies suggest that EPA lowers blood fats (triglycerides) and decreases the chance of a blood clot. Foods rich in EPA include flaxseed, legumes, green leafy vegetables, fish, and other seafood. Other studies are being conducted on the anti-inflammatory effects of omega-3 fatty acids. Omega-3 fatty acids could prove beneficial for the prevention of rheumatoid arthritis and systemic lupus erythematosus. Omega-3 fatty acids also demonstrate the ability to prevent fatal heart arrhythmia.

For more information on omega-3 fatty acids or DHA and EPA, search the Web or visit these Web sites:

<hcrc.org/faqs/omega-3.html>
<www.flaxcouncil.ca/flaxnutT.htm>
<www.purdue.edu/UNS/html4ever/970926.Watkins.omega3.html>

Measurement of Protein Quality. Measurement of protein quality is usually done with growing rats. Three main indices are used:

1. Protein efficiency ratio (**PER**)
2. Net protein utilization (**NPU**)
3. Biological value (**BV**)

BV, NPU, and PER are determined from feeding studies (usually with rats). BV and NPU are very similar and are determined by measuring the amount of nitrogen fed and the amount of nitrogen excreted. Thus, they are a measure of nitrogen retained in the body. BV and NPU range from 0 to 1.0, and BV is always equal to or greater than NPU.

PER is determined by feeding rats for 28 days and measuring the weight gain. In this case, weight gain is used as a measure of growth instead of nitrogen retention. PER is the weight gained divided by the amount of protein consumed. PER values range from 0 to about 3.5 with the standard being 2.5 for casein (the major milk protein).

Complementary Relationships. By combining a protein that is deficient in a given amino acid with a protein that has an excess of that amino acid, the protein quality can be increased. Complementary groups include:

- Grains + milk products
- Grains + legumes
- Seeds + legumes

Factors Affecting Protein Use. Four factors affect protein use in the body:

1. Ratios of essential amino acids
2. Amount of protein in the diet
3. Physiological state of the subject
4. Digestibility

If a protein is of poor quality, can a person eat more of it to satisfy needs? This can only be done to a certain extent. Adults must obtain 4 percent of their calories from high-quality protein, whereas an infant requires that 8 percent of its calories be from high-quality protein. For example, corn would be an adequate source of protein for an adult, but not for a growing child. A child fed corn as the sole source of protein would develop kwashiorkor. No matter how much food it was fed, it would not be able to meet its protein requirements.

Lipids

In food, lipids (fats) provide a source of essential fatty acids, add caloric density (energy), act as carriers for flavors, carry fat-soluble vitamins, contribute to texture and mouthfeel, become precursors of flavor, and provide heat transfer medium (in frying).

Essential Fatty Acids. The body can produce most of the fatty acids that it requires. It cannot make some fatty acids that contain double bonds. Given an eighteen carbon fatty acid containing two double bonds called linoleic acid, humans can synthesize all the other fatty acids they require. Thus, linoleic acid is considered as an essential nutrient. The requirements for this essential fatty acid are not well established, and most adults have large amounts of it stored in their adipose tissue. It is estimated that adults should consume about 1 percent and infants 2 percent of their calories from linoleic acid.

Vitamins

Table 4-1 lists the fat- and water-soluble vitamins and their functions.

TABLE 4-1 Functions of Some Vitamins

Vitamins	Some Functions
	Fat-Soluble Vitamins
Vitamin A	Growth and development of bone and epithelial cells, vision
Vitamin D	Absorption of dietary calcium and phosphorus
Vitamin E	Antioxidant in tissues
Vitamin K	Aids in blood clotting
	Water-Soluble Vitamins
Thiamin	**Coenzyme** in energy metabolism
Riboflavin	Coenzyme in many enzyme systems
Niacin	Coenzyme for cell respiration; release of energy from fat, carbohydrates, and proteins
Vitamin C	Metabolism of amino acids, fats, lipids, folic acid, and cholesterol control; collagen formation
Vitamin B_{12}	Coenzyme for red blood cell maintenance and nerve tissue; carbohydrate, fat, and protein metabolism

Minerals

Table 4-2 lists some of the macrominerals and microminerals and their functions.

FOOD PYRAMID

Recently, the FDA has proposed the Food Pyramid to help consumers determine what they should eat to meet their dietary needs and to prevent disease. The following is suggested:

- Bread, cereal, rice, and pasta–6 to 11 servings
- Vegetables–3 to 5 servings
- Fruits–2 to 4 servings
- Milk, yogurt, and cheese–2 to 3 servings
- Meat, poultry, fish, nuts, and beans–2 to 3 servings
- Fats, oils, and sweeteners–use sparingly

TABLE 4-2 Functions of Some Minerals

Mineral (Requirement)	Some Functions
Calcium	Bone mineral; blood clotting; nerve, muscle, and gland function
Phosphorus	Bone mineral, part of many proteins involved in metabolism
Iron	Part of hemoglobin and some enzymes, oxygen transport
Copper	Iron absorption, hemoglobin synthesis, skin pigments, collagen metabolism
Magnesium	Bone mineral, enzyme activator; energy metabolism
Sodium, Potassium, Chloride	Tissue fluid pressure and acid-base balance, passage of nutrients and water into cells, nerve and muscle function
Zinc	Activator of many enzymes
Iodine	Thyroid function
Manganese	Synthesis of bone and cartilage components, cholesterol metabolism
Selenium	Removal of peroxides from tissues, enzyme activation

The USDA also has made some recommendations and provides a set of guidelines for healthy eating. These last two are generally much easier for the average person to follow and ensure adequate nutrition in the diet.

Figure 4-4 shows the Food Pyramid.

For information on specific nutrients supplied by foods, refer to Table A-8.

DIGESTIVE PROCESSES

The processing of food takes place in four stages:

1. Ingestion
2. Digestion
3. **Absorption**
4. Elimination

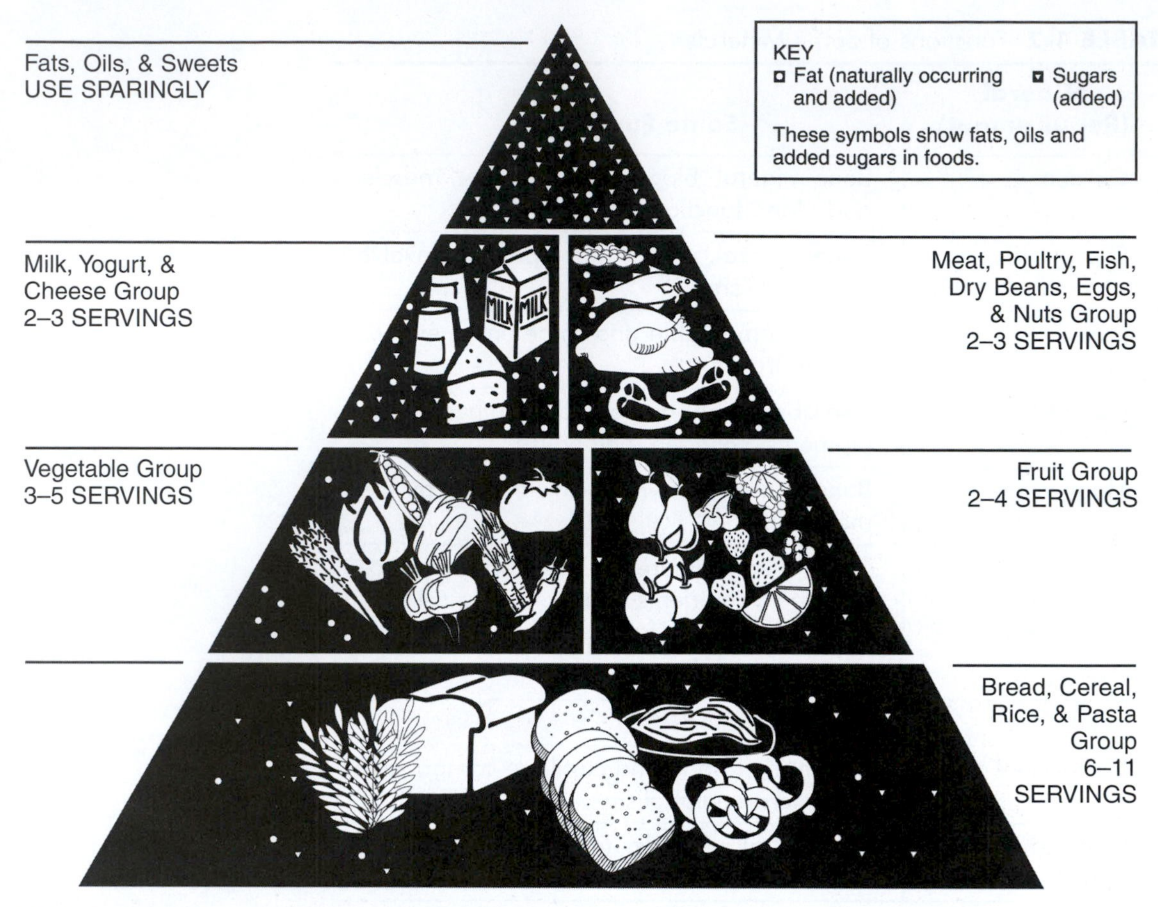

FIGURE 4-4

Food pyramid. (*Source:* USDA)

Ingestion

The act of eating. This is the first of four main stages of food processing.

Digestion

Digestion breaks down food into molecules small enough to be absorbed. It breaks polymers into monomers that are easier to absorb and that can be used to synthesize new polymers required by the organism.

Absorption

Cells that line the digestive tract take up the nutrients. Nutrients are transported to the cells where they are incorporated into the cells and converted to energy that may be used immediately or stored until needed.

Elimination

In the last stage of food processing–elimination–undigested wastes pass out of the digestive tract.

Components of the Human Digestive System

Figure 4-5 illustrates the human digestive system. The following structures are considered parts of the digestive system:

- Mouth
- Tongue
- Pharynx
- Salivary glands
- Esophagus
- Stomach
- Liver
- Gall bladder
- Pancreas
- Small intestine
- Large intestine
- Rectum
- Anus

Mouth. Food enters the mouth and is reduced in size by teeth and tongue. Salivary glands secrete saliva, which lubricates, buffers, contains antimicrobial substances, and contains amylase to digest starch.

Swallowing. Food passes to the pharynx, which contains both the trachea and the esophagus. The epiglottis prevents food from entering the trachea. Food passes through the esophagus into the stomach.

Stomach. The stomach stores and digests food. It contains pits leading to gastric glands with three types of cells:

1. Mucous cells–produce mucus, which lubricates and protects lining
2. Parietal cells–secrete hydrochloric acid
3. Chief cells–secrete pepsinogen

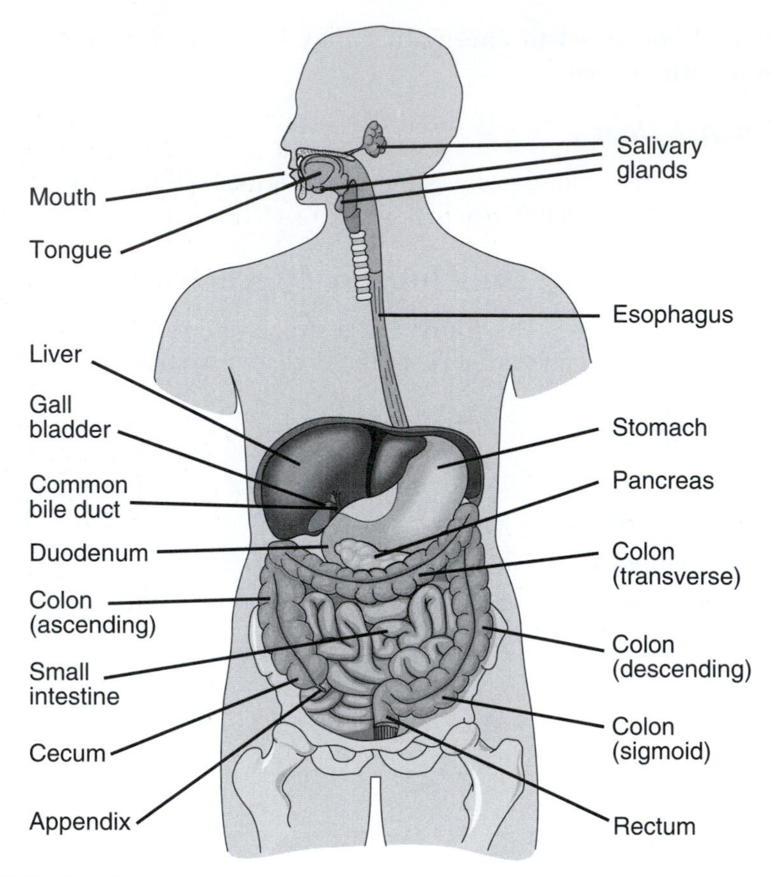

FIGURE 4-5

Parts of the digestive system.

Small Intestine. The small intestine receives food from the stomach. Bile enters from the liver via the gall bladder. **Enzymes** in the small intestine come from the pancreas. The small intestine is the site of most digestion and the site of most absorption.

Pancreas. The pancreas secretes many digestive enzymes into the small intestine. It also produces and secretes the hormones insulin and glucagon.

Liver. The liver produces bile, but no digestive enzymes. Bile contains bile salt, which emulsifies fats. Bile is made from cholesterol and stored in the gall bladder.

Duodenum. The duodenum is the first 10 inches (25 cm) of small intestine. It receives enzymes from the pancreas, and it neutralizes acid from the stomach. The duodenum is the site of most chemical changes in food.

- Enzymatic action. Pancreatic amylases break down starch into the disaccharide maltose. Disaccharides are converted to monosaccharides. Maltose is hydrolyzed by maltase to give glucose + glucose. Sucrose is hydrolyzed by sucrase to give glucose + fructose. Lactose is hydrolyzed by lactase to give glucose + galactose.

  Not all simple saccharides can be easily digested. Lactose intolerance is a very common problem. Lactose intolerance is caused by a deficiency of lactase, the enzyme that catalyzes the hydrolysis of lactose to glucose and galactose. Lactose cannot be absorbed unless it is broken down first. If insufficient lactase is present, then some of the lactose will travel to the large intestine and cause problems. It will increase the flow of water into the intestine due to increased osmolality of the contents of the intestine. This results in osmotic diarrhea.

  Emulsified fat is broken down by lipase into fatty acids and glycerol. Nucleic acid is converted by nucleases into nucleotides, which are converted to nitrogenous bases, sugars, and phosphates.

  Proteins or polypeptides are converted to smaller peptides by trypsin and chymotrypsin, and small polypeptides are converted into amino acids by enzymes such as aminopeptidases, carboxypeptidases, and dipeptidases.

- **Absorption.** Following enzymatic action in the duodenum, food is absorbed in the remainder of the small intestine. This portion of the small intestine has a very large surface area. It contains villi and microvilli and is rich in capillaries and lymph vessels.

Large Intestine. Material not digested or absorbed passes into the large intestine. Most of the water (90 percent) is absorbed into the blood from the large intestine. Some vitamins are produced by bacteria in the large intestine and are absorbed. The residue is stored in the rectum, and then eliminated through the anus.

VEGETARIAN DIETS

Some people eat vegetarian diets for reasons of culture, belief, or health. Most vegetarians eat milk products and eggs, and as a group, these lacto-ovo-vegetarians enjoy excellent health. Vegetarian diets are consistent with the Dietary Guidelines for

Americans and can meet Recommended Dietary Allowances for nutrients. Individuals can get enough protein from a vegetarian diet as long as the variety and amounts of foods consumed are adequate. Meat, fish, and poultry are major contributors of iron, zinc, and B vitamins in most American diets, and vegetarians should pay special attention to these nutrients.

Vegans eat only food of plant origin. Because animal products are the only food sources of vitamin B_{12}, vegans must supplement their diets with a source of this vitamin. In addition, vegan diets, particularly those of children, require care to ensure adequacy of vitamin D and calcium, which most Americans obtain from milk products.

BIOAVAILABILITY OF NUTRIENTS

Chemical analysis of a food may determine the presence of a nutrient, but this can be misleading. Though the nutrient may be in the food, whether it is available in a form that can be used by the metabolic processes of the body is another question. If the nutrient is in a form that can be used, it said to be bioavailable. Factors determining the **bioavailability** of a nutrient include digestibility, absorption, nutrient-to-nutrient interactions, binding to other substances, processing and cooking procedures. Also, age, gender, health, nutritional status, drugs, and food combinations influence the bioavailability of carbohydrates, proteins, fats, vitamins, and minerals.

STABILITY OF NUTRIENTS

The nutritive value of food starts with the genetics of the plants or animals. Fertilization, weather, and the maturity at harvest also influence the composition of the plant or animal being used for food. Storage before processing affects nutrient levels. Then all of the processing steps continue to affect the nutrient levels in a food. Finally, preparation in the home or at the restaurant can reduce the final nutritive value of a food before the digestive process.

A primary goal of food science is to preserve the nutrients through all phases of food harvesting, processing, storage, and preparation. To do this, the food scientist needs to know what the **stability** of the nutrients is under varying conditions of pH, air, light, heat, and cold. Nutrient losses are small in most modern food processing operations, but when nutrient losses are unavoidably high, the law allows enrichment.

DIET AND CHRONIC DISEASE

Food choices also can help to reduce the risk for chronic diseases, such as heart disease, certain cancers, diabetes, stroke, and osteoporosis, which are leading causes of death and disability among Americans. Good diets can reduce major risk factors for chronic diseases–factors such as obesity, high blood pressure, and high blood cholesterol.

Healthful diets contain the amounts of essential nutrients and calories needed to prevent nutritional deficiencies and excesses. Healthful diets also provide the right balance of carbohydrate, fat, and protein to reduce risks for chronic diseases.

Summary

The Food Nutrition Board of the National Research Council establishes Recommended Dietary Allowances and Dietary References Intakes. These guidelines provide daily nutrient levels for maintaining normal functions and health. The RDA lists recommendations for energy; protein; vitamins A, E, K, C, and the B vitamins; and the minerals iron, zinc, iodine, and selenium. These recommendations vary according to age, gender, pregnancy, and lactation. Protein requirements for humans should consider the essential amino acids. The food pyramid provided by the FDA is easy for the average person to follow to ensure adequate nutrition.

The digestive process includes ingestion, digestion, absorption, and elimination. Nutrients in the diet are progressively broken into smaller components by mechanical, chemical, and enzymatic means. Small molecules resulting from digestion are absorbed to supply the body with energy, protein, vitamins, and minerals.

Review Questions

Success in any career requires knowledge. Test your knowledge of this chapter by answering these questions or solving these problems.

1. Name six minerals required by the body.
2. Identify the protein requirement for a 19-year-old male and female.
3. Describe the function of protein in the diet.
4. How many calories are in 1 gram of protein, carbohydrate, fat, and alcohol?
5. ___________ acid is an essential fatty acid.

6. Identify the organ of digestion that receives enzymes from the pancreas.
7. During digestion, enzymes such as aminopeptidases, carboxypeptidases, and dipeptidases convert polypeptides into ___________ ___________.
8. What nutritional deficiency causes kwashiorkor and marasmus?
9. List five essential amino acids.
10. What factor determines protein quality?

Student Activities

1. Use a log to track your diet for five days. Analyze how closely your diet conforms to the recommendations of the Food Pyramid. Estimate your average daily consumption of calories, fat, and protein by using Table A-8. Report your findings to the class.
2. Using Figure 4-1 (page 66), look up the RDA for all the nutrients for your age and gender. Report your findings.
4. Develop a short presentation on one of these topics: bioavailability, fiber, or nutrient stability in foods.
4. Collect labels from food products and report on the use of the RDAs on the label.
5. Develop a mnemonic that will help you remember the essential amino acids.
6. If possible, obtain a digestive tract of a pig and use this to discuss and describe the digestive process in humans. Dissect the tract and trace the passage of food.
7. Plan a meal that meets the dietary guidelines of the food pyramid or RDA. Then describe how each item on the menu is digested and in what form it is absorbed.

Resources

Brody, J. E. 1981. *Jane Brody's nutrition book.* New York: Bantam Book.

Drummond, K. E. 1994. *Nutrition for the food service professional.* 2nd ed. New York: Van Nostrand Reinhold.

Ensminger, A. H., M. E. Ensminger, J. E. Konlande, and J. R. Robson. 1994. *Foods and nutrition encyclopedia.* 2 Vols. Boca Raton, FL: CRC Press.

Gardner, J. E., Ed. 1982. *Reader's digest. Eat better, live better.* Pleasantville, NY: Reader's Digest Association, Inc.

Potter, N. N., and J. H. Hotchkiss. 1995. *Food science,* 5th ed. New York: Chapman and Hall.
Vaclavik, V. A., and E. W. Christina. 1999. *Essentials of food science.* Gaithersburg, MD: Aspen Publishers, Inc.
Vieira, E. R. 1996. *Elementary food science,* 4th ed. New York: Chapman and Hall.

Internet

Internet sites represent a vast resource of information. The URLs (uniform resource locator) for the World Wide Web sites can change. Using one of the search engines on the Internet such as Yahoo!, HotBot, AltaVista, Excite, Dogpile, About, or Google, find more information by searching for these words or phrases: food pyramid, Recommended Dietary Allowances, Dietary Reference Intakes, protein deficiencies, protein quality, amino acids, protein efficiency ratio, net protein utilization, biological value, human digestive system, vegetarian diets, vitamins, minerals, fiber. Also, Table A-7 provides a listing of some useful Internet sites that can be used as a starting point.

Chapter 5

Food Composition

Objectives

After reading this chapter, you should be able to:

- Find foods in a food composition table and describe their nutritional value
- List three methods of determining the composition of foods
- Describe the method for determining the caloric content of foods
- Explain the difference between Calorie and calorie
- Identify common abbreviations and terms used in a food composition table
- Discuss the use of food composition tables
- List four factors that affect the nutrient content of foods

Key Terms

bomb calorimeter
Calorie
chromatography
energy
ether extract
proximate analysis
spectrophotometry

Food composition tables are used to evaluate the nutritional value of food supplies, to develop food distribution programs, to plan and evaluate food consumption surveys, to provide nutritional counseling, and to estimate the nutritional content of individual diets.

DETERMINING THE COMPOSITION OF FOODS

Nutrient content of foods is influenced by variety, season, geographical differences, stage of harvesting, handling, commercial processing, packaging, storage, display, home preparation, cooking, and serving. The composition of foods is determined by a variety of scientifically sound, standardized methods. The first system of approximating the value of a food or feed for nutritional purposes was developed at the Weende Experiment Station in Germany more than 100 years ago. This system separates a food into nutritive fractions through a series of chemical determinations. These determinations reflect a food's nutritive value. The different fractions included water or dry matter, crude protein, **ether extract** or fat, crude fiber, nitrogen-free extract (sugars and starches), and ash or total mineral. This system became known as **proximate analysis**. Newer methods of determining the composition of foods have replaced or supplemented the old proximate analysis and allowed determination of more specific nutrients in foods. Some of these newer methods include **spectrophotometry**, liquid **chromatography**, and gas chromatography (Figure 5-1). These new methods allow the determination of fatty acids, cholesterol, amino acids, specific minerals, and vitamins.

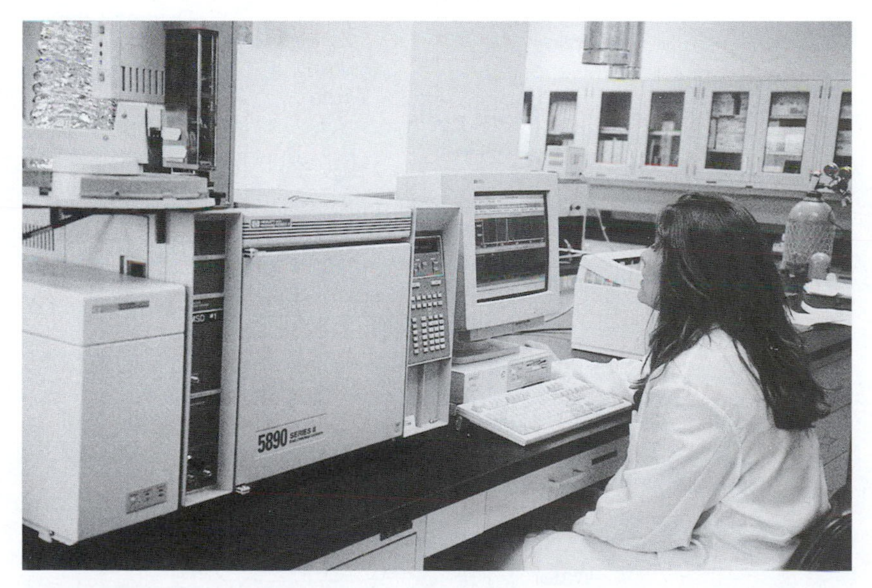

FIGURE 5-1

High-tech equipment such as the gas chromatograph used to determine some nutrients in foods.

ENERGY IN FOOD

Energy in food is measured in terms of calories. A **Calorie** is a metric unit of heat measurement. The small calorie (cal) is the amount of heat required to raise the temperature of 1 gram of water from 14.5° to 15.5°C. The definition now generally accepted in the United States, and is standard in thermochemistry, is that 1 cal is equal to 4.1840 joules (J).

A large calorie, or kilocalorie (Cal), usually referred to as a calorie and sometimes as a kilogram calorie, equals 1,000 cal. This unit is used to express the amount of energy that a food provides when consumed.

Calorimeters measure the heat developed during the combustion of food. **Bomb calorimeters** have been used to determine the calorie content of foods. Basically, a bomb calorimeter consists of an enclosure in which the reaction takes place, surrounded by a liquid, such as water, that absorbs the heat of the reaction and thus increases in temperature. Measurement of this temperature rise for a known weight of food permits the total amount of heat generated to be calculated.

The design of a typical bomb calorimeter is shown in the Figure 5-2. The food to be analyzed is placed inside a steel reaction

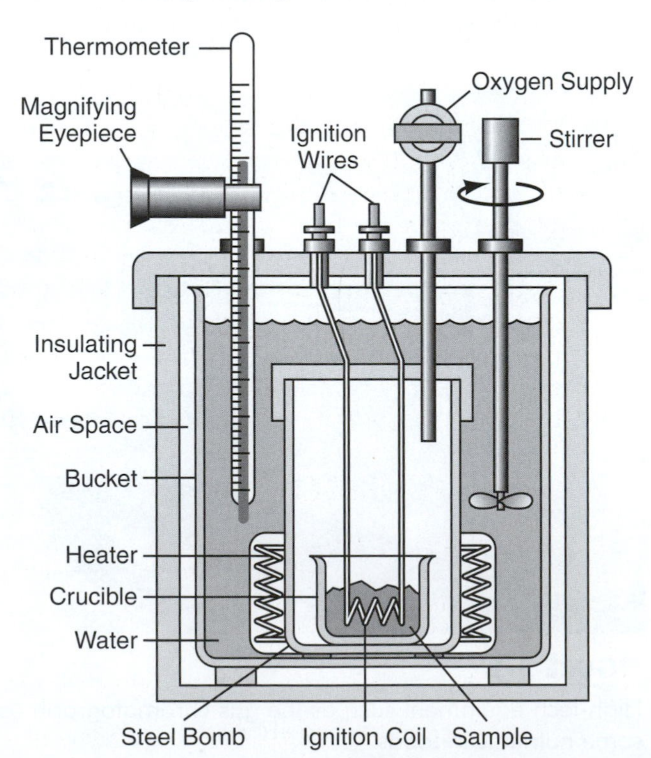

FIGURE 5-2

Bomb calorimeter determines the energy content of foods.

MUSCLE CRAMPS AND DIET

Almost everyone has experienced a muscle cramp. The muscle becomes contracted and rigid, and this is usually quite painful. The contracted muscle gets locked into a self-sustaining knot, which can last for a few minutes, hours, or even days. The informal name for this involuntary muscle cramp is charley horse. The term charley horse is most often used in connection with athletes, especially baseball players.

Many muscle cramps are associated with exercise. These cramps are often due to a depletion or imbalance of salts in the muscle tissue, especially calcium, sodium, and potassium, which are lost through perspiration (sweat). A buildup of lactic acid, one of the byproducts of heavy exercise, also can contribute to cramping. Drinking "electrolyte" drinks that restore the salt balance can often relieve cramps. Also, eating foods high in sodium and potassium can relieve cramps.

Check out the foods in the Food Composition Table, Table A-8, and suggest some foods that would be good for muscle cramps.

To learn more about muscle cramps and diet, search the Web or start by visiting this site:

<www.covenanthealth.com/features/health/sports/spor3206.htm>

vessel called a bomb. The steel bomb is placed inside a bucket filled with water, which is kept at a constant temperature relative to the entire calorimeter by use of a heater and a stirrer. The temperature of the water is monitored with a thermometer fitted with a magnifying eyepiece, which allows accurate readings to be taken. Inserting an air space between the bucket and an exterior insulating jacket minimizes heat losses. Slots at the top of the steel bomb allow ignition wires and an oxygen supply to enter the vessel. When an electric current passes through the ignition coil, a combustion reaction occurs. The heat released from the sample is largely absorbed by the water, which results in an increase in temperature.

FOOD COMPOSITION TABLES

The Food Composition Table, Appendix Table A-8, is from the U.S. Department of Agriculture, Agricultural Research Service (1990. Data Set 72-1, Release 3.2). Other food composition tables can be found at their Web site:

<www.nal.usda.gov/fnic/foodcomp/Data/index.html>

Table 5-1 explains the abbreviations used in Table A-8.

TABLE 5-1 Explanation of Abbreviations Used in the Food Composition Table, Table A-8.

Description	Abbreviation
Food item number	No.
Description of food and measure	Description of Food
Percentage of water	Water (%)
Food energy in kilocalories	Energy (kcal)
Protein in grams	Prot (g)
Fat in grams	Fat (g)
Saturated fatty acid in grams	Sat (g)
Monounsaturated fatty acid in grams	Mono (g)
Polyunsaturated fatty acid in grams	Poly (g)
Cholesterol in milligrams	Chols (mg)
Carbohydrate in grams	Carb (g)
Calcium in milligrams	Ca (mg)
Phosphorus in milligrams	P (mg)
Iron in milligrams	Fe (mg)
Potassium in milligrams	K (mg)
Sodium in milligrams	Na (mg)
Vitamin A in International Units	Vit A (IU)
Vitamin A in Retinol Equivalents	Vit A (RE)
Thiamin in milligrams	Thmn (mg)
Riboflavin in milligrams	Ribof (mg)
Niacin in milligrams	Niacin (mg)
Ascorbic acid in milligrams	Vit C (mg)

Summary

Food composition tables are used to evaluate diets and food supplies. Methods such as spectrophotometry, liquid chromatography, and gas chromatography determine the composition of foods. The bomb calorimeter measures the caloric content of foods. Many food composition tables are available, but the USDA maintains and updates data on the composition of foods.

Review Questions

Success in any career requires knowledge. Test your knowledge of this chapter by answering these questions or solving these problems.

Note: The answers to some of these questions will be found by using Table A-8.

1. How many Calories and grams of protein are in 3 oz. of Froot Loops® cereal?
2. How many grams of fat are in one slice of cheese pizza?
3. Describe item #4270.
4. List three methods for determining the composition of foods.
5. A ___________ is defined as the amount of heat required to raise the temperature of one gram one °C.
6. Describe two uses of a food composition table.
7. Name four factors that affect the nutrient content of foods.
8. Explain the relationship between Calorie, Kcal, calorie, and cal.
9. Identify the following abbreviations: oz, mg, IU, RE, mono, sat, poly, carb, chols.
10. In terms of energy and protein, what is the difference between a slice of white bread and a slice of whole wheat bread?

Student Activities

1. Check out USDA's Nutrient Database for Standard Reference on the World Wide Web. Containing nearly 6,000 foods and over 70 components, it is the nation's primary source of food composition data at this URL:

   <www.nal.usda.gov/fnic/foodcomp>

   Use this Web site or use Table A-8 to answer these questions:

   a. Energy (calories) is reported in both ___________ and ___________. Search the home page to find the difference between these units.

b. Another name for fat is ___________.

c. Another name for vitamin C is ___________.

d. What minerals are listed in the database?

e. The three major classes of fatty acids are ___________, ___________, and ___________.

f. How much sodium is in a teaspoon of salt and in a large double cheeseburger with everything?

g. Deep yellow and dark green leafy vegetables are among the best sources of vitamin A. List three.

h. Compare the fat in several popular snack foods. List them from most to least fat.

i. Find three fruits low in fat.

j. Compare the vitamin C content in five different beverages. Use 1-cup portions.

k. Which of the following is highest in cholesterol–2 tbsp chunky peanut butter, 1 cup orange juice, a batter-fried chicken drumstick, 3 cups rice, or ½ cup salsa?

l. Provide product descriptions and dietary fiber values for NDB No. 09200 (1 large) and NDB No. 16005 (1 cup).

m. Which has more calcium–a cup of 1% low fat milk or ½ cup 1% low fat cottage cheese? Record values and item descriptions.

2. Develop a report that describes other ways you could use the USDA's Nutrient Database.

3. Create a form for tracking the foods you eat for five days.

Resources

Brody, J. E. 1981. *Jane Brody's nutrition book.* New York: Bantam Books.

Cremer, M. L. 1998. *Quality food in quantity. Management and science.* Berkeley, CA: McCutchan Publishing Corporation.

Drummond, K. E. 1994. *Nutrition for the food service professional,* 2nd ed. New York: Van Nostrand Reinhold.

Ensminger, A. H., M. E. Ensminger, J. E. Konlande, and J. R. Robson. 1994. *Foods and nutrition encyclopedia.* 2 Vols. Boca Raton, FL: CRC Press.

Gardner, J. E., Ed. 1982. *Reader's digest. Eat better, live better.* Pleasantville, NY: Reader's Digest Association, Inc.

Horn, J., J. Fletcher, and A. Gooch. 1997. *Cooking a to z. The complete culinary reference source.* Glen Ellen, CA: Cole Publishing Group, Inc.

Wagner, S. Ed. 1999. *The recipe encyclopedia: The complete illustrated guide to cooking.* San Diego, CA: Thunder Bay Press.

Internet

Internet sites represent a vast resource of information. The URLs (uniform resource locator) for the World Wide Web sites can change. Using one of the search engines on the Internet such as Yahoo!, HotBot, AltaVista, Excite, Dogpile, About, or Google, find more information by searching for these words or phrases: food composition, proximate analysis, calorimetry, nutritional value, dietary intake. Also, Table A-8 provides a listing of some useful Internet sites that can be used as a starting point.

Chapter 6

Quality Factors in Foods

Objectives

After reading this chapter, you should be able to:

- Describe the influence of color on food quality
- Identify the instrument that could be used to measure food texture
- Discuss the influence of color, texture, size, and shape on consumer acceptance
- Describe how water changes texture
- Identify six words used to describe food flavor
- Describe sensory methods humans use to determine food flavor
- Discuss three factors that can affect food flavor
- Explain three means for maintaining or assessing quality in foods
- Describe the role the USDA plays in food quality

Key Terms

astringency
chroma
GMP
HACCP
hue
Maillard reaction
phenolic compounds
pigment
rheology
standards
texture
TQM
value
volatile

Quality of a food product involves maintenance or improvement of the key attributes of the product–including color, flavor, texture, safety, healthfulness, shelf life, and convenience. To maintain quality, it is important to control microbiological spoilage, enzymatic degradation, and chemical degradation. These components of quality depend upon the composition of the food, processing methods, packaging, and storage.

APPEARANCE FACTORS

Of the sensory attributes of food, those related to appearance are the most susceptible to objective measurement, but appearance is important to the consumers. They have certain expectations of how food should look. Two separate categories of appearance include:

1. Color attributes
2. Geometric attributes (size and shape)

Color

Of these two, color is by far the most important. Consumers expect meat to be red, apple juice to be light brown and clear, orange juice to be orange, egg yolks to be bright yellow-orange, and so on.

Food color measurements provide an objective index of food quality. Color is an indication of ripeness or spoilage. The end point of cooking processes is judged by color. Changes in expected colors can also indicate problems with the processing or packaging.

Browns and blackish colors can be either enzymatic or nonenzymatic reactions. The major nonenzymatic reaction of greatest interest to scientists is the **Maillard reaction**, which is the dominant browning reaction. Other less explained reactions include blackening in potatoes or the browning in orange juice. The enzymatic browning found widespread in fruits and selected vegetables is due to the enzymatic catalyzed oxidation of the **phenolic compounds**.

Naturally occurring **pigments** play a role in food color. Water-soluble pigments may be categorized as anthocyanins and anthoxthanins. Lesser known water-soluble pigments include the leucoanthocyanins. Fat-soluble plant pigments are primarily categorized into the chlorophyll and carotenoid pigments. These green and orange-yellow pigments considerably impact the color. Myoglobins contribute to the color of meat.

Measuring Color

In order to maintain quality, the color of food products must be measured and standardized. If a food is transparent, like a juice or a colored extract, colorimeters or spectrophotometers can be used for color measurement. The color of liquid or solid foods can be measured by comparing their reflected color to defined (standardized) color tiles or chips. For a further measurement of color, reflected light from a food can be divided into three components: **value**, **hue**, and **chroma**. The color of a food can be precisely defined with numbers for these three components with tri-stimulus colorimetry. Instruments such as the Hunterlab Color and Color Difference Meter measure the value, hue, and chroma of foods for comparisons.

Size and Shape

Depending on the product, consumers expect foods to have certain sizes and shapes (Figure 6-1). For example, consumers have some idea of what an ideal french fry should look like, or an apple, or a cookie, or a pickle. Size and shape are easily measured. Fruits and vegetables are graded based on their size and shape, and this is done by the openings they will pass through during grading. Now computerized electronic equipment can determine the size and shape of foods.

TEXTURAL FACTORS

Consumers expect gum to be chewy, crackers to be crisp, steak to be tender, cookies to be soft, and breakfast cereal to be crunchy. The **texture** of food refers to the qualities felt with the fingers, the

FIGURE 6-1

Consumers expect foods to have a particular shape.

tongue, or the teeth. Textures in food vary widely, but any departure from what the consumer expects is a quality defect.

Texture is a mechanical behavior of foods measured by sensory (physiological/psychological) or physical (**rheology**) means. Rheology is the study of the science of deformation of matter. The four main reasons for studying rheology include:

1. Insight into structure
2. Information used in raw material and process control in industry
3. Applications to machine design
4. Relation to consumer acceptance

Regardless of the reason for studying texture, classification and understanding are difficult because of the enormous range of materials. Moreover, food materials behave differently under different conditions (Figure 6-2).

Texture testing in foods is based upon the action of stress and strain. Many of the methods are based upon compression, shearing, shear-pressure, cutting, or tensile strength. For example, the compressimeter was used to determine the compressibility of cakes and other "spongelike" products. Historically, the penetrometer, has been used to measure gel strength. The Warner-Bratzler shear apparatus has been the standard method of evaluating meat tenderness. The Instron has adapted many of the historical texture

FIGURE 6-2

Texture is important to consumers, and scientists measure the texture of different foods. (*Source:* USDA, ARS Image Gallery)

measuring instruments. It measures elasticity. The Brookfield viscometer will measure the viscosity in terms of Brookfield units. Other instruments used to measure texture include a succulometer and a tenderometer.

Changes in texture are often due to water status. Fresh fruits and vegetables become soggy as cells break down and lose water. On the other hand if dried fruits take on water, their texture changes. Bread and cake lose water as they become stale. If crackers, cookies, and pretzels take up water, they become soft and undesirable.

Various methods are used to control the texture of processed foods. Lipids (fats) are softeners and lubricants used in cakes. Starch and gums are used as thickeners. Protein can also be a thickener, or if coagulated as in baked bread, it can form a rigid structure. Depending on its concentration in a product, sugar can add body as in soft drinks or in other products add chewiness, or in greater concentrations it can thicken and add chewiness or brittleness.

FLAVOR FACTORS

Food flavor includes taste sensations perceived by the tongue–sweet, salty, sour, and bitter–and smells perceived by the nose. Often the terms flavor and smell (aroma) are used interchangeably. Food flavor and aroma are difficult to measure and difficult to get people to agree on. A part of food science called sensory science is dedicated to finding ways to help humans accurately describe the flavors and other sensory properties of their food.

Flavor, like color and texture, is a quality factor. It influences the decision to purchase and to consume a food product. Food flavor is a combination of taste and smell, and it is very subjective and difficult to measure. People differ in their ability to detect tastes and odors. People also differ in their preferences for these.

Besides the tastes of sweet, salty, sour, and bitter (see Figure 6-3), an endless number of compounds give food characteristic aromas, such as

- Fruity
- **Astringency**
- Sulfur
- Hot

Sweetness may result from sugars like arabinose, fructose, galactose, glucose, riboses, xylose, and other sweetners. Organic acids may be perceived on the bottom of the tongue. Some of these

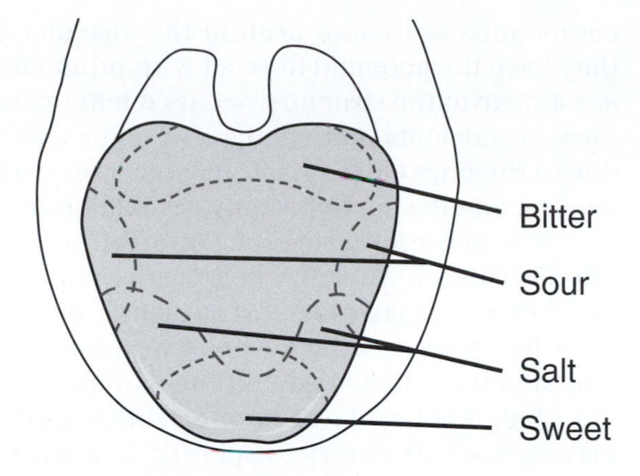

FIGURE 6-3
Taste areas on our tongues help perceive flavor.

common acids are citric, isocitric, malic, oxalic, tartaric, and succinic acids. The fruity flavors are often esters, alcohols, ethers, or ketones. Many of these are **volatile** and are associated with acids.

Phenolic compounds are closely related to the sensory and nutritional qualities of plants. They are found in many fruits, including apples, apricots, peaches, pears, bananas, and grapes; and vegetables such as avocado, eggplant, and potatoes, and contribute to color, astringency, bitterness, and aroma. Most phenolic

TASTE OF HOT CHILI PEPPERS

The active ingredient in hot chili peppers is a substance called capsaicin. It is so potent that even a minute amount has a strong effect. Why does capsaicin taste so "hot"? There are receptor molecules in the membranes of certain pain sensor nerves that respond strongly to capsaicin molecules. Heat, acids, and other various chemical or physical stimuli, including injuries, also activate the same pain receptors.

Capsaicin's "hot" taste is created by directly activating the heat/pain sensory system. Once capsaicin triggers a receptor, the same receptor becomes even more sensitive to heat, making warm soup taste even hotter.

For more information about capsaicin, search the Web or visit these Web sites:

<www.apnet.com/inscight/10221997/graphb.htm>
<www.steveweiss.com/products.html>

compounds are found around the vascular tissues in plants, but they have the potential to react with other components in the plant as damage to the structure occurs during handling and processing. Loss of nutrients and changes in color and flavor occur in foods due to the phenolic compounds' reaction with polyphenol oxidase, or PPO, an enzyme that catalyzes oxidation.

The sense of taste is a powerful predictor of food selection. The four main tastes the body experiences are sweet, sour, salty, and bitter. Humans like sweet-tasting foods. Possibly this preference for sweet is a holdover from ancient ancestors, who found that sweetness indicated that the food provided energy.

Judgment of flavor is often influenced by color and texture. Flavors such as cherry, raspberry, and strawberry are associated with the color red. Beef flavor is brown. Actually, the flavor essences are colorless. As for texture, people expect potato chips to be crunchy and gelatin to be soft and cool.

Depending on the food, flavor can also be influenced by:

- Bacteria
- Yeasts
- Molds
- Enzymes
- Heat/cold
- Moisture/dryness
- Light
- Time
- Additives

Finally, depending on the product, the influence these factors have on a food flavor can be positive or negative and sometimes differs depending on the person.

Taste Panels

For consumer quality acceptance, the best method of measuring taste is to have people taste the products. Taste panels may be a group of professionals, or they may be a group of customers. Typically taste panels are in separate booths so that they cannot influence each other. Food samples are coded with letters and numbers, and tasters are given an evaluation form to complete as they taste the product and evaluate it (Figure 6-4).

FIGURE 6-4
Taste test panels help determine the acceptance of new products. (*Source:* USDA, ARS Image Gallery)

ADDITIONAL QUALITY FACTORS

Additional quality factors include shelf life, safety, healthfulness, and convenience. The extension of storage life of products generally involves heat treatments, irradiation, refrigeration or freezing, or reduction of water activity by either addition of water-binding agents, like sugars, or drying. In many cases compromises are made to achieve desired shelf life or convenience. Such processes, though improving shelf life, almost always have some effect on the components of the food. The factors that influence changes of various ingredients in foods include the following: proteins, lipids, carbohydrates, vitamins, chemicals, and microbiological characteristics.

Proteins

Heat denaturation changes solubility and texture of foods; light oxidation of protein causes off flavors. Enzymatic degradation of protein can cause changes in body and texture and also bitter flavors. Freezing can alter protein conformation and solubility in some cases.

Lipids

Enzymatic hydrolysis of lipids can cause off flavors, such as soapy or goaty, depending on type of oil. This also makes frying oils unsuitable for use. It can change functionality and crystallization properties. Oxidation of unsaturated fatty acids causes off flavors.

Carbohydrates

High-heat treatments cause interactions between reducing sugars and amino groups to give Maillard browning and changes in flavor. Hydrolysis of starch and gums can change texture of food systems. Some starches can be degraded by enzymes or under acidic conditions.

Vitamins

Depending upon the vitamin, losses can occur when the food is heated, exposed to light, or to oxygen.

Chemicals and Microbiological Characteristics

Ensuring the safety of food involves careful control of the process from the farm gate to the consumer. Safety includes control of both chemical and microbiological characteristics of the product. Most processing places emphasis on microbial control, and often has as its objective the elimination of organisms or prevention of their growth.

Processes that are aimed at prevention of growth include:

- Irradiation
- Refrigeration
- Freezing
- Drying
- Control of water activity (addition of salt, sugars, polyols, and so forth)

Processes that are aimed at minimizing organisms include:

- Pasteurization
- Sterilization (canning)
- Cleaning and sanitizing
- Membrane processing

A further method of processing that is aimed at the control of undesirable microflora is the deliberate addition of microorganisms and the use of fermentation.

Safety from a chemical viewpoint generally relates to keeping undesirable chemicals, such as pesticides, insecticides, and antibiotics, out of the food supply. Making sure that food products are free from extraneous matter (metal, glass, wood, etc.) is another facet of food safety.

Today's consumers want food products that are convenient to use and still have all the qualities of a fresh product.

QUALITY STANDARDS

Quality **standards** help ensure food quality (see Figure 6-5). Types of standards include research standards, trade standards, and government standards. Research standards are set up by a company to help ensure the quality of its products in a competitive market. Trade standards are established by members of an industry. These are voluntary and assure at least minimum acceptable quality. As for government standards, some are mandatory and some are optional. Grade standards established by the government provide a common language for producers, dealers, and consumers for buying and selling.

Quality Standards USDA/AMS

In cooperation with industry, the Agricultural Marketing Service (AMS) of the USDA develops and maintains official U.S. quality standards and grades for hundreds of agricultural products (see Figure 6-6). These standards are based on attributes which describe the value, utility, and entire range of quality for each product in the following categories:

- Nuts and specialty crops
- Dairy
- Poultry and eggs (including rabbits)
- Fresh fruits and vegetables (including fresh fruits and vegetables for processing)
- Processed fruits and vegetables (including juices and sugar products)
- Livestock (including wool and mohair)

The USDA/AMS Web site is <www.ams.usda.gov/standards/index.htm>.

Grading and Certification

Quality grading (a user-fee service) is based on the standards developed for each product (see Figure 6-7). Grading services are often operated cooperatively with state departments of agriculture. Quality grades provide a common language among buyers and sellers, which in turn assures consistent quality for consumers.

Beef Programs Page 1 of 5

Characteristic	Certified Angus Beef	Sterling Silver Excel Corp.	SYSCO Supreme Angus Beef	SYSCO Imperial Angus Beef	Certified Hereford Beef	Farmland Angus Beef	Wal-Mart Angus Beef	Packerland Angus Beef[1]	Omaha Steaks Angus Beef	Excel Corp. Angus Pride
Live Requirements										
GLA-phenotype (51% black)	X		X	X		X	X	X	X	X
GLA-genotype			X (Red Angus)	X (Red Angus)						X (Red Angus)
Quality Factors										
U.S. Prime	X	X	X	X		X	X	X	X	X
U.S. Choice	X	X	X	X	X	X	X	X	X	X
U.S. Select					X					
Marbling requirements	Modest^{00} or higher	Modest^{00} or higher	Modest^{00} or higher	Small^{00} or higher	Slight^{00}– Moderate^{99}	Small^{50} or higher	Modest^{00} or higher	Modest^{00} or higher	Small^{00} or higher	Small^{50} or higher
Medium or fine marbling texture	X	X	X	X	X		X	X		
Maturity[a]	A	A or B	A	A	A	A	A	A		A
Yield Factors										
Yield grade	3.9[b] or lower				3.9 or lower	3.9 or lower	3.9[b] or lower	3.9[b] or lower		3.9 or lower
Fat thickness (inches)										
Ribeye area (square inches)										
Muscling[c]	X	X	X	X		X	X	X	X	X
Hot carcass weight (pounds)					600–950					
Carcass Characteristics										
No ribeye muscle internal hemorrhages	X	X	X	X	X	X	X	X	X	X
Free of "dark cutting" characteristics	X	X	X	X	X	X	X	X	X	X
Hump height (inches)	≤ 2	≤ 2	≤ 2	≤ 2		≤ 2	≤ 2	≤ 2	≤ 2	≤ 2
Steer and heifer beef carcasses	X	X	X	X	X	X	X	X	X	X
USDA Information										
Schedule number	G1	G2	G9	G9	G10	G14	G16	G17	G18	G19
Initial release date	1978	Jul 98	Dec 96	Dec 96	Jan 96	Dec 96	Mar 96	Jun 98	Feb 97	May 98
Effective date	May 94	May 99	Dec 96	Dec 96	Jan 99	Dec 96	Mar 96	Jun 98	Feb 97	May 98
USDA Certified	X	X	X	X	X	X	X	X	X	X
USDA Process Verified										
Management Claims										
Contact program for requirements										
Breed claim										

a—Lean color, texture, firmness, and overall skeletal characteristics, each must meet the requirements for the designated maturity, or younger
b—A yield grade of 3.9 or lower, except carcasses evaluated after removal of all or part of the kidney, pelvic, and heart fat may not have a yield grade higher than 3.5
c—Moderately thick or thicker muscling and tend to be moderately wide and thick in relation to their length
X—Indicates program requirement
1—Replaced Ada Angus Beef

May 17, 2000 USDA Certified & Process Verified Programs

FIGURE 6-5 Standards ensure the production of consistent products. (*Source:* USDA)

TABLE II
ALLOWANCES FOR DEFECTS IN RAISINS WITH SEEDS EXCEPT LAYER OR CLUSTER

Defects	U.S. Grade A	U.S. Grade B	U.S. Grade C
	Maximum count (per 32 ounces)		
Pieces of stem	1	2	3
	Maximum count (per 16 ounces)		
Capstems in other than uncapstemmed types	10	15	20
Seeds in seeded types	12	15	20
Loose capstems in uncapstemmed types	20	20	20
	Maximum (percent by weight)		
Sugar	5	10	15
Discolored, damaged, or moldy raisins	5	7	9
Provided these limits are not exceeded:			
Damaged	3	4	5
Moldy	2	3	4
Substandard Development and Undeveloped	2	5	8
	Appearance or edibility of product:		
Slightly discolored or damaged by fermentation or any other defect not described above	May not be affected.	May not be more than slightly affected.	May not be more than materially affected.
Grit, sand, or silt	None of any consequence may be present that affects the appearance or edibility of the product.		Not more than a trace may be present that affects the appearance or edibility of the product.

FIGURE 6-6
Agriculture Marketing Service of the USDA provides guidelines for grade standards. (*Source:* USDA, American Marketing Service)

Table I.—Classification of Flavor

Identification of flavor characteristics	U.S. extra grade	U.S. standard grade
Cooked	Definite	Definite.
Feed	Slight	Definite.
Bitter	—	Slight.
Oxidized	—	Slight.
Scorched	—	Slight.
Stale	—	Slight.
Storage	—	Slight.

Table II.—Classification of Physical Appearance

Identification of physical appearance characteristics	U.S. extra grade	U.S. standard grade
Dry product:		
Unnatural color	None	Slight.
Lumps	Slight pressure	Moderate pressure.
Visible dark particles	Practically free	Reasonably free.
Reconstituted product:		
Grainy	Free	Reasonably free.

Table III.—Classification According to Laboratory Analysis

Laboratory tests	U.S. extra grade	U.S. standard grade
Bacterial estimate, SPC/gram.	50,000	100,000.
Coliform estimate/gram.	10	10.
Milkfat content, percent.	Not less than 26.0, but less than 40.0.	Not less than 26.0, but less than 40.0.
Moisture content, percent.[1]	4.5	5.0.
Scorched particle content, mg:		
Spray proc.	15.0	22.5.
Roller proc.	22.5	32.5.
Solubility index, ml:		
Spray proc.	1.0	1.5.
Roller proc.	15.0	15.0.

[1] Milk solids not fat basis.

FIGURE 6-7

Tables provided by USDA indicate differences between extra grade and standard grade. (*Source:* USDA)

Certification services, which facilitate ordering and purchasing of products used by large-volume buyers, assure these buyers that the products they purchase will meet the terms of the contract–with respect to quality, processing, size, packaging, and delivery.

- Fresh fruits, vegetables, and specialty crops
- Processed fruits and vegetables
- Milk and other dairy products
- Livestock and meat
- Poultry
- Eggs
- Cotton
- Tobacco

Mission

The U.S. Department of Agriculture's Agricultural Marketing Service (AMS) facilitates the strategic marketing of agricultural products in domestic and international markets by grading, inspecting, and certifying the quality of these products in accordance with official USDA standards or contract specifications.

U.S. Grade Standards are quality driven and provide a foundation for uniform grading of agricultural commodities nationwide. Uniform standards provide identification, measurement, and control of quality characteristics important to the marketing function. In addition, they provide a common language for marketing, a means of establishing the value or basis for prices, and a gauge of consumer acceptance.

USDA Grade Standards also form the basis for quality certification services that buyers and sellers of agricultural products use in domestic and international contracting. AMS provides the following services upon request for a fee:

- Quality standards for more than 200 agricultural commodities to help buyers and sellers trade on agreed-upon quality levels
- Grading, inspection, quality assurance, and acceptance services to certify the grade or quality of products for buyers and sellers
- Inspection of facilities involved in the processing of agricultural commodities
- Assessment and registration of product and service quality management systems to established internationally recognized standards for some commodities

A variety of quality management services for some commodities are based on the International Organization for Standardization's audit-based quality assurance standards. These services are designed to provide additional and alternative approaches to verifying compliance with voluntary standards or contractual requirements.

Food Quality Assurance

The AMS Food Quality Assurance Staff manages the Federal food product description system, as well as associated quality assurance policies and procedures for food procured by federal agencies using appropriated funds. The Food Quality Assurance Staff works with user agencies, research and development groups, and industry on food specification issues. This work leads to the development of Commercial Item Descriptions (CID) and quality assurance procedures that will better serve government user needs.

Commercial Item Descriptions

In cooperation with industry, AMS develops and maintains commercial item descriptions (CIDs) for hundreds of food items. A CID is a simplified product description that concisely describes key product characteristics of an available, acceptable, commercial product. These CIDs are based on attributes that describe the odor, flavor, color, texture, analytical requirements, and so on, for each product (see Figure 6-8). The product areas include:

- Meat, poultry, fish, and shellfish
- Dairy foods and eggs
- Fruit, juices, nectars, and vegetables
- Bakery and cereal products
- Confectionery, nuts, and sugar
- Jams, jellies, nectars, and preserves
- Bouillions and soups
- Dietary foods and food specialty preparations
- Fats and oils
- Condiments and related products
- Coffee, tea, and cocoa
- Beverages, nonalcoholic
- Composite food packages

COMMERCIAL ITEM DESCRIPTION

CHICKEN NUGGETS, FINGERS, STRIPS, FRITTERS, AND PATTIES, FULLY COOKED, INDIVIDUALLY FROZEN

The U.S. Department of Agriculture (USDA) has authorized the use of this Commercial Item Description.

1. SCOPE

1.1 This Commercial Item Description (CID) covers individually frozen, fully cooked, solid muscle, chunked and formed or ground/chopped and formed, breaded or unbreaded, seasoned or unseasoned, chicken nuggets, fingers, strips, fritters, and patties (chicken products) packed in commercially acceptable containers, suitable for use by Federal, State, local governments, and other interested parties.

2. CLASSIFICATION

2.1 The frozen, fully cooked chicken products shall conform to the classifications in the following list and shall comply with USDA, Food Safety and Inspection Service (FSIS), Meat and Poultry Inspection Regulations, (9 CFR Part 381) and applicable State regulations. When applicable, the frozen, fully cooked chicken products shall comply with the USDA, Food and Nutrition Service (FNS), Child Nutrition Programs, National School Lunch Program

FIGURE 6-8

Numerous food items purchased by the government have commercial item descriptions to ensure uniform products. (*Source:* USDA)

QUALITY CONTROL

Regardless of government, research, or trade standards, most food manufacturing plants have some type of internal, formal, quality control or quality assurance department. These departments perform a wide variety of functions to ensure that a consistent, quality product is produced. Quality control may perform inspection duties, laboratory tests, oversee sanitation and microbiological aspects, and guide research and development. Total Quality Management (**TQM**) and Hazard Analysis and Critical Control Point (**HACCP**) are two newer ideas for controlling quality and safety.

TQM seeks to continuously improve the quality of products by making small changes in ingredients, manufacturing, handling, or storage, resulting in an overall improvement. All workers at a plant are involved in and responsible for the quality improvements in a product. HACCP is a preventative food safety system. First a step-by-step analysis of the process for manufacturing, storing, and distributing a food product is conducted. Then tight control of the process is established at potential problem points. Control measures are put in place before problems occur.

HOW TO READ A MARKET REPORT

The following terms and definitions are frequently used in Fruit and Vegetable Market News reports:

QUALITY includes size, color, shape, texture, cleanness, freedom from defects, and other more permanent physical properties of a product which can affect its market value.

The following terms, when used in conjunction with "quality," are interpreted as meaning:

FINE: Better than good. Superior in appearance, color, and other quality factors.

GOOD: In general, stock which has a high degree of merchantability with a small percentage of defects. This term includes U.S. No. 1 stock, generally 85 percent U.S. No. 1 or better quality on some commodities, such as tomatoes.

FAIR: Having a higher percentage of defects than "good." From a quality

<http://www.ams.usda.gov/standards/>

FIGURE 6-9
Knowledge of government standards is necessary to read market reports. (*Source:* USDA)

Good Manufacturing Practices (**GMPs**) are guidelines that a company uses to evaluate the design and construction of food processing plants and equipment. These standards require that all stainless steel and plastics used during the processing steps must meet food-grade specifications. Agencies such as the USDA and the FDA will help food companies select appropriate equipment.

The GMPs also require that hygiene and food contact procedures must be met. These procedures include wearing white uniforms, hair-nets, disposable gloves, face masks, and other protective gear. Standards for cleaning and sanitizing practices in food processing plants and equipment are also outlined in the GMPs.

The treatment of water to make it of drinkable quality, the filtering of air, and the treatment of food processing wastes are also addressed in the GMPs. The management of unavoidable pests in food processing plants and warehouses is also done to ensure that GMPs are used.

In effect, the GMPs cover every aspect of the processing of food (see Figure 6-9). The FDA and the USDA use these guidelines when inspecting a plant to ensure that it is in compliance with the regulations set forth in the Federal Food, Drug, and Cosmetic Act. The FDA provides copies of the GMP regulations.

Summary

Consumers expect certain qualities from their food. These include color, flavor, texture, and even size. When these are missing or different than expected, the food is rejected. Food science determines and uses methods to measure food-quality factors. These methods ensure a consistent, reliable product. Some evaluation methods use chemical and mechanical techniques. Others are completely human, such as taste panels. The USDA-AMS establishes quality and grading standards. Also, in cooperation with industry, the AMS develops and maintains commercial item descriptions for hundreds of items. Within the food industry, methods such as HACCP, TQM, and GMP monitor quality.

Review Questions

Success in any career requires knowledge. Test your knowledge of this chapter by answering these questions or solving these problems.

1. List the three components of reflected light used to define colors.
2. Name one instrument used to measure texture.
3. Discuss what humans can taste and what they smell and how this forms food flavor.
4. Identify the following acronyms: AMS, HACCP, TQM, GMP, CID.
5. Industry and ___________ develop and maintain CIDs.
6. List six factors that can influence the flavor of food.
7. Changes in the texture of food are often due to _______.
8. What qualities do consumers expect of their food?
9. The study of the science of the deformation of matter is called _______.
10. How do fats or lipids affect the texture of food?

Student Activities

1. Cut an apple or a potato and time how long it takes for browning to occur on the cut surface.
2. Make a list of foods you eat and describe their color. Discuss what would happen to your consumption if the food color was changed.
3. Leave a slice of bread on a plate for a couple of days. Describe the textural changes.
4. Conduct a taste test. This revolves around taste alone or taste and the appearance of the food. For example, find

out how red color affects food choice, or compare the taste of a name brand product with a generic product. A taste test could also be designed around the preferred texture of a food.

5. Many charts are available that visually explain the government grading standards. Obtain one of these charts and display it.
6. Remove potato chips from their packaging and place them in a plastic bag exposed to light. Explain the changes after few days.
7. Visit the USDA/AMS Web site on the Internet and describe the quality standards for one of the product groups or find a CID for a food item. Report your findings to the class.

Resources

Corriher, S. O. 1997. *Cookwise: The hows and whys of successful cooking.* New York: William Morrow and Company, Inc.

Cremer, M. L. 1998. *Quality food in quantity. Management and science.* Berkeley, CA: McCutchan Publishing Corporation.

Drummond, K. E. 1994. *Nutrition for the food service professional,* 2nd ed. New York: Van Nostrand Reinhold.

Gardner, J. E., Ed. 1982. *Reader's digest. Eat better, live better.* Pleasantville, NY: Reader's Digest Association, Inc.

McGee, H. 1997. *On food and cooking. The science and lore of the kitchen.* New York: Simon and Schuster Inc.

Vaclavik, V. A., and E. W. Christina. 1999. *Essentials of food science.* Gaithersburg, MD: Aspen Publishers, Inc.

Internet

Internet sites represent a vast resource of information. The URLs (uniform resource locator) for the World Wide Web sites can change. Using one of the search engines on the Internet such as Yahoo!, HotBot, AltaVista, Excite, Dogpile, About, or Google, find more information by searching for these words or phrases: HACCP, Maillard reaction, rheology, phenolic compounds, taste panels, pasteurization, quality food standards, quality grading, Food Quality Assurance, commercial items descriptions, quality control. Also, Table A-7 provides a listing of some useful Internet sites that can be used as a starting point.

Chapter 7

Unit Operations in Food Processing

Objectives

After reading this chapter, you should be able to:

- Describe materials handling in the food industry
- Name three methods of reducing the size of a food product
- Name three methods for separating food products
- Identify two general types of pumps
- Describe four factors that affect mixing
- Describe five factors influencing heat transfer
- Identify five unit processes that include heat transfer
- Discuss the uses of three common methods of drying
- List two examples of a formed food
- Describe the purposes of concentration
- Identify two reasons for packaging food products
- Discuss why some unit operations overlap

Key Terms

agglomeration
aggregation
centrifuge
concentration
conduction
convection
extrusion
freeze drying
gelation
gravity flow
heat transfer
impeller
laminar
microfiltration
micron
permeate
reciprocating
retentate
reverse osmosis
solutes
specific heat
ultrafiltration
viscosity

Most food processing is comprised as a series of physical processes that can be broken down into a number of basic operations. These unit operations can stand alone and depend upon logical physical principles. Unit operations include materials handling, cleaning, separating, size reduction, fluid flow, mixing, heat transfer, concentration, drying, forming, packaging, and controlling.

MATERIALS HANDLING

Materials handling includes the variety of operations from harvesting on the farm or ranch, refrigerated trucking of perishable produce, transportation of live animals, to conveying a product such as flour from a railcar or truck to a bakery storage bin (see Figure 7-1). During all of these operations, sanitary conditions must be maintained, losses minimized, quality maintained, and bacterial growth minimized. Also, all transfers and deliveries of materials must be on time and that time kept to a minimum for efficiency and quality.

Materials handling involves trucks and trailers, harvesting equipment, railcars, a variety of conveyors, forklifts, storage bins, pneumatic (air) lift systems, and so on.

FIGURE 7-1

Transporting grain, a materials handling operation. (*Source:* USDA, Photography Library)

FIGURE 7-2
Cleaning carrots, a materials handling operation. (*Source:* USDA, Photography Library)

CLEANING

The way foods are grown or produced in open environments on the farm or ranch often requires cleaning before use. Cleaning ranges from the removal of dirt to the removal of bacteria from a liquid food. Brushes, high-velocity air, steam, water, vacuum, magnets, **microfiltration**, and mechanical separation are all used to clean foods. Cleaning methods are prescribed according to the surface of the food product (see Figure 7-2).

Aside from the food itself, food processing equipment requires frequent, thorough, and special cleaning to maintain the quality of the product. Also, the floors and walls of the processing facility must be cleaned.

SEPARATING

Separations can be achieved on the basis of density or size and shape. Separations that are based on density differences include the separation of cream from milk, recovery of solids from suspensions, and removal of bacteria from fluids.

Cream Separator

Milk can be separated into skim milk and cream based on the density difference between fat and nonfat solids of milk. A cream separator is used to obtain the cream from milk and is a disc-type **centrifuge** in which the fluid is separated into low- and high-density fluid streams, which permits the separate collection of cream and skim milk.

Clarification

Sediment and microorganisms can be removed centrifugally in a clarifier, which is generally a disc-type centrifuge that applies forces of 5,000 to 10,000 times gravity and forces the denser material to the outside. By periodically opening the bowl, the solids can be continuously removed from the remainder of the fluids. This same principle has been used to recover yeast cells from spent fermentation broths and to continuously concentrate bakers cheese from whey.

Membrane Processes

Reverse osmosis (RO), **ultrafiltration** (UF), and microfiltration (MF) are processes that use membranes with varying pore sizes to separate on the basis of size and shape (see Figure 7-3). Reverse osmosis uses membranes with the smallest pore and is used to separate water from other **solutes**. Ultrafiltration uses membranes with larger pores and will retain proteins, lipids, and colloidal salts, while allowing smaller molecules to pass through to the **permeate** phase. Microfiltration, with pores less than 0.1 **micron**, is used to separate fat from proteins and to reduce microorganisms from fluid food systems. High-pressure pumps are required for RO and low-pressure pumps for UF and MF.

FIGURE 7-3

Reverse osmosis unit, a separation process. *(Courtesy Osmonics, Inc., Minnetonka, MN)*

SIZE REDUCTION

Size reduction can be through the use of high-shear forces, graters, cutters, or slicers (see Figure 7-4). Emulsions with very small fat globule droplets are frequently made with a homogenizer, which is a high-shear positive pump that forces fluid through a very small opening at very high pressure to form or reduce the size of an emulsion. The positive pump uses a **reciprocating** or rotating cavity between two lobes, or gears between a stationary cavity and a rotor. The fluid forms the seal between the rotating parts.

Typical equipment for size reduction in meat products and their component parts include:

- Grinder
- Bacon slicer
- Sausage stuffer
- Vertical chopper

Ball mills grind products into fine particles.

Sometimes this process is better thought of as size adjustment either through size reductions through such methods as slicing, dicing, cutting, or grinding, or size increase by **aggregation**, **agglomeration**, or **gelation**.

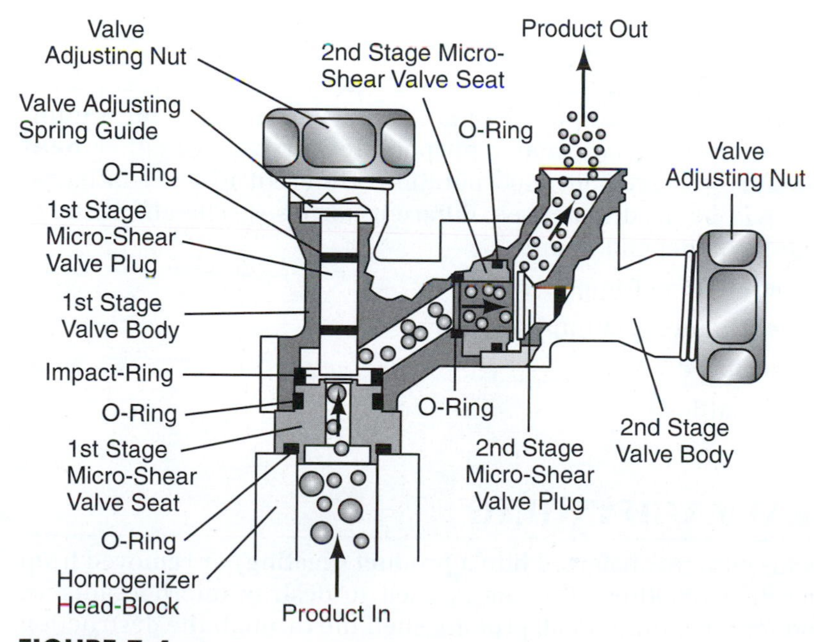

FIGURE 7-4

Homogenization, a size reduction operation.

PUMPING (FLUID FLOW)

Transport of fluids is achieved either by **gravity flow** or through the use of pumps (see Figure 7-5). In gravity flow, the flow is **laminar**, where the flow is transferred from the fluid to the wall between adjacent layers. Adjacent molecules do not mix. Often, fluids are transported from one unit operation or process to another by pumps and in turbulent flow where there is mixing of adjacent particles. Two different types of pumps are commonly used for different purposes:

1. The centrifugal pump uses a rotating **impeller** to create a centrifugal force within the pump cavity. The flow is controlled by the choice of impeller diameter and rotary speed of the pump drive. The capacity of a centrifugal pump is dependent upon the speed, impeller length, and the inlet and outlet diameters.
2. A positive pump consists of a reciprocating or rotating cavity between two lobes or gears and a rotor. Fluid enters by gravity or a difference in pressure, and the fluid forms the seals between the rotating parts. The rotating movement of the rotor produces the pressure to cause the fluid to flow.

MIXING

An agitation (mixing) device may be placed in a tank for a number of purposes. Two major purposes of mixing are either **heat transfer** or ingredient incorporation. Different mixer configurations will be used to achieve different purposes. The efficiency of mixing will depend upon:

- Design of impeller
- Diameter of impeller
- Speed
- Baffles

HEAT EXCHANGING

Heat is either transferred into a product (heating) or removed from a product (cooling). Heating is used to destroy microorganisms, produce a healthful food, prolong shelf life through the destruction of certain enzymes, and to promote a product with acceptable taste, odor, and appearance.

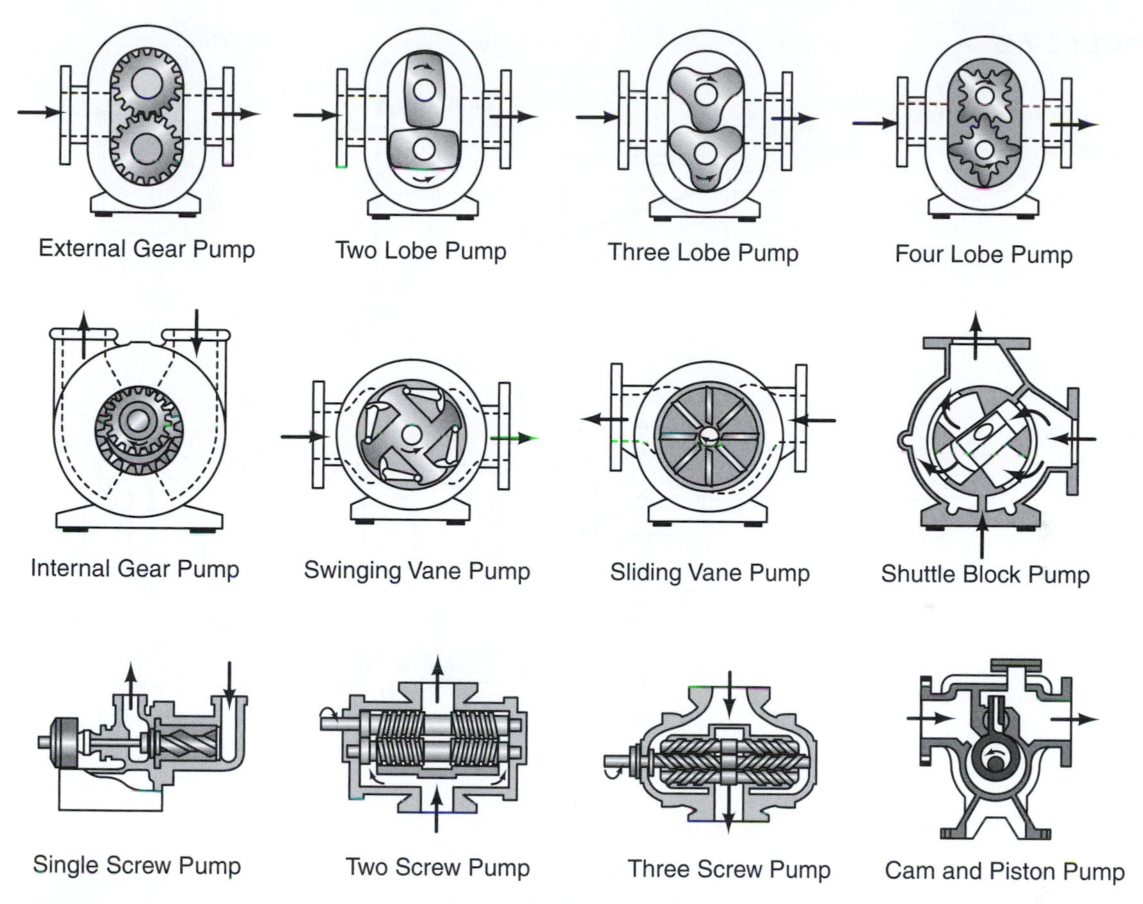

FIGURE 7-5

Examples of various pumps to control fluid flow.

Five factors influence the heat transfer into or out of the product:

1. Heat exchanger design (see Figure 7-6)
2. Heat transfer properties of the product, such as:
   - **specific heat** (amount of heat required to change the temperature of a unit mass of product a specific temperature without change in state of the material)
   - thermal conductivity (rate by which heat is transferred through a material)
   - latent heat (heat required to change the state of a material)
3. Density (weight per unit volume)

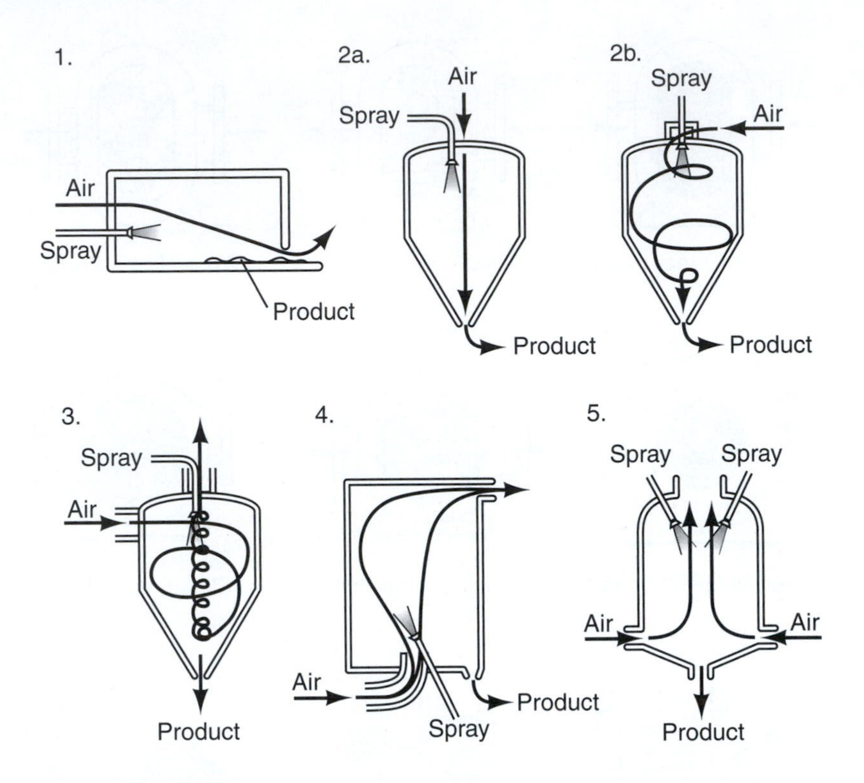

FIGURE 7-6

Examples of spray driers, a heat exchange operation.

4. Method of heat transfer, such as:
   - **conduction** (transfer from molecule to molecule through the material)
   - radiation (transfer from electromagnetic radiation of a body due to the vibration of its molecules)
   - **convection** (transfer through movement of mass)
5. **Viscosity** (related to the amount of force required to move the fluid product)

A variety of heat exchanges are used in the food industry, which include:

- Plate heat exchanges
- Tubular heat exchanges
- Swept surface heat exchangers

Plate heat exchanges pass fluid over a plate where a heating or cooling medium is being passed up or down on the other side of the plate. The thin film makes for rapid heat transfer and is the most efficient method of heating fluids of low viscosity.

HISTORICAL FREEZE DRYING

The Incas of Peru, who stored their vegetables near the peaks of high mountains, first used the process of freeze-drying. There, the vegetables froze solid. Over time, the frozen water sublimated into the thin mountain air (converted directly to vapor without passing through the liquid state), leaving behind the perfectly preserved, desiccated vegetables.

Modern freeze-drying started during World War II to preserve blood plasma for use at the front lines. Today, freeze-drying is done using flash-freezing and vacuum dehydration. Freeze-drying preserves almost all the nutrients of foods, as well as the important flavor elements.

To learn more about freeze-drying, search the Web or visit these sites:

<www.sciam.com/0996issue/0996working.html>
<forums.cosmoaccess.net/forum/survival/prep/freeze.htm>

Tubular heat exchanges generally are composed of a tube within a tube, in which product and heating or cooling medium are flowing in opposite (countercurrent) directions. This a low-cost method of heating or cooling and is applied to fluids of higher viscosities that generally passed through a plate heat exchanger.

Swept surface heat exchanges have blades that scrape the surface of the heat exchanger and bring new product continuously to the heat or cooling surface. They are used for fluids of very high viscosity. An ice cream freezer is an example of a swept surface heat exchanger.

Common unit processes that include heat transfer as a unit operation include:

- Pasteurization (heat)
- Sterilization (heat)
- Drying (heat)
- Evaporation (heat)
- Refrigeration (cold)
- Freezing (cold)

CONCENTRATION

Concentration can be achieved through evaporation and through reverse osmosis. Evaporation generally involves heating the fluid in a vessel under a vacuum to cause a change in state of water from liquid to vapor and then recovers water by passing the vapor through a condenser. In some products, evaporation causes the loss of flavor volatiles. When this happens, a low-temperature unit is added to recover the flavor volatiles so that they can be added back to the product.

To reduce operating costs, multiple effect evaporators are used, which have two or more evaporators placed in a series to provide a means for the continuous concentration of a fluid product. This increases the efficiency of the evaporation process.

Reverse osmosis (RO) is a process where the fluid is passed through a semipermeable membrane with very small pores that permit only the transfer of water. Most systems consist of a membrane cast on a solid porous backing–usually in the form of a tube. High pressure is applied to force the water (called permeate) through the membrane. The concentrated fluid (called **retentate**) is retained in the tubing. The rate of water removal decreases as the fluid is concentrated, until it is no longer economically feasible to remove more water.

Concentration is often used as a pre-step to drying to reduce the amount of water that needs to be removed during drying, thus reducing drying costs. Evaporation can achieve higher solids economically than can reverse osmosis. RO is preferred over evaporation for heat-sensitive fluids.

Contact equilibrium processes or mass transfer may or may not require a change in state. Generally, a type of molecule is transferred to or from a product. Processes that use mass transfer include distillation, gas absorption, crystallization, membrane processes, drying, and evaporation (see Figure 7-7).

FIGURE 7-7

Washing and cooling cabbages. (*Source:* USDA Photography Library)

DRYING

Three common methods of drying are (1) sun or tray drying, (2) spray drying, and (3) **freeze drying**. Sun or tray drying is least expensive, followed by spray drying and freeze drying. The drying method of choice is generally based on the characteristics of the product.

Products that are already solid lend themselves to sun or tray drying. These include fruits and vegetables. The products may be dried by exposure to sun or placed in trays and dried in a current of warm or hot air. This method is used to make raisins from grapes.

Products that are very heat sensitive are freeze dried. Commercially, only instant coffee is widely freeze dried. Some freeze dried fruits are beginning to reach the market, but these are in limited quantities. In freeze drying, the moisture is removed without a phase change (sublimation).

The most common drying method is spray drying, which is applied to fluid products. The bulk density (weight per unit volume) is controlled to a large extent by the solids that are sent to the dryer. Several different designs of spray nozzles are used to atomize the fluid into the heated air. These generally are either centrifugal nozzles or high-pressure spray nozzles. The type of nozzles will vary with the product being dried.

For some products that are very hygroscopic (take on water from the air), the dried product may be partially rewetted and then redried. This produces agglomerated products that are easily dispersed in solution. Spray dried powders with a surfactant are also a method for improving dispersion.

An older method is roller drying. The product was allowed to flow over a hot, rotating drum and the dried product scraped off. This was a low-cost method of drying, but it created a lot of heat damage to the product.

FORMING

Often foods need to be formed into specific shapes. For example, hamburger patties, chocolates, jellies, tablets, snack foods, breakfast cereals, butter and margarine bars, cheeses, variety breads, and sausages. Processes used to form foods can include compacting, pressure **extrusion**, molds, powders and binding agents, heat and pressure, and extrusion cooking.

PACKAGING

Packaging is used for a variety of purposes including shipping, dispensing, improving the usefulness of the product, and protection from microbial contamination, dirt, insects, light, moisture, drying, flavor changes, and physical alterations. Attractive packaging also helps with marketing of the food product.

Foods are packaged in metal cans, glass and/or plastic bottles, paper, cardboard, and plastic and metallic films. Many foods are packaged in a combination of materials. In the past, rigid containers of glass and metal were commonly used. Now, more products are being packaged in flexible and formable containers, such as retortable pouches used for fruit juices and chewy bars.

Machines that automatically package food products operate at high speeds, and the complete process is step-wise and automated from the forming of the container, filling of the container, to the sealing, labeling, and stacking.

CONTROLLING

Producing a food often requires that the unit operations are combined into a complex processing operation. To ensure the quality of a food product, food processors need to measure and control these operations. The tools used in controlling and measuring may include valves, thermometers, scales, thermostats, and other instruments to measure and control pressure, temperature, fluid flow, acidity, weight, viscosity, humidity, time, and specific gravity.

In modern processing plants, the instrumentation and controls are automatic and computer controlled. An operator oversees and controls the processing from a remote console (see Figure 7-8).

FIGURE 7-8

Controls help combine unit operations into a complex process. (*Source:* USDA Photography Library)

OVERLAPPING OPERATIONS

Some food processing operations may use a single unit operation, but most food processing includes a combination of unit operations to achieve the total process. For example, the manufacture of a dried coffee creamer from a combination of fluid and dry ingredients includes the following unit operations in sequence:

- Mixing
- Fluid flow
- Size reduction (homogenization)
- Heat transfer (heating)
- Fluid flow
- Heat transfer (cooling)
- Mass transfer (conversion of water to vapor during drying)
- Pasteurization of milk (includes the unit operations of fluid flow and heat transfer both heating and cooling)

Some other examples of unit processes, and their associated unit operations, include the following:

- Freeze drying involves heat transfer and mass transfer.
- Extrusion requires fluid flow, heat transfer, mass transfer, and size reduction in the case of cereals and snack foods.
- Ice cream manufacturing is comprised of two processes: (1) mix-making, which uses mixing, fluid flow, heat transfer, and size reduction, and (2) freezing, which involves fluid flow, heat transfer, and mass transfer of air into the ice cream.

CONSERVING ENERGY

Food processing is energy intensive, so the energy represents a significant share of the cost of the final product. Food processors are always looking for new ways to optimize the use of energy. For example, heat that is used or removed is captured and used somewhere else in the process. Times and temperatures for processes like dehydration, concentration, freezing, and sterilization are reevaluated. Processors continually monitor the energy required for all the unit processes and look for more efficient ways. Also, such simple measures as improved temperature control, insulation, controlling ventilation rates, reducing lighting, and checking sensors save energy.

NEW PROCESSES

Major goals of food scientists and food processing engineers are to develop new methods that improve quality and/or increase efficiency. New processes are constantly being tried in the unit operations. New processes to watch include ohmic heating, irradiation, supercritical fluid extraction, and high hydrostatic pressure.

Ohmic heating uses alternating current to rapidly increase the temperature in a product. This process destroys the microorganisms and maintains the quality of the food. Supercritical fluid extraction uses a gas such as carbon dioxide at high pressure to extract or separate the food components. High hydrostatic pressure is used in liquid foods to inactivate microorganisms and some enzymes.

Summary

Unit operations make up the basics of food processing. These include materials handling, cleaning, separating, size reduction, fluid flow, mixing, heat transfer, concentration, drying, forming, packaging, and controlling. Most food processing involves a combination or an overlap of these unit operations. Where unit operations overlap or are combined, complex controls ensure the proper function of each operation. Many of the unit processes discussed in this chapter are discussed in more detail in chapters that follow.

Review Questions

Success in any career requires knowledge. Test your knowledge of this chapter by answering these questions or solving these problems.

1. The manufacture of ice cream is an example of a/an ___________.
2. Why are foods packaged?
3. ___________ is the amount of heat required to change the temperature of a unit mass of product a specific temperature without changing the material.
4. Name the three methods for separating foods.
5. What are the two types of fluid flow pumps?
6. ___________ heat exchanges pass fluid over a plate where a heating or cooling medium is being passed up or down on the other side of the plate.
7. List the four factors affecting the mixing of food products.
8. Why is it important to handle food materials carefully?

9. Explain the three common methods of drying foods.
10. List three membrane processes for separating food products.

Student Activities

1. Bring in seeds (beans or wheat) direct from a combine. Clean the product and separate the broken seeds. Develop a report on this activity.
2. Develop a demonstration of conduction, radiation, and convection.
3. Build a simple food drier and dry some fruits.
4. Bring a collection of food packages to class and discuss the purposes of the packaging.
5. Identify a food product you eat. List the unit processes required to produce that food.
6. Pick a food product that has been formed. Describe to the class how is it formed.

Resources

Cremer, M. L. 1998. *Quality food in quantity. Management and science.* Berkeley, CA: McCutchan Publishing Corporation.

Potter, N. N., and J. H. Hotchkiss. 1995. *Food science,* 5th ed. New York: Chapman and Hall.

Vaclavik, V. A., and E. W. Christina. 1999. *Essentials of food science.* Gaithersburg, MD: Aspen Publishers, Inc.

Vieira, E. R. 1996. *Elementary food science,* 4th ed. New York: Chapman and Hall.

Internet

Internet sites represent a vast resource of information. The URLs (uniform resource locator) for the World Wide Web sites can change. Using one of the search engines on the Internet such as Yahoo!, HotBot, AltaVista, Excite, Dogpile, About, or Google, find more information by searching for these words or phrases: food processing, unit operations, freeze dried, specific heat, ultrafiltration, food processing equipment, pasteurization, food sterilization, food evaporation, refrigeration of food, freezing food, food forming, food packaging, ohmic heating of food, supercritical fluid extraction, heat exchanges (plate, tubular, swept surface). Also, Table A-7 provides a listing of some useful Internet sites that can be used as a starting point.

Chapter 8

Food Deterioration

Objectives

After reading this chapter, you should be able to:

- List three general categories of food deterioration
- Discuss shelf life and dating
- Name six factors that cause food deterioration
- Identify six preservation techniques that prevent deterioration
- Describe normal changes in food products following harvest or slaughter
- Identify four food enzymes and describe their function

Key Terms

denature
emulsion
facultative
mesophilic
mycotoxins
obligative
organoleptic
osmosis
pyschrophilic
radiation
shelf life
thermophilic
water activity (Aw)

Deterioration includes changes in organoleptic quality, nutritional value, food safety, aesthetic appeal, color, texture, and flavor. To some degree, all foods undergo deterioration after harvest. The role of food science is to minimize negative changes as much as possible.

TYPES OF FOOD DETERIORATION

The three general categories of food deterioration are:

1. Physical
2. Chemical
3. Biological

Factors that cause food deterioration are many, including light, cold, heat, oxygen, moisture, dryness, other types of **radiation**, enzymes, microorganisms, time, industrial contaminants, and macroorganisms (insects, mice, and so on).

SHELF LIFE AND DATING OF FOODS

All foods have a time limit of their usefulness. This time limit depends on the type of food, the storage conditions, and other factors. If food is held at about 70°F (21°C), its useful life varies as shown in Table 8-1.

Shelf life is the time required for a food product to reach an unacceptable quality. It depends on the food item (Table 8-1), the processing method, packaging, and storage conditions. Food manufacturers put code dates on their products (see Figure 8-1). "Pack date" is the date of manufacture. The date of display is called the "display date," and the "sell by date" is the last day to sell. Some foods have a "best used by date," or the last date of maximum quality. The "expiration date" indicates when the food is no longer acceptable.

TABLE 8-1 Useful Life at 70°F

Food	Days
Meat	1 to 2
Fish	1 to 2
Poultry	1 to 2
Dried, smoked meat	360+
Fruits	1 to 7
Dried fruits	360+
Leafy vegetables	1 to 2
Root crops	7 to 20
Dried seeds	360+

CAUSES OF FOOD DETERIORATION

Specific causes of food deterioration include the following:

- Microorganisms such as bacteria, yeast, and molds
- Activity of food enzymes
- Infestations by insects, parasites, and rodents
- Inappropriate temperatures during processing and storage
- Gain or loss of moisture
- Reaction with oxygen
- Light
- Physical stress or abuse
- Time

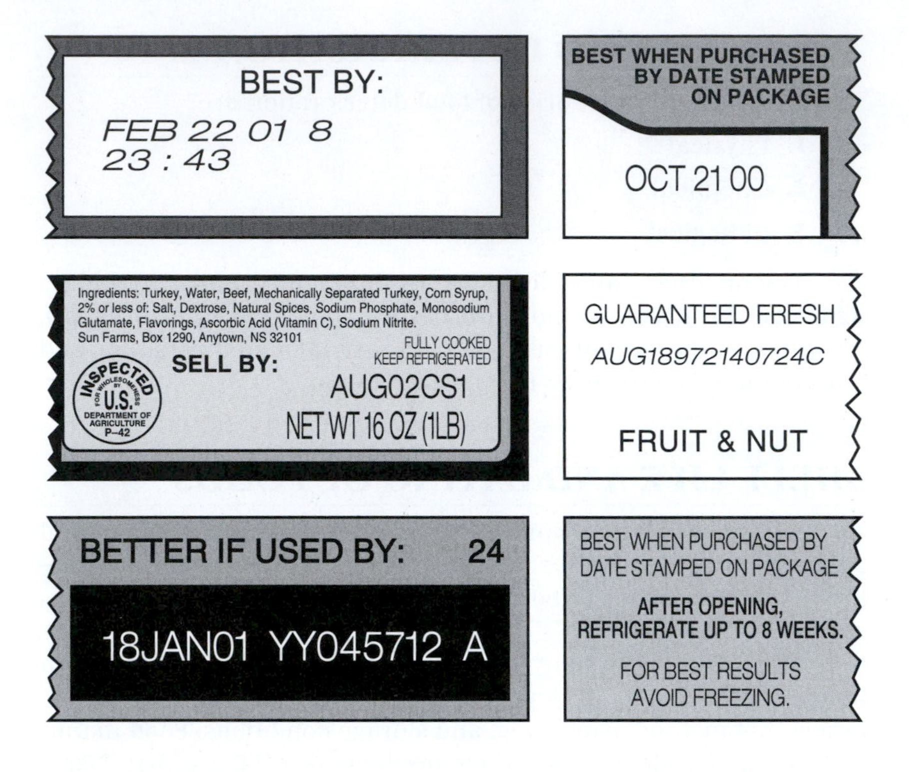

FIGURE 8-1
Dates on labels help ensure quality.

Deterioration can be caused by these items individually or in any combination.

Bacteria, Yeast, Mold

Thousands of species of microorganisms exist, and a few hundred are associated with foods. Not all are bad because some are desirable in food preservation. Microorganisms are found in the soil, water, and air; on animal skins, plant surfaces, and in digestive tracts; but they are usually not in healthy tissue.

Bacteria are single-celled organisms occurring in three shapes: round (cocci), rod (bacilli), and spiral (spirilla and vibrios). Some produce spores, and these spores are resistant to heat, chemicals, and other adverse conditions.

Yeasts are the largest of the microorganisms but are still single cells, and some produce spores. Molds are larger than bacteria. They are often filamentous, and they all produce spores.

In foods these microorganisms attack basically all the food components–sugars, starches, cellulose, fats, and proteins. Depending on the food and the microorganism, the action on food could be to produce acids, making the food sour, or produce

FUNGI AS A LIFE FORM

Although life-forms in the Fungus Kingdom (molds, mushrooms, and yeasts) may seem like plants, they are actually more closely related to animals. In the history of life, fungi and primitive animals branched apart after plants had evolved.

Like some kinds of animal cells, walls made out of chitin surround the cells of fungi. Chitin is a complex molecule made of sugar and nitrogen. Plants use cellulose for their cell walls.

Fungi get their energy by breaking down organic molecules that they soak up directly from their environment. They emit substances that chop complex organic molecules into smaller ones that are easy to absorb. Fungi are vital to the global ecosystem because they are so good at breaking down anything organic.

To learn more about fungi, search the Web or visit this site:

<www.perspective.com/nature/fungi/>

alcohol. Some microorganisms produce gas, making the food foamy; still others produce unwanted pigments or toxins.

Environmental conditions that affect microbial growth include temperature and oxygen. Microbes that prefer cold temperatures are said to be **pyschrophilic**. **Mesophilic** microorganisms prefer normal temperatures; **thermophilic** microorganisms prefer hot temperatures. Bacteria or molds that require atmospheric oxygen are said to be aerobic, and those yeast and bacteria that do not require atmospheric oxygen are called anaerobic. **Facultative** microorganisms are both aerobic and anaerobic; **obligative** microorganisms are either.

Food-Borne Disease

Food infections are caused when a microorganism is present in the food and it causes an infection in the human when the food is consumed. Infections can be caused by *Clostridium perfringen, Salmonella sp., Escherichia coli* (*E. coli* 0157), and several others (see Chapter 25). Food intoxication occurs when a food is consumed that contains a chemical toxic to humans. *Staphylococcus aureus* and *Clostridium botulinum* produce toxins. Molds in foods produce **mycotoxins** like aflatoxin. These toxins are not destroyed by heat.

Insects

Insect damage can be minor, but this wounds the tissue for additional damage by microorganisms. Insect damage and infestation can also be so complete as to render the food inedible. Pesticides control insects, as does an inert atmosphere and cold storage.

Food Enzymes

All foods from living tissues have enzymes. Most of these enzymes will survive harvest or slaughter. At the time of harvest or slaughter, enzymes that control digestion and respiration proceed uncontrolled and cause tissue damage. Some of the post-harvest enzymatic reactions are desirable–for example, the ripening of tomatoes and the aging or tenderizing of beef. Enzyme action can be controlled by heat, chemicals, and radiation. Food enzymes are also used in food processing (Table 8-2).

Heat and Cold

Normal harvest temperatures range from 50° to 100°F. The higher the temperature, the faster the biochemical reactions occur. In fact, the rate of chemical reactions doubles with each 10-degree rise in temperature. On the other hand, subfreezing temperatures damage tissues. Cold temperatures may also cause discoloration, change the texture, break an **emulsion**, and **denature** protein. Chilling can injure the tissue of fruits.

Oxygen

Chemical oxidation reactions can destroy vitamins (especially A and C), alter food colors, cause off-flavors, and promote the growth of molds.

FIGURE 8-2

Environmentally controlled spud cellars maintain quality of potatoes during storage.

POST-HARVEST BIOCHEMICAL CHANGES

Fruits and vegetables are high-water-content foods. This promotes bacterial, yeast, and mold growth. If fruits and vegetables become partially dehydrated or wilted because of bruising or rough handling, their economic value sharply decreases (Figure 8-2). Processors must practice superb techniques in handling and transporting these commodities. Processing techniques must also minimize natural enzymatic deterioration.

Processors realized the need for immediate cooling of freshly picked produce. This led to the development of mobile processing units that supercool produce immediately after harvest. Jet streams wash the product and begin to remove internal heat, and the supercoolers remove the remaining heat.

TABLE 8-2 Important Food Processing Enzymes

Enzyme	Source	Importance	Action
Ascorbic acid oxidase	Citrus fruit, vegetables (squash, cabbage, cucumbers)	Browning and off flavor in fruits; destruction vitamin C activity	Oxidizes ascorbic acid to dehydro form destroying the browning prevention ability.
Beta-amylase	Grains, wheat, barley, sweet potatoes	Produces maltose and glucose to support yeast growth and/or sweeten	Beta-amylase in flour with fungal glucoamylase produces mixtures of fermentable sugars: glucose, maltose.
Bromelain	Pineapple	Meat tenderizing; chill-proofing beef	Acts on collagen to hydrolyze peptides, amides, and esters from the non-reducing end.
Catalase	Meat, liver, blood, molds, bacterial, milk	Removes excess H_2O_2 from treated milk in cheese making; cheese making	Removes residual H_2O_2 (hydrogen peroxide) treated foods, converts H_2O_2 to H_2O and oxygen. Used when pasteurization is not feasible.
Cathepsins	Animal tissue, liver, muscle	Aging of meat or game when hung	Autolytic reaction during aging of meat (on protein substrate).
Cellulase	Digestive juice of invertebrates, snails	Hydrolysis cell wall breaks glucosidic link in cellulose; clarification of fruit juices; removes graininess of pears	Converts native cellulose soluble, low-molecular-weight products such as maltose and glucose.
Chlorophyllase	Spinach; other selected vegetables, green	Chlrophyll is made water soluble; lose color	Makes chlorophyll water soluble.
Collagenase	Clostridia bacterial	Tenderizing of meat—collagen	Hydrolyzes native collagen and denatured collagen such as gelatin and hide powder.

(continued)

TABLE 8-2 Important Food Processing Enzymes *(continued)*

Enzyme	Source	Importance	Action
Elastase	Mammalian organs (hog, pancreas), microorganisms	Breaks down meat muscle fiber	Hydrolyzes elastin, hemoglobin, fibrin, albumin casein, soy protein.
Ficin	Fig trees	Meat tenderizer	Hydrolyzes muscle fiber proteins and both types of connective tissue.
Glucoxidase	Molds, fungi	Removal of glucose and oxygen in foods, impaired by residual reducing sugars such as egg white	Will catalyze the oxidation of glucose to gluconic acid with the formation of H_2O_2 (hydrogen peroxide) in presence of moisture and atmospheric oxygen.
Lactase	Plants; almonds, peaches; calf intestine	Cheese making; removing lactose from products for allergy sensitive persons	Splits lactose and other galactosides yielding glucose and galactose.
Lysozyme	Egg whites	Protects eggs from microbacterial action	Dissolves cells of certain bacteria.
Maltase	Yeast and other micoorganisms, grain, GI tract	Baking—lean dough with little added sugar	Hydrolyzes glucose of starch.
Papain	Papaya melons	Chill-proofing beer; meat tenderizing	Acts on muscle fiber proteins and on elastin and lessens extent on collagen.
Pectinase	Citrus fruit, Fungi-Aspergillus niger	Production of clear jellies and juices; greater yields (reduces viscosity)	Hydrolyzes pectins or pectic acids; lowers viscosity, eliminates protective colloid action of pectin.
Pectin methyl esterase	Citrus, tomatoes, apples	Commercial pectinase to remove pectin from fruit juice used in diabetic low-sugar jellies	Demethylation of pectins resulting in methanol and pectate (polygalacturonic acid).

TABLE 8-2 Important Food Processing Enzymes *(concluded)*

Enzyme	Source	Importance	Action
Peroxidase	Liver, milk, horseradish	Presence or absence can be used to judge adequacy of pasteurization	Splits H_2O_2 to H_2O and O_2 when there is an oxygen acceptor.
Polyphenol oxidase, phenolase, catecholase	Plants, mushrooms	Tea fermentation; production of colored end products and flavors associated with browning reaction in fruits and vegetables	Catalyzes reactions in which molecular oxygen is the H acceptor and phenols act as H donor; oxidizes phenolic compounds to quinines.
Phosphatase	Baby food, milk, brewing	Detects effectiveness of pasteurization as denatured slightly higher temperature than tubercle bacillus.	Liberates inorganic phosphate from phenyl phosphate.
Rennin	Salt extract of calves' fourth stomach	Coagulation of milk in cheese preparation; gel desserts	Hydrolyzes peptides. Acts on the surface of the kappa-casein molecule.
Sucrase	Yeast	Slow crystallization in candy; hydrolyzes sucrose; artificial honey; development of creamy centers in candy	Splits sucrose to the invert sugars.
Zymase	Yeast	Fermentation yeast, maltozymase	Zymase is the heat-labile fraction of the enzymes system responsible for alcoholic fermentation. Fermentation converts glucose into CO_2 and ethanol.

Much of today's harvesting is accomplished by mechanical methods. Because fruits and vegetables are delicate items, harvesting is usually conducted before maturity is reached. Ripening chambers containing ethylene gas are an important part of the processing of certain fresh fruits and vegetables. Other controlled atmospheric conditions, like temperature, humidity, oxygen levels, and light, are used to regulate shelf life.

POST-SLAUGHTER BIOCHEMICAL CHANGES

Rigor mortis is an essential process in the conversion of live muscle to meat. After death, the biochemistry of muscle tissue changes. The muscle will use up energy from glycogen–a complex carbohydrate found in animal tissue. Because the blood is no longer flowing to remove the by-products of metabolism, lactic acid builds up in the muscle. This reduces the pH, causing a complex series of reactions that results in the contraction of the muscle fibers. This contraction makes the muscle feel hard or stiff, thus the name *rigor mortis.*

After more time, the muscle fibers will begin to relax. The relaxation of muscle post-rigor is sometimes called the resolution of rigor. This process is greatly influenced by temperature, being faster at higher temperatures. In processing meats, the time and temperature during rigor are carefully controlled to maximize tenderness. This part of the process is sometimes called aging.

PRINCIPLES OF FOOD PRESERVATION

Food preservation involves the use of heat, cold, drying (**water activity**, or **Aw**), acid (pH), sugar and salt, smoke, atmosphere, chemicals, radiation, and mechanical methods (Table 8-3).

Heat

Most bacteria are killed at 180° to 200°F, but spores are not. To ensure sterility requires wet heat at 250°F for 15 minutes. High-acid foods require less heat. (See Chapter 9 for a complete discussion.)

Cold

Microbial growth slows at temperatures under 50°F, but some psychrophiles will continue slow growth. Foods frozen at less than 14°F usually do not have any free water, so these foods also benefit from low water activity to help protect against microbial growth. Freezing may kill some but not all of the microorganisms (see Tables 8-4 and 8-5). (See Chapter 10 for more information.)

TABLE 8-3 Summary of Processing Methods to Control Factors Affecting Food Safety, Quality, and Convenience

To Be Controlled	Method and Effect				
	Heat	**Cold**	**Chemicals**	**Water Activity (Aw)**	**Mechanical**
Micro-organisms	Prevents growth	Reduces growth rate	Preservatives retard growth	Do not grow below Aw of 0.6	Reduces numbers
Enzymes	Destroyed by heat activity	Decreases reaction rate	Modify activity	Alters rate of enzyme activity	Increases enzyme-substrate complex formation
Chemical Reactions	Increases chemical rate; browning oxidation	Reduces reaction rate	May inhibit or activate	Can alter rate of reaction, especially oxidation	Not applicable
Physical Structure	Increases effects	Decreases effects	May modify structure	High Aw may cause caking	Can destroy structures

TABLE 8-4 Approximate Storage Life of Frozen Foods (in months unless indicated otherwise)

Product	**0°F**	**10°F**	**20°F**
Orange juice (heated)	27	10	4
Peaches	12	<2	6 days
Strawberries	12	2.4	10 days
Cauliflower	12	2.4	10 days
Green beans	11–12	3	1
Peas	11–12	3	1
Spinach	6–7	3	21 days
Chicken, raw	27	15½	<8
Fried chicken	<3	<1	18 days
Turkey pies	>30	9½	2½
Beef, raw	13–14	5	<2
Pork, raw	10	<4	<1½
Lean fish, raw	3	<2½	<1½
Fat fish, raw	2	1½	24 days

Source: USDA

TABLE 8-5 Maximum Cold Storage Time for Meat

Meat Product	Refrigerator (38° to 40°F)	Freezer (at 0°F or lower)
Beef (fresh)	2 to 4 days	6 to 12 months
Veal (fresh)	2 to 4 days	6 to 9 months
Pork (fresh)	2 to 4 days	3 to 6 months
Lamb (fresh)	2 to 4 days	6 to 9 months
Poultry (fresh)	2 to 3 days	3 to 6 months
Ground beef, veal, lamb	1 to 2 days	3 to 4 months
Ground pork	1 to 2 days	1 to 3 months
Variety meats	1 to 2 days	3 to 4 months
Luncheon meats	7 days	not recommended
Sausage, fresh pork	7 days	2 months
Sausage, smoked	3 to 7 days	1 month
Frankfurters	4 to 5 days	1 month
Bacon	5 to 7 days	1 month
Smoked ham, whole	1 week	2 months
Smoked ham, slices	3 to 4 days	1 month
Beef, corned	7 days	2 weeks
Leftover cooked meat	4 to 5 days	2 to 3 months
Frozen combination foods:		
Meat pies (cooked)		3 months
Swiss steak (cooked)		3 months
Stews (cooked)		3 to 4 months
Prepared meat dinners		2 to 6 months

Drying

Drying reduces the water activity (Aw) in a food. Because microorganisms contain about 80 percent moisture, drying or dehydrating the food also dehydrates the microorganism. Changing the amount of water in a food also alters the rate of enzyme activity and other chemical reactions. (See Chapter 11 for a more complete discussion.)

Acid

As the food becomes more acid (lower pH), the heat required for sterilization is reduced. For example, the pH of corn is about 6.5. At 226°F, 15 minutes are required to destroy *C. botulinum* spores. The pH of pears is about 3.8, and only 5 minutes are necessary to destroy *C. botulinum* at 226°F. Acid may occur naturally in foods, may be produced by fermentation, or may be added artificially.

Sugar and Salt

Sugar, salt, and smoke are chemical means of controlling food deterioration. The addition of sugar or salt to a food item increases the affinity of the food for water. This removes the water from the microorganism through **osmosis**.

Smoke

Smoke contains formaldehyde and other preservatives. The heat involved with adding the smoke helps reduce the microbial populations, and it dries the food somewhat. Chapter 14 provides more details on the use of chemicals.

Atmosphere

Changing the storage atmosphere reduces food deterioration. The growth of aerobes is slowed by removing the oxygen; providing oxygen limits the growth of anaerobes. Adding carbon dioxide or nitrogen also slows deterioration. (See Chapter 20 for more details.)

Chemicals

Chemical additives such as sodium benzoate, sorbic acid, sodium or calcium propionate, and sulphur dioxide retard the growth of microorganisms, modify enzyme activity, inhibit chemical reactions, or modify the structure of foods. (See Chapter 14 for a discussion on the use of chemicals.)

Radiation

Radiation includes X rays, microwave, ultraviolet light, and gamma rays. Radiation can destroy the microorganisms and inactivate enzymes. Chapter 12 provides complete details on irradiation.

Summary

All foods undergo deterioration–physical, chemical, or biological. There are many ways to control this deterioration–from proper handling in the initial stages of harvesting to correct food preservation techniques. Some deterioration produces toxins that are not destroyed by heat. Some of these toxins can cause infections in humans. All foods from living tissues have enzymes. Some of the post-harvest enzymes are desirable and are controlled by heat, chemicals, and radiation. Food processors realize the importance of controlling deterioration through such means as heat, cold, drying, acids, sugar, atmosphere, chemicals, and radiation.

Since product dates aren't a guide for safe use of a product, how long can the consumer store the food and still use it at top quality? Follow these tips:

- Purchase the product before the date expires.
- Follow handling recommendations on product.
- Keep beef in its package until using.
- It is safe to freeze beef in its original packaging. If anything freezing longer than 2 months, overwrap these packages with airtight heavy-duty foil, plastic wrap, or freezer paper or place the package inside a plastic bag.
- For poultry or fish, select them just before checking out at the register. Put in a disposable plastic bag (if available) to contain any leakage which could cross contaminate cooked foods or produce. Make the grocery your last stop before going home.

Review Questions

Success in any career requires knowledge. Test your knowledge of this chapter by answering these questions or solving these problems.

1. Name the two environmental conditions that affect microbial growth on food.
2. Name the three general categories of food deterioration.
3. Some ___________ are desirable in food preservation.
4. Why do foods have a shelf life?
5. The growth of aerobes is slowed by removing the ___________; while providing ___________ limits the growth of anaerobes.

6. List four factors that cause food deterioration.
7. What is a food-borne disease?
8. Give four preservation techniques to prevent food deterioration.
9. Why are some fruits and vegetables washed immediately after being picked?
10. Name four food enzymes and describe their function.

Student Activities

1. Make a list of the date codes on five different foods. List the "sell by date," the "best used by date," and the "expiration date." If possible, bring the labels and discuss these in class.
2. Leave a food such as meat, bread, fruit, and so on, at room temperature and describe the changes in food quality over the course of two weeks. Try to categorize the changes and their causes.
3. Develop a report or presentation on why the occurrence of *E. coli* 0157 in food is such a concern.
4. What is the chemical make up of enzymes and how many enzymes exist?
5. Describe the **organoleptic** properties of six common foods. Organoleptic is a term describing the major quality factors of appearance, texture, and flavor as perceived by the senses. Find an enzyme or a food product produced by an enzyme described in Table 8-2. Develop a report or presentation on this enzyme or food product.
6. Obtain foods that are processed with several methods–for example, dried, canned, fresh, and frozen. Describe what method extends their shelf life. Refer to Table 8-3.

Resources

Cremer, M. L. 1998. *Quality food in quantity. Management and science.* Berkeley, CA: McCutchan Publishing Corporation.

Ensminger, A. H., M. E. Ensminger, J. E. Konlande, and J. R. Robson. 1994. *Foods and nutrition encyclopedia.* 2 Vols. Boca Raton, FL: CRC Press.

Potter, N. N., and J. H. Hotchkiss. 1995. *Food science,* 5th ed. New York: Chapman and Hall.

Vaclavik, V. A,. and E. W. Christina. 1999. *Essentials of food science.* Gaithersburg, MD: Aspen Publishers, Inc.
Vieira, E. R. 1996. *Elementary food science,* 4th ed. New York: Chapman and Hall.

Internet

Internet sites represent a vast resource of information. The URLs (uniform resource locator) for the World Wide Web sites can change. Using one of the search engines on the Internet such as Yahoo!, HotBot, AltaVista, Excite, Dogpile, About, or Google, find more information by searching for these words or phrases: shelf life of foods, food deterioration, food dating, food-borne disease, food enzymes, chemical oxidation, mycotoxins, organoleptic, pyschrophilic, thermophilic. Also, Table A-7 provides a listing of some useful Internet sites that can be used as a starting point.

SECTION *Two*

Preservation

Chapter 9

Heat

Objectives

After reading this chapter, you should be able to:

- Name four degrees of preservation achieved by heating
- Describe how specific heat treatments are selected
- Identify how the heat resistance of microorganisms is determined
- Discuss methods of heating foods before or after packaging
- Describe one type of retort
- Compare conduction and convection heating
- Explain time-temperature combinations
- List three factors that influence how foods heat
- Describe a thermal death curve
- Compare the acidity of various foods to the heat treatment required

Key Terms

aseptic packaging
batch
blanching
commercial sterility
D value
hot-pack
hot-fill
hydrostatic retort
logarithmic
pasteurization
retort
spores
sterilization
still retort
Z value

Heating (cooking) foods kills some microorganisms, destroys most enzymes, and improves shelf life. It does not preserve a food indefinitely. Heating methods create various degrees of preservation depending on the product.

HEAT

Heat, in physics, is the transfer of energy from one part of a substance to another by virtue of a difference in temperature. Heat is energy in transit; it always flows from a substance at a higher temperature to the substance at a lower temperature, raising the temperature of the latter and lowering that of the former substance, provided the volume of the bodies remains constant. The sensation of warmth or coldness of a substance on contact is determined by the property known as temperature. Adding heat to a substance, however, not only raises its temperature, causing it to impart a more acute sensation of warmth, but also expands or contracts a substance and alters its electrical resistance properties, and heat increases the pressure exerted by gases. Temperature depends on the average kinetic energy (motion) of the molecules of a substance.

DEGREES OF PRESERVATION

Depending on the product and the use of the product, heat can be used to create varying degrees of preservation including **sterilization**, **commercial sterility**, **pasteurization**, and **blanching**. Table 9-1 categorizes and compares these heat treatments into mild and severe.

TABLE 9-1 Mild and Severe Heat Treatments

Comparisons	Mild	Severe
Aims	Kill pathogens; reduce bacterial count	Kill all bacteria; food will be commercially sterile
Advantages	Minimal damage to flavor, texture, nutritional quality	Long shelf life; no other preservation method is necessary
Disadvantages	Short shelf life; another preservation method must be used, such as refrigeration or freezing	Food is overcooked; major changes in texture, flavor, nutritional quality
Examples	Pasteurization; blanching	Canning

Sterilization

Sterilization refers to the complete destruction of microorganisms. It often requires at least 250°F (121°C) for 15 minutes to destroy all **spores**. In practice, food would require much longer times because there is a lag between exposure to heat and when the product reaches the desired temperature.

Commercial Sterility

This describes the condition where all pathogenic and toxin-forming organisms have been destroyed, as well as other organisms capable of growth and spoilage under normal handling and storage conditions. Such products may contain viable spores, but they will not grow under normal conditions.

Pasteurization

Pasteurization is a comparatively low-energy thermal process with two main objectives: (1) destroy all pathogenic microorganisms that might grow in a specific product; and (2) extension of shelf life by decreasing number of spoilage organisms present. The product is not sterile and will be subject to spoilage.

Blanching

Blanching is a mild heat treatment, generally applied to fruits and vegetables to inactivate enzymes that might decrease product quality. It may also destroy some microorganisms and thus lead to increased product shelf life. The primary objective, however, is enzyme inactivation.

SELECTING HEAT TREATMENTS

A heat treatment sufficient to destroy all microorganisms and enzymes is detrimental to the food qualities of color, flavor, texture, nutrition, and consistency. To pick the right heat treatment severity for a specific food, two factors must first be determined:

1. Time-temperature combination required to inactivate the most resistant microbe
2. Heat penetration characteristics of the food and the container

Heat penetration characteristics of the food vary with the consistency and particle size; the heat penetration characteristics of the container vary with the size, shape, and material.

HEAT RESISTANCE OF MICROORGANISMS

The most resistant microbe in canned foods is *Clostridium botulinum* (botulism), so food processors must use a time-temperature combination adequate to kill this microbe.

Heat kills any bacteria **logarithmically**. For example, if 90 percent are killed in the first minute at a certain temperature, then 90 percent of those remaining alive will die during the second minute, and 90 percent of those remaining alive will die during the third minute, and so on. (**Note:** Spores are more heat resistant than bacterial cells.)

In foods where the type of contamination is unknown, a margin of safety is applied. Especially in foods that are low acid, processors assume *C. botulinum* to be present and treat the food accordingly.

HEAT TRANSFER

Food processors cannot assume that a product heated in an environment at 250°F (121°C) for 10 minutes has been exposed to that temperature for that time. Product at the edge of the container will reach temperature much sooner than will the product at the center of the container.

Many factors will influence the time required for the coldest portion of the product to obtain the required temperature. The size and shape of the container are important as are the physical properties of the product within the container (Figure 9-1). To overcome some of the problems, different methods of heating are used. These include the use of conduction, convection, and radiation, and combinations of these.

Conduction is the transfer of heat from one particle to another by contact. Food particles in a can do not move. Convection heating

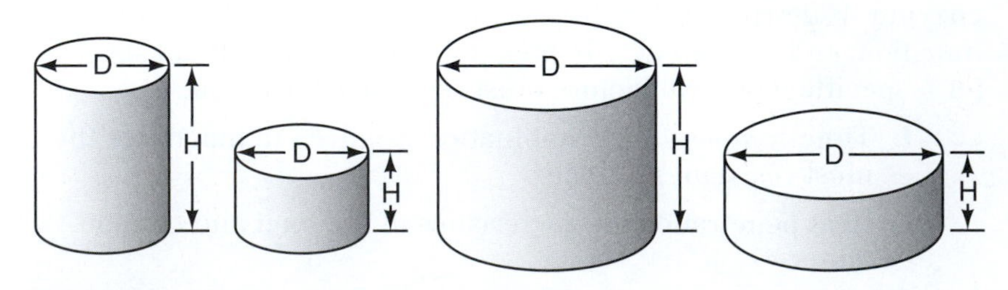

FIGURE 9-1

Size and shape of container (can) expressed by diameter (D) and height (H) determines how the product will heat.

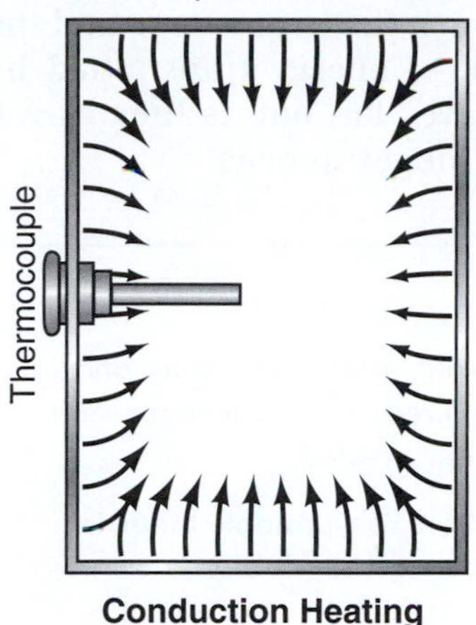

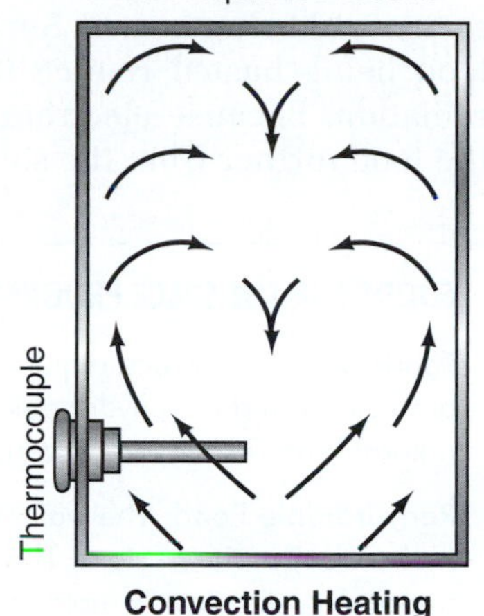

FIGURE 9-2
Container content determines how heating occurs.

means that the movement inside a can distributes the heat (Figure 9-2). Radiation heating occurs when the energy transfers through a medium that itself is not heated.

Convection-Conduction

Conduction heating is thermal transfer due to collisions of hot food particles with cooler ones. Convection heating involves circulation of warm molecules and results in more effective thermal transfer. Foods that heat by convection require less time to reach temperature than foods that heat by conduction.

Heating starts out rapidly by convection. A change in the texture of the food occurs that may cause further heating to be by conduction. This often involves the gelatinization of starch in products like pork and beans. The longer that heating is by convection, the shorter the total process time.

Conduction-Convection

Heating starts as conduction, then a change in food makes the heating convective. An example would be the heating of foods containing large pieces of meat. At first the heating occurs as conduction. Then as juices are released by the meat, heating switches more to convection.

Radiation

Radiation is the transfer of energy in the form of electromagnetic waves. It is the fastest method of heat transfer. Radiation transfers heat directly from the radiant heat source, such as a broiler plate, to the food being heated. Surfaces between the heat source and the food being heated reduce the amount of energy transmitted by radiation. Because electromagnetic waves fan out as they travel, the food further from the source takes longer to heat.

FOODS FOR THE SPACE PROGRAM

Foods going into space require special packaging and processing and are categorized as rehydratable, thermostabilized, intermediate moisture, natural form, irradiated, frozen, fresh, and refrigerated.

Rehydratable Food: The water is removed from rehydratable foods to make them easier to store. This process of dehydration is also known as freeze-drying. Water is replaced in the foods before they are eaten. Rehydratable items include beverages as well as food items. Hot cereal such as oatmeal is a rehydratable food.

Thermostabilized Food: Thermostabilized foods are heat processed so that they can be stored at room temperature. Most of the fruits and fish (tuna fish) are thermostabilized in cans. The cans open with easy-open pull-tabs similar to fruit cups that can be purchased in the local grocery store. Puddings are packaged in plastic cups.

Intermediate Moisture Food: Taking some water out of the product while leaving enough in to maintain the soft texture preserves intermediate moisture foods. This way, it can be eaten without any preparation. These foods include dried peaches, pears, apricots, and beef jerky.

Natural Form Food: These foods are ready to eat and are packaged in flexible pouches. Examples include nuts, granola bars, and cookies.

Irradiated Food: Beefsteak and smoked turkey are the only irradiated products being used at this time. These products are cooked and packaged in flexible foil pouches and sterilized by ionizing radiation so that they can be kept at room temperature.

Frozen Food: These foods are quick frozen to prevent a buildup of large ice crystals. This maintains the original texture of the food and helps it taste fresh. Examples include quiches, casseroles, and chicken potpie.

Fresh Food: These foods are neither processed nor artificially preserved. Examples include apples and bananas.

Refrigerated Food: These foods require cold or cool temperatures to prevent spoilage. Examples include cream cheese and sour cream.

For more information on space foods, visit the NASA Web site at:

<spacelink.nasa.gov/space.food>

PROTECTIVE EFFECTS OF FOOD CONSTITUENTS

Some food constituents protect bacterial spores. Sugar protects bacterial spores in canned fruit. Starch, protein, fats, and oils also protect bacterial spores. This makes some foods difficult to sterilize–for example meat products, fish packed in oil, and ice cream.

DIFFERENT TEMPERATURE-TIME COMBINATIONS

Food processors must determine the time-temperature combination required to kill the most heat resistant pathogen and or spoilage organism in the product of interest. Also, they must know the heat penetration characteristics of the food and its container.

Organisms

The most heat resistant of all pathogens is *Clostridium botulinum.* Some nonpathogenic spore-formers are even more heat resistant–for example, *Putrefactive anaerobe* 3679 and *Bacillus stearothermophilus* 1518. If these organisms are destroyed, the pathogens will also be killed.

Thermal Death Curves

Microbial death, like microbial growth, is generally described by a logarithmic equation. Food processors are concerned with two values: the **D value** and the **Z value** (Figure 9-3). These numbers relate the death of microorganisms to the time and temperature of heating.

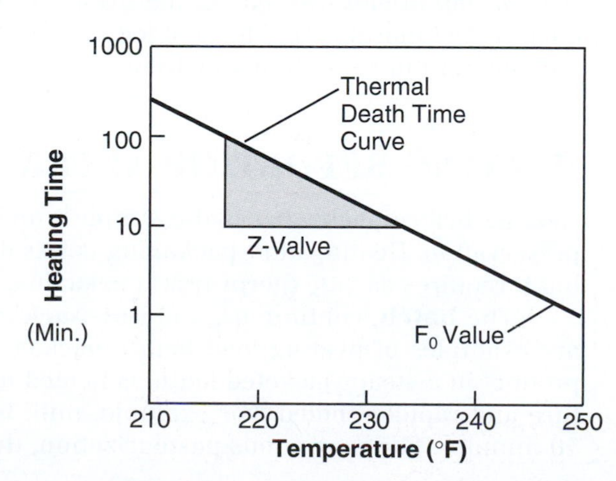

FIGURE 9-3
Graphing heating time and temperature provides a thermal death curve.

D Value

The D value is the time in minutes at a specified temperature to reduce the number of microorganisms by one log cycle. The units of time (minutes or seconds) and a temperature must be specified–for example, D121 = 3.5 minutes (D250 on F scale).

Z Value

The Z value is the temperature required to decrease the time necessary to obtain a one log reduction in cell numbers to one tenth of the original value. For example, if 100 minutes at 220°F (104°C) killed 90 percent of the organisms and 10 minutes at 238°F (114°C) also killed 90 percent of the organisms, the Z value would be equal to 18°F (10°C).

Time-Temperature Combinations

Many combinations of heating times and temperatures will give the same lethality. For *C. botulinum*, the following are equivalent:

0.78 min. at 261°F (127°C)	10 min. at 241°F (116°C)
1.45 min. at 255°F (124°C)	36 min. at 230°F (110°C)
2.78 min. at 250°F (121°C)	150 min. at 219°F (104°C)
5.27 min. at 244°F (118°C)	330 min. at 212°F (100°C)

Low-Acid Foods

The exact number of spores in a food is not known. In the United States, the legal processing required for low-acid foods is exposure to a temperature for a period equal to 12 D values for *Clostridium botulinum*. This is sufficient to decrease any population of *Clostridium botulinum* through 12 log cycles, thus providing an adequate safety margin. Acid level of food affects the processing times and temperatures as shown in Table 9-2.

HEATING BEFORE OR AFTER PACKAGING

Heating before packaging is the simplest and oldest form of heat preservation. Heating after packaging is less damaging to the food, but it requires aseptic (germ-free) packaging.

The **batch**, continuous, and **hot-pack** and **hot-fill** methods are examples of heating food before packaging. In the batch, the product in a steam-jacketed kettle is heated to a specific temperature and rapidly cooled. For example, milk is heated to 145°F for 30 minutes. For continuous pasteurization, the food is heated to a

TABLE 9-2 Processing Requirements and pH

Heat and Processing Requirements	Acidity	pH	Food
High temperature processing 240°–250°F (116°–121°C)	Low acid	7.0	Hominy, ripe olives, crabmeat, eggs, oysters, milk, corn, duck, chickens, codfish, beef, sardines
		6.0	Corned beef, lima beans, peas, carrots, beets, asparagus, potatoes
		5.0	Figs, tomato soup
	Medium acid	4.5	Ravioli, pimentos
Boiling water processing 212°F (100°C)	Acid	—	Potato salad, tomatoes, pears, apricots, peaches, oranges
		3.7	Sauerkraut, pineapple, strawberry, grapefruit
	High acid	3.0	Pickles, relish, cranberry juice, lemon juice, lime juice

high temperature for a short time (HTST). For example, milk is heated to 161°F for 15 seconds. In the hot-pack, hot-fill method, food processors fill unsterilized containers with sterilized food that is still hot enough to render the package commercially sterile. In **retorts**, the product is heated after being placed in a container. Several types of retorts process food this way.

Still Retort

In the **still retort** process, the product is placed in a container and then heated in a steam atmosphere without agitation (Figure 9-4). If the temperature is above 250°F (121°C), considerable burning of the product on the side of the can will occur. Heating times in still retorts are often in the range of 30 to 45 minutes.

Agitating Retort

In some retorts the product is agitated during cooking. This allows for the use of high temperatures during processing. The agitation will allow for convection heating and reduce the time necessary to reach final temperature. Cooking times are only 10 to 20 percent of those in still retorts.

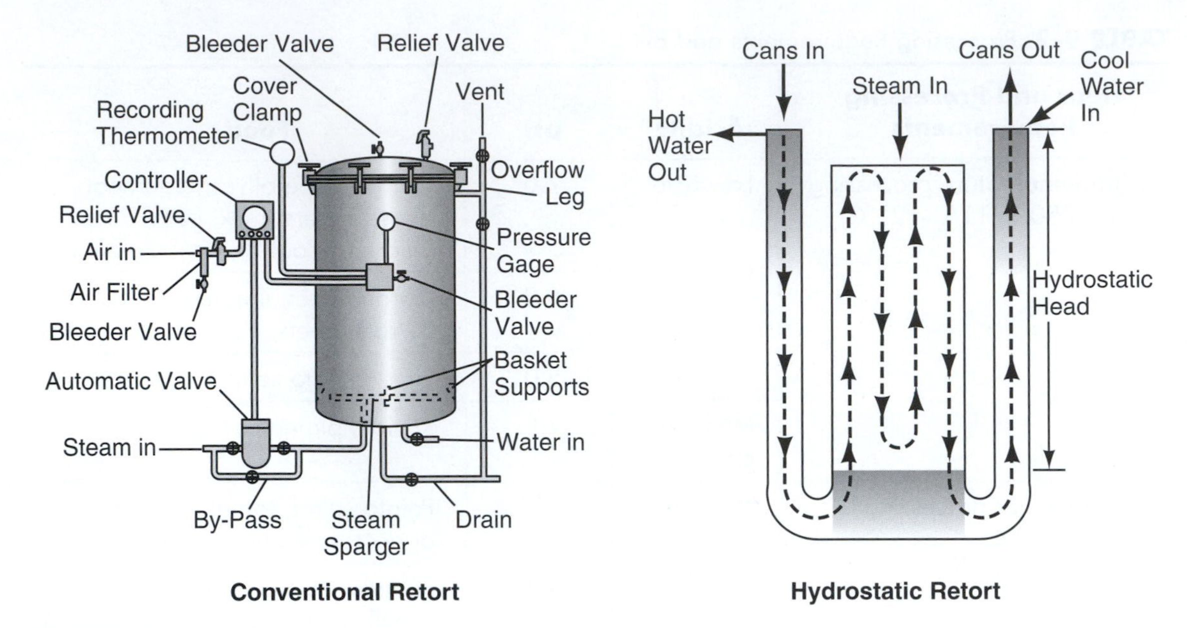

FIGURE 9-4

Retorts are used to heat cans.

Hydrostatic Retort

In a **hydrostatic retort** the cans flow continuously (Figure 9-4). It uses hydrostatic heat to control pressure, and it is an agitating system.

Aseptic Packaging

For **aseptic packaging** the food is sterilized outside of the container (Figure 9-5). Then it is placed into a sterile container and sealed under aseptic conditions. Paper and plastic packaging materials are most commonly used, and this method is most suitable for liquid-based food products.

HOME CANNING

The high percentage of water in most fresh foods makes them very perishable. They spoil or lose their quality for several reasons:

- Growth of undesirable microorganisms–bacteria, molds, and yeasts
- Activity of food enzymes
- Reactions with oxygen
- Moisture loss

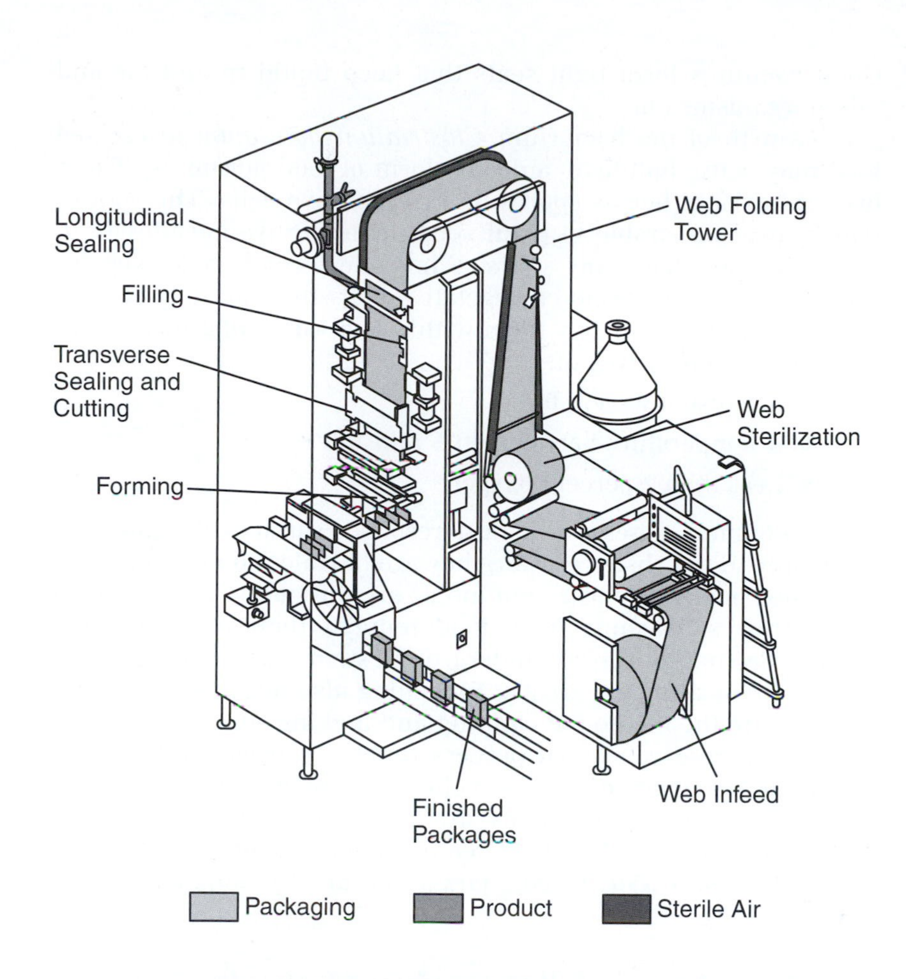

FIGURE 9-5

Aseptic packaging.

Microorganisms live and multiply quickly on the surfaces of fresh food and on the inside of bruised, insect-damaged, and diseased food. Oxygen and enzymes are present throughout fresh food tissues. Proper home canning practices include:

- Carefully selecting and washing fresh food
- Peeling some fresh foods
- Hot-packing many foods
- Adding acids (lemon juice or vinegar) to some foods
- Using acceptable jars and self-sealing lids
- Processing jars in a boiling-water or pressure canner for the correct period of time

Together, these practices remove oxygen; destroy enzymes; prevent the growth of undesirable bacteria, yeasts, and molds; and help form a high vacuum in jars (similar to commercial canning).

Good vacuums form tight seals that keep liquid in and air and microorganisms out.

Growth of the bacterium *Clostridium botulinum* in canned food may cause botulism–a deadly form of food poisoning. These bacteria exist either as spores or as vegetative cells. The spores, which are comparable to plant seeds, can survive harmlessly in soil and water for many years. When ideal conditions exist for growth, the spores produce vegetative cells that multiply rapidly and may produce a deadly toxin within 3 to 4 days of growth in an environment consisting of:

- A moist, low-acid food
- A temperature between 40°F and 120°F
- Less than 2 percent oxygen

Botulinum spores are on most fresh food surfaces. Because they grow only in the absence of air, they are harmless on fresh foods.

Most bacteria, yeasts, and molds are difficult to remove from food surfaces. Washing fresh food reduces their numbers only slightly. Peeling root crops, underground stem crops, and tomatoes reduces their numbers greatly. Blanching also helps, but the vital controls are the method of canning and making sure that the recommended research-based process times are used. Correct processing times ensure destruction of the largest expected number of heat-resistant microorganisms in home-canned foods. Properly sterilized canned food will be free of spoilage if lids seal and jars are stored below 95°F. Storing jars at 50° to 70°F enhances retention of quality.

Food Acidity and Processing Methods

Whether food should be processed in a pressure canner or boiling-water canner to control botulinum bacteria depends on the acidity in the food. Acidity may be natural, as in most fruits, or added, as in pickled food. Low-acid canned foods contain too little acidity to prevent the growth of these bacteria. Acid foods contain enough acidity to block their growth, or destroy them more rapidly when heated. The acidity level in foods can be increased by adding lemon juice, citric acid, or vinegar.

Low-acid foods have pH values higher than 4.6. They include red meats, seafood, poultry, milk, and all fresh vegetables except for most tomatoes. Most mixtures of low-acid and acid foods also have pH values above 4.6 unless their recipes include enough lemon juice, citric acid, or vinegar to make them acid foods. Acid foods have a pH of 4.6 or lower. They include fruits, pickles, sauerkraut, jams, jellies, marmalades, and fruit butters.

Although tomatoes usually are considered an acid food, some are now known to have pH values slightly above 4.6. Figs also have pH values slightly above 4.6. If they are to be canned as acid foods, these products must be acidified to a pH of 4.6 or lower with lemon juice or citric acid. Properly acidified tomatoes and figs are acid foods and can be safely processed in a boiling-water canner.

Botulinum spores are very hard to destroy at boiling-water temperatures–the higher the canner temperature, the more easily the spores are destroyed. All low-acid foods should be sterilized at temperatures of 240° to 250°F. These temperatures are possible with pressure canners operated at 10 to 15 PSIG. PSIG means pounds per square inch of pressure as measured by gauge. At temperatures of 240° to 250°F, the time needed to destroy bacteria in low-acid canned food ranges from 20 to 100 minutes. The exact time depends on the kind of food being canned, the way it is packed into jars, and the size of jars. The time needed to safely process low-acid foods in a boiling-water canner ranges from 7 to 11 hours; the time needed to process acid foods in boiling water varies from 5 to 85 minutes.

Process Adjustments at High Altitudes

Using the process time for canning food at sea level may result in spoilage if you live at altitudes of 1,000 feet or more. Water boils at lower temperatures as altitude increases. Lower boiling temperatures are less effective for killing bacteria. Increasing the process time or canner pressure compensates for lower boiling temperatures.

For more information about home canning, check out the USDA Complete Guide to Home Canning on the Web at <www.foodsafety.ufl.edu/cmenu/preserve.htm>.

Summary

Heat produces varying degrees of preservation depending on the product and the use of the product. Heat treatments can be selected on the basis of time and temperature combination to inactivate the most resistant microbe and the heat penetration characteristics of the food and the container. The most heat resistant microbe in canned foods is *Clostridium botulinum.* Thermal death curves describe the time/temperature relationships for preservation of food.

Depending on the type of food and its acidity, the heat treatment can be mild or severe. Convection, conduction, and radiation transfer heat to the processed foods. Commercially, foods are heated before or after packaging. Different types of retorts efficiently process commercial foods. Aseptic packaging is used when the food is sterilized outside the container. Home canning follows the same principles as commercial heat preservation methods.

Review Questions

Success in any career requires knowledge. Test your knowledge of this chapter by answering these questions or solving these problems.

1. The most heat resistant microbe in canned foods is ___________ ___________.
2. What are the two main objectives of pasteurization?
3. Name four types of preservatives achieved by heating.
4. In the thermal death curve, the D value relates to the ___________ to reduce the number of microorganisms, and the Z value relates to the ___________ required to decrease the microorganisms.
5. Heating after packaging requires what type of packaging?
6. ___________ heating is thermal transfer due to collisions of hot food particles with cooler ones.
7. What is the difference between a still retort and an agitating retort?
8. Identify the two factors to pick the right heat treatment severity for a specific food.
9. Define conduction heating.
10. ___________ is the transfer of energy in the form of electromagnetic waves.

Student Activities

1. Using an electronic spreadsheet, develop a chart that shows logarithmic growth of bacteria.
2. Develop a report or presentation on botulism.
3. Create a drawing that describes a spore.
4. Search the Internet for the use of heat in processing a food product or write to a company such as Del Monte® or Campbell's® and request information on the use of heat in some of their products.
5. Culture some bacteria in petri dishes on agar, or obtain some prepared microscope slides of bacteria and observe the different types through a microscope.

6. Choose a food that is easy to process and demonstrate home canning.

Resources

Asimov, I. 1988. *Understanding physics.* New York: Hippocrene Books, Inc.
Potter, N. N., and J. H. Hotchkiss. 1995. *Food science,* 5th ed. New York: Chapman and Hall.
Vaclavik, V. A., and E. W. Christina. 1999. *Essentials of food science.* Gaithersburg, MD: Aspen Publishers, Inc.
Vieira, E. R. 1996. *Elementary food science,* 4th ed. New York: Chapman and Hall.

Internet

Internet sites represent a vast resource of information. The URLs (uniform resource locator) for the World Wide Web sites can change. Using one of the search engines on the Internet such as Yahoo!, HotBot, AltaVista, Excite, Dogpile, About, or Google, find more information by searching for these words or phrases: aseptic packaging, hydrostatic retort, agitating retort, still retort, food preservation, conduction/convection heating, radiation transfer of heat, low-acid foods, home canning. Also, Table A-7 provides a listing of some useful Internet sites that can be used as a starting point.

Chapter 10

Cold

Objectives

After reading this chapter, you should be able to:

- Compare cooling, refrigeration, and freezing
- Identify four requirements for refrigeration
- Correlate storage temperature to length of storage
- Compare requirements for refrigeration to those of freezing
- List three methods of freezing
- Describe changes in food quality that may occur during refrigeration or freezing
- Compare home freezing to commercial freezing

Key Terms

cool storage
cryogenic
hydrocooling
hypobaric
immersion freezing
refrigeration
vacuum cooling

Cold (cool) storage, refrigeration, and frozen storage are methods of food preservation and processing differing in temperature and time. Various methods take food to temperatures necessary for cold storage or frozen storage.

REFRIGERATION VERSUS FREEZING

Cool storage is considered any temperature from 68° to 28°F (16° to –2°C). Refrigerator temperatures range from 40° to 45°F (4.5° to 7°C); frozen storage temperatures range from 32° to 0°F (0° to –18°C).

Microbes grow more rapidly at temperatures above 50°F (10°C). Some growth occurs at subfreezing temperatures as long as water is available. Little growth occurs below 15°F (–9.5°C).

Historically, the first mechanical ammonia **refrigeration** system was invented in 1875. In the 1920s Clarence Birdseye started frozen food packaging.

REFRIGERATION AND COOL STORAGE

Cool storage is the gentlest of the food preservation methods. It affects taste, texture, nutritive value, and color the least, but it is a short-time preservation method.

CLARENCE BIRDSEYE

Clarence Birdseye was born in Brooklyn, New York, on December 9, 1886. He attended Amherst College, majoring in biology, but Clarence did not graduate. Instead, he pursued a career as a field naturalist for the U.S. government. The job took him far north, near the Arctic, where he made a discovery that changed the history of the food industry. He noticed that freshly caught fish, when placed onto the Arctic ice and exposed to the icy wind and frigid temperatures, froze solid almost immediately. Clarence found too that the fish, when thawed and eaten, still had all its fresh characteristics.

In September 1922, Clarence organized his own company, Birdseye Seafoods, Inc., where he began processing chilled fish fillets at a plant near the Fulton Fish Market in New York City. In 1924, he developed the process of packing dressed fish or other food in cartons, then freezing the contents between two flat, refrigerated surfaces under pressure.

On July 3, 1924, he organized the General Seafood Corporation, with the financial help of Wetmore Hodges, Basset Jones, I. L. Rice, William Gamage, and J. J. Barry. This was the beginning of the wholesale frozen foods industry. The retail frozen foods business began March 6, 1930, in Springfield, Massachusetts, when the Springfield Experiment Test Market produced and packaged twenty-six different vegetables, fruits, fish, and meats.

For more information, visit the Birdseye Web site at:

<www.birdseye.com/index_new.html>

Refrigeration is also a gentle method of food preservation (Figure 10-1). It minimally effects the taste, texture, and the nutritional value of foods. Refrigeration has a limited contribution toward preserving food. For most foods, refrigeration extends the shelf life by a few days. In many cases, refrigeration is not the sole means of preserving the food. For example, milk is first heat processed (pasteurized), and then refrigerated. The heat treatment destroys the pathogenic microorganisms and reduces the total microbial load as well as inactivates the enzymes in the milk. Refrigeration will keep the spoilage reactions (microbial or enzymatic) to a minimum. Refrigeration does not kill microorganisms or inactivate enzymes, rather it slows down their deteriorative effects.

Refrigeration temperature is a key factor in predicting the length of the storage period (refer to Tables 8-4 and 8-5). For example, meat will last 6 to 10 days at 32°F (0°C), one day at 72°F (22°C), and less than one day at 100°F (38°C). Household refrigerators usually run at 40.5° to 44.6°F (4.7° to 7°C). Commercial refrigerators are operated at a slightly lower temperature. A group of microorganisms called psychrophilic will grow at refrigerated temperatures.

When cooling is used as a preserving technique, it is essential that the food be maintained at the proper cold temperature during manufacturing, transport, display, and home storage. Signs of spoilage in refrigerated foods vary with the food product. In fruits and vegetables, loss of firmness or crispness takes place. Red meat will change in color; fish will get softer, with noticeable drippage.

FIGURE 10-1

Commercial refrigeration units.

Refrigerated food will last longer if it is cleaned and properly packaged prior to refrigeration and, of course, maintained at the proper temperature with minimum exposure to surrounding temperatures.

In relation to the timelines of refrigeration and the processing of food, heat must be rapidly removed from foods at the time of harvest or slaughter. To do this, such methods as moving air, **hydrocooling** (water), **vacuum cooling**, and liquid nitrogen are used.

Because refrigeration is an energy-demanding process that must be maintained throughout the life of the product, refrigerated foods tend to cost more than nonrefrigerated foods. Yet, today's consumer is buying more refrigerated foods for their fresh quality. When attempting to preserve the high quality of refrigerated foods, the consumer needs to keep cold food at temperatures between 33°F and 45°F.

Requirements of Refrigerated Storage

Refrigerated storage requires low temperatures, air circulation, humidity control (80 to 95 percent), and modified gas atmosphere. Insulation, frequency of door opening, quantity of hot product added daily, and the respiration rate of the food product can affect the requirements.

In controlled atmosphere storage, standard cold storage temperatures and humidity are maintained but the oxygen levels are reduced and the carbon dioxide levels are elevated to reduce respiration in the food product. Some controlled atmosphere storage units are **hypobaric**, meaning the pressure is also reduced. This reduces the availability of oxygen.

Changes in Food during Refrigerated Storage

During refrigerated storage, foods can experience chill injury, flavor (odor) absorption, and loss of firmness, color, flavor, and sugar.

FREEZING AND FROZEN STORAGE

Freezing can be thought of as a continuation of refrigeration. The freezing point for pure water is 32°F (0°C). For food, however, the freezing point is below that. Chemistry and physics teach that the presence of solutes in water will lower its freezing point. The addition of one mole (molecular weight in grams) of any nonionic substance (solute) to one liter of water lowers the freezing point by 1.885 degrees C (Figure 10-2). Water contained in foods has many solutes, mainly salts and sugars.

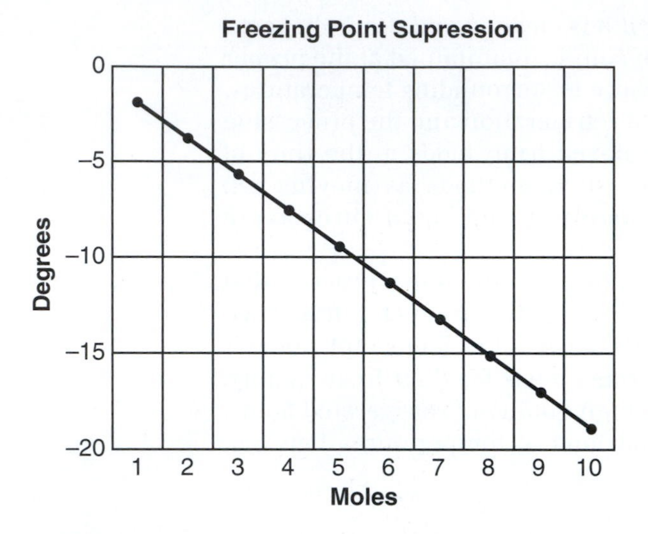

FIGURE 10-2

Each mole of a solute in water suppresses the freezing point by 1.885°C.

Freezing is similar to refrigeration because it will not destroy microorganisms or inactivate enzymes but rather slow their deteriorative effect. In some cases, it will completely inhibit the activity of microorganisms, even though they will still be alive. Enzymes, on the other hand, will maintain a certain level of activity during freezing. For this reason, food processors blanch vegetables prior to freezing them. Blanching is a mild heat treatment designed to inactivate enzymes.

Freezing has been a key technology in bringing convenience foods to homes and restaurants. It causes minimal changes in the quality of food in terms of size, shape, texture, color, flavor, and microbial load. This is assuming that the freezing process is carried out properly, and assuming that the food can be frozen.

Food processors freeze foods to an internal temperature of 0°F (–18°C). The food must be maintained at this temperature or slightly lower during transport and storage. Many fruits and vegetables will retain good quality at the above temperature for up to 12 months or even longer. The expected frozen storage will vary with temperature. For example, frozen orange juice will last 27 months at 0°F (–18°C), 10 months at 10°F (–12°C), and only 4 months at 20°F (–6.7°C). Table 10-1 provides a list of products and their frozen storage time.

TABLE 10-1 Suggested Maximum Home-Storage Periods for Commercially Frozen Foods

Food	Approximate Months Storage (0° F)
Fruits and Vegetables	
Fruits:	
Cherries	12
Peaches	12
Raspberries	12
Strawberries	12
Fruit Juice concentrates:	
Apple	12
Grape	12
Orange	12
Vegetables:	
Beans	8
Corn	8
Peas	8
Spinach	8
Baked Goods	
Bread and yeast rolls:	
White bread	3
Plain rolls	3
Cakes:	
Angel or chiffon	2
Chocolate layer	4
Fruit	12
Pound or yellow	6
Danish pastry	3
Doughnuts:	
Cake type	3
Yeast raised	3
Pies, Fruit (unbaked)	8
Meat	
Beef:	
Ground beef	4
Roast	12
Steaks	12
Lamb:	
Steaks	9
Roast	9
Pork, cured	2
Pork, fresh:	
Roasts	8
Sausage	2
Veal:	
Cutlets, chops	9
Roasts	9
Cooked meat:	
Meat dinners	3
Meat pie	3
Swiss steak	3
Poultry	
Chicken:	
Cut-up	6
Livers	3
Whole	12
Duck, whole	6
Goose, whole	6
Turkey:	
Cut-up	6
Whole	12
Cooked chicken and turkey:	
Chicken or turkey (sliced meat and gravy)	6
Chicken or turkey pies	6
Fried chicken	3
Fried chicken dinners	3
Fish and Shellfish	
Fish Fillets:	
"Lean" fish:	
Cod, flounder, haddock, halibut	6
"Fatty" fish:	
Mullet, ocean perch, sea trout, striped bass	3
Shellfish:	
Clams, shucked Crabmeat	3
King or Dungeness	2
Oyster, shucked	1
Shrimp (unbreaded)	12
Cooked fish and shellfish:	
Fish with cheese sauce	3
Fish with lemon butter sauce	3
Fried fish dinner, fried fish sticks, scallops, or shrimp	3
Shrimp creole	3
Tuna pie	3
Frozen desserts	
Ice cream	1

Source: Bulletin 989, "So Easy to Preserve," produced by the Cooperative Extension Service, the University of Georgia, College of Agricultural and Environmental Sciences. Third Edition published in 1993.

Chemical Changes during Freezing

Enzymes in fruits and vegetables are slowed down, but not destroyed during freezing. If not inactivated, these enzymes can cause color and flavor changes as well as loss of nutrients.

Enzymes in Vegetables. Blanching inactivates enzymes in vegetables. Blanching is the exposure of the vegetable to boiling water or steam for a brief period of time. The vegetable must then be rapidly cooled in ice water to prevent cooking. Contrary to statements in some publications on home freezing, blanching is essential for top-quality frozen vegetables. Blanching also helps to destroy microorganisms on the surface of the vegetables. It makes vegetables such as broccoli and spinach more compact, so they do not take up as much room in the freezer. Overblanching results in a cooked product and a loss of flavor, color, and nutrients. Underblanching stimulates enzyme activity and is worse than no blanching at all.

Enzymes in Fruits. Enzymes in fruits can cause browning and loss of vitamin C. Because fruits are usually served raw and people like their uncooked texture, they are not usually blanched. Instead, the use of chemical compounds controls enzymes in frozen fruits. The most common control chemical is ascorbic acid (vitamin C). Ascorbic acid may be used in its pure form or in commercial mixtures of ascorbic acid and other compounds.

Some directions for freezing fruits also include temporary measures to control browning. Such temporary measures include placing the fruit in citric acid or lemon juice solutions, or in a sugar syrup. These methods do not prevent browning as effectively as treatment with ascorbic acid.

Rancidity in Foods. Another type of chemical change that can take place in frozen products is the development of rancid, off flavors. This can occur when fat, such as in meat, is exposed to air over a period of time. Using a wrapping material that does not permit air to reach the product can control rancidity. Removing as much air as possible from the freezer container reduces the amount of air in contact with the product.

Textural Changes during Freezing

Freezing actually consists of freezing the water contained in the food. When the water freezes, it expands and the ice crystals formed cause the cell walls to rupture. Consequently the texture of the product will be much softer when the product thaws.

These textural changes are more noticeable in fruits and vegetables that have a higher water content and those that are usually eaten raw. For example, when a frozen tomato is thawed, it turns into mush and liquid. This explains why celery, lettuce, and tomatoes are not usually frozen. For this reason frozen fruits, usually consumed raw, are best served before they have completely thawed. In the partially thawed state, the effect of freezing on the fruit tissue is less noticeable.

Textural changes due to freezing are not as apparent in products that are cooked before eating because cooking also softens cell walls. These changes are less noticeable in high-starch vegetables, such as peas, corn, and lima beans.

Rate of Freezing. Freezing products as quickly as possible can control the extent of cell wall rupture. In rapid freezing, a large number of small ice crystals are formed. These small ice crystals cause less cell wall rupture than slow freezing, which produces only a few large ice crystals. This is why some home freezer manuals recommend that the temperature of the freezer be set at the coldest setting several hours before foods will be placed in the freezer.

Changes Caused by Fluctuating Temperatures. Storing frozen foods at temperatures higher than 0°F increases the rate at which deterioration can take place and can shorten the shelf life of frozen foods. For example, the same loss of quality in frozen beans stored at 0°F for one year will occur in three months at 10°F, in three weeks at 20°F, and in five days at 30°F.

Fluctuating temperatures can cause the ice in the foods to thaw slightly and then refreeze. Each time this happens, the smaller ice crystals form larger ones, further damaging cells and creating a mushier product. Fluctuating temperatures can also cause water to migrate from the product. This defect may also be seen in commercially frozen foods that have been handled improperly.

Moisture Loss. Moisture loss, or ice crystals evaporating from the surface area of a product, produces freezer burn–a grainy, brownish spot where the tissues become dry and tough. This surface freeze-dried area is very likely to develop off-flavors, but it will not cause illness. Packaging in heavyweight, moisture-resistant wrap will prevent freezer burn.

Microbial Growth in the Freezer

The freezing process does not actually destroy the microorganisms that may be present on fruits and vegetables. Although blanching destroys some microorganisms and there is a gradual decline in the number of microorganisms during freezer storage, sufficient numbers are still present to multiply and cause spoilage of the product when it thaws. For this reason, it is necessary to carefully inspect any frozen products that have accidentally thawed by the freezer going off or the freezer door being left open.

Clostridium botulinum, the microorganism that causes the greatest problem in canning low-acid foods, does not grow and produce toxin at 0°F. Therefore, freezing provides a safe and easy alternative to pressure canning low-acid foods.

Freezing Methods

Freezing methods that freeze the product in air include the still-air sharp freezer, the blast freezer, and the fluidized-bed freezer (Figure 10-3). Single plate, double plate, pressure plate, and slush freezer all are methods of freezing where the food or food packages directly contact a surface that is cooled by a refrigerant. In **immersion freezing** intimate contact occurs between the food or package and the refrigerant. Types of immersion freezing include a heat exchange fluid, compressed gas, or refrigerant spray. Very fast freezing occurs using liquid nitrogen–a **cryogenic** liquid.

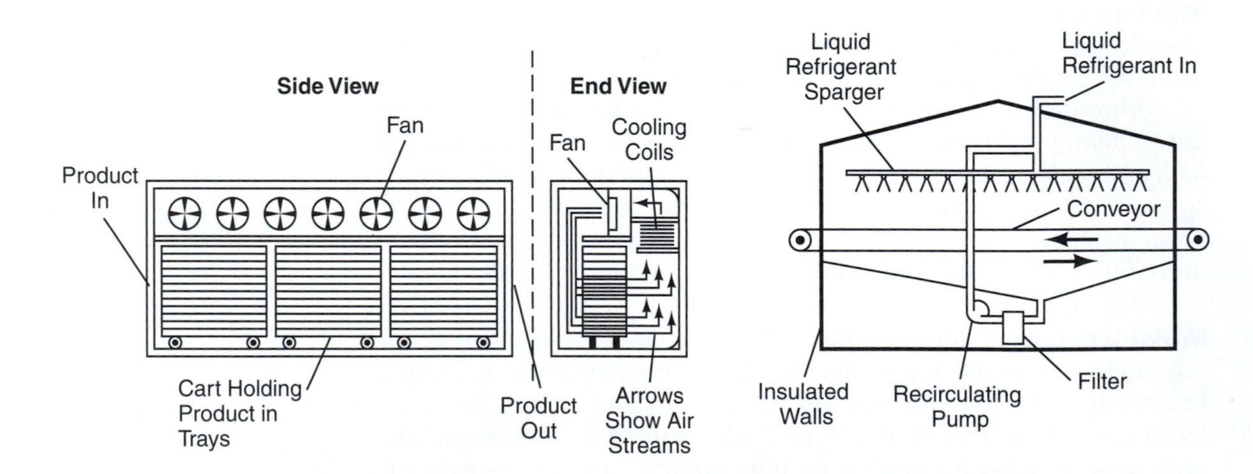

FIGURE 10-3

Tunnel blast freezers and continuous liquid-refrigerant freezers preserve large quantities of product.

Packaging

Packaging for frozen foods protects against dehydration, light, and air. It needs to be strong, flexible, and liquid tight.

NEW DEVELOPMENTS

The success of freezing technology has opened a new field for food processors. They can prepare complete meals and freeze them until the consumer is ready to thaw and heat them. Many of these meals are sold in their serving dishes. Other popular frozen foods are pot pies, fish sticks, desserts, and potatoes. At present, no other form of food preservation offers the convenience of frozen foods.

HOME FREEZING

Freezing is one of the easiest, most convenient, and least time-consuming methods of preserving foods. Freezing does not sterilize foods; the extreme cold simply retards the growth of microorganisms and slows down chemical changes that affect quality or cause food to spoil. By following the directions for freezing food, high-quality, nutritious frozen food can be enjoyed year round.

Foods for the home freezer must have proper packaging materials to protect their flavor, color, moisture content, and nutritive value from the dry climate of the freezer. The selection of containers depends on the type of food to be frozen, personal preference, and types that are readily available. Fruits and vegetables should not be frozen in containers with a capacity over one-half gallon. Foods in larger containers freeze too slowly to result in a satisfactory product. In general, packaging materials must have these characteristics:

- Moisture-vapor resistant
- Durable and leak-proof
- Not become brittle and crack at low temperatures
- Resistant to oil, grease, or water
- Protect foods from absorption of off flavors or odors
- Easy to seal
- Easy to mark

Two types of packaging materials are used for home use–rigid containers and flexible bags or wrappings.

Rigid Containers

Rigid containers made of plastic or glass are suitable for all packs and are especially good for liquid packs. Straight sides on rigid containers make the frozen food much easier to get out. Rigid containers are often reusable and make the stacking of foods in the freezer easier. Cardboard cartons for cottage cheese, ice cream, and milk are not sufficiently moisture-vapor resistant to be suitable for long-term freezer storage, unless they are lined with a freezer bag or wrap.

Regular glass jars break easily at freezer temperatures. If using glass jars, wide-mouth dual-purpose jars made for freezing and canning must be used. These jars have been tempered to withstand extremes in temperatures. The wide-mouth allows easy removal of partially thawed foods. If standard canning jars (those with narrow mouths) are used for freezing, extra head space allows for expansion of foods during freezing. Expansion of the liquid could cause the jars to break at the neck. Some foods will need to be thawed completely before removal from the jar is possible. Covers for rigid containers should fit tightly. If they do not, freezer tape seals and reinforces the covers. This tape is especially designed to stick at freezing temperatures.

Flexible Bags or Wrappings

Bags and sheets of moisture-vapor resistant materials and heavy-duty aluminum foil are suitable for dry packed vegetables and fruits, meats, fish, or poultry. Bags can also be used for liquid packs.

Protective cardboard cartons may be used to protect bags and sheets against tearing and to make stacking easier.

Laminated papers made of various combinations of paper, metal foil, glassine, cellophane, and rubber latex are suitable for dry packed vegetables and fruits, meats, fish, and poultry. Laminated papers are also used as protective overwraps.

Freezer Pointers

The following freezer pointers will help any individual enjoy the convenience of home freezing of foods:

- Freeze foods at 0°F or lower. To facilitate more rapid freezing, set the temperature control at –10°F or lower about 24 hours in advance.
- Freeze foods as soon as they are packaged and sealed.

- Do not overload the freezer with unfrozen food. Add only the amount that will freeze within 24 hours, which is usually 2 to 3 pounds of food per cubic foot of storage space. Overloading slows down the freezing rate, and foods that freeze too slowly may lose quality.
- Place packages in contact with refrigerated surfaces in the coldest part of the freezer.
- Leave a little space between packages so that air can circulate freely. When the food is frozen, store the packages close together.

Foods to Freeze for Quality

Freezing cannot improve the flavor or texture of any food, but when properly done, it can preserve most of the quality of the fresh product. Only the best-quality fruits and vegetables at their peak of maturity should be frozen.

Fruit should be firm yet ripe–firm for texture and ripe for flavor. Vegetables should be young, tender, unwilted, and garden fresh. If fruits or vegetables must be held before freezing, they should be stored in the refrigerator to prevent deterioration. Some varieties of fruits or vegetables are more suitable for freezing than others.

Foods found in Table 10-2 do not freeze well and are best preserved by another method or left out of mixed dishes that are to be frozen.

Effect of Freezing on Spices and Seasonings

Freezing spices and seasonings can create some undesirable effects, including the following:

- Pepper, cloves, garlic, green pepper, imitation vanilla, and some herbs tend to get strong and bitter.
- Onion and paprika change flavor during freezing.
- Celery seasonings become stronger.
- Curry may develop a musty off-flavor.
- Salt loses flavor and has the tendency to increase rancidity of any item containing fat.

When using seasonings and spices, foods can be seasoned lightly before freezing, and additional seasonings added when reheated or served.

TABLE 10-2 Foods Showing Reduced Quality after Freezing

Foods	Usual Use	Condition after Thawing
Cabbage[1], celery, cress, cucumbers[1], endive, lettuce, parsley, radishes	As raw salad	Limp, water-logged, quickly develops oxidized color, aroma, and flavor
Irish potatoes, baked or boiled	In soups, salads, sauces with butter	Soft, crumbly, water-logged, mealy
Cooked macaroni, spaghetti, or rice	When frozen alone for later use	Mushy, tastes warmed over
Egg whites, cooked	In salads, creamed foods, sandwiches, sauces, gravy, or desserts	Soft, tough, rubbery, spongy
Meringue	In desserts	Toughens
Icings made from egg whites	Cakes, cookies	Frothy, weep
Cream or custard fillings	Pies, baked goods	Separates, watery, lumpy
Milk sauces	For casseroles or gravies	May curdle or separate
Sour cream	As topping, in salads	Separates, watery
Cheese or crumb toppings	On casseroles	Soggy
Mayonnaise or salad dressing	On sandwiches (not in salads)	Separates
Gelatin	In salads or desserts	Weeps
Fruit jelly	Sandwiches	May soak bread
Fried foods	All except French fried potatoes and onion rings	Lose crispness, become soggy

Source: Bulletin 989, "So Easy to Preserve," produced by the Cooperative Extension Service, the University of Georgia, College of Agricultural and Environmental Sciences. Third Edition, 1993 (Web URL: <www.ces.uga.edu/ Family/soeasy/soeasy.html>)

[1]Cucumbers and cabbage can be frozen as marinated products such as "freezer slaw" or "freezer pickles." These do not have the same texture as regular slaw or pickles.

Freezer Management

A full freezer is most energy-efficient, and refilling your freezer several times a year is most cost-efficient. If the freezer is filled and emptied only once each year, the energy cost per package is very high. You can lower the cost for each pound of stored food by filling and emptying your freezer two, three, and even more times each year.

Posting a frozen foods inventory near the freezer and keeping it up-to-date by listing the foods and dates of freezing keeps foods from being forgotten. This inventory should show the exact amounts and kinds of foods in the freezer at all times. Also, foods in a freezer should be organized into food groups for ease in locating. Packages in the freezer the longest need to be the first ones used.

Storage temperature must be maintained at 0°F or lower. At higher temperatures, foods lose quality much faster. A freezer thermometer in the freezer allows frequent checking of the temperature.

For more details on home freezing, search the Web or visit these Web sites:

<www.ces.uga.edu/Family/soeasy/freeze1.html>

<www.oznet.ksu.edu/dp_fnut/foodpreservation/freezing.htm>

Summary

Freezing and refrigeration preserve foods not because they destroy microbes but because they slow or stop microbial growth. Refrigeration temperatures range from 40° to 45°F; frozen storage temperatures range from 32° to 0°F. Refrigerated storage also requires air circulation, humidity control, and a modified gas atmosphere. During refrigerated storage, foods can experience the absorption of flavor, a loss of firmness, color, flavor, and sugar.

Freezing technology has been key to the development of convenience foods. Foods properly frozen and stored experience minimal changes in food quality. Frozen storage varies with temperature and the type of food being stored. General freezing methods include still air, blast freezer, and fluidized-bed freezer. Single plate, double plate, and slush freezers freeze foods or packages that directly contact a cold surface. Immersion freezing is the direct contact of the food or package and the refrigerant such as liquid nitrogen. To maintain quality, frozen foods must be packaged in airtight and liquid-tight strong and flexible containers.

Home freezing of foods follows the same general principles as commercial freezing.

Review Questions

Success in any career requires knowledge. Test your knowledge of this chapter by answering these questions or solving these problems.

1. Name the three methods of freezing.
2. List the four requirements of refrigerated storage.
3. Identify four changes in food during refrigeration.

4. A key factor in food freezing is how __________ the food is frozen.
5. Describe the temperature difference between cooling, refrigeration, and freezing.
6. Why do food processors blanch vegetables prior to freezing them?
7. Name the two types of containers for home freezing use.
8. Freezing cannot improve the __________ or texture of any food.
9. Explain why a freezer should not be overloaded with unfrozen food.
10. List the three things packaging for frozen foods protects against.

Student Activities

1. Freeze one of the following and report what you observe when the food thaws: basil, artificial vanilla, egg custard, fresh apples or peaches, lettuce. Use your senses of taste, smell, and sight to describe what you observe.
2. Develop a report or presentation on what happens as water freezes.
3. Obtain some liquid nitrogen and/or dry ice. Freeze some food products using these and compare the speed of freezing. Also, compare the speed of freezing to that in a normal freezer.
4. Develop a report or presentation on the use of refrigeration and freezing in other countries.
5. Keep some of the following foods at refrigerator temperatures for extended times and describe changes in food quality: hamburger, celery, apple, tomato, potato, strawberries, and cheese.

Resources

National Council for Agricultural Education. 1993. *Food science, safety, and nutrition.* Madison, WI: National FFA Foundation.

Potter, N. N., and J. H. Hotchkiss. 1995. *Food science,* 5th ed. New York: Chapman and Hall.

Vaclavik, V. A., and E. W. Christina. 1999. *Essentials of food science.* Gaithersburg, MD: Aspen Publishers, Inc.

Vieira, E. R. 1996. *Elementary food science*, 4th ed. New York: Chapman and Hall.

Internet

Internet sites represent a vast resource of information. The URLs (uniform resource locator) for the World Wide Web sites can change. Using one of the search engines on the Internet such as Yahoo!, HotBot, AltaVista, Excite, Dogpile, About, or Google, find more information by searching for these words or phrases: air cooling, cool storage, hydrocooling, hypobaric, immersion freezing, vacuum cooling, frozen food storage, freezing point, freezing methods. Also, Table A-7 provides a listing of some useful Internet sites that can be used as a starting point.

Chapter 11

Drying and Dehydration

Objectives

After reading this chapter, you should be able to:

- Discuss two reasons for dehydrating foods
- Describe changes that occur during dehydrating
- List three factors affecting dehydration
- Identify chemical changes that can occur in food during drying
- List two problems that can occur during drying
- Identify reasons for food concentration
- Describe three drying methods
- Describe the methods of food concentration
- Compare home drying of foods to commercial drying

Key Terms

atmospheric pressure
bars
caramelization
case hardening
dehydration
enzymatic browning
rehydrated
sublimation
surface area
vacuum

Drying and dehydration both remove water from foods. Dehydration occurs under natural conditions in the field and during cooking. Dehydrated and dried food are lighter, take up less space, and cost less to ship.

DEHYDRATION

Water is removed from foods under natural conditions in the field in the case of grains, raisins, and seeds. Water is also removed during cooking and as part of a controlled **dehydration** process. Dehydration is the almost complete removal of water. It is commonly used to produce dried milk, coffee, soups, and corn flakes (Figure 11-1). Dehydration also results in:

- Decreased weight
- Increased amount of product per container
- Decreased shipping costs

The purpose of drying is to remove enough moisture to prevent microbial growth. Figure 11-2 shows some types of commercial driers. In some cases, such as sun drying, the rate may be too slow and organisms may cause spoilage before the product can be thoroughly dried. In these cases, salt or smoke may be added to the product prior to drying.

The lower limit of moisture content by the sun drying method is approximately 15 percent. Sun drying can have problems with contamination, and it only works in areas of low humidity. Dried foods are not sterile. Many spores survive in dry areas of food. Drying never completely removes all water.

Four factors affect heat and liquid transfer in food products during drying:

1. Surface area
2. Temperature
3. Humidity
4. Atmospheric pressure

FIGURE 11-1

Wide varieties of dried/dehydrated soups are available at many grocery stores.

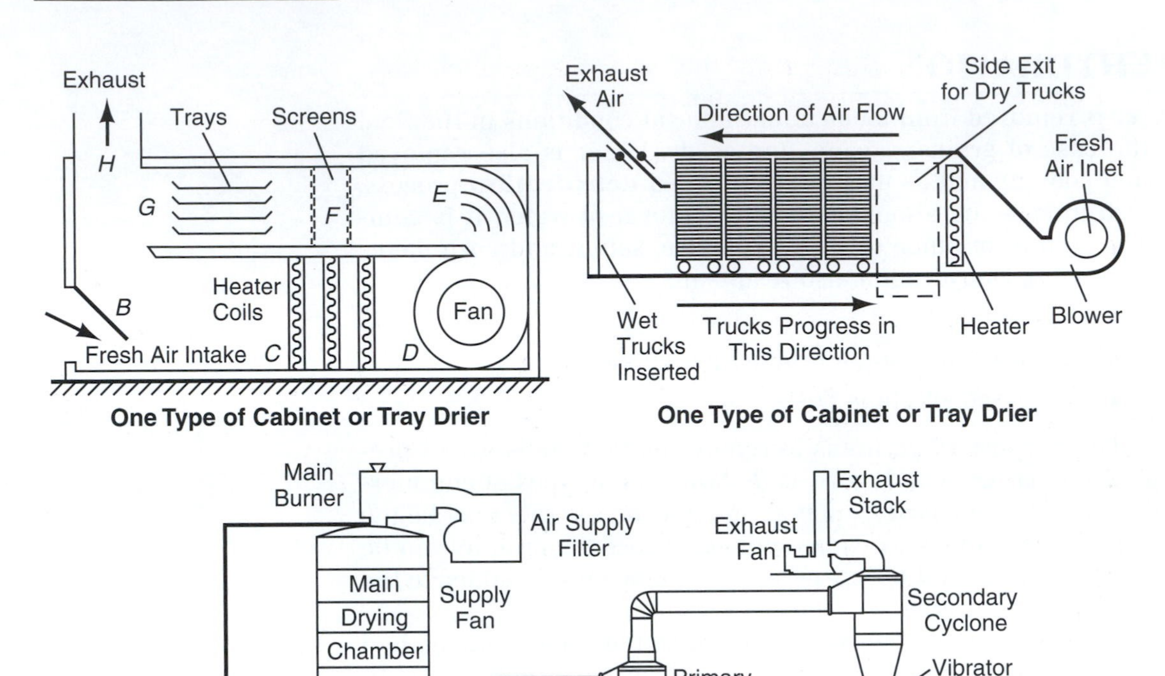

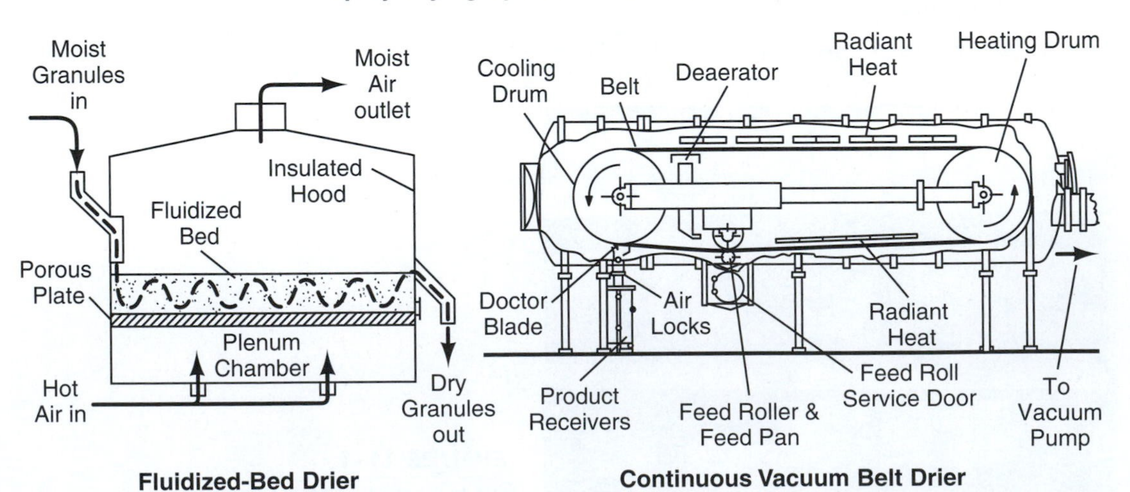

FIGURE 11-2 Diagrams of typical commercial dehydrators.

The greater the **surface area**, the faster the product dries. The greater the temperature differential between the product and the drying medium, the greater the rate of drying. The higher the humidity, the slower the drying will be. **Atmospheric pressure** controls the temperature at which water boils–the lower the pressure, the lower the temperature required to remove water. To overcome this, a vacuum can be used. So, dehydration can be enhanced by increasing the surface area of the food product; increasing the temperature and air velocity; and reducing the humidity and pressure (vacuum).

Drying Curve

When foods are dehydrated, they lose water at a changing rate. Water is lost rapidly at first because it is being lost from the surface of the food. As the food develops a dried layer on the outside, the remaining moisture is confined to the center. The outer dry layer creates an insulation barrier preventing rapid heat transfer into the food. Also, the moisture in the center has farther to travel. Eventually, the food piece reaches its normal equilibrium in relative humidity. It is picking up moisture from the drying atmosphere as fast as it loses moisture. So, when the moisture content of a food is graphed against the time of drying (11-3), a curve with a typical shape develops. The graph shows moisture being lost rapidly at first and then being lost very slowly toward the end of the drying period.

Solute Concentration

Foods high in sugar or other solutes dry more slowly. Also, as drying progresses the concentration of solutes becomes greater in the water that remains. This is another reason the drying rate slows.

Binding of Water

As a product dries, its free water is removed. This water is the easiest to remove, and it evaporates first. Some water is held by absorption to food solids. Water that is in colloidal gels, such as starch, pectin, or other gums, is more difficult to remove. The most difficult water to remove is that chemically bound in the form of hydrates, like glucose monohydrate, or inorganic salt hydrates.

Chemical Changes

Several chemical changes can occur during drying, including:

FIGURE 11-3

A typical drying curve shows a rapid water loss at first that slows as the product dries.

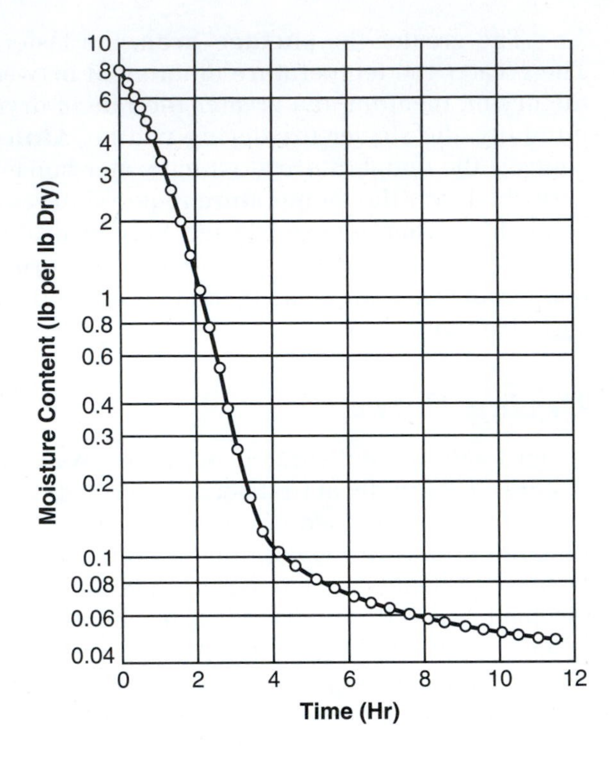

- Caramelization
- Enzymatic browning
- Nonenzymatic browning
- Loss of ease of rehydration
- Loss of flavor

The extent of these chemical changes depends on the foods and the type of drying method used.

Caramelization of sugars occurs if the temperature is too high. **Enzymatic browning** is caused by enzymes and can be prevented by inactivating the enzymes before drying. This is usually done by pasteurizing or blanching the food before drying. Nonenzymatic browning or Maillard browning is controlled by dehydrating the food rapidly through the moisture ranges that are optimal for the Maillard browning. Physical changes, denatured proteins, and loss of sugars and salts during dehydration make the reabsorption of water by the dried product less than equal that of the original. So, **rehydrated** dried products have an altered texture. Preventing a loss of flavor during dehydration is almost impossible. Food processors develop ways of trapping the vapors carrying the flavor and then adding them back after dehydration.

DRINKS FOR THE SPACE PROGRAM

At the beginning of the space program, the astronauts indicated that they wanted to drink orange juice and milk in space. Pure orange juice or whole milk cannot be dehydrated. Orange drink crystals, when rehydrated, just make orange "rocks" in water. A freeze-dried orange juice is available, but it is difficult to rehydrate. Whole dried milk does not dissolve properly. It floats around in lumps and has a disagreeable taste. Therefore, nonfat dry milk must be used in space packaging. To solve the orange juice "problem," a new product had to be developed.

During the 1960s, General Foods developed a synthetic orange-flavored juice called Tang for the space program. It can be used in place of orange juice. Today, there are over 30 flavors of Tang that were developed and used throughout the world. Tang is now a product of Kraft Foods.

For more information on space food or Tang, search the Web or visit these Web sites:

<spacelink.nasa.gov/space.food>
<www.kraftfoods.com/cgi-bin/product.cgi?PRODUCT_ID=26>

Drying Methods

Common drying methods are:

- Air convection
- Drum
- Vacuum
- Freeze

Air Convection. A typical air convection drier has an insulated enclosure, a way of circulating air through the enclosure, and a way to heat this air. The food is supported within the enclosure, and the movement of air is controlled by fans, blowers, and baffles. Dried product is collected by some specially designed devices. Types of air driers include kiln, tunnel, continuous conveyor belt, air lift, fluidized bed, spray, cabinet, tray or pan, and belt trough. These driers can dehydrate food pieces, purees, liquid, and granules.

Drum. Drum or roller driers are used for drying liquid foods, purees, pastes, and mashes. These food products are applied in a thin layer onto the surface of a revolving heated drum. Food is applied continuously, and as the drum rotates, the thin layer of food dries. The speed of the drum is regulated so that when the food reaches a point where a scraper is located, it will be dry. This

FIGURE 11-4

A drum drier processing potato mash into dried potato flakes.

type of drier can have one or two drums. Milk, potato mash (Figure 11-4), tomato paste, and animal feeds are typically dried on drums. The high surface temperature of the drum restricts the foods that can be dried this way.

Vacuum. Vacuum drying produces the highest quality of product but is also the most costly. Essential elements of vacuum drying systems include a vacuum chamber, a heat source, a device for maintaining a vacuum, and a component to collect (condense) water vapor as it leaves the food. Shelves or other supports suspend the food in the vacuum chamber. Vacuum shelf driers (batch driers) and the continuous vacuum belt drier are two main types of vacuum driers. Fruit juices, instant tea, milk, and delicate liquid foods are dried in vacuum driers.

Freeze-Drying. Freeze-drying is used to dehydrate sensitive, high-value foods such as coffee, juices, strawberries, whole shrimp, diced chicken, mushrooms, steaks, and chops. Freeze-drying protects the delicate flavors, colors, texture, and appearance of foods. To dry foods with this method, they are first frozen so that the frozen food does not shrink or distort while giving up its moisture. The principle of freeze-drying is that under conditions of low vapor pressure (vacuum), water evaporates from ice without the ice melting. Water goes from a solid to a gas without passing through the liquid phase. This is called **sublimation**.

FOOD CONCENTRATION

Evaporation concentrates food (Figure 11-5) by the removal of about one third to two thirds of the water present. It has some preservative effects, but mainly it reduces weight and volume. During concentration and depending upon the method, food may take on a cooked flavor, darken somewhat, change in nutritional value, and some microbial destruction is possible. Methods of concentration include:

- Solar (sun)
- Open kettles
- Flash evaporators
- Thin film evaporators
- Vacuum evaporators
- Freeze concentration
- Ultrafiltration and reverse osmosis

Reduced Weight and Volume

Reducing volume and weight saves money. For example, most all liquid foods that are to be dehydrated are first concentrated because the early stages of water removal are more economical with some form of evaporation. Reducing the volume and weight saves money by removing water that would have to be contained and shipped. For example, a soup producer needing tomato solids has little need for the original tomato pulp, which is only about 6 percent solids. By reducing the water content, the solids can be

FIGURE 11-5

Concentrated products produced by evaporation.

increased to 32 percent. Food concentration saves container costs, transportation costs, warehousing costs, and handling costs. Foods that are commonly concentrated include evaporated and sweetened condensed milks, fruit and vegetable juices, sugar syrups, jams and jellies, tomato paste, and other types of purees, buttermilk, whey, and yeast. Some food by-products are concentrated and used as animal feed. The chart in Figure 11-6 indicates the method used according to the size of material.

Solar Evaporation

Solar evaporation is the oldest method of food concentration. It is slow and only used today to concentrate salt solutions in human-made lagoons.

Open Kettles

Open, heated kettles are used for jellies, jams, and some soups. High temperatures and long concentration times damage many foods. Kettles are still used to make maple syrup, but the high heat produces the desirable color and typical flavor of maple syrup.

Flash Evaporators

Flash evaporators subdivide the food and bring it into direct contact with the steam. The concentrated food is drawn off at the bottom of the evaporator, and the steam and water vapor are removed out of a separate outlet.

Thin-Film Evaporators

In thin-film evaporators (Figure 11-7), the food is pumped onto a rotating cylinder and spread into a thin layer. Steam quickly removes the water from the thin layer, and the concentrated food is then wiped from the cylinder wall. The concentrated food and water vapor are continuously removed to an external separator. Here the food product is taken and the water vapor condensed.

Vacuum Evaporators

Low-temperature vacuum evaporators are used for heat-sensitive foods. With a vacuum, lower temperatures can be used to remove water from the food. Often vacuum chambers are in a series, and the food product becomes more concentrated as it moves through the chambers.

The Filtration Spectrum

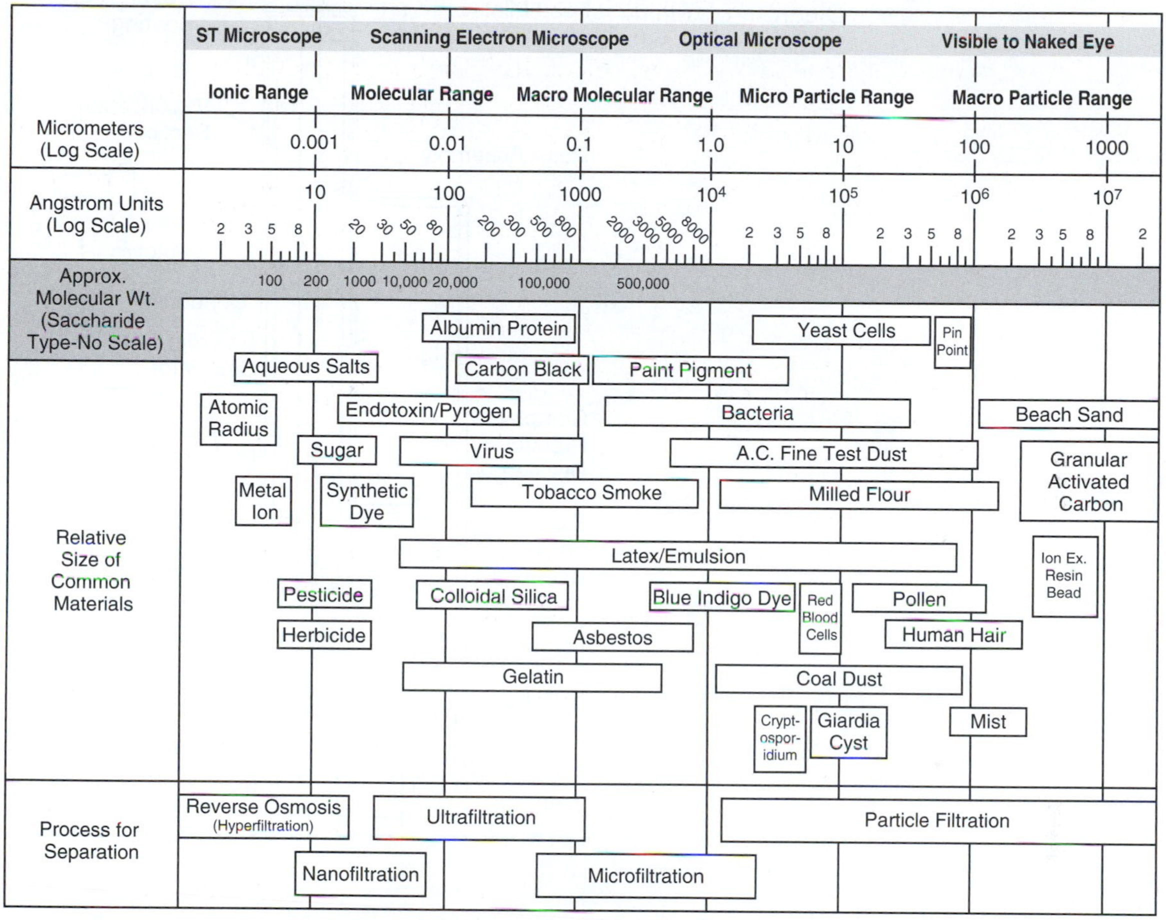

ST Microscope
Scanning Electron Microscope
Optical Microscope
Visible to Naked Eye

Ionic Range
Molecular Range
Macro Molecular Range
Micro Particle Range
Macro Particle Range

Micrometers (Log Scale)
0.001 0.01 0.1 1.0 10 100 1000

Angstrom Units (Log Scale)
10 100 1000 10^4 10^5 10^6 10^7
2 3 5 8 20 30 50 80 200 300 500 800 2000 3000 5000 8000 2 3 5 8 2 3 5 8 2 3 5 8 2

Approx. Molecular Wt. (Saccharide Type-No Scale)
100 200 1000 10,000 20,000 100,000 500,000

Relative Size of Common Materials
Albumin Protein
Yeast Cells
Pin Point
Aqueous Salts
Carbon Black
Paint Pigment
Atomic Radius
Endotoxin/Pyrogen
Bacteria
Beach Sand
Sugar
Virus
A.C. Fine Test Dust
Granular Activated Carbon
Metal Ion
Synthetic Dye
Tobacco Smoke
Milled Flour
Latex/Emulsion
Ion Ex. Resin Bead
Pesticide
Colloidal Silica
Blue Indigo Dye
Red Blood Cells
Pollen
Herbicide
Asbestos
Human Hair
Gelatin
Coal Dust
Cryptosporidium
Giardia Cyst
Mist

Process for Separation
Reverse Osmosis (Hyperfiltration)
Ultrafiltration
Particle Filtration
Nanofiltration
Microfiltration

Note: 1 Micron (1×10^{-6} Meters) ≈ 4×10^{-5} Inches (0.00004 Inches)
1 Angstrom Unit = 10^{-10} Meters = 10^{-4} Micrometers (Microns)

FIGURE 11-6

Chart showing the relative size of materials and the method for filtration separation. For example, note that protein requires ultrafiltration and sugar (lactose) requires reverse osmosis.

Freeze Concentration

All of the components of a food do not freeze at once. First to freeze is the water. It forms ice crystals in a mixture. So, before an entire mixture freezes, it is possible to separate the initial ice crystals. To do this, the partially frozen mixture is centrifuged, then the frozen slush is put through a fine-mesh screen. Frozen water crystals are held back by the screen and discarded. Freeze concentration is used commercially in orange juice production.

FIGURE 11-7

Diagram of a thin-film evaporator.

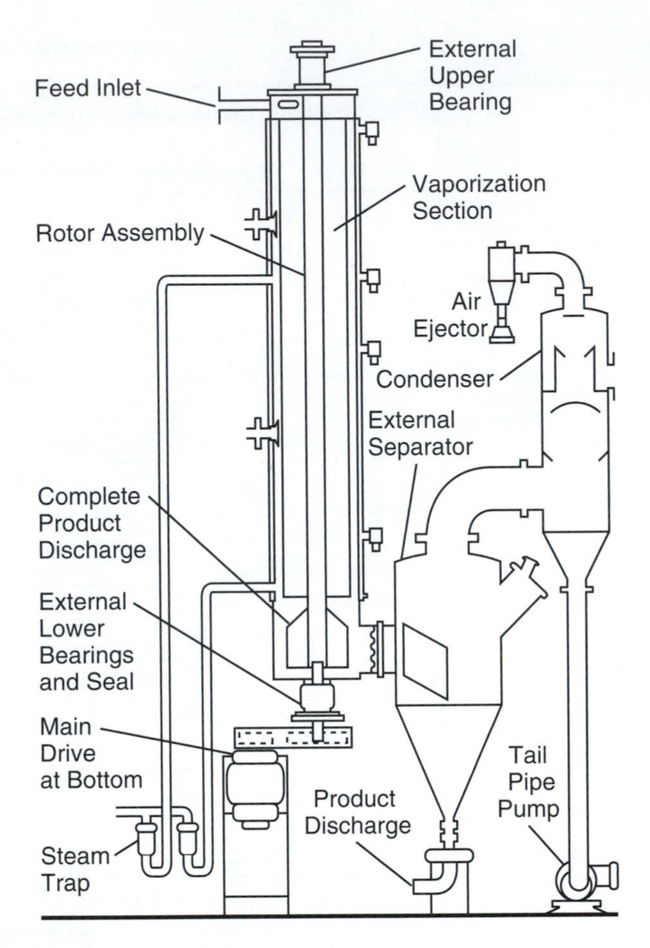

Ultrafiltration and Reverse Osmosis

Ultrafiltration (Figure 11-8) is a membrane filtration process operating at 2 to 10 **bars** (international unit of pressure equal to 29.531 in. of mercury at 32°F) pressure and allowing molecules the size of salts and sugars to pass through the membrane pores, while molecules the size of proteins are rejected. Typical ultrafiltration applications include: milk for protein standardization, cheeses, yogurt, whey, buttermilk, eggs, gelatin, and fruit juice.

The reverse osmosis membrane filtration process uses the tightest membranes and operates at 10 to 100 bars pressure, allowing only water to pass through the pores of the membrane. It is used to concentrate whey and reduce transportation costs of milk by removing the water. Reverse osmosis can also be used to recover rinsing water by concentration of rinsing water from tanks,

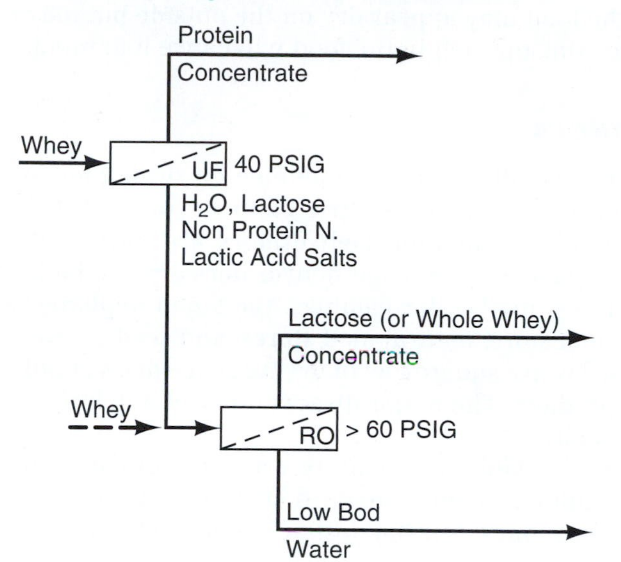

FIGURE 11-8
Schematic showing the use of ultrafiltration and reverse osmosis to separate the components of cheese whey.

pipes, and so on for recovery of milk solids, and it is used to concentrate eggs, blood, gelatin, and fruit juices.

Besides the reduction in water and the recovery of valuable food components, ultrafiltration and reverse osmosis discharge water that is low in organic matter. This decreases the potential for pollution from discharge water.

HOME DRYING

Most foods can be dried indoors using modern food dehydrators, countertop convection ovens, or conventional ovens. Microwave ovens are recommended only for drying herbs, because not enough airflow can be created in them. Some foods can be dried outdoors in the sun.

Foods can be dried in an oven, or in a food dehydrator by using the right combination of warmth, low humidity, and air current. In drying, a warm temperature allows the moisture to evaporate. Air current speeds up drying by moving the surrounding moist air away from the food. Low humidity allows moisture to move from the food to the air.

Drying food is a slow process. It will take 6 or more hours in a dehydrator; in an oven, it will take 8 or more hours. Drying time depends on type of food, thickness, and type of dryer. Turning up the oven temperature cannot speed up drying. This cooks the food on the outside before it dries on the inside. This is called **case hardening**. Also, the food may appear dry on the outside but may be wet on the inside. Moisture left in the food will cause it to mold.

Food Dehydrators

A food dehydrator is a small electrical appliance for drying foods indoors. A food dehydrator has an electric element for heat and a fan and vents for air circulation. Dehydrators are efficiently designed to dry foods fast at 140°F. Food dehydrators are available from department stores, mail-order catalogs, the small appliance section of a department store, natural food stores, and seed or garden supply catalogs. Twelve square feet of drying space dries about one-half bushel of produce. The major disadvantage of a dehydrator is its limited capacity.

Instructions are available from county extension offices and various books for building a homemade dehydrator. Building a dehydrator could save money, but the design is not as efficient as commercial dehydrators.

Oven Drying

Everyone who has an oven has a food dehydrator. By combining the factors of heat, low humidity, and air current, an oven can be used as a dehydrator. An oven is ideal for occasional drying of meat jerkies, fruit leathers, banana chips, or for preserving excess produce like celery or mushrooms. Because the oven may also be needed for everyday cooking, it may not be satisfactory for preserving abundant garden produce.

Oven drying is slower than dehydrators because the oven does not have a built-in fan for air movement. Drying food in an oven takes twice as long as in a dehydrator.

Room Drying

The room drying method takes place indoors in a well-ventilated attic, room, car, camper, or screened-in porch. Herbs, hot peppers, nuts in the shell, and partially dried, sun-dried fruits are the most common air dried items.

Herbs and peppers can be strung on a string or tied in bundles and suspended from overhead racks in the air until dry. Enclosing them in paper bags, with openings for air circulation, protects them from dust, loose insulation, and other pollutants. Nuts are spread on papers, a single layer thick. Partially sun-dried fruits should be left on their drying trays.

Sun Drying

The high sugar and acid content of fruits make them safe to dry outdoors when conditions are favorable for drying. Vegetables (with the exception of vine dried beans) and meats are not recommended for outdoor drying. Vegetables are low in sugar and acid. This increases the risks for food spoilage. Meats are high in protein, making them ideal for microbial growth when heat and humidity cannot be controlled.

Sun-dried raisins are the best known of all dried foods. California produces much of the world's supply of raisins (Figure 11-9).

To dry fruits outdoors, hot, dry, breezy days are best. A minimum temperature of 85°F is needed with higher temperatures being better. Drying foods outdoors requires several days. Because the weather is uncontrollable, drying fruits outdoors can be risky. Rain in California while the grapes are drying destroys an entire production of raisins. High humidity in the South is a problem for

FIGURE 11-9

Dried fruits, especially raisins and prunes, are familiar products.

drying fruits outdoors. Humidity below 60 percent is best. Often these ideal conditions are not available when the fruit ripens, and other alternatives to dry the food are needed.

Racks or screens placed on blocks allow for better air movement around the food. Because the ground may be moist, it is best to place the racks or screens on a concrete driveway or if possible over a sheet of aluminum or tin. The reflection of the sun on the metal increases the drying temperature.

Because birds and insects are attracted to dried fruits, two screens are best for drying food. One screen acts as a shelf and the other as a protective cover. Cheesecloth could also be used to cover the food. Solar dryers may need turning or tilting throughout the day to capture the direct, full sun. Food on the shelves needs to be stirred and turned several times a day.

More information about drying can be found at these Web sites:

<www.ext.nodak.edu/extnews/askext/foodss/4214.htm>

<www.msue.msu.edu/msue/imp/mod01/master01.html>

<muextension.missouri.edu/xplor/hesguide/foodnut/gh1562.htm>

Summary

Besides preservation, drying and dehydration decrease the weight and volume of a product and thereby decrease shipping costs. Drying is affected by surface area, temperature, humidity, and atmospheric pressure. As foods dry, they demonstrate a typical drying curve. Some chemical changes that can occur during dehydration include caramelization, browning, loss of rehydration, and loss of flavor. Foods can be dried by air convection, drum, vacuum, or freeze-drying.

Food concentration removes one third to two thirds of the water and reduces weight and volume. Methods of concentration include solar, open-kettle, flash evaporators, thin-film evaporators, freeze concentration, and ultrafiltration or reverse osmosis.

Home drying of foods follows the same general principles as commercial processes. Small home dehydrators are available or can be built. The kitchen oven (or microwave) can also be used for drying foods.

Review Questions

Success in any career requires knowledge. Test your knowledge of this chapter by answering these questions or solving these problems.

1. List the three drying methods.
2. Dehydration results in decreased ____________ and ____________ and shipping costs.
3. Vacuum drying produces the highest quality of product but is also very ____________.
4. What is ultrafiltration?
5. The principle of ____________ is that under conditions of low vapor pressure (vacuum), water evaporates from ice without the ice melting.
6. The purpose of drying is to remove enough moisture to prevent ____________ growth.
7. Define sublimation.
8. What types of foods are dried using a drum or roller driers?
9. Discuss the two problems with drying of a food product.
10. List three chemical changes that occur during drying.

Student Activities

1. Develop a report or presentation on the dehydration of foods in underdeveloped countries.
2. Identify a product that has been freeze-dried. If possible bring a sample to the class and reconstitute it.
3. Develop a report or presentation on the use of reverse osmosis or ultrafiltration in the food industry.
4. Using a microwave, a small food dehydrator, or a conventional oven at a low temperature, dehydrate a variety of fruits and vegetables. Describe any desirable or undesirable changes in the food, including taste. Then rehydrate some of the foods. Also calculate the water percentage of the foods by using the original wet weight and the dry weight.

Resources

Ensminger, A. H., M. E. Ensminger, J. E. Konlande, and J. R. Robson. 1994. *Foods and nutrition encyclopedia.* 2 Vols. Boca Raton, FL: CRC Press.

Potter, N. N., and J. H. Hotchkiss. 1995. *Food science,* 5th ed. New York: Chapman and Hall.

Vaclavik, V. A., and E. W. Christina. 1999. *Essentials of food science.* Gaithersburg, MD: Aspen Publishers, Inc.

Vieira, E. R. 1996. *Elementary food science,* 4th ed. New York: Chapman and Hall.

Internet

Internet sites represent a vast resource of information. The URLs (uniform resource locator) for the World Wide Web sites can change. Using one of the search engines on the Internet such as Yahoo!, HotBot, AltaVista, Excite, Dogpile, About, or Google, find more information by searching for these words or phrases: dehydration, food drying, freeze-drying, caramelization, enzymatic/nonenzymatic browning, air convection drying, food/freeze concentration, solar drying, ultrafiltration, reverse osmosis, evaporation, sublimation. Also, Table A-7 provides a listing of some useful Internet sites that can be used as a starting point.

Chapter 12

Radiant and Electrical Energy

Objectives

After reading this chapter, you should be able to:

- Describe ionizing radiation
- Name two requirements for the irradiation process
- Discuss the four areas in which irradiation is most useful
- List the three specific ways the FDA has approved irradiation uses
- Explain how microwaves heat food
- Describe ohmic (electrical) heating and its major advantage
- Explain why salt and water content are important in microwave heating

Key Terms

electromagnetic energy
ionization
ions
irradiation
microwaves
ohmic heating
polar
radioisotopes

Ionizing radiation and microwaves are invisible energy waves moving through space. Food processors use these two forms of radiation. Electrical or ohmic heating of foods is a relatively new method for heating and preserving foods.

FOOD IRRADIATION

Radiation is broadly defined as energy moving through space in invisible waves (Figure 12-1). Radiant energy has differing wavelengths and degrees of power. Light, infrared heat, and **microwaves** are forms of radiant energy. So are the waves that bring radio and television broadcasts into our homes. Broiling and toasting use low-level radiant energy to cook food.

The radiation of interest in food preservation is ionizing radiation, also known as **irradiation**. These shorter wavelengths are capable of damaging microorganisms such as those that contaminate food or cause food spoilage and deterioration. Scientists have experimented with irradiation as a method of food preservation since 1950. They found irradiation to be a controlled and very predictable process.

As in the heat pasteurization of milk, the irradiation process greatly reduces but does not eliminate all bacteria. Irradiated poultry, for example, still requires refrigeration, but would be safe longer than untreated poultry. Strawberries that have been irradiated will last two to three weeks in the refrigerator compared to only a few days for untreated berries. Irradiation complements, but does not replace, the need for proper food handling practices by producers, processors, and consumers.

Two requirements for the irradiation process include:

1. A source of radiant energy
2. A way to confine that energy

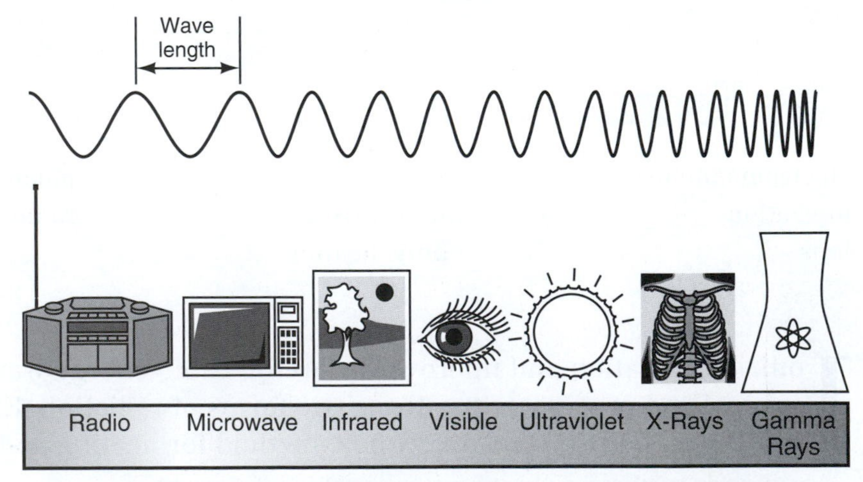

FIGURE 12-1
A representation of the electromagnetic spectrum showing wavelengths from radio through visible light to invisible gamma rays.

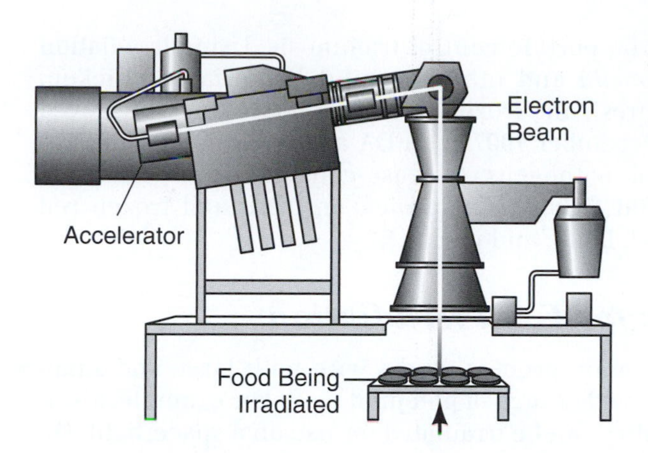

FIGURE 12-2

Diagram of irradiation equipment that uses an accelerator.

For food irradiation, the two sources of radiant energy are (1) **radioisotopes** (radioactive materials), and (2) machines that produce high-energy beams. Specially constructed containers or compartments are used to confine the beams so that people will not be exposed. Radioisotopes are used in medical research and therapy in many hospitals and universities. They require careful handling, tracking, and disposal. Machines that produce high-energy beams offer greater flexibility (Figure 12-2). For example, they can be turned on and off unlike the constant emission of gamma rays from radioisotopes.

Food Irradiation Process

Irradiation is known as a cold process. It does not significantly increase the temperature or change the physical or sensory characteristics of most foods. An irradiated apple, for example, will still be crisp and juicy. Fresh or frozen meat can be irradiated without cooking it.

During irradiation, the energy waves affect unwanted organisms but are not retained in the food. Similarly, food cooked in a microwave oven, or teeth and bones that have been X-rayed do not retain those energy waves.

Approved Uses for Food Irradiation

Irradiation has been approved for many uses in about 36 countries, but only a few applications are presently used because of consumer concern and because the facilities are expensive to build.

In the United States, the Food and Drug Administration (FDA) has approved irradiation for eliminating insects from wheat, potatoes, flour, spices, tea, fruits, and vegetables. Irradiation also can be used to control sprouting and ripening. Approval was given in 1985 to use irradiation on pork to control trichinosis. Using irradiation to control *Salmonella* and other harmful bacteria in chicken, turkey, and other fresh and frozen uncooked poultry was approved in May 1990. In December 1997, the FDA approved the use of irradiation to control pathogens (disease-causing microorganisms such as *E. coli* and *Salmonella* species) in fresh and frozen red meats such as beef, lamb, and pork.

Applications for Food Irradiation

Because the irradiation process works with both large and small quantities, it has a wide range of potential uses. For example, a single serving of poultry can be irradiated for use on a spaceflight. Or, a large quantity of potatoes can be treated to reduce sprouting during warehouse storage.

Irradiation cannot be used with all foods. It causes undesirable flavor changes in dairy products, and it causes tissue softening in some fruits, such as peaches and nectarines.

Irradiation is most useful in four areas: preservation; sterilization; control of sprouting, ripening, and insect damage; and control of foodborne illness.

1. Irradiation can be used to destroy or inactivate organisms that cause spoilage and decomposition, thereby extending the shelf life of foods. It is an energy-efficient food preservation method that has several advantages over traditional canning. The resulting products are closer to the fresh state in texture, flavor, and color. Using irradiation to preserve foods requires no additional liquid, nor does it cause the loss of natural juices. Both large and small containers can be used, and food can be irradiated after being packaged or frozen.
2. Foods that are sterilized by irradiation can be stored for years without refrigeration–just like canned (heat sterilized) foods. With irradiation it will be possible to develop new shelf-stable products. Sterilized food is useful in hospitals for patients with severely impaired immune systems, such as some patients with cancer or AIDS. These foods can be used by the military and for spaceflights.
3. In the role of the control of sprouting, ripening, and insect damage, irradiation offers an alternative to chemicals for use with potatoes, tropical and citrus fruits, grains, spices, and seasonings. However, because no residue is left in the food, irradiation does not protect against reinfestation like insect sprays and fumigants do.

4. Irradiation can be used to effectively eliminate those pathogens that cause foodborne illness, such as *Salmonella.*

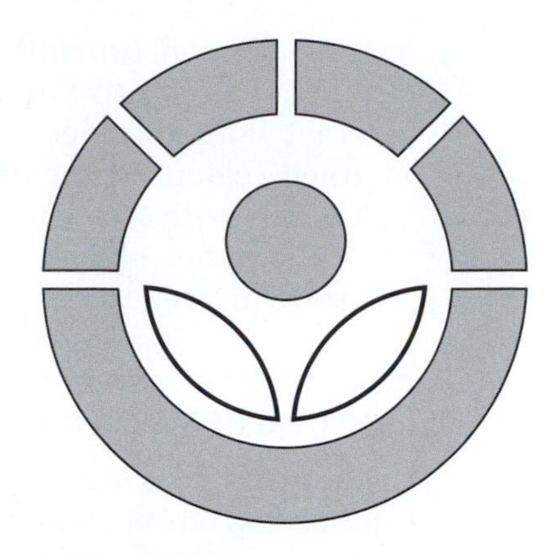

FIGURE 12-3

International symbol placed on irradiated food.

Nutritional Quality of Irradiated Foods

Scientists believe that irradiation produces no greater nutrient loss than what occurs in other processing methods, such as canning.

Regulation of Food Irradiation

Since 1986, all irradiated products must carry the international symbol called a radura, which resembles a stylized flower (Figure 12-3). Accurate records are essential to regulation because there is no way to verify or detect if a product has been irradiated, or how much radiation it has received.

MICROWAVE HEATING

Microwaves are another method of heating foods. Conventional methods heat foods by conducting heat from an external source to the inside. Microwaves generate heat inside a food due to water friction. Conventional methods brown or crust foods on the surface, microwaves do not.

A microwave oven generates microwaves from electricity. Microwaves are **electromagnetic energy**–that is, they have an electric and a magnetic component. (Originally, microwave ovens were called "radar" ranges.) Other forms of electromagnetic energy are radio waves, sunlight, and electricity.

Scientists refer to microwave energy as radiation because, similar to other radiation energy, it travels through space. Microwave radiation is often called nonionizing radiation to distinguish it from other forms of radiation, like X-rays, which are called ionizing radiation. **Ionization** refers to breaking atoms or molecules into two electrically charged groups.

Microwaves themselves are generated by a magnetron tube that converts electrical energy at 60 cycles/second into an electromagnetic field with positive and negative charges that change direction millions of times per second (915+ million times per second). Essentially, it consists of **polar** and nonpolar molecules and functions by switching back and forth. This switching back and forth causes friction and makes heat. This friction is caused by the disruption of hydrogen bonds between neighboring water molecules, and the heat is really "molecular friction." In order to have microwave heating of foods, there must be a polar substance available. Water is such a polar substance.

Other constituents in foods are also a factor in microwave heating. The positive and negative **ions** of table salt in foods will interact with the electrical field by migrating to the oppositely charged regions of the electrical field and also disrupt the hydrogen bonds with water to generate heat.

A number of factors will influence the speed of cooking of a microwave oven. The electrical and physical properties of foods

PAST, PRESENT, AND POTENTIAL OF IRRADIATION

Research on food irradiation dates back to the 1920s. The U.S. Army used the process on fruits, vegetables, dairy products, and meat during World War II, and NASA routinely sends irradiated food on U.S. spaceflights.

Irradiation to control microorganisms on beef, lamb, and pork is safe and eventually could mean that consumers will have less risk of becoming ill from contaminated meat. Irradiation of fruits and vegetables will also mean longer shelf life.

Research shows that specific doses of radiation can kill rapidly growing cells such as those of insect pests and spoilage bacteria. Irradiation kills foodborne pathogens such as *E. coli* 0157:H7 and *Salmonella*.

A recent survey conducted by the International Food Information Council (IFIC) in Dallas, New York, and Los Angeles found that consumers were willing to buy irradiated foods for themselves and their families, including children. Food irradiation is one food safety tool whose time has come. The informed consumer will make choices based on irradiated products identified with clear labeling. The IFIC survey found that consumers appreciated the safety benefit of irradiation to eliminate harmful bacteria rather than to extend shelf life.

In addition to the Food and Drug Administration and the USDA, the list of irradiation endorsers includes the U.S. Public Health Service, the American Medical Association, and the National Food Processors Association. The World Health Organization (WHO) also sanctions irradiation, which is currently being used in about 40 countries. The WHO in 1992 called irradiation a "perfectly sound food-preservation technology." In 1997, WHO again endorsed food irradiation, joined this time by the United Nations Food and Agriculture Organization (FAO).

For more information about the irradiation of food, search the Web, or visit these Web sites:

<www.exnet.iastate.edu/foodsafety/rad/irradhome.html>
<www.food-irradiation.com/>
<www.iaea.or.at/icgfi/>

themselves will determine microwave penetration depth, the dielectric constant, heat capacity, density, and the conventional heat transfer and heating rate. The food dielectric properties of interest in microwave heating are determined by the moisture and heating.

Microwave energy, like all electromagnetic radiation, travels in a wave pattern. The waves are reflected by metals; pass through air, glass, paper, and plastic; and are absorbed by food. Most microwave containers are designed to transmit microwave energy without reflecting or absorbing it, and thus are made of paper or plastic. The microwaves will travel through the container to the food.

When food is exposed to microwaves, it absorbs that energy and converts it to heat. Food composition (mainly water content) is a key factor that determines how fast it will heat in a microwave environment. The higher the water content of a food, the higher its loss factor, thus the faster it will heat. Solutes, like sugar and salt, also influence the loss factor of foods exposed to microwave energy. For foods to heat in a microwave oven, the electromagnetic waves must penetrate the food. There are limits to the depth of their penetration. This makes container geometry a very important factor in heating foods by microwave energy.

Food composition does not only influence the loss factor, but also penetration depth. For example, when salt is added to water, it will change its microwave heating characteristics in two different directions. On one hand, it will increase the water's loss factor causing the water to heat faster. On the other hand, salt will decrease the penetration depth of microwaves into the salt/water solution, decreasing the heating rate. If salt is added to water to enhance its loss factor and, at the same time, the container geometry changes to minimize the drop in penetration depth, the heating rate will be greatly increased.

Food Processing Applications

Food scientists are dedicated to developing microwavable foods that can heat quickly and evenly while maintaining high quality. Heating foods evenly in a microwave oven is difficult at best, particularly with solid foods of different composition. A frozen dinner tray is an example of a solid food with varying composition.

Understanding how food composition influences microwave heating is a skill needed by those developing microwavable foods. Developing new microwavable meals and snacks is a growing segment of the food industry. Aside from using microwave energy to heat meals at home, many attempts have been made to use microwave energy to heat food at processing plants. Some of these uses

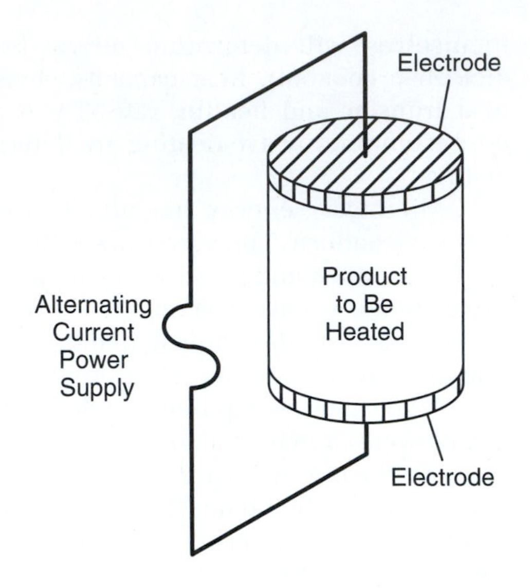

FIGURE 12-4

Diagram of the principle of ohmic heating. An alternating electrical current is passed through a food in a conducting fluid.

include baking, concentrating, cooking, curing, drying, enzyme inactivation (blanching), finish drying, freeze-drying, heating, pasteurizing, precooking, puffing and foaming, solvent removal, sterilizing, tempering, and thawing. The best use of microwaves depends on product quality and cost.

OHMIC (ELECTRICAL) HEATING

Ohmic heating is the heating of a food product by using an alternating current flowing between two electrodes (Figure 12-4). It is one of the newest methods of heating foods. The major advantage of ohmic heating is that solid pieces and liquids in a food are heated at the same time. Also, the ohmic method works well as a continuous heating system for foods with particles, like beef stew. And it is suitable for aseptic packaging methods.

Summary

Ionizing radiation (irradiation) and microwaves are both used to preserve food. Although irradiation has been in use since 1950, the FDA must approve its uses on each additional food. The latest approval for irradiation occurred in 1997 for fresh and frozen red meats. Irradiation serves four uses in the food industry: preservation; sterilization; control of sprouting, ripening, and insect damage; and control of foodborne illness. Though irradiation demonstrates little effect on nutrient content, the greatest challenge will be consumer acceptance. Microwaves are also used to heat foods by generating heat inside the food due to water friction. Scientists refer to microwaves as nonionizing radiation.

Microwave development has led to a whole new group of convenience foods. Attempts to use microwave energy in food processing include baking, concentrating, cooking, curing, drying, heating, puffing, foaming, tempering, and thawing. The best use of microwaves depends on the product quality and the cost.

Ohmic heating heats foods between electrodes using alternating current. It is one of the newest methods of heating, and it is useful because solid and liquid in a food are heated at the same time.

Review Questions

Success in any career requires knowledge. Test your knowledge of this chapter by answering these questions or solving these problems.

1. Describe ohmic heating.
2. Name the two requirements for irradiation.
3. Radiation is broadly defined as ___________ moving through space in invisible waves.
4. Explain ionizing radiation.
5. List the four ways in which irradiation is most useful.
6. Describe how microwaves heat food.
7. When ___________ is added to water, it changes the microwave heating characteristics in two different directions.
8. List three specific ways irradiation has been approved for use by the FDA.
9. Food composition influences microwave heating of food in what two ways?
10. Irradiation cannot be used on what two specific products?

Student Activities

1. Research the use of irradiation on food. Develop a report or presentation and take either the pro or con to the use of irradiation.
2. Collect news articles from newspapers, TV, magazines, or the Internet that tell of some outbreak of a foodborne disease. Use these articles for a classroom discussion on how irradiation could have prevented the outbreak.

3. Develop a report or presentation on the use of microwaves in U.S. households. Include how microwaves were introduced and the growth in sales.
4. Compare a microwave cookbook with a traditional cookbook. Report on your findings.
5. Using a microwave, demonstrate heating/cooking of small quantities of different types of food, including liquids and different shapes. Also demonstrate some of the unique features of microwaves.
6. Contact an appliance repair shop and ask if they have any microwaves that cannot be repaired. Disassemble these old microwaves and identify the parts.

Resources

Asimov, I. 1988. *Understanding physics.* New York: Hippocrene Books, Inc.

FDA. 2000. *Food irradiation: A safe measure.* Publication No. 00-2329. Washington, DC: USDA.

Potter, N. N., and J. H. Hotchkiss. 1995. *Food science,* 5th ed. New York: Chapman and Hall.

Internet

Internet sites represent a vast resource of information. The URLs (uniform resource locator) for the World Wide Web sites can change. Using one of the search engines on the Internet such as Yahoo!, HotBot, AltaVista, Excite, Dogpile, About, or Google, find more information by searching for these words or phrases: food irradiation, microwaves, ohmic heating, ionizing food. Also, Table A-7 provides a listing of some useful Internet sites that can be used as a starting point.

Chapter 13

Fermentation, Microorganisms, and Biotechnology

Objectives

After reading this chapter, you should be able to:

- Discuss the use of fermentation in food preservation
- Provide the general reactions for fermentation
- Name three methods for controlling fermentation
- List six foods produced by fermentation
- Identify the uses of acetic acid bacteria and lactic acid bacteria
- Describe fermentation use in bread making
- Identify four uses of acetic acid
- Describe the use of microorganisms as food
- Discuss a role of biotechnology in the food industry

Key Terms

anaerobic
biosensors
brewing
brine
coagulation
hops
inoculation
leavening
malt
mashing
microsensors
pathogenic
single-celled protein (SCP)
starter culture
syneresis
yeast

Fermentation allows the growth of microorganisms in order to produce a stable product. Products commonly produced, at least in part, by fermentation include: beer, pickles, olives, some meat products, bread, cheese, coffee, cocoa, soy sauce, sauerkraut, and wine. Microorganisms can become food, and biotechnology changes the way microorganisms are used.

FERMENTATIONS

Fermentation is the oldest form of food preservation. The principle of fermentation is the breakdown of carbohydrate materials by bacteria and **yeasts** under **anaerobic** (without atmospheric oxygen) conditions. It produces acids and alcohols, with some aldehydes, ketones, and flavorings. Products produced by fermentation help preserve foods against microbial degradation. Fermentation by lactic acid bacteria produces:

- Pickles
- Olives
- Some meat products (sausage and salami)
- Sour cream
- Cottage cheese
- Cheddar cheese
- Coffee

Acetic acid bacteria produces cooking wine and cider. Lactic acid bacteria with propionic acid bacteria produces swiss cheese. Molds produce blue cheese. Yeasts are involved in the production of beer, wine, rum, whiskey, and bread. Yeasts with acetic acid bacteria act on cacao beans (chocolate).

Fermentation follows these general reactions:

Glucose
↓
pyruvic acid
↓
acetaldehyde + carbon dioxide
↓
alcohol (ethanol)

A **starter culture** is a concentrated number of the organisms desired to start the fermentation. This will assure that at the beginning the proper organisms are growing.

Benefits

The main benefit of fermentations is preservation of the product. For example, acid produced may prevent spoilage by some microorganisms. Fermentation may add flavor as in wine. It may remove or alter existing flavors as in soya. Or fermentation can alter the chemical characteristics of the food as in sugar to ethanol, ethanol to acetic acid, or sugar to lactic acid.

Fermented food can be more nutritious than the unfermented foods from which they were derived. Fermentation microorganisms produce vitamins and growth factors in the food. They also may liberate nutrients locked in plant cells and structures by indigestible materials. Finally, fermentation can enzymatically split polymers like cellulose into simpler sugars that are digestible by humans.

Control

To encourage the growth of certain microorganisms, fermentation can be controlled by:

- pH
- Salt content
- Temperature

Acid or a low pH has an inhibitory effect. Acid may be added to the food, or it is produced by the microorganisms causing fermentation. It must be added or formed quickly to prevent spoilage or the growth of undesirable microorganisms.

Microorganisms exhibit differing tolerance for salt. Lactic acid-producing microorganisms used to ferment olives, pickles, sauerkraut, and some meats are tolerant of salt concentrations from 10 to 18 percent. Many of the microorganisms that cause spoilage cannot tolerate a salt concentration above 2.5 percent, and they are especially intolerant of a salt and acid combination. This gives lactic acid-producing microorganisms an advantage. Salt in a solution is called **brine**. Salt also serves another purpose when fermenting vegetables. It draws water and sugar out of the vegetables. The sugar entering the brine is available to the microorganisms for continued fermentation.

Depending on the temperature, various microorganisms dominate fermentation. Temperature can affect the final acid concentration and the time it takes to reach different acidities. For example, in sauerkraut fermentation, three different types of microorganisms are used, and the actions of each depends on the temperature, so the temperature starts low and is gradually increased to allow each microorganism to grow. Controlling fermentation is an attempt to favor desired organisms. Fermentation is stopped by pasteurizing and cooling.

USES OF FERMENTATION

Fermentation produces dairy products, bread, pickles, processed meats, vinegar, wine, and beer (Figure 13-1). The following paragraphs describe a few specific uses of fermentation in foods.

Fermented Dairy Products

Cheese making relies on the fermentation of lactose (milk sugar) by lactic acid bacteria. This bacteria produces lactic acid, which lowers the pH and in turn assists **coagulation** (clot formation), promotes **syneresis** (expulsion of water from the clot), helps prevent spoilage and **pathogenic** bacteria from growing, and contributes to cheese texture, flavor, and keeping quality. Lactic acid bacteria also produce growth factors, which encourages the growth of nonstarter organisms and provides enzymes that act on fats (lipases) and proteins (proteases) necessary for flavor development during curing.

Yogurt is a semisolid fermented milk product that originated centuries ago in Bulgaria. Its popularity has grown and is now consumed in most parts of the world. Although the consistency, flavor, and aroma may vary from one region to another, the basic ingredients of yogurt are milk and a starter culture. The fermentation products of lactic acid, acetaldehyde, acetic acid, and diacetyl contribute to the flavor.

FIGURE 13-1
Representation of products produced by fermentation.

Cultured buttermilk was originally the fermented by-product of butter manufacture, but today it is more common to produce cultured buttermilks from skim or whole milk. The fermentation is allowed to proceed for 16 to 20 hours, to an acidity of 0.9 percent lactic acid. This product is frequently used as an ingredient in the baking industry, in addition to being packaged for sale in the retail trade.

Acidophilus milk is a traditional milk fermented with *Lactobacillus acidophilus (LA)*, which has been thought to have therapeutic benefits in the gastrointestinal tract.

Sour cream usually has a fat content between 12 and 30 percent, depending on the required properties. The starter is similar to that used for cultured buttermilk. **Inoculation** and fermentation conditions are also similar to those for cultured buttermilk, but the fermentation is stopped at an acidity of 0.6 percent.

Other fermented dairy products include such products as kefir and kumiss. Many of these have developed in regional areas and, depending on the starter organisms used, have various flavors, textures, and components from the fermentation process.

Dairy products are covered more completely in Chapter 16.

Bread Making

Bread is leavened with yeast. Baker's yeast is composed of the living cells of *Saccharomyces cerevisiae*, a unicellular microorganism. Yeast performs its **leavening** function by fermenting carbohydrates such as the sugars, glucose, fructose, maltose, and sucrose. (Yeast cannot metabolize lactose, the predominant sugar in milk.) The principal products of fermentation are carbon dioxide, which produces the leavening effect, and ethanol. Yeast also produces many other chemical substances that flavor the baked product and change the dough's physical properties.

Although most breads and rolls are leavened by yeast, some breadlike products like corn bread and certain kinds of muffins are leavened by chemicals such as baking powder (Figure 13-2).

Chapter 19 contains more information about bread making.

Pickling

Pickles are made by covering the fruit with a sweetened vinegar solution containing a spice or spices, such as cloves, added for flavor. Some vegetable pickles are produced by fermentation. The vegetable is placed in a covered crock and allowed to ferment in a brine solution for a period of time ranging from a few days to several weeks. Because of the acidity of the pickling solution, heat processing under pressure in order to kill microorganisms is not required either commercially or in the home.

QUOTES ABOUT BIOTECHNOLOGY AND THE GMO DEBATE

Is the debate about biotechnology and GMOs about science versus pseudoscience? Consider these quotes from reliable sources:

"Scientists are gaining the ability to insert genes that give biological defense against diseases and insects, thus reducing the need for chemical pesticides, and convey genetic traits that enable crops to better withstand drought conditions. With this powerful new genetic knowledge, scientists have the capability to pack large amounts of technology into a single seed."

Norman Borlaug, Ph.D.,
Nobel Peace Prize Laureate
(July 31, 1997, testimony before the U.S. Senate Agriculture Committee).

"Biotechnology differs from crossbreeding in that one gene is inserted into an organism to achieve the desired effect. With traditional crossbreeding, every gene of an organism is potentially mixed with another. The one desirable trait that breeders want can be passed on, but so will some undesirable ones. With biotechnology, scientists are able to focus on the desired gene and subject it to extensive testing before and after it is inserted into the new organism."

Dr. Steve Taylor,
Department of Food Science and Technology,
University of Nebraska, Lincoln
(as stated in Letter to the Editor, Wall Street Journal, *July 21, 1999).*

"I have absolutely no anxiety. . . . I am worried about a lot of things, but not about modified food. To argue that you don't know what is going to occur is true about everything in life. People wouldn't get married, have children, do anything . . ."

James Watson, Ph.D.,
codiscoverer of DNA structure and Nobel Laureate
(from the Daily Telegraph *of the United Kingdom, February 25, 1999).*

"I can imagine some of biotechnology's most vocal critics saying this—but in fact, this was a criticism of Luther Burbank's genetic research in 1906—technology that we now accept and benefit from."

U.S. Grains Council President and CEO Kenneth Hobbie
(speech embargoed until its presentation at the
International Grains Council Grains Conference in London on June 10, 1999).

"We have confidence in the findings of our Food and Drug Administration that these [biotech] foods are safe. And if we didn't believe that, we wouldn't be selling them and we certainly wouldn't be eating them. . . . I would never permit an American child to eat anything I thought was unsafe."

President Clinton, conference call with farm radio broadcasters
from Hermitage, Arkansas (as reported by Reuters on November 5, 1999).

FIGURE 13-2
When yeast ferments sugars, bread rises.

FIGURE 13-3
Some vegetable pickles are produced by fermentation.

Vegetables, fruit, meat, eggs, and even nuts can be pickled. Some well-known pickled foods include sauerkraut (cabbage fermented in brine), dill or sweet pickles made from cucumbers, peach pickles, and pickled watermelon rind. Favorite mixtures of pickled vegetables include piccalilli, chowchow, and assorted relishes (Figure 13-3).

Processed Meats

Some processed meats have microbial starter cultures added to achieve fermentation to enhance preservation and create a unique "tangy" flavor from the production of lactic acid. Fermentation inhibits the growth of spoilage and pathogenic microorganisms.

Vinegar

Vinegar (from the French vinaigre, meaning "sour wine") is an acidic liquid obtained from the fermentation of alcohol and used either as a condiment or as a preservative. Vinegar usually has an acetic acid content of between 4 and 8 percent. Its flavor may be sharp, rich, or mellow, depending on the original product used. Vinegar is made by combining sugary materials (or materials produced by the breakdown of starches) with vinegar or acetic acid bacteria and air. The sugars or starches are converted to alcohol by yeasts, and the bacteria make enzymes that cause oxidation of the alcohol to acetic acid.

Several varieties of vinegar are manufactured. Wine vinegars, produced in grape-growing regions, are used for salad dressings and relishes and may be either reddish or white, depending on the wine used in the fermentation process. Tarragon vinegar has the distinctive flavor of the herb. **Malt** vinegar, popular in Great Britain, is known for its earthy quality. White vinegar, also called distilled vinegar, is made from industrial alcohol; it is often used as a preservative or in mayonnaise because of its less distinctive flavor and clear, untinted appearance. Rice vinegar, which has an agreeably pungent quality, is often used in Oriental countries for marinades and salad dressings.

Vinegar may be used as an ingredient of sweet-and-sour sauces for meat and vegetable dishes, as a minor ingredient in candies, or as an ingredient in baking, as a part of the leavening process. Vinegar is also added to milk, if sour milk is needed in a home recipe. Commercially and in the home, the most common use of vinegar is in the making of salad dressings.

Wine Making

For red wine, the grapes are crushed immediately after picking and the stems generally removed. The yeasts present on the skins come into contact with the grape sugars, and fermentation begins naturally. Cultured yeasts, however, are sometimes added. During fermentation the sugars are converted by the yeasts to ethyl alcohol and carbon dioxide. The alcohol extracts color from the skins. Glycerol and some of the esters, aldehydes, and acids that contribute to the character, bouquet (aroma), and taste of the wine are by-products of fermentation. Maturation of the wine may take years in 50-gallon oak casks. During this time the wine is drawn off three or four times into fresh casks to avoid bacterial spoilage. Further aging usually occurs after bottling.

Brewing

Brewing is a centuries-old technique. It involves four steps: **mashing**, boiling, fermentation, and aging.

1. Mashing. Infusion of malt, water, and crushed cereal grains at temperatures that encourage the complete conversion of the cereal starch into sugars (Figure 13-4).
2. Boiling. Concentration of the resulting "wort"(liquid) and the addition of **hops**.
3. Fermentation. Addition of yeast to the wort, resulting in the production of alcohol and carbon dioxide gas, byproducts of the action of yeast on sugar.

FIGURE 13-4
Copper kettles used for mashing during the fermentation process of making beer.

4. Aging. Proteins settle out of beer or are "digested" by enzymatic action. The aging process may last from 2 to 24 weeks.

The uniform clarity of modern beers results from filtration systems that use such agents as cellulose and diatomaceous earth (deposits of cell walls of diatoms). Additives are frequently used to stabilize foam and to maintain freshness. With few exceptions, bottled and canned beer is pasteurized in the container in order to ensure that the yeast that may have passed through the filters is incapable of continued fermentation. Genuine draft beer is not pasteurized and must be stored at low temperature.

Chapter 23 provides more details on brewing and wine making.

In the future, technology will design more systems that call for continuous fermentations. Also, genetic engineering of organisms will give better control of the process.

MICROORGANISMS AS FOODS

Besides being used to produce desirable changes in food, some microorganisms are used for animal feed and human food. These microorganisms are selected for their rapid growth, nutritional value, and other properties that make them a good food or feed. Sometime these are called **single-celled protein (SCP)**. Examples of microorganisms used as food or feed include brewer's yeast and baker's yeast. The practice of growing yeasts for food goes back many years.

GENETIC ENGINEERING AND BIOTECHNOLOGY

In the food industry, genetic engineering is being used to improve the yields and efficiency of traditional fermentation products and to convert unused or underused raw materials to useful products. Researchers now can isolate a known trait from any living species –plant, animal, or microbe–and incorporate it into another species. These traits are contained in genes–segments of the DNA molecules found in all living cells. The process of recombining genes bearing a chosen trait into the DNA molecules of a new host is called recombinant DNA technology (Figure 13-5).

Another difference with recombinant DNA, which can be a benefit but which concerns some, is the power of genetic engi-

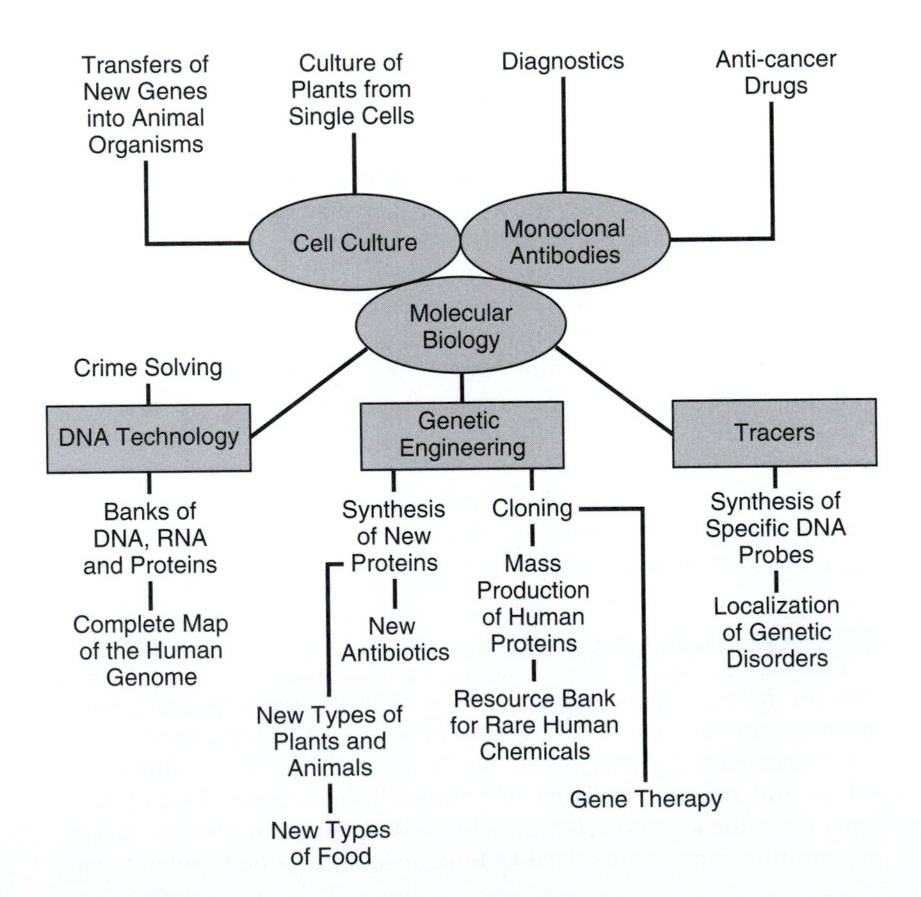

FIGURE 13-5

Recombinant DNA technology offers hope for new food products.

neering–the ability to transfer genes from a wide variety of species. Because the chemical makeup of DNA is similar in all living things, desirable genes from any organism can be inserted into any other animal, plant, or microorganism. This provides the developer with a much larger selection of valuable traits.

Genetic engineering and biotechnology are also being used to produce new and improved enzymes, flavors, sweeteners, and other food ingredients.

Chymosin (Rennin)

Rennin is used in cheese manufacture. It used to be extracted from the stomachs of calves, and the demand far exceeded supply. Scientists inserted the gene for producing rennin into microorganisms. Rennin became the first genetically engineered product approved for use in food.

Bovine Somatotropin (BST)

Bovine (cattle) somatotropin (BST) or growth hormone is produced by genetically engineered microorganisms. When administered to dairy cows, it increases milk production by 10 to 25 percent, whereas feed intake increases by only about 6 percent, thus increasing the efficiency of the dairy cow.

Tomatoes

Flavr Savr tomatoes produced by Calgene contains an antisense gene for the enzyme, polygalacuronase. These tomatoes ripen normally with full flavor, but the rate of softening is much slower. For example, the fruit can be harvested ripe in California and be shipped to Ohio and still be consumable for two weeks.

Cloning

In theory, every cell contains the genetic information to produce a complete new organism identical to the original. Plants can be cloned. Soon it will be possible to produce many identical, genetically, engineered animals from somatic (body) cells. These animals and plants might provide our food in the future.

GMO Debate

Currently, the debate rages on about the use of GMOs–genetically modified organisms–in our food supply. Though genetic engineering promises better and more plentiful products, genetically engineered foods may encounter a few obstacles to widespread public acceptance. Some consumers, along with a few advocacy groups, have voiced concern about the safety and environmental impact of these new food products. Some urge an outright ban on any genetically engineered foods. Others support mandatory labeling that discloses the use of genetic engineering. Still others advocate more stringent testing of these products before marketing.

From the standpoint of the Food and Drug Administration, the important thing for consumers to know about these new foods is that they will be every bit as safe as the foods now on store shelves. All foods, whether traditionally bred or genetically engineered, must meet the provisions of the federal Food, Drug, and Cosmetic Act. Whether genetically engineered foods succeed or fail ultimately depends on public acceptance. FDA scientists and others in the field blame some negative consumer reaction on the recombinant DNA technique's complexity. The technology is difficult to understand, so there is a fear of the unknown.

The Future

Biotechnology is also used in some food processing related areas including processing aids, ingredients, rapid detection systems, and **biosensors**. Enzymes acting as protein catalysts are used extensively in the food processing industry to control texture, appearance, and nutritive value, and for the generation of desirable flavors and aromas. Because they are isolated from plants, animals, or microorganisms, their availability is dependent upon the availability of the source material. Using genetically engineered microorganisms for the production of enzymes eliminates the need to rely on source materials while ensuring a continuous supply of enzymes.

New technologies allow researchers to target the genetics of plants, animals, and microorganisms and to manipulate them to our food production advantage. Predictions for the future include the following:

- Environmentally hardy food-producing plants that are naturally resistant to pests and diseases and capable of growing under extreme conditions of temperature, moisture, and salinity
- An array of fresh fruits and vegetables, with excellent flavor, appealing texture, and optimum nutritional content, that stay fresh for several weeks
- Custom designed plants with defined structural and functional properties for specific food-processing applications

- Cultures of microorganisms that are programmed to express or shut off certain genes at specific times during fermentation in response to environmental triggers
- Strains engineered to serve as delivery systems for digestive enzymes for individuals with reduced digestive capacity
- Cultures capable of implanting and surviving in the human gastrointestinal tract for delivery of antigens to stimulate the immune response or protect the gut from invasion by pathogenic organisms
- Microbially derived, high-value, "natural" food ingredients with unique functional properties
- **Microsensors** that accurately measure the physiological state of plants; temperature-abuse indicators for refrigerated foods; and shelf-life monitors built into food packages
- On-line sensors that monitor fermentation processes or determine the concentration of nutrients throughout processing
- Biotechnologically designed foods to supply nutritional needs; meat with reduced saturated fat, eggs with decreased levels of cholesterol, and milk with improved calcium bioavailability

But the future depends primarily on how the GMO controversy is resolved.

Summary

Fermentation is the breakdown of carbohydrate materials by bacteria and yeasts under anaerobic (without atmospheric oxygen) conditions. The main benefit of fermentation is preservation of the product. For example, acid produced may prevent spoilage by some microorganisms. Fermentation may add flavor–for example, wine. It may remove or alter existing flavors–for example, soy. Or fermentation can alter the chemical characteristics of the food as in sugar to ethanol, ethanol to acetic acid, or sugar to lactic acid. To encourage the growth of certain microorganisms, fermentation can be controlled by pH, salt content, and temperature. Controlling fermentation is an attempt to favor desired organisms. Fermentation is stopped by pasteurizing and cooling. Microorganisms, like SCP, can become food. Biotechnology can bring great changes to food production but not without controversy.

Review Questions

Success in any career requires knowledge. Test your knowledge of this chapter by answering these questions or solving these problems.

1. Describe SCP.
2. Explain fermentation.
3. Show the general reactions that fermentation follows.
4. List six foods produced by fermentation.
5. Name three factors that control fermentation.
6. BST, ___________ ___________, is a growth hormone and is produced by genetically engineered microorganisms.
7. Acetic acid produces ___________; while lactic acid bacteria produce ___________ and ___________.
8. What action does yeast perform in bread making?
9. ___________ became the first genetically engineered product approved for use in food.
10. Define recombinant DNA technology.

Student Activities

1. Make a list of the names of 10 cheeses. Compare with other members of the class and see who can have the least duplicated list.
2. Describe how yeast is prepared for bread making.
3. Develop a list of the chemical products of fermentation.
4. Conduct an in-class taste test of some of the variety of foods produced by fermentation.
5. Demonstrate the use of fermentation on a product such as bread, cucumber pickles, or sauerkraut. Instructions for producing these foods are easy to find.
6. Develop a report or presentation around the controversies of using genetic engineering and biotechnology (genetically modified organisms or GMO) in foods.

Resources

Brody, J. E. 1981. *Jane Brody's nutrition book.* New York: Bantam Books.

Mathewson, P. R. 1998. *Enzymes.* St. Paul, MN: Eagan Press.

McGee, H. 1997. *On food and cooking. The science and lore of the kitchen.* New York: Simon and Schuster Inc.

Vieira, E. R. 1996. *Elementary food science*, 4th ed. New York: Chapman and Hall.

Internet

Internet sites represent a vast resource of information. The URLs (uniform resource locator) for the World Wide Web sites can change. Using one of the search engines on the Internet such as Yahoo!, HotBot, AltaVista, Excite, Dogpile, About, or Google, find more information by searching for these words or phrases: fermentation, food preservation, lactic acid bacteria, acetic acid bacteria, coagulation, syneresis, biotechnology in foods, single-celled protein (SCP), wine making, beer making, bread making, genetic engineering of food. Also, Table A-7 provides a listing of some useful Internet sites that can be used as a starting point.

Chapter 14

Chemicals

Objectives

After reading this chapter, you should be able to:

- Describe three reasons for using food additives
- Discuss how food additives are monitored and controlled
- List five general categories of intentional food additives
- Identify the five specific uses of food additives and give examples
- Discuss the use of nutritional additives
- Identify five uses of food additives that would be considered an abuse
- Name two methods used to reduce fat intake
- Identify three color additives exempt from certification
- Read food labels and identify the additives and provide a reason for their use

Key Terms

acidulants
antimicrobial agents
antioxidants
chelating agents
Delaney clause
emulsifier
GRAS
lake
sequestrants
stabilizer
surface active agents

Food additives (chemicals) are used only to maintain or to improve quality of food or to give it some added quality that consumers want. The Food and Drug Administration (FDA) monitors the use of additives and allows them to be used only if proven information has shown that the additive will accomplish the intended effect in the food. Also, the amount used cannot be more than is needed to accomplish the intended effect in the food.

REASONS FOR USE

Food additives are any substance used intentionally in food and that may reasonably be expected to, directly or indirectly, become a component of food or affect the characteristic of any food. This includes any substance intended for use in producing, manufacturing, packaging, processing, preparing, treating, transporting, or holding food. The uses of food additives are governed by the Food, Drug and Cosmetic Act.

Intentional food additives include the following:

- Flavors
- Colors
- Vitamins
- Minerals
- Amino acids
- Antioxidants
- Antimicrobial agents
- Acidulants
- Gums
- Sequestrants
- Surface active agents
- Sweeteners

The use of food additives is controlled by the **Delaney clause**. This clause basically states that the food industry cannot add any substance to food if it induces cancer when ingested by man or animal or if it is found, after tests that are appropriate for the evaluation of the safety of food additives, to induce cancer in man or animals.

Additives are used to achieve one or a combination of four purposes:

1. To maintain or improve nutritional value
2. To maintain freshness
3. To aid in processing or preparation
4. To make food more appealing

Without additives or preservatives, some food would soon spoil or it would taste bland. If some foods were not made storable, food would be wasted. Also, if some foods are not stored properly, they can cause illnesses. Some of the general categories of additives (chemicals) that benefit food include preservatives, nutritional additives, color modifiers, flavoring agents, texturing agents, and aids to processing.

HOW MUCH IS ONE TRILLIONTH?

The analytical capabilities of the 1950s and 1960s could detect approximately 100 parts per billion (0.00001 percent) of a chemical in food, and any amount less than 100 ppb was then equal to zero. With the vastly improved detection methods in analytical chemistry, it is now possible to detect amounts as low as 2 ppb (0.0000002 percent) and, in some cases, parts per trillion (ppt). These small concentrations are difficult to even imagine, but in one analogy, one part per million equals about 1/32 ounce (or about 1 gram) in 1 ton of food.

One part per billion is about one drop in a 10,000-gallon tank or can. One part per trillion is one grain of sugar in an Olympic-size swimming pool. This increase in the sensitivity of detection has produced evidence of trace contaminants in food that was before unsuspected. This presents a dilemma. Chasing an ever-receding zero level in foods as analytical methods continually improve could bring us eventually to the question: "Does the presence of one molecule of a carcinogen constitute grounds for removing a food from the marketplace?"

To find out more about the Delaney clause, search the Web or visit these Web sites:

<www.pmac.net/sine.htm>
<www.cfsan.fda.gov/list.html>

TORTILLAS: ENRICHED FLOUR (FLOUR, NIACIN, REDUCED IRON, THIAMIN MONONITRATE, RIBOFLAVIN), WATER, HYDROGENATED SOYBEAN OIL WITH BHT ADDED TO PROTECT FLAVOR, GLYCERIN, SALT, DEXTROSE, CALCIUM PROPIONATE (PRESERVATIVE), WHEY, MONO- AND DIGLYCERIDES, SORBIC ACID (PRESERVATIVE), CITRIC ACID, L-CYSTEINE (DOUGH CONDITIONER).

SAUCE MIX: TOMATO POWDER, MALTODEXTRIN, SALT, CHILI PEPPER, FRUCTOSE, SUGAR, MODIFIED CORN STARCH, MONOSODIUM GLUTAMATE, ONION POWDER, PAPRIKA, GARLIC POWDER, SPICE, SILICON DIOXIDE, HYDROGENATED SOYBEAN OIL WITH BHT ADDED TO PROTECT FLAVOR, MALIC ACID, AUTOLYZED YEAST EXTRACT, ARTIFICIAL COLOR, ASCORBIC ACID, NATURAL FLAVOR.

RICE PACKET: RICE

FIGURE 14-1

Food label with the additives circled.

Additives have many other functions too. These uses include hardening, drying, leavening, antifoaming, firming, crisping, antisticking, whipping, creaming, clarifying, and sterilizing (Tables 14-1 and 14-2). Without additives to help in the processing of food, today's grocery stores would need a lot less room to sell what foods that would keep on the shelf for any length of time. Tested and approved food additives are a part of today's modern food technology.

Over the past few years, the food industry has worked to reduce the use of food additives with special emphasis on certain groups of additives like artificial colors and preservatives (Figure 14-1). This effort has been driven by consumer desires. Food companies spend millions of dollars to find out what consumers want, and they adjust their products to attempt to meet these demands. The group of food additives that add the most value to foods are the flavors, and these are generally used in very low levels. Sweeteners on the other hand are the most heavily used additives. The most heavily used additives are sucrose, high fructose corn syrup, dextrose, and salt. The per capita consumption of the others is less than one pound per capita per year.

TABLE 14-1 Additives Classes and Functions

Additive Class	Function
Curing/pickling agents	Impart a unique flavor and/or color to food, often increase shelf life and stability
Dough conditioners	Modify starch and gluten to improve baking quality of yeast-leavened dough
Drying agents	Absorb moisture
Enzymes	Improve food processing and the quality of finished foods
Firming agents	Maintain the shape or crispness in fruits and vegetables
Flavor enhancers	Supplement, enhance, or modify the original flavor or aroma of a food without contributing flavors of their own
Flour treating agents	Improve the color and/or baking qualities of flour
Formulation aids	Promote or produce a desired physical state or texture
Fumigants	Control insects or pests
Leavening agents	Produce or stimulate CO_2 production in baked goods
Lubricants/release agents	Prevent sticking of food to contact surfaces
Nutritive sweeteners	Provide greater than 2% of the caloric value of sucrose per equivalent unit of sweetening capacity when used to sweeten food
Oxidizing/reducing agents	Produce a more stable product
pH control agents	Change or maintain active acidity or alkalinity
Processing aids	Enhance the appeal or utility of a food or food components
Propellants, aerating agents and gases	Supply force to expel a product
Solvents and vehicles	Extract or dissolve another substance
Surface active agents	Modify surface properties of liquid food components
Surface-finishing agents	Increase palatability, preserve gloss, and inhibit discoloration
Synergists	Produce a total effect different or greater than the sum of the effects produced by the individual food ingredients
Texturizers	Affect the feel or appearance of the food

TABLE 14-2 Additives Classes and Functions

Additives	Example	Function in Food
Anti-caking free-flowing agents	calcium silicate, magnesium carbonate	Used to keep food dry and to prevent caking as moisture is absorbed in foods. This situation will occur in foods such as salt, powdered sugar, and baking powder.
Antimicrobial agents	salt, sugar, sodium nitrate, sodium propionate, potassium sorbate	Prevent growth of micro-organisms in foods such as breads, carbonated beverages, and margarine. Nitrites preserve color, enhance flavor, and prevent rancidity.
Antioxidants	BHT, BHA, ascorbic acid	Retard rancidity of unsaturated oils; prevents browning in fruits and vegetables that occurs during exposure to oxygen.
Buffers	sodium bicarbonate, malic acid, citric acid, delactose whey	Cause leavening (rising) of starch products; imparts tart taste in carbonated drinks and candies.
Coloring agents and adjuncts	beta-carotene, Red #3, yellow #6	Used to enhance or correct the colors of foods, making them more eye-appealing; most criticized nutritive additive.
Flavoring agents	natural extracts, essential oils, MSG, benzaldehyde	2000 kinds are now added to many types of food to enhance flavor; largest group of additives. These substances modify the original flavor or aroma without contributing flavors of their own.
Emulsifiers	lecithin, monocyglycerides	Used to evenly distribute the fat- and water-soluble ingredients throughout food products; used in margarine, bakery products, and chocolate.

TABLE 14-2 Additives Classes and Functions *(concluded)*

Additives	Example	Function in Food
Humectants	glycerine, propylene glycol, sorbitol	Retain moisture, keep foods soft; used in marshmallows, flaked coconut, cake icings, soft and chewy cookies.
Maturing and bleaching agents	chlorine, benzoyl peroxide, acetone peroxide	Improve baking properties and whiten appearance of wheat flour and cheese.
Sequestrants	EDTA, citric acid	Added to bind with metals such as iron, calcium, and copper to prevent color, flavor, and appearance changes in food products such as wine and cider.
Stabilizers and thickeners	gums, pectins, alginates, modified starch, carrageenan, dextrins, gelatin, guar, protein derivatives, cellulose	Used to maintain the texture and body of many food products such as ice cream, pudding, candy, and milk products.
Nonnutritive sweeteners	aspartame, saccharin	Used to give sweet flavor to items such as beverages, cereals, and dietetic foods without adding the calories of ordinary sweeteners.

Over the years, chemicals have been added to food for unacceptable reasons, including:

- To disguise inferior products
- To deceive the consumer
- To provide otherwise desirable results that lower the nutritional value
- To replace good manufacturing practices
- To use in amounts greater than are necessary

Tables 14-1 and 14-2 provide complete descriptions of food additives, their use, and some examples.

PRESERVATIVES

Preservatives include **antioxidants**, sequesterants, and **antimicrobial agents** (see Table 14-3). Common antioxidants are BHA, BHT, TBHQ, erythorbic acid, sodium erythorbate, tocopherols, and ascorbic acid. Some of the common antimicrobial agents include benzoic acid or sodium benzoate, calcium propionate, potassium, and sorbate or sorbic acid. Sequesterants are **chelating agents**. They are organic compounds that react with metallic ions to bind in a relatively inactive structure. **Sequestrants** prevent metals from catalyzing reactions of fat oxidation, pigment discoloration, flavor loss, and odor loss. Common sequesterants include EDTA, citric acid and its salts, and phosphoric acid and its salts.

NUTRITIONAL ADDITIVES

Vitamins and minerals are added to foods to make them more nutritious and sometimes to replace those nutrients lost during

TABLE 14-3 Common Preservatives and Their Uses

Preservative	Uses
Salt	Retards bacterial growth on cheese and butter
Nitrites and nitrates of sodium and potassium	Adds to flavor, maintains pink color in cured meats, and prevents botulism in canned foods
Sulfur dioxide and sulfites	Used as bleaches and antioxidants to prevent browning in alcoholic beverages, fruit juices, dried fruits and vegetables, and to prevent yeast growth
Benzoic acid and sodium benzoate	In oyster sauce, fish sauce, ketchup, alcoholic beverages, fruit juices, margarine, salads, jams, and pickled products
Propionic acid and propionates	In bread, chocolate products, cheese
Sorbic acid and sorbates	Prevent mold formation in cheese and flour confectioneries
Butylated hydroxyanisole (BHA) and Butylated hydroxytoluene (BHT)	Prevent oxidative spoilage of unsaturated fats and oils in potato chips, cheese balls, and so on
Ascorbic acid and ascorbates	In pork sausages
Sodium citrate	In cooked cured meat and canned baby foods

TABLE 14-4 Some Common Nutritional Additives

Name or Type	Nutrient and Use
Alpha tocopherols	Vitamin E; antioxidant; nutrient; used in vegetable oil
Ascorbic acid	Vitamin C; antioxidant reacts with unwanted oxygen; stabilizing colors, flavors, oily foods; nutrient added to beverages, breakfast cereals, cured meats; prevents formation of nitrosamines
Beta carotene	Vitamin A; coloring agent added to butter, margarine, shortening
Calcium pantothenate	Added for calcium
Ferrous gluconate, ferric orthophosphate, ferric sodium pyrophosphate, ferrous fumarate, ferrous lactate	Added to supply iron; some black color
Minerals	Added to improve nutritional value; zinc and iodine
Vitamins	Added to improve nutritional value; thiamin, riboflavin, niacin, pyridoxine, biotin, folate, and vitamins A and D

processing. Some examples include bread that has been enriched, milk with vitamin D added, and margarine with vitamins A and D added. Table 14-4 provides a list of some of the common nutritional additives.

COLOR MODIFIERS

Colors include both natural and synthetic colorants. FD&C stands for foods, drugs, and cosmetics, and this means that this colorant can be used in all three of these. If the color is marked D&C or C colorants, it is used in household products, like shampoos.

The Food and Drug Administration (FDA) is responsible for controlling all color additives used in the country. All color additives permitted for use in foods are classified either as "certifiable" or "exempt from certification" (Figure 14-2).

Color additives permitted for direct addition to human food in the United States includes these certifiable colors:

- FD&C Blue No. 1 (dye and **lake**)
- FD&C Blue No. 2 (dye and lake)

FIGURE 14-2
Scientists measure the color of foods because it is important to the food's acceptance. (*Source:* USDA, ARS Image Gallery)

- FD&C Green No. 3 (dye and lake)
- FD&C Red No. 3 (dye)
- FD&C Red No. 40 (dye and lake)
- FD&C Yellow No. 5 (dye and lake)
- FD&C Yellow No. 6 (dye and lake)
- Orange B (restricted to specified use)
- Citrus Red No. 2 (restricted to specified use)

Colors exempt from certification include the following:

- Annatto extract
- Beta-carotene
- Beet powder
- Canthaxanthin

- Caramel color
- Carrot oil
- Cochineal extract (carmine)
- Cottonseed flour, toasted partially defatted, cooked
- Ferrous gluconate (restricted to specified use)
- Fruit juice
- Grape color extract or enocianina (restricted to a specified use)
- Paprika and paprika oleoresin
- Riboflavin
- Saffron
- Titanium dioxide (restricted to specified use)
- Turmeric and turmeric oleoresin
- Vegetable juice

Certifiable color additives are available for use in food as either dyes or lakes. Dyes dissolve in water and are made as powders, granules (small hard pieces), liquids, or other special-purpose forms. They can be used in beverages, dry mixes, baked goods, confections (food made with sweet ingredients), dairy products, pet foods, and a variety of other products. Lakes are the water-insoluble form of the dye. Lakes are more stable than dyes and are ideal for coloring products containing fats and oils or items lacking sufficient moisture to dissolve dyes. Typical uses include coated tablets, cake and donut mixes, and hard candies. Table 14-5 describes some certifiable colors and indicates their uses in food.

Colors are used in food products for the following reasons:

- To offset color loss due to exposure to light, air, extremes of temperature, moisture, and storage conditions.
- To correct natural variations in color. Off-colored foods are often incorrectly associated with poor quality. For example, some oranges are often sprayed with Citrus Red No. 2 to correct the natural orangy-brown or the patches of green color of their peels. However, masking poor quality is an unacceptable use of colors.
- To strengthen colors that occur naturally but at levels weaker than those usually associated with a given food.
- To provide a color identity to foods that would otherwise be colorless. For example, red colors provide a pleasant identity to strawberry ice cream, and lime sherbet is known by its bright green color.

TABLE 14-5 Color Additives Certifiable for Food Use

Name/Common Name	Hue	Common Food Uses
FD&C Blue No. 1/ Brilliant Blue FCF	Bright blue	Beverages, dairy products, dessert powders, jellies, confections, condiments, icings, syrups, extracts
FD&C Blue No. 2/ Indigotine	Royal blue	Baked goods, cereals, snack foods, ice cream, confections, cherries
FD&C Green No. 3/ Fast Green FCF	Sea green	Beverages, puddings, ice cream, sherbet, cherries, confections, baked goods, dairy products
FD&C Red No. 40/ Allura Red AC	Orange-red	Gelatins, puddings, dairy products, confections, beverages, condiments
FD&C Red No. 3/ Erythrosine	Cherry red	Cherries in fruit cocktail and in canned fruits for salads, confections, baked goods, dairy products, snack foods
FD&C Yellow No. 5/ Tartrazine	Lemon yellow	Custards, beverages, ice cream, confections, preserves, cereals
FD&C Yellow No. 6/ Sunset Yellow	Orange	Cereals, baked goods, snack foods, ice cream, beverages, dessert powders, confections

- To provide a colorful appearance to certain “fun foods”. For example, many candies and holiday treats are colored to create a festive appearance.
- To protect flavors and vitamins that may be affected by sunlight during storage.
- To provide an appealing variety of healthy and nutritious foods that meet consumers’ demands.

FLAVORING AGENTS

Some flavoring agents, such as spices and liquid derivatives of onion, garlic, cloves, and peppermint, enhance flavor. Synthetic flavorings that resemble natural flavors have been developed, and these have the advantage of being more stable than natural flavors. Some of these are described in Table 14-6. Flavors cost the most and add the most value to products.

TEXTURING AGENTS

Emulsifiers are sometime called **surface active agents**. These improve the uniformity of a food–the fineness of grain–the smoothness and body of foods such as bakery goods, ice creams, and confectionery products. Mono- and diglycerides, polysorbate 60 and 80, lecithin as well as proteins, are all considered emulsifiers.

Stabilizers and thickeners add smoothness, color uniformity, and flavor uniformity to such foods as ice creams, chocolate milk, and artificially sweetened beverages. Stabilizers and thickeners are labeled as pectin, vegetable gums, and gelatins. Table 14-7 lists some common emulsifiers, stabilizers, thickeners, and their uses.

ACIDULANTS

Acidulants make a food acid or sour. They are added to foods primarily to change the taste and to control microbial growth. These include citric acid, acetic acid, phosphoric acid, hydrochloric acid, and others.

TABLE 14-7 Emulsifiers, Stabilizers, Thickeners, and Their Uses

Texturing Agent	Use
Carboxymethylcellulose	Batter coating, frozen chips, and fish sticks
Xanthan gum	Seafood dressings, frozen pizza, packet dessert topping
Pectin	Jams, marmalades, and jellies
Dextrins	Icings, frozen desserts, confectioneries, whipped cream, cake mixtures, mayonnaise, and salad dressings
Lecithins	Milk chocolate and powdered milk
Sodium alginate	Ice cream, yogurt, sauces, and syrups
Mono- and diglycerides of stearic acid	Low cholesterol margarine, hot chocolate mix, dehydrated mashed potato, aerosol cream
Potassium dihydrogencitrate	Processed cheese, condensed and evaporated milk

TABLE 14-6 Some Common Flavorings Used as Food Additives

Flavor	Food Additive
Camphor	Bornyl acetate
Cinnamon	Cinnam aldehyde
Ginger	Ginger oil
Grape	Methyl anthranilate
Lemon	Citral
Orange	Orange oil
Pear	Amyl butyrate
Peppermint	Menthol
Rum	Ethyl formate
Spearmint	Carvone
Spicy	Ethyl cinnamate
Vanilla	Ethyl vanillin
Wintergreen	Methyl salicylate

FAT REPLACERS

In an attempt to reduce the fat intake, the food industry is attempting to modify fat itself. These approaches generally fall into the following categories:

- Decreasing fat content
- Using fat replacers, substitutes, extenders, mimetics, or synthetic fat

Fat replacers include various carbohydrate-based, protein-based, and fat-based replacers for different food categories. Consumers constantly demand new and improved fat replacers. Some of these are:

- Olean (sucrose polyester)
- Olestra
- Amalean I and Amalean II (modified high-amylose corn)
- Cellulose and hemicelluloses
- Chitosan (fiber of crustaceans)
- Hydrocolloids

IRRADIATION

The FDA considers irradiation of food as an additive. Irradiated food is safe and will last longer. Irradiation to control microorganisms on beef, lamb, and pork is safe and eventually could mean that consumers will have less risk of becoming ill from contaminated meat.

Irradiation of fruits and vegetables means longer shelf life for those items. Irradiation passes through food without leaving any residue. The ionizing radiation kills bacteria and other pathogens in food, but the food never comes in contact with radioactive materials. The process does not make the food radioactive.

Research on food irradiation dates back to the 1920s. The U.S. Army used the process on fruits, vegetables, dairy products, and meat during World War II, and NASA routinely sends irradiated food on U.S. spaceflights. Moreover, research shows that any changes in irradiated food are similar to the effects of canning, cooking, or freezing. Nutritionally, irradiated food is virtually identical to nonirradiated food.

Chapter 12 covers more details of irradiation.

HAZARDS

Additives remain a public concern. Contrary to popular perception, the majority of direct food additives are Generally Recognized

TABLE 14-8 Common Foodborne Toxicants

Food Type	Substance	Effect
Peaches, lima beans, apple seeds, apricot pits, certain grains, legumes	Cyanide	Neurological degenerative disease, death
Nutmeg, sassafras tea	Safrole	Liver cancer
Coffee, tea, red wine	Tannin	Liver cancer
Coffee, tea, colas	Caffeine	Birth defects
Green potato skins	Solanine	Nervous system disorders
Cabbage, mustard brussel sprouts, cauliflower, turnips	Glucosinolates	Goiter

as Safe (**GRAS**) substances. GRAS substances are ingredients that may be added to food without extensive prior testing and were established to avoid the burden of proving the safety of substances already regarded as safe. The banning of cyclamates in 1969 produced a presidential directive to review the safety of the GRAS substances and led to the Food and Drug Administration's (FDA's) cyclic review of all direct and indirect additives.

A review of Generally Recognized as Safe substances revealed that about 90 percent present no significant hazard with normal human food uses. Most of those remaining to be tested have not been associated with hazards to humans. The other direct food additives have been approved, and their uses are regulated by the FDA. Indirect additives, such as those used in production, processing, and packaging and that might migrate to food, are numerous but normally occur in foods at trace levels, if at all. Many may occur at parts per billion or less. Examination of severity, incidence, and onset of effects indicates that additives are the lowest-ranking hazard.

Some chemicals are naturally a part of the food, but they are at such low levels they are harmless (Table 14-8).

Summary

Food additives are used to maintain or improve the quality of food. The Food and Drug Administration (FDA) monitors the use of additives. According to the FDA, additives are any substance intentionally or indirectly a component of food. Intentional food additives include the general categories of flavor, colors, vitamins, minerals, amino acids, antioxidants, antimicrobial agents, acidulants, sequestrants, gums (thickeners), sweeteners, and surface active agents. Some additional additives include fat replacers and irradiation. The majority of direct food additives fall into the category of Generally Recognized As Safe (GRAS). Unacceptable uses of food additives include any uses to deceive, disguise, lower nutritional value, or avoid good manufacturing practices. For years the use of food additives has been controlled by the Delaney clause.

Review Questions

Success in any career requires knowledge. Test your knowledge of this chapter by answering these questions or solving these problems.

1. Name the two methods used to reduce fat intake.
2. List three reasons for using food additives.
3. Using Table 14-1, choose three additives and describe their function.
4. BHT is a/an ___________ that retards rancidity of unsaturated oils and prevents browning in fruits and vegetables.
5. Nitrates and nitrites of sodium and potassium add to ___________ and maintain ___________ in cured meats.
6. What organization is responsible for controlling all color additives used on foods in the United States?
7. Why are vitamins and minerals added to foods?
8. Green potato skins and apple seeds are two examples of common foods containing foodborne ___________.
9. List five categories of intentional food additives.
10. List three colors that are exempt from FDA certification.

Student Activities

1. Conduct a contest where everyone collects five food labels. Declare the winner as the person who identifies the most additives and their functions from their labels.
2. Find an example for each of the following uses of color additives: identify for otherwise colorless foods, fun foods, and correction of natural variations.

3. Identify at least two food products that use one of the flavors listed in Table 14-6.
4. Visit the Web site for the FDA (<www.fda.gov/>) and develop a report or presentation on food additives based on information from the site.

Resources

Alexander, R. J. 1998. *Sweeteners: Nutritive.* St. Paul, MN: Eagan Press.
Cremer, M. L. 1998. *Quality food in quantity. Management and science.* Berkeley, CA: McCutchan Publishing Corporation.
Ensminger, A. H., M. E. Ensminger, J. E. Konlande, and J. R. Robson. 1994. *Foods and nutrition encyclopedia.* 2 Vols. Boca Raton, FL: CRC Press.
Francis, F. J. 1998. *Colorants.* St. Paul, MN: Eagan Press.
McGee, H. 1997. *On food and cooking. The science and lore of the kitchen.* New York: Simon and Schuster Inc.
Stauffer, C. E. 1999. *Emulsifiers.* St. Paul, MN: Eagan Press.
Thomas, D. J., and W. A. Atwell. 1999. *Starches.* St. Paul, MN: Eagan Press.

Internet

Internet sites represent a vast resource of information. The URLs (uniform resource locator) for the World Wide Web sites can change. Using one of the search engines on the Internet such as Yahoo!, HotBot, AltaVista, Excite, Dogpile, About, or Google, find more information by searching for these words or phrases: food additives, antioxidants (BHA, BHT, TBHQ, erythorbic acid, sodium erythorbate, tocopherols, ascorbic acid), antimicrobial agents, acidulants, sequestrants, humectant, Food and Drug Administration, food coloring, food preservatives, nutritional additives, FD&C (food, drug, and cosmetics), certifiable color additives, flavoring agents, irradiation, GRAS (Generally Recognized As Safe). Also, Table A-7 provides a listing of some useful Internet sites that can be used as a starting point.

Chapter 15

Packaging

Objectives

After reading this chapter, you should be able to:

- Identify three types of food packaging
- Name and describe the use of four basic packaging materials
- List ten features or requirements of packaging material
- Describe tests that measure the properties of packaging material
- Identify packages with special features
- Discuss how packaging addresses environmental concerns
- Identify and describe a packaging innovation

Key Terms

aseptically
copolymer
extratries
hermetically
ionomer
PET
primary
printability
recycled
secondary
tertiary

Food packaging development started with humankind's earliest beginnings. Early forms of packaging ranged from gourds to seashells to animal skins. Later came pottery, cloth, and wooden containers. These packages were created to facilitate transportation and trade.

Aside from protecting the food, the package serves as a vehicle through which the manufacturer can communicate with the consumer. Nutritional information, ingredients, and often recipes are found on a food label. The package is also used as a marketing tool designed to attract your attention at the store. This makes printability an important property of a package.

TYPES OF CONTAINERS

Food packaging can be divided into three general types:

1. Primary
2. Secondary
3. Tertiary

Primary containers come in direct contact with the food. A **secondary** container is an outer box or wrap that holds several primary containers together. **Tertiary** containers group several secondary holders together into shipping units.

Many containers used in the food industry are part of form-fill-seal packaging. Containers may be preformed at another site and then filled at the processing plant. Or containers may be formed in the production line just ahead of the filling operation. This is called form-fill-seal, and it is one of the most efficient ways to package food.

To protect the food against exchange of gases and vapors, and contamination from bacteria, yeasts, molds, and dirt, containers are **hermetically** sealed. The most common hermetic containers are cans and glass bottles.

FOOD-PACKAGING MATERIALS AND FORMS

The food industry uses four basic packaging materials: metal, plant matter (paper and wood), glass, and plastic. A number of basic packaging materials are often combined to give a suitable package. The fruit drink box is an example where plastic, paper, and metal are combined in a laminate to give an ideal package. Another example is a peanut butter jar. The main package containing the food (primary package) is made of glass (or plastic); the lid is made of metal lined with plastic; and the label is made of paper.

Cans

Cans (Figure 15-1) are formed at the food processing factory or shipped with their bottoms attached with separate can lids. Lids are seamed onto the cans. The outside of the steel can is protected from rust by a thin layer of tin (.025 percent by weight). The inside of the can is protected by a thin layer of tin or baked-on enamel. Tin-free steel and thermoplastic adhesive-bonded seams have become more common. These do away with the need for solder, which can contribute to traces of lead in food. Factory equipment

FIGURE 15-1
Various types and sizes of cans are used by the food industry.

allows hermetically sealed sanitary steel cans to be manufactured and later sealed at the rate of 1,000 units per minute. Rigid aluminum, tin plate, and tin free containers also can be readily formed without side seams or bottom end seams by pressure extrusion. Aluminum is used as a packaging metal because of its light weight, low levels of corrosion (rust), recyclability, and ease of shipping. However, aluminum has less structural strength than metal cans. This limitation has been overcome by the injection of a small amount of liquid nitrogen into the can prior to closure. This gas provides for internal pressure that adds rigidity.

Glass

Glass provides a chemically inert and noncorrosive recyclable food packaging material. Glass breaks, and it is too heavy for some processing uses. Also, recycling is not easy, except in the case of home canning use.

Paper

Paper (Figure 15-2) used as a primary container must be treated, coated, or laminated. Paper from wood pulp and reprocessed waste paper is bleached and coated or impregnated with waxes, resins, lacquers, plastics, and laminations of aluminum to improve its water strength and gas impermeability, flexibility, tear resistance, burst strength, wet strength, grease resistance, sealability, appearance, and **printability** of advertising or labels. Papers treated for

FIGURE 15-2
Various types of paper packaging are used by the food industry.

primary contact with food are reduced in their ability to be **recycled**. Paper that comes in contact with foods must meet FDA standards for chemical purity. Paper used for milk cartons must come from sanitary virgin pulp. The major safety concern is with the puncturability or tearability that will allow for the outside environment to enter and contaminate the food.

Plastics

Some popular plastics include cellophane, cellulose acetate, nylon, mylar, saran, and polyvinyl chloride. **Copolymer** plastics extend the range of useful food-packaging applications. **Ionomer** (ionic bonds) plastic materials are improved food-handling materials that function under greater oil, grease, solvent resistance, and they have a higher melting strength. Newer plastic materials contain cornstarch, which makes them more biodegradable.

Laminates

Commercial laminates with as many as eight different layers can be custom-designed for a specific product. In the case of prepackaged dry beverages, the laminate (from outside of package to inside) may have a special cellophane that is printable, polyethylene for a moisture barrier, treated paper for stiffness, a layer for bonding, an aluminum foil (prime gas barrier) inside a layer of polyethylene for an additional water vapor barrier (Figure 15-3).

FIGURE 15-3

Diagram of a laminate cross section.

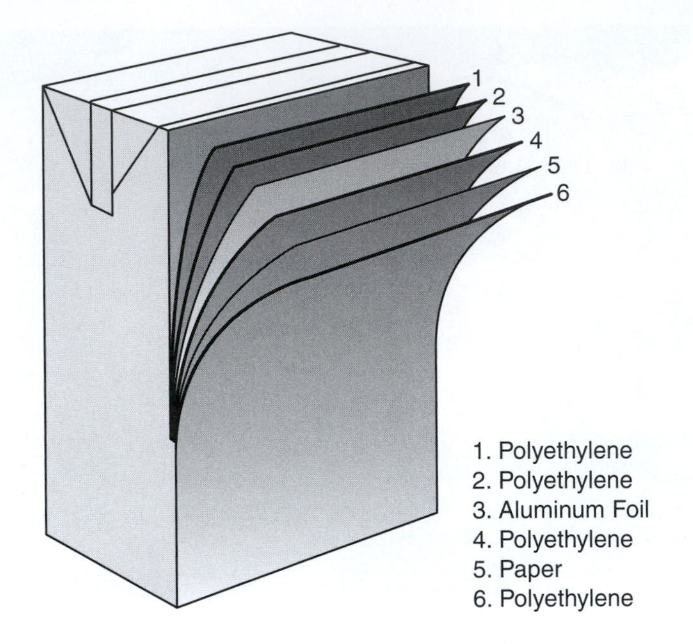

Each basic packaging material has advantages and disadvantages. Metal is strong and a good overall barrier, but it is heavy and prone to corrosion. Paper is economical and has good printing properties; however, it is not strong, and it absorbs water. Because paper absorbs water, it gains moisture from the milk, gets weaker, and fails. Glass is transparent allowing the consumer to see the product, but it is breakable. Plastics are versatile but often expensive. Paper makes a good economical material and provides a good printing surface.

Overall, the requirements, functions, and considerations of food containers include:

1. Nontoxic and compatibilty with food
2. Sanitary protection
3. Moisture protection
4. Resistance to impact
5. Light protection
6. Gas and odor protection
7. Ease of opening and closing
8. Tamper-resistance and tamper evident
9. Pouring features
10. Size, shape, and weight limitations

11. Reseal features
12. Ease of disposal
13. Appearance and printability
14. Transparency
15. Affordability

Plastics can also serve as secondary packages. The milk case in which a number of milk cartons are delivered to the supermarket is a good example.

Retortable Pouches

Twenty years of development went into the technology to ensure that the materials used in the production of retortable pouches (Figure 15-4) would protect the food and not contribute harmful **extratries** (chemical interaction with foods) to foods. The three layers consist of an outer layer of polyester film for strength, temperature resistance, and printability; a middle layer of aluminum film for barrier properties; and an inner layer of polypropylene film that provides for a heat-seal.

Edible Films

Sausage casings are an example of an edible film. By spraying gelatin, gum arabic, starch, monoglycerides, proteins, or other edible materials, a thin protective coating can be formed around food

FIGURE 15-4
This retortable pouch is used by the military.

particles. For example, raisins in breakfast cereals are sprayed to prevent them from moistening the cereal in the box, and nuts are coated to protect them from oxidative rancidity. An edible wax film is used to coat the surface of vegetables to reduce moisture loss and provide increased resistance to the growth of molds.

All edible films must be approved by the FDA for human consumption.

PACKAGE TESTING

Many tests measure the protective properties of packaging materials. Basically, the tests can be divided into chemical and mechani-

FAKE FOODS

According to FDA investigations, some companies and individuals have made hundreds of thousands, even millions, of dollars off of their fraudulent foods. These early cases helped lead to passage of the 1938 Federal Food, Drug, and Cosmetic Act, which, among other things, specifically bars economic adulteration of food.

1922—The Bureau of Chemistry in the U.S. Department of Agriculture—FDA's precursor—investigated reports that various makers of Eskimo Pie coatings were illegally substituting coconut oil and cottonseed sterine for the required cocoa butter and milk fat. One report claimed that the imitation coating, unlike the real coating, did not snap off when bitten into but instead would break in such a way that it allowed melting ice cream to leak out.

1923—Two businesses dealing in adulterated olive oil were convicted and fined in Federal Court of the Southern District of New York. This was the second conviction for one of the companies, which was fined in 1922 for selling so-called pure olive oil adulterated with peanut oil.

1926—Chicago police arrested two Chicago men following a complaint from New York bakers that the two men were making and selling a butter substitute for genuine butter. Their substitute butter included melted low-grade butter and lard that was churned with buttermilk. The federal government seized 32 tubs of the product, and the case was turned over to the Internal Revenue Bureau of the U.S. Treasury Department.

1933—A San Francisco man was fined for adulterating dried egg yolks with artificial color and skim milk powder. He imported the dried egg yolks from China, unpacked them and added the adulterants, and then repacked them in containers closely matching those in which they were imported.

cal. Chemical tests are used to determine if any of the packaging material, such as plastic, migrated into the food, and to measure resistance to greases, acids, alkalies, and other solvents. Mechanical tests measure barrier properties, strength, heat-seal ability, and clarity.

Actual tests consist of subjecting a few samples of the food-filled packages through the complete processing, shipping, warehousing, and merchandising sequence. This allows the packaging material to be subjected to all the normally occurring abuses. These packages are recovered and analyzed to see how they withstood the vibrations, humidity, temperature, and handling. Similar tests are sometimes performed in simulations.

An Atlanta company was fined for selling so-called vanilla extract that contained insufficient vanilla but was artificially colored to make it look like real vanilla extract. It was sold to Army posts in Kansas, Nebraska, and Oklahoma.

1936—The new Food and Drug Administration recorded the seizure of 62 cases of sardines labeled as "packed in olive oil." An inspector learned that the sardines were actually packed in a mixture of one-third sesame and two-thirds olive oil.

1938—A complaint from the assistant to the quartermaster supply officer at Fort Sam Houston in Texas led to the seizure of more than 500 bottles of a product labeled "Vanilla Extract." It contained artificial color and little, if any, true vanilla. Fortunately, the product had not been paid for.

Under the supervision of a state inspector, 60 cases of a product labeled tomato catsup were poured down a sewer after the catsup was found to consist largely of water, starch, and cochineal—a red dye made from the dried bodies of female cochineal insects.

In recent years, FDA has sought and won convictions against companies and individuals engaged in making and selling bogus orange juice, apple juice, maple syrup, honey, cream, olive oil, and seafood.

1990s—An orange juice manufacturer defrauded consumers of more than $45 million during an estimated 20-year period. Another orange juice company netted $2 million in two years by substituting invert beet sugar for frozen orange juice concentrate. Still another orange juice manufacturer saw its earnings rise from zero in the company's second year of operation to $57 million in its fifth year before being convicted and sentenced for adulterating orange juice concentrate with liquid beet sugar.

PACKAGES WITH SPECIAL FEATURES

With the new foods being produced, packages frequently have some type of added convenience feature. Often a package must withstand freezer temperatures as well as boiling or the temperature of steam without bursting. The properties of polyester and nylon films allow this type of packaging.

Microwave oven packaging presents another challenge to packaging. Microwave oven packaging must meet all the standard requirements for packaging, and it also must be transparent to microwaves and able to withstand high temperatures.

Squeezable plastic bottles are used for all types of packaging. These bottles have the high barrier properties of glass with less than one fourth the weight, and they do not break.

Composite paper cartons can be sterilized and then **aseptically** filled with sterile liquid products. The process is called aseptic packaging. This type of packaging allows foods like milk and fruit juices to be packaged in inexpensive flexible containers that require no refrigeration. From the outside of the container inward, this packaging material is made from lamination of polyethylene, paper, polyethylene, aluminum foil, polyethylene, and a coating of ionomer resin.

Supplying food to the military has always created special problems. The packaging must provide protection, and it must simplify preparation and consumption under adverse circumstances.

ENVIRONMENTAL CONSIDERATIONS

Packaging waste can adversely affect the environment. Recycling is a sound approach. However, the problem often lies in feasibility of collection, separation, and purification of the consumer's disposed food packages. This mode of recycling is called post-consumer recycling. Though it offers a logistic challenge, recycling is gaining in popularity, and the packaging industry is cooperating in that effort. Aluminum cans are the most recycled container at this time. Plastic recycling is increasing, yet most plastic is recycled during manufacturing of the containers–not as post-consumer recycling (Figure 15-5). For example, trimmings from plastic bottles are reground and reprocessed into new ones.

The plastics industry facilitates consumer recycling by identifying the type of plastic from which the container is made. A number from 1 to 7 is placed within the recycling logo on the container's bottom. For example, 1 refers to **PET** (Polyethylene Terephthalate), the plastic used for the large 2-liter soft drink bottles.

FIGURE 15-5

Crushed and baled cans at a recycling center. Environmental concerns require many packaging materials to be recycled.

Plastics have the advantage of being light. This helps to conserve fuel during transport and also reduces the amount of package waste.

Environmental issues have gained importance because of regulatory requirements. Increasingly, it is not possible to sell a new packaging material without covering all the environmental issues. Possibly if someone feels that a package does not meet environmental standards, the brand name could suffer.

Environmental regulations play an important part in beverage packaging as well. An important trend in beverage packaging is the use of nonreturnable bottles as well as cans. The PET bottle is also used more and more in both returnable and nonreturnable applications. With the increased use of plastic containers worldwide, recycling and return concepts become an absolute necessity.

INNOVATIONS IN PACKAGING

Packaging innovation is becoming more about convenience than cost. Consumers want convenience, and food companies are developing packages that provide it. Globalization and continued environmental pressures provide new challenges to the food packaging industry.

Cost is no longer the main driver behind packaging innovations. Consumers want convenience, whether it is the elderly consumer who needs to read the label on a box of cookies or the working mother who does not have time to cook a fresh meal. More households are small, and the number of elderly consumers is increasing. This means that packaging should not only be easy to open but able to be closed again so that food will keep longer.

Ready meals are catching on. This phenomenon has swept across the United States and will become more important in the European market. Food processors see a growing demand for convenience food–for example, kitchen-ready preparations of pasta products, oven-ready preparations of pizza, and related products.

Along with the desire for convenience, the popularity of microwavable food is growing. For these foods, some packaging companies developed an expanded polypropylene tray that is convenient to handle. These trays provide foods with extended shelf life, fresh product presentation, and environmental savings. The material can also be heated safely in the microwave, and the insulating feel of the foam material allows the consumer to comfortably handle the package after heating.

Consumer attitudes toward packaging have also changed. Though consumers want convenience, they also demand higher levels of package security. A similar conflict arises with child-resistant packages, which must be too difficult for children but possible for infirm or elderly adults.

Processors recognize that foods must meet consumer quality standards and also appeal to the different palates in the various regions of the world. For example, McDonald's recognized that many consumers in India look on cows differently than in Westernized countries. The company adjusted the menu to meet local needs. This is happening in other areas as well, such as Japan. Although the Japanese have developed a taste for beef products, they do not eat large portions so the package size is adjusted. In China, where they have a lot of pigs, more sophisticated pork products are being produced.

Summary

Using modern technology, society created an overwhelming number of new packages containing a multitude of food products. A modern food package has many functions, its main purpose being to physically protect the product during transport. The package also acts as a barrier against potential spoilage agents, which vary with the food product. Practically all foods should be protected from filth, microorganisms, moisture, and objectionable odors. Consumers rely on the package to offer that protection.

Aside from protecting the food, the package serves as a vehicle for the manufacturer to communicate with the consumer. Nutritional information, ingredients, and often recipes are found on a food label. The package is also used as a marketing tool designed to attract attention at the store. This makes printability an important property of a package. Globalization of the food industry and the consumers are driving the development of innovations in packaging.

Review Questions

Success in any career requires knowledge. Test your knowledge of this chapter by answering these questions or solving these problems.

1. What are the two basic tests to measure the protective properties of packaging materials?
2. Name the three general types of food packaging.
3. Explain the three layers in a retortable package.
4. List ten features of packaging materials.
5. Name the four basic packaging materials used by the food industry.
6. The collection, separation, and purification of the consumer's disposed food packages is the main problem with ___________.
7. Packaging ___________ is of concern because of small children and the elderly.
8. Packages with special features have what requirements?
9. ___________ ___________ are an example of edible film packaging.
10. Describe the functions of primary, secondary, and tertiary packaging.

Student Activities

1. Form teams to collect and display 10 different types of food packaging. Indicate on the display whether the packaging is primary, secondary, or tertiary and indicate the type of material used.
2. Identify a raw fruit or vegetable and list all possible types of packaging for fresh and processed forms.
3. Identify a food product using each of the following for primary packaging: metal, glass, plastic, wood, paper.
4. Search the Internet for information about the unique

packaging challenges presented by the space program and the military. Report on your findings.

5. Conduct a contest to see who can bring to class an actual example of the most innovative packaging for a product. Innovation includes not only bright colors and design but also includes flexibility and convenience. Defend your choice.
6. Let potato chips set outside their package for a few days. Then describe the changes in the chips and relate this to the function of the packaging.

Resources

Potter, N. N., and J. H. Hotchkiss. 1995. *Food science,* 5th ed. New York: Chapman and Hall.

Vaclavik, V. A., and E. W. Christina. 1999. *Essentials of food science.* Gaithersburg, MD: Aspen Publishers, Inc.

Vieira, E. R. 1996. *Elementary food science,* 4th ed. New York: Chapman and Hall.

Internet

Internet sites represent a vast resource of information. The URLs (uniform resource locator) for the World Wide Web sites can change. Using one of the search engines on the Internet such as Yahoo!, HotBot, AltaVista, Excite, Dogpile, About, or Google, find more information by searching for these words or phrases: food packaging, recycling, paper packaging, extratries, post-consumer recycle, edible film packaging. Also, Table A-7 provides a listing of some useful Internet sites that can be used as a starting point.

SECTION *Three*

Foods and Food Products

Chapter 16

Milk

Objectives

After reading this chapter, you should be able to:

- Define the term "milk"
- Describe quality control during the production of milk and milk products
- Explain pasteurization and homogenization
- Identify three methods of pasteurization
- Describe the "solids" composition of milk
- Discuss the separation of butterfat and its uses
- List four beverage milk products
- Describe butter
- Name five concentrated or dried dairy products
- List the steps in cheese making
- Identify three bacteria used to produce dairy products
- Name five fermented dairy products
- List the steps in making ice cream
- Describe three USDA quality grade shields

Key Terms

buttermilk
churning
coalesce
curd
HTST
lipolysis
LTLT
rennet
ripening
ropey
solids-not-fat
standard plate count (SPC)
standardized
thermization
UHT
ultrapasteurization
vacuum evaporation
whey

Milk is the first food of young mammals. It provides a high-quality protein, a source of energy, and vitamins and minerals. Worldwide, many mammalian species are used to produce milk and milk products. Some of these include goats, sheep, horses, and yaks. The focus of this chapter is milk from dairy cows.

Most milk in the United States is produced by cows. The dairy industry produces milk as a fluid product and in a variety of manufactured products including butter, cheeses, condensed, dry, and cultured.

FLUID MILK

Milk is composed primarily of water (87 to 89 percent). Like other foods, if water is removed, the shelf life is extended. The term total milk solids describes the remaining 12 to 13 percent of milk. This includes the carbohydrates, lactose, fat, protein, and minerals of milk. The term milk **solids-not-fat**, excludes the fat and includes the lactose, caseins, **whey**, proteins, and minerals (calcium, phosphorus, magnesium, potassium, sodium, chloride, and sulphur). Table A-8 provides the composition of milk and milk products.

Milk also contains water-soluble vitamins (thiamin and riboflavin) and mineral salts. It is described as a colloidal dispersion of the protein casein and the whey proteins. Finally, fluid milk is an emulsion with fat globules suspended in the water phase of milk.

Legal Description

The standard of identity under CFR 131.110(a) describes milk as follows:

> *Milk is the lacteal secretion, practically free from colostrum, obtained by the complete milking of one or more healthy cows. Milk that is in the final package form for beverage use shall have been pasteurized or ultrapasteurized, and shall contain not less than 8¼ percent milk solids not fat and no less than 3¼ percent milk fat. Milk may have been adjusted by separating part of the milk fat therefrom, or by adding thereto cream, concentrated milk, dry whole milk, skim milk, concentrated skim milk or nonfat dry milk. Milk may be homogenized.*

Production Practices

Fewer cows produce more milk in the United States, and the dairies are becoming larger. In major production areas, dairies of 1,000 cows or more are common (Figure 16-1).

Milk fresh from the cow is virtually a sterile product. All post-milking handling must maintain the milk's nutritional value and prevent deterioration caused by numerous physical and biological factors. In addition, equipment on the farm must be maintained to government and industry standards. Most cows are milked twice a day, although some farms milk three or four times per day. The milk is immediately cooled from the body temperature of the cow to below 41°F (5°C), then stored at the farm under refrigeration until picked up by insulated tanker trucks at least every other day (Figure 16-2). When the milk is pumped into the tanker, a sample is collected for later lab analysis.

Quality Control

At the dairy farm, inspectors monitor herd health, farm water supplies, sanitation of milking equipment, milk temperature, holding times, and bacteria counts of milk. Violations of health standards result in heavy penalties up to and including suspension from business. Inspections, whether at the farm or at the processing plant, and regulatory inspections occur on an on-going basis. Inspectors have full authority to suspend plant operations in order

FIGURE 16-1
Modern dairies milk 1,000 cows or more in a clean, convenient environment. (*Source:* USDA Photography Library)

FIGURE 16-2
Large milk tanker trucks haul milk from a dairy to a processing plant. (*Source:* USDA Photography Library)

to allow detailed examination of all equipment, facilities, and products. The dairy industry works hard to ensure that they comply with or exceed all regulations.

All finished dairy products are tested regularly by state inspectors to ensure compliance with each of the following: (1) Standards of Identity, which refers to such criteria as moisture, butterfat, and protein content; (2) Purity, which refers to such criteria as pathogens and residues. The Food and Drug Administration (FDA) sets standards of identity for beverage milk products.

Processing

When the milk arrives at the milk plant it is checked to make sure that it meets the standards for temperature, total acidity, flavor, odor, tanker cleanliness, and the absence of antibiotics. The butterfat and solids-not-fat contents of this raw milk are also analyzed. The amounts of butterfat (BF) and solids-not-fat (SNF) in the milk will vary according to time of year, breed of cow, and feed supply. Butterfat content, solids-not-fat content, and volume are used to determine the amount of money paid the producer.

Once the load passes these receiving tests, it is then pumped into large refrigerated storage silos (nearly half-million pounds capacity) at the processing plant.

Pasteurizing

All raw milk must be processed within 72 hours of receipt at the plant. Milk is such a nutritious food that numerous naturally occurring bacteria are always present. The milk is pasteurized, which is a process of heating the raw milk to kill all pathogenic microorganisms that may be present. Pasteurization is not sterilization (sterilization eliminates all viable life-forms, whereas pasteurization does not). After pasteurization, some harmless bacteria may survive the heating process. These bacteria will cause milk to "go sour." Keeping milk refrigerated is the best way to slow the growth of these bacteria.

The batch method of pasteurization heats the milk to at least 145°F and holds it at that temperature for at least 30 minutes. This method is called the Low-Temperature Longer Time (**LTLT**) pasteurization. Because it can cause a "cooked" flavor, this process is not used by some milk plants for fluid milk products.

High Temperature/Short Time (**HTST**) pasteurization heats the milk to at least 161°F for at least 15 seconds. The milk is immediately cooled to below 40°F and packaged into plastic jugs or plastic-coated cartons.

Heating milk to 280°F or higher for 2 seconds, followed by rapid cooling to 45°F or lower is called **ultrapasteurization**.

Sterilization of milk occurs when it is heated to 280° to 302°F for 2 to 6 seconds. This is called ultrahigh temperature (**UHT**) processing. With this treatment the sterilized milk is aseptically packaged, and the milk does not require refrigeration until it is opened.

Butterfat

Butterfat content accounts for several different types of products. Whole milk, 2%, 1%, nonfat, and Half & Half are some examples. A machine called a separator separates the cream and skim portions of the milk. During the separation of whole milk, two streams are produced: the fat-depleted stream, which produces the beverage milks as described; skim milk for evaporation and possibly for subsequent drying, and the fat-rich stream, the cream. This usually comes off the separator with fat contents in the 35 to 45 percent range.

Cream is used for further processing in the dairy industry for the production of ice cream or butter, or can be sold to other food processing industries. These industrial products normally have higher fat contents than creams for retail sale, normally in the range of 45 to 50 percent fat. A product known as "plastic" cream can be produced from certain types of milk separators. This product has a fat content approaching 80 percent fat, but it remains as an oil-in-water emulsion (the fat is still in the form of globules and the skim milk is the continuous phase of the emulsion), unlike butter which also has a fat content of 80 percent but which has been churned so that the fat occupies the continuous phase and the skim milk is dispersed throughout in the form of tiny droplets (a water-in-oil emulsion).

Homogenization

Milk is homogenized to prevent the cream portion from rising to the top of the package. Cream is lighter in weight than milk. The cream portion of unhomogenized milk would form a cream layer at the top of the carton. A "homogenizer" forces the milk under high pressure through a valve that breaks up the butterfat globules to such small sizes they will not **coalesce** (stick together). Homogenization does not affect the nutrition or quality of the product.

Beverage Milk

While the fat content of most raw milk is 4 percent or higher, the fat content in most beverage milks has been reduced to 3.4 percent. Lower-fat alternatives, such as 2 percent fat, 1 percent fat, or skim milk (less than 0.1 percent fat), are also available in most markets. These products are either produced by partially skimming the whole milk, or by completely skimming it and then adding an appropriate amount of cream back to achieve the desired final fat content.

Nutritional Qualities

Vitamins may be added to both full fat and reduced fat milks. Vitamins A and D (the fat-soluble ones) are often added to milk. Vitamin A is lost during fat separation and heating, and vitamin D is not present in milk. These are supplemented in the form of a water-soluble emulsion. Many states have milk standards that require the addition of milk solids. These solids represent the natural mineral (calcium and iron), protein (casein), and sugar (lactose) portion of nonfat dry milk.

Table A-8 provides the details on the nutritional qualities of beverage milk.

Quality Control

Quality control personnel conduct numerous tests on the raw and pasteurized products to ensure optimum quality and nutrition. A sample is analyzed for the presence of microbiological organisms with a **standard plate count (SPC)** and **ropey** milk test. The equipment used to analyze butterfat and solids-not-fat is calibrated on a regular basis to ensure a consistent, quality product that meets or exceeds government requirements.

All milk products have a sell-by date printed on the package. This is the last day the item should be offered for sale. However, most companies guarantee the quality and freshness of the product for at least 7 days past the date printed on the package. Samples of each product packaged each day are saved to confirm that they maintain their freshness 7 days after the sell-by date.

Packaging

Once the milk has been separated, **standardized**, homogenized, and pasteurized, it is held below 40°F in insulated storage tanks, then packaged into gallon, half-gallon, quart, pint, and half-pint containers. The packaging machines are maintained under strict sanitation specifications to prevent bacteria from being introduced into the pasteurized product. All equipment that comes into contact with product (raw or pasteurized) is washed daily. Sophisticated automatic Clean-in-Place (CIP) systems guarantee consistent sanitation with a minimum of manual handling, reducing the risk of contamination.

Once packaged, the products are quickly conveyed to a cold storage warehouse. They are stored there for a short time and shipped to the supermarket on refrigerated trailers. Once at the store, the milk is immediately placed into a cold storage room or refrigerated display case.

MILK PRODUCTS AND BY-PRODUCTS

Milk products and by-products include butter, concentrated and dried products, cheese, whey products, yogurt and other fermented products, and ice cream.

Butter

Butter is made by **churning** pasteurized cream. Churning breaks the fat globule membrane so the emulsion breaks, fat coalesces, and water (**buttermilk**) escapes. Federal law requires that it contain at least 80 percent milkfat. Salt and coloring may be added. Nutritionally, butter is a fat (see Table A-8). Whipped butter is regular butter whipped for easier spreading. Whipping increases the amount of air in butter and increases the volume of butter per pound.

Today's commercial butter making is a product of the knowledge and experience gained over the years in such matters as hygiene, bacterial acidifying, and heat treatment, as well as the rapid technical development that has led to the advanced machinery now used.

The principal constituents of a normal salted butter are fat (80 to 82 percent), water (15.6 to 17.6 percent), salt (about 1.2 percent) as well as protein, calcium, and phosphorous (about 1.2 percent). Butter also contains fat-soluble vitamins A, D, and E.

Butter should have a uniform color, be dense, and taste clean. The water content should be dispersed in fine droplets so that the butter looks dry. The consistency should be smooth so that the butter is easy to spread and melts readily on the tongue.

The butter-making process involves quite a number of stages. The continuous butter maker has become the most common type of equipment used.

From the storage tanks, the cream goes to pasteurization. This destroys enzymes and microorganisms that would impair the keeping quality of the butter.

In the **ripening** tank, the cream is subjected to a program of heat treatment designed to give the fat the required crystalline structure when it solidifies on cooling. The program is chosen to accord with factors such as the composition of the butterfat, expressed, for example, in terms of the iodine value, which is a measure of the unsaturated fat content. The treatment can even be modified to obtain butter with good consistency despite a low iodine value, that is, when the unsaturated proportion of the fat is low.

As a rule, ripening takes 12 to 15 hours. From the ripening tank, the cream is pumped to the churn or continuous butter maker via a plate heat exchanger that brings it to the requisite temperature. In the churning process the cream is violently agitated to break down the fat globules, causing the fat to coagulate into butter grains, while the fat content of the remaining liquid, the buttermilk, decreases.

Thus, the cream is split into two fractions: butter grains and buttermilk. In traditional churning, the machine stops when the grains have reached a certain size, whereupon the buttermilk is drained off. With the continuous butter maker the draining of the buttermilk is also continuous.

After draining, the butter is worked to a continuous fat phase containing a finely dispersed water phase. It used to be common practice to wash the butter after churning to remove any residual buttermilk and milk solids, but this is rarely done today. If the butter is to be salted, salt is spread over its surface, in the case of batch production. In the continuous butter maker, a salt slurry is added to the butter.

After salting, the butter must be worked vigorously to ensure even distribution of the salt. The working of the butter also influences the characteristics by which the product is judged–aroma, taste, keeping quality, appearance, and color.

The finished butter is discharged into the packaging unit, and from there to cold storage.

Continuous Butter Making. Methods of continuous butter making were introduced at the end of the nineteenth century but their

ORGANIC DAIRY: PERCEPTION AND MARKETING

Organic dairying represents the organic food industry's largest and fastest growing segment, showing a 50 percent increase in sales in recent years. Organic dairy sales are predicted to reach $2 billion by 2005. This is rapid growth compared to the general food industry, which typically demonstrates a 3 percent annual growth.

Figures indicate that 30 percent of U.S. homes are purchasing some type of organic product. Whereas 58 percent of organic products are sold through natural food stores, more grocery stores are beginning to offer organic products. Organic foods seemingly are becoming a mainstream idea, not a niche anymore. This says something about a changing U.S. consumer base.

With products from organic dairies pushing beyond specialty stores and moving into supermarkets, large organic dairy operations attribute their success and growth to seven major trends:

1. Mainstreaming of organic consumers, products, and retailers
2. Increased knowledge of the relationship between diet and health
3. Environmental awareness
4. Search for alternatives to genetically modified organisms (GMOs)
5. Declining organic food-production costs
6. Development and distribution of organic standards
7. Capital investments in "organic" from the financial community

Organic dairy products—milk, blended yogurts, and cheese—require that cows in the dairy not be given antibiotics or hormones for a minimum of one year prior to milking. Organic dairies often forbid pesticide or GMO use.

Some organic dairies also market eggs and calcium-enriched orange and grapefruit juices. These companies market their products by emphasizing "organic" and by pushing brand identity.

For more information about organic, visit these Web sites:

<www.horizonorganic.com/>
<www.ams.usda.gov/nop/>

application was very restricted. In the 1940s the work was resumed and resulted in three different processes, all based on the traditional methods–churning, centrifuging, and concentration or emulsifying. One of the processes based on conventional churning was the Fritz method, which is the one now used predominantly in Western Europe. In machines based on this method, butter is made in generally the same way as by traditional methods. The butter is basically the same, except that it is somewhat more dense as a result of uniform and fine water dispersion.

The Manufacturing Process. The cream is prepared in the same way as for conventional churning before being fed continuously from the ripening tanks to the butter maker.

The cream is first fed ito a churning cylinder fitted with beaters that are driven by a variable speed motor.

Rapid conversion takes place in the cylinder and, when finished, the butter grains and buttermilk pass on to a draining section. The first washing of the butter grains sometimes takes place en route–either with water or recirculated chilled buttermilk. The working of the butter commences in the draining section by means of a screw that also conveys it to the next stage. On leaving the working section, the butter passes through a conical channel to remove any remaining buttermilk. Immediately afterward, the butter may be given its second washing, this time by two rows of adjustable high-pressure nozzles.

The water pressure is so high that the ribbon of butter is broken down into grains, and consequently any residual milk solids are effectively removed. Following this stage, salt may be added through a high-pressure injector.

The third section in the working cylinder is connected to a vacuum pump. Here it is possible to reduce the air content of the butter to the same level as conventionally churned butter.

In the final or mixing section the butter passes a series of perforated disks and star wheels. There is also an injector for final adjustment of the water content. Once regulated, the water content of the butter deviates less than ±0.1 percent, provided the characteristics of the cream remain the same.

The finished butter is discharged in a continuous ribbon from the end nozzle of the machine and then into the packaging unit.

Concentrated and Dried Dairy Products

Fluid milk contains approximately 88 percent water. Concentrated milk products are obtained through partial water removal. Dried dairy products have even greater amounts of water removed to usually less than 4 percent. The benefits of both these processes include an increased shelf life, convenience, product flexibility, decreased transportation costs, and storage. Concentrated dairy products include:

- Evaporated skim or whole milk
- Sweetened condensed milk
- Condensed buttermilk
- Condensed whey

Dried dairy products include:

- Milk powder
- Whey powder
- Whey protein concentrates

The principles of evaporation and dehydration can be found in Chapter 11.

Evaporated Skim or Whole Milk. After the raw milk is clarified and standardized, it is given a preheating treatment. Milk is then concentrated at low temperatures by **vacuum evaporation**. The vacuum lowers the boiling point to approximately 104° to 113°F (40° to 45°C). This results in little to no cooked flavor.

The milk is concentrated to 30 to 40 percent total solids. A second standardization is done at this time to ensure that the proper salt balance is present. The ability of milk to withstand intensive heat treatment depends to a great degree on its salt balance.

The product at this point is quite perishable. The fat is easily oxidized. The evaporated milk at this stage is often shipped by the tanker for use in other products.

In order to extend the shelf life, evaporated milk can be packaged in cans and then sterilized in an autoclave. Continuous flow sterilization is followed by packaging under aseptic conditions. Though the sterilization process produces a light brown coloration, the product can be successfully stored for up to a year.

Sweetened Condensed Milk. Where evaporated milk uses sterilization to extend its shelf life, sweetened condensed milk has an extended shelf life due to the addition of sugar. Sucrose, in the form of crystals or solution, increases the osmotic pressure of the liquid. This prevents the growth of microorganisms. The only real heat treatment (185° to 194°F [85° to 90°C] for several seconds) this product receives is after the raw milk has been clarified and standardized. This treatment destroys osmiophilic and thermophilic microorganisms, inactivates lipases and proteases, decreases fat separation, and inhibits oxidative changes.

The milk is evaporated in a manner similar to the evaporated milk. Although sugar may be added before evaporation, addition after evaporation is recommended to avoid undesirable viscosity changes during storage. Enough sugar is added so that the final concentration of sugar is approximately 45 percent.

The sweetened evaporated milk is then cooled and lactose crystallization is induced. The milk is inoculated, or seeded, with powdered lactose crystals, then rapidly cooled while being agitated. The product is packaged in smaller containers, such as cans, for retail sales and bulk containers for industrial sales.

Condensed Buttermilk. Buttermilk is a by-product of the butter industry. It can be evaporated on its own, or it can be blended with skim milk and dried to produce skim milk powder. This blended product may oxidize readily due to the higher fat content. Condensed buttermilk is perishable and must be stored cool.

Condensed Whey. The process of cheese making creates a lot of whey that needs disposal. One of the ways of using cheese whey is to condense it by evaporation. The whey contains fat, lactose, lactoglobulin, lactalbumin, and water. The fat is generally removed by centrifugation and churned as whey cream or used in ice cream. Evaporation is the first step in producing whey powder.

Milk Powder. Milk used in the production of milk powder is first clarified, standardized, and then given a heat treatment. This heat treatment is usually more severe than that required for pasteurization. Besides destroying all the pathogenic and most of the spoilage microorganisms, it also inactivates the enzyme lipase that could cause **lipolysis** during storage.

The milk is then evaporated prior to drying. Homogenization may be applied to decrease the free fat content. Spray drying is the most used method for producing milk powders. After drying, the powder must be packaged in containers able to provide protection from moisture, air, and light. Whole milk powder can then be stored for long periods (up to about 6 months) of time at ambient temperatures. Skim milk powder (SMP) processing is similar.

Instant milk powder is produced by partially rehydrating the dried milk powder particles causing them to become sticky and agglomerate. The water is then removed by drying resulting in an increased amount of air incorporated between the powder particles.

Whey Powder. Converting whey into powder has led to a number of products into which it can be incorporated (Figure 16-3).

FIGURE 16-3
Many food products contain whey or whey products.

Whey powder is essentially produced by the same method as other milk powders. Reverse osmosis can be used to partially concentrate the whey prior to vacuum evaporation. Before the whey concentrate is spray dried, lactose crystallization is induced to decrease the hygroscopicity. This is accomplished by quick cooling in flash coolers after evaporation. Crystallization continues in agitated tanks for 4 to 24 hours.

A fluidized bed may be used to produce large agglomerated particles with free-flowing, nonhygroscopic, no caking characteristics.

Whey Protein Concentrates. Both whey disposal problems and high-quality animal protein shortages have increased worldwide interest in whey protein concentrates. After clarification and pasteurization, the whey is cooled and held to stabilize the calcium phosphate complex. The whey is commonly processed using ultrafiltration, although reverse osmosis, microfiltration, and demineralization methods can be used. During ultrafiltration, the low-molecular-weight compounds such as lactose, minerals, vitamins, and nonprotein nitrogen are removed in the permeate while the proteins become concentrated in the retentate. After ultrafiltration, the retentate is pasteurized, may be evaporated, then dried. Drying, usually spray drying, is done at lower temperatures than for milk in order that large amounts of protein denaturation may be avoided.

Cheese

Traditionally, cheese was made as a way of preserving the nutrients of milk. In a simple definition, cheese is the fresh or ripened product obtained after coagulation and whey separation of milk, cream, or partly skimmed milk, buttermilk, or a mixture of these products. It is essentially the product of selective concentration of milk (Figure 16-4). Thousands of varieties of cheeses have evolved that are characteristic of various regions of the world.

Some common cheese-making steps include:

1. Treatment of milk
2. Additives
3. Inoculation and milk ripening
4. Coagulation
5. Enzyme
6. Acid
7. Heat-acid
8. Curd treatment
9. Cheese ripening

Like most dairy products, cheese milk must first be clarified, separated, and standardized. An initial **thermization** treatment results in a reduction of high initial bacteria counts before storage. It must be followed by proper pasteurization. HTST pasteurization is often used.

FIGURE 16-4

Milk vats where cheese is produced. (*Courtesy* Glanbia Foods, Inc., Twin Falls, ID)

Homogenization is not usually done for most cheese milk. It disrupts the fat globules and increases the fat surface area where casein particles adsorb. This results in a soft, weak **curd** at **renneting** and increased hydrolytic rancidity.

Calcium chloride is added to replace calcium lost during pasteurization. The calcium assists in coagulation and reduces the amount of rennet required.

Because milk color varies from season to season, color may added to standardize the color of the cheese throughout the year. Annato, Beta-carotene, and paprika are used.

The addition of hydrogen peroxide is sometimes used as an alternative treatment for full pasteurization.

Lipases, normally present in raw milk, are inactivated during pasteurization. The addition of lipases are common to ensure proper flavor development through fat hydrolysis.

The basis of cheese making relies on the fermentation of lactose by lactic acid bacteria (LAB). LAB produce lactic acid, which lowers the pH and in turn assists coagulation, promotes syneresis, helps prevent spoilage and pathogenic bacteria from growing, contributes to cheese texture, flavor, and keeping quality. LAB also produce growth factors, which encourages the growth of nonstarter organisms, and provides lipases and proteases necessary for flavor development during curing.

After inoculation with the starter culture, the milk is held for 45 to 60 minutes at 77° to 86°F (25° to 30°C) to ensure that the bacteria are active, growing, and have developed acidity. This stage is called ripening the milk and is done prior to renneting.

Coagulation is essentially the formation of a gel by destabilizing the casein micelles causing them to aggregate and form a network that partially immobilizes the water and traps the fat globules in the newly formed matrix. This may be accomplished with enzymes, acid treatment, or heat-acid treatment. Chymosin, or rennet, is most often used for enzyme coagulation. Lowering the pH of the milk results in aggregation. Acid curd is more fragile than rennet curd due to the loss of calcium. Acid coagulation can be achieved naturally with the starter culture, or artificially with the addition of gluconodeltalactone. Acid coagulated fresh cheeses may include cottage cheese, quark, and cream cheese.

Heat causes denaturation of the whey proteins (Figure 16-5). The denatured proteins then interact with the caseins. With the addition of acid, the caseins precipitate with the whey proteins. In rennet coagulation, only 76 to 78 percent of the protein is recovered, whereas in heat-acid coagulation, 90 percent of protein can be recovered. Examples of cheeses made by this method include Paneer, Ricotta, and Queso Blanco.

FIGURE 16-5
Liquid (whey) being separated from cheese curds. (*Courtesy* Glanbia Foods, Inc., Twin Falls, ID)

After the milk gel has been allowed to reach the desired firmness, it is carefully cut into small pieces with knife blades or wires. Cutting shortens the distance and increases the available area for whey to be released. The curd pieces immediately begin to shrink and expel the greenish liquid called whey (syneresis). This syneresis process is further driven by a cooking stage. The increase in temperature causes the protein matrix to shrink due to increased hydrophobic interactions, and also increases the rate of fermentation of lactose to lactic acid. The increased acidity also contributes to shrinkage of the curd particles. The final moisture content is dependent on the time and temperature of the cook stage.

When the curds have reached the desired moisture and acidity, they are separated from the whey. The whey may be removed from the top or drained by gravity. The curd-whey mixture may also be placed in molds for draining. Some cheese varieties, such as Colby, Gouda, and Brine Brick include a curd washing that increases the moisture content, reduces the lactose content and final acidity, decreases firmness, and increases openness of texture.

Curd handling from this point on is very specific for each cheese variety. Salting may be achieved through brine as with Gouda, surface salt as with Feta, or vat salt as with Cheddar. To achieve the characteristics of Cheddar, cheddaring (curd manipulation), milling (cutting into shreds), and pressing at high pressure are crucial.

Except for fresh cheese, the curd is ripened, or matured, at various temperatures and times until the characteristic flavor, body, and texture profile is achieved. During ripening, degradation of lactose, proteins, and fat is carried out by ripening agents. The ripening agents in cheese include: bacteria and enzymes of the milk, lactic culture, rennet, lipases, added molds or yeasts, and environmental contaminants.

The microbiological content, the biochemical composition of the curd, as well as temperature and humidity affect the final product. This final stage–aging–varies from weeks to years according to the cheese variety.

Yogurt

Yogurt (also spelled yoghurt) is a semisolid fermented milk product that originated centuries ago in Bulgaria. Yogurt flavor and aroma varies from one region to another but, the basic ingredients and manufacturing are essentially consistent. Although milk of various animals has been used for yogurt production in various parts of the world, most of the industrialized yogurt production uses cow's milk. Whole milk, partially skimmed milk, skim milk, or cream may be used.

The starter culture for most yogurt production in North America is a blend of *Streptococcus salivarius thermophilus (ST)* and *Lactobacillus delbrueckii bulgaricus (LB).* Although they can grow independently, the rate of acid production is much higher when used together than either of the two organisms grown individually. *ST* grows faster and produces both acid and carbon dioxide. These microorganisms are ultimately responsible for the formation of typical yogurt flavor and texture. The yogurt mixture coagulates during fermentation due to the drop in pH. The streptococci are responsible for the initial pH drop of the yogurt mix to approximately 5.0. The lactobacilli are responsible for a further decrease to pH 4.0. The fermentation products of lactic acid, acetaldehyde, acetic acid, and diacetyl contribute to flavor.

The milk is clarified and separated into cream and skim milk, then standardized to achieve the desired fat content. The various ingredients are then blended together in a mix tank equipped with a powder funnel and an agitation system. The mixture is then pasteurized. Once the homogenized mix has cooled to an optimum growth temperature, the yogurt starter-culture is added.

Other Fermented Milk Beverages. Cultured buttermilk was originally the fermented by-product of butter manufacture, but today it is more common to produce cultured buttermilks from skim or whole milk. Milk is usually heated to 203°F (95°C) and cooled to 68° to 77°F (20° to 25°C) before the addition of the starter culture. Starter is added, and the fermentation is allowed to proceed for 16 to 20 hours, to an acidity of 0.9 percent lactic acid. This product is frequently used as an ingredient in the baking industry, in addition to being packaged for sale in the retail trade.

Acidophilus Milk. Acidophilus milk is a traditional milk fermented with *Lactobacillus acidophilus (LA)*, which has been thought to have therapeutic benefits in the gastrointestinal tract. Skim or whole milk may be used. The milk is heated to a high temperature, such as 203°F (95°C) for 1 hour, to reduce the microbial load and favor the slow growing *LA* culture. Milk is inoculated with *LA* and incubated at 99°F (37°C) until coagulated. Some acidophilus milk has an acidity as high as 1 percent lactic acid, but for therapeutic purposes 0.6 to 0.7 percent is more common. Another variation has been the introduction of a sweet acidophilus milk, one in which the *LA* culture has been added but there has been no incubation.

Sour Cream. Cultured cream usually has a fat content between 12 and 30 percent, depending on the required properties. The starter is similar to that used for cultured buttermilk. The cream after standardization is usually heated to 167° to 176°F (75° to 80°C) and is homogenized to improve the texture. Inoculation and fermentation conditions are also similar to those for cultured buttermilk, but the fermentation is stopped at an acidity of 0.6 percent.

Other Fermented Products. Many other fermented dairy products include kefir, koumiss, beverages based on bulgaricus or bifidus strains, labneh, and a host of others. Many of these have developed in regional areas and, depending on the starter organisms used, have various flavors, textures, and components from the fermentation process.

Ice Cream

Ice cream is greater than 10 percent milkfat by legal definition, and as high as 16 percent fat in some premium ice creams. It contains 9 to 12 percent milk solids-not-fat. This component, also known as the serum solids, contains the proteins (caseins and whey proteins) and carbohydrates (lactose) found in milk. Ice cream also contains 12 to 16 percent sweeteners, which are usually a combination of sucrose and glucose-based corn syrup sweeteners, and

0.2 to 0.5 percent stabilizers and emulsifiers. Finally, 55 to 64 percent of ice cream is water that comes from the milk or other ingredients (Figure 16-6). These percentages are by weight, either in the mix or in the frozen ice cream. However, when frozen, about one half of the volume of ice cream is air. All ice cream is made from a basic white mix.

The basic steps in the manufacturing of ice cream generally include the following:

1. Blending of the mix ingredients
2. Pasteurization
3. Homogenization
4. Aging the mix
5. Freezing
6. Packaging
7. Hardening

FIGURE 16-6

Ice cream contains more than 10 percent milkfat and sweeteners in the form of sugars and corn syrup.

Butterfat in ice cream increases the richness of flavor, produces a characteristic smooth texture by lubricating the palate, helps to give body to the ice cream, aids in good melting properties, and aids in lubricating the freezer barrel during manufacturing. Nonfat mixes are extremely hard on the freezing equipment.

Milk solids-not-fat (MSNF) contain the lactose, caseins, whey proteins, minerals, and ash content of the product from which they were derived. They are an important ingredient for the following reasons: they improve the texture of ice cream; they help to give body and chew resistance to the finished product; they allow higher incorporation of air; they provide an inexpensive source of total solids.

Consumers like a sweet ice cream. As a result, sweetening agents are added to ice cream mix at a rate of usually 12 to 16 percent by weight. Sweeteners improve the texture and palatability of the ice cream, enhance flavors, and are usually the cheapest source of total solids. In addition, the sugars, including the lactose from the milk components, contribute to a depressed freezing point so that the ice cream has some unfrozen water associated with it at very low temperatures typical of their serving temperatures, 5° to 0°F (−15° to −18°C). Without this unfrozen water, the ice cream would be too hard to scoop.

The stabilizers in ice cream are a group of compounds, usually polysaccharides, that are responsible for adding viscosity to the unfrozen portion of the water and thus holding this water so that it cannot migrate within the product. This produces an ice cream that is firmer. Without the stabilizers, the ice cream would become coarse and icy very quickly due to the migration of this free water and the growth of existing ice crystals.

The emulsifiers are a group of compounds in ice cream that aid in developing the appropriate fat structure and air distribution necessary for the smooth eating and good meltdown characteristics desired in ice cream. Because each molecule of an emulsifier contains a hydrophilic portion and a lipophilic portion, they reside at the interface between fat and water. They act to reduce the interfacial tension or the force that exists between the two phases of the emulsion. The emulsifiers actually promote a destabilization of the fat emulsion, which leads to a smooth, dry product with good meltdown properties.

The original ice cream emulsifier was egg yolk, which was used in most of the original recipes. Today, two emulsifiers predominate most ice cream formulations: mono- and diglycerides, which are derived from the partial hydrolysis of fats or oils of animal or vegetable origin.

QUALITY PRODUCTS

The USDA establishes U.S. grade standards to describe different grades of quality in butter, cheese (Cheddar, Colby, Monterey, and Swiss), and instant nonfat dry milk. The FDA established the Grade A designation for fluid milk products, yogurt, and cottage cheese. Manufacturers use the grade standards to:

- Identify levels of quality
- Provide a basis for establishing prices at wholesale
- Supply consumers with a choice of quality levels

The USDA also provides inspection and grading services that manufacturers, wholesalers, or other distributors may request. A fee is charged to cover the cost of the service. Only products that are officially graded may carry the USDA grade shield.

The U.S. Grade AA or Grade A shield (Figure 16-7) is most commonly found on butter and sometimes on Cheddar cheese. U.S. Extra Grade is the grade name for instant nonfat dry milk of high quality. Processors who use the USDA's grading and inspection service may use the official grade name or shield on the package. The "Quality Approved" shield may be used on other dairy products (like cottage cheese) or other cheeses for which no official U.S. grade standards exist if the products have been inspected for quality under USDA's grading and inspection program.

Grade Shields
Marks of Quality

FIGURE 16-7

USDA grade shields commonly found on butter and cheese.

Milk available in stores today is usually pasteurized and homogenized. Very little raw milk is sold today. Federal, state, and local laws or regulations control the composition, processing, and handling of milk. Federal laws apply when packaged or bottled milk is shipped interstate. Raw milk is prohibited from being sold interstate.

The Pasteurized Milk Ordinance of the Food and Drug Administration (FDA) requires that all packaged or bottled milk shipped interstate be pasteurized to protect consumers. Milk can be labeled "Grade A" if it meets FDA or state standards under the Pasteurized Milk Ordinance.

The Grade A rating designates wholesomeness or safety rather than a level of quality. According to the standards recommended in the ordinance, Grade A pasteurized milk must come from healthy cows and be produced, pasteurized, and handled under strict sanitary controls that are enforced by state and local milk sanitation officials.

Once the consumer purchases milk and milk products, proper storage conditions and time maintain quality and safety. Guidelines of storage times for maintaining the quality of some products in the home refrigerator include the following:

- Fresh milk–5 days
- Buttermilk–10 to 30 days
- Condensed or evaporated milk–opened 4 to 5 days
- Half & half, light cream, and heavy cream–10 days
- Sour cream–2 to 4 weeks

MILK SUBSTITUTES

One well-known substitute for a milk product is margarine. It is made from vegetable fat. Other substitutes include frozen desserts, coffee whiteners, whipped toppings, and imitation milk (soy milk). These substitutes are made by combining nondairy fats or oils with certain classes of milk components.

REDUCED FAT PRODUCTS

In an effort to reduce calories, saturated fat, and cholesterol in the diet, a number of low-fat, reduced-fat, and no-fat products have appeared on the market. Where fat is replaced in a milk product, the replacement must perform the same function as the original fat. In other words, fat replacements must give the product the same texture (mouthfeel). Fat substitutes have been made of proteins processed into very small particles and of carbohydrates that bind large amounts of water. New fat substitutes being developed and marketed are fats that are nonabsorbable or digestible. These do not contribute to the calories or fat intake of the individual. One such product is Olestra.

Summary

Milk provides high-quality protein, energy, vitamins, and minerals to human nutrition. Besides fluid milk the dairy industry produces a variety of milk products including butter, cheeses, condensed and dried products, and cultured products. The USDA and the FDA maintain quality standards. To protect the consumer against pathogenic microorganisms in milk, it is pasteurized. Butterfat globules in homogenized milk are reduced in size to prevent coalescence. Butterfat is also separated from milk and added back to produce beverage milk with specific fat content or to be used in the production of butter and creams. Butter is produced by churning butterfat, and a by-product is buttermilk.

Concentrated or dried dairy products such as evaporated milk, sweetened condensed milk, condensed whey, milk powder, or whey powder increase shelf life and convenience and decrease transportation costs. Traditionally, cheese developed as a way to preserve the nutrients of milk. Today, many varieties of cheese have evolved. Production of cheese basically involves the coagulation of the milk and separation of the whey. Coagulation can be accomplished with enzymes, acid, or heat. Yogurt is a fermented dairy product, as are acidophilus milk, sour cream, and kefir. Fermented dairy products require a starter culture for fermentation. By legal definition, ice cream contains 10 percent or more butterfat. It also relies on sweeteners at a level of 12 to 16 percent.

The USDA established grade standards for butter, cheese, and instant nonfat dry milk. The FDA established grade designations for fluid milk, yogurt, and cottage cheese. Only officially graded products may carry the grade shield.

In an effort to meet new consumer demands, the food industry developed milk and milk product substitutes such as coffee whiteners, whipped toppings, imitation milk, and reduced fat products.

Review Questions

Success in any career requires knowledge. Test your knowledge of this chapter by answering these questions or solving these problems.

1. ___________ heats the milk to at least 161°F for at least 15 seconds. ___________ eliminates all viable life-forms.
2. Name four beverage milk products.
3. ___________ is made by churning pasteurized cream.
4. Define “milk.”
5. List the steps in cheese making.
6. Why is milk homogenized?
7. Name the three reasons manufacturers use the grade standards.
8. List the six areas of inspection that occur at the dairy farm for quality control of milk production.
9. After water, what are the components of milk solids?
10. Name three dried milk products.
11. LAB or lactic acid bacteria aid in making ___________.
12. List the steps in making ice cream.

Student Activities

1. Track all the milk and milk products you consume in a week. Using Table A-8 develop a report or presentation on the amount of nutrition they provided.
2. Develop a report or presentation to compare butter to margarine. Use Table A-8 as a resource.
3. How is advertising used to encourage consumers to drink more milk? Find examples of current milk or milk product advertising campaigns and post these in the classroom.
4. Make butter, cheese, yogurt, or ice cream as an in-class project. These products are easy to make, and instructions can be found in many resources.
5. Develop a taste test to identify various types of cheese.
6. Taste and describe various fermented dairy products.
7. Develop a report or presentation on the competition that dairy products face from nondairy substitutes.
8. Compare the taste of a dairy product to that of one of the substitutes such as margarine, dairy creamer, or soy milk.
9. Develop a report or presentation on the uses of whey.

Resources

Bartlett, J. 1996. *The cook's dictionary and culinary reference.* Chicago: Contemporary Books.

Ensminger, A. H., M. E. Ensminger, J. E. Konlande, and J. R. Robson. 1994. *Foods and nutrition encyclopedia.* 2 Vols. Boca Raton, FL: CRC Press.

Horn, J., J. Fletcher, and A. Gooch. 1997. *Cooking a to z. The complete culinary reference source.* Glen Ellen, CA: Cole Publishing Group, Inc.

Medved, E. 1973. *The world of food.* Lexington, MA: Ginn and Company.

National Council for Agricultural Education. 1993. *Food science, safety, and nutrition.* Madison, WI: National FFA Foundation.

Vaclavik, V. A., and E. W. Christina. 1999. *Essentials of food science.* Gaithersburg, MD: Aspen Publishers, Inc.

Internet

Internet sites represent a vast resource of information. The URLs (uniform resource locator) for the World Wide Web sites can change. Using one of the search engines on the Internet such as Yahoo!, HotBot, AltaVista, Excite, Dogpile, About, or Google, find more information by searching for these words or phrases: milk, buttermilk, ice cream, names of specific cheese types or brands, milk industry, milk production, specific brand names of milk and milk products, pasteurization, quality control of milk, whey protein concentrates, homogenization, cheese making, milk grading, milk substitutes. Also, Table A-7 provides a listing of some useful Internet sites that can be used as a starting point.

Chapter 17

Meat, Poultry, and Eggs

Objectives

After reading this chapter, you should be able to:

- Describe the production of meat from cattle, pigs, and poultry
- Identify meat products from cattle, pigs, and poultry
- Discuss the general composition of meat and meat products
- List five factors affecting meat tenderness
- Describe the cooking of meat
- Discuss the production of meat substitutes
- Identify quality grading of meat
- Describe egg production
- Identify factors affecting egg quality
- Discuss egg grading

Key Terms

aging
albumen
antemortem
blood spot
bromelin
by-products
cold shortening
curing
deboning
electrical stimulation
eviscerated
ficin
integrated
Julian date
marinading
mechanically separated
myoglobin
offal
papain
postmortem
processed meats
rigor mortis
smoking
textured protein
vitelline membrane
yield grade

The first meat packers in the United States were the colonial New England farmers, who packed meats in salt as a means of preservation. The beef industry moved from the large metropolitan areas to be near commercial feedlots in the central United States in such states as Texas, Oklahoma, Kansas, and Nebraska. The pork industry remains centrally located in the Midwest, principally in Iowa, Illinois, Minnesota, Michigan, and Nebraska, but it is making a move to the west and southeast. Rapid growth and vertical integration characterize the poultry industry.

MEAT AND MEAT PRODUCTS

Livestock slaughter occurs in federally inspected plants that do most all of the processing. A few large packers dominate the industry. Only about 62 percent of beef is consumed as beef cuts; 24 percent is ground for hamburger, and 14 percent is processed into meat products. In the case of pork, more than 65 percent of the total is consumed in the form of processed meat such as ham, bacon, and sausage. In addition, the meatpacking industry produces such valuable **by-products** as cosmetics, glues, gelatins, tallow, variety meats, and meat and bonemeal.

Livestock marketing and prices are affected by weather, feed prices, federal import policies, and consumer demands.

Traditionally, meatpackers sold carcasses as sides, quarters (hinds or fores), or as wholesale cuts (large cuts such as entire rounds, loins, ribs, or chucks). Most beef today is sold as "boxed beef." Boxed beef is prepared at the packing plant by removing more of the bone and fat from the carcass as it is cut into smaller portions, vacuum-packed to reduce spoilage and shrinkage, and placed into boxes that are easier to ship and handle than quarters (Figure 17-1). Boxed beef reduces shipping costs, labor costs, and the increased value of the fat and bone to the packer. Some large packers prepare consumer-ready meat cuts such as precut steaks and roasts in vacuum packages to be placed directly in the supermarket meat case.

Government Surveillance

Inspection takes place at practically every step of the livestock procurement and meatpacking processes. Inspection attempts to ensure that harmful additives and ingredients are kept out of manufactured meat products, that sick and diseased animals are

FIGURE 17-1
Much of the beef today is sold as boxed beef.

excluded from the market, that misleading labeling and packaging are eliminated, and that contaminated and unwholesome meats are prevented from reaching consumers.

Federal meat inspection was authorized by the Meat Inspection Act (1906) and is administered by the Food Safety and Inspection Service (FSIS) of the Department of Agriculture (USDA). Meat that is to be used entirely within a given state may be inspected only by that state's department of agriculture. All meats entering interstate commerce must be federally inspected.

Grading

Unlike inspection, which is mandatory, meat grading is a service offered to packers on a voluntary basis by the Agricultural and Marketing Service (AMS) of the USDA. Grading is operated on a self-supporting basis and is funded from fees paid by the users. Grading establishes and maintains uniform trading standards and aids in the determination of the value of various cuts of meat.

Carcasses are given both a quality and a **yield grade** (Figure 17-2 and Figure 17-3). Quality grades for beef carcasses are prime,

Relationship Between Marbling, Maturity, and Carcass Quality Grade*

Maturity**

Degrees of Marbling	A***	B	C	D	E
Slightly Abundant	Prime				
Moderate			Commercial		
Modest	Choice				
Small					
Slight	Select			Utility	
Traces					Cutter
Practically Devoid	Standard				

* Assumes that firmness of lean is comparably developed with the degree of marbling and that the carcass is not a "dark cutter".
** Maturity increases from left to right (A through E).
*** The maturity portion of the Figure is the only portion applicable to bullock carcasses.

FIGURE 17-2

Graphical representation of the USDA quality grades for beef. (*Source:* USDA)

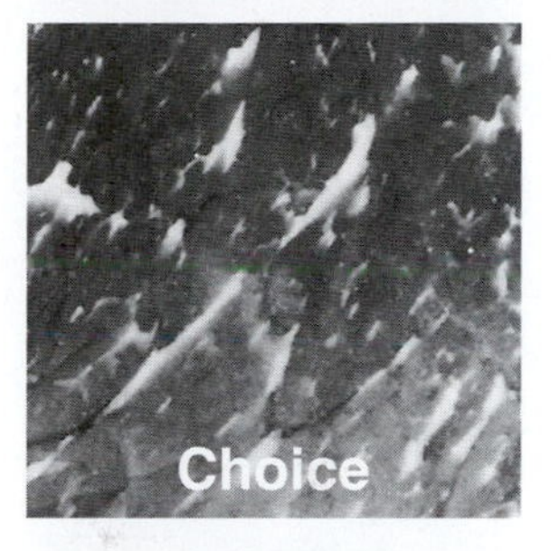

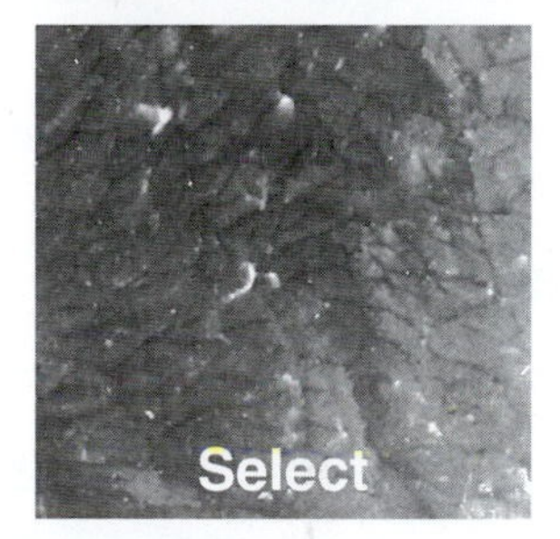

FIGURE 17-3

Degree of marbling is the primary difference between the quality grades: prime, choice, and select. (*Source:* USDA)

choice, good, standard, commercial, utility, cutter, and canner and are assigned in terms of carcass characteristics associated with palatability. These grades are assigned on the basis of carcass marbling (fat flecks or streaks within the lean), color and texture of the lean, and maturity, which is determined according to the color, size, and texture of the cartilage bones. Although grading was not originally intended to provide estimates of palatability (taste and tenderness) for consumers, it has become a consumer rating for beef.

There are five Yield Grades applicable to all classes of beef, denoted by 1 through 5, with a Yield Grade 1 representing the highest degree of cutability.

Carcasses below the choice grade have rarely been stamped with a grade because they were thought to be less palatable. The belief held by some consumers that leaner meats (those with a lower fat content) are more healthful has led to an increased demand for the select grade. Yield grades classify carcasses on the basis of the proportion of usable meat to bone and fat, and are used in conjunction with quality grades to determine the monetary value of a carcass.

For pork, there are different considerations for barrows and gilts carcasses versus sow. Grades for barrow and gilt carcasses are based on two general considerations:

1. Quality–which includes characteristics of the lean and fat
2. The expected yield of the four lean cuts (ham, loin, picnic shoulder, and Boston butt)

The standards for grades of sow carcasses are based on:

1. Differences in yields of lean cuts and of fat cuts
2. Differences in quality of cuts

Lamb falls into four classifications: Prime, Choice, Good, and Utility. Carcasses are evaluated by muscle tone, fat streaking, size of flat rib bone, and firmness.

The USDA/AMS Web site for quality and yield grade standards is <www.ams.usda.gov/standards/index.htm>.

Grading Formulas

The following are specifications for official United States standards of grades for beef, pork and lamb.

- Beef: The yield grade of a beef carcass is determined on the basis of the following: Yield grade–2.50 + (2.50 × adjusted fat thickness, inches) + (0.20 x precent kidney, pelvic, and heart fat) + (0.0038 × hot carcass weight, pounds)–(0.32 × area ribeye, square inches).
- Pork: The grade of a barrow or gilt carcass is determined on the basis of the following equation: Carcass grade = (4.0 × backfat thickness over the last rib, inches) (1.0 × muscling score). To apply this equation, muscling should be scored as follows: thin muscling = 1, average muscling = 2, and thick muscling = 3. Carcasses with thin muscling cannot grade U.S. No. 1.
- Lamb: The yield grade of lamb is determined on the basis of the adjusted fat thickness over the ribeye muscle between the 12th and 13th ribs. The adjusted fat thickness range for each yield grade is as follows: Yield Grade 1–0.00 to0.15 inch; Yield Grade 2–0.16 to 0.25; Yield Grade 3–0.26 to 0.35 inch; Yield Grade 4–0.36 to 0.45 inch; and Yield Grade 5–0.46 inch and greater.

Slaughtering Practices

The Humane Slaughter Act (1960) requires that prior to slaughter, animals be rendered completely unconscious with a minimum of excitement and discomfort, by mechanical, electrical, or chemical (carbon dioxide gas) methods.

After being bled, skinned, and **eviscerated** (removal of the internal organs), the carcasses are chilled for 24 to 48 hours before being graded and processed. Meat items, such as brains, kidneys, sweetbreads (calf thymus glands), the tail, and the tongue, do not accompany a carcass and are considered by-products to be sold separately as specialty items. These parts, and all other items removed from the carcass, such as feet, hide, and intestines, are called **offal** and are an important source of income for meatpackers.

Structure and Composition of Meat

The term "meat" generally refers to the skeletal muscle from the carcasses of animals–beef and veal (cattle), pork (hogs), and lamb (sheep). The general composition of such meat is approximately 70 percent water, 21 percent protein, 8 percent fat, and 1 percent ash (mineral). For the specific composition of meats, refer to Table A-8.

Meat and processed meat products, and other foods of animal origin, provide a complete protein source that contains, in favorable quantities, all the essential amino acids. Proteins from plant sources are frequently deficient in one or more of the essential amino acids. Overall, use of protein by the body is more efficient when a complete protein is consumed.

Meats, including processed meat products, are an excellent source of iron, which is more biologically available than iron from plant sources or that added through fortification. The iron in meat also improves the absorption and utilization of iron from other sources.

Fat contributes to product juiciness, tenderness, and flavor of meat and **processed meats**. Additionally, fat reduces their formulation costs. The fat content of processed meat products is regulated by the USDA. For example, products such as wieners cannot contain more than 30 percent fat. Some specialty loaf items may contain more than 30 percent fat. However, many processed products contain considerably less fat. The meat industry is responding to consumer demands for leaner products by providing new reduced fat options.

Chilling

Immediately after slaughter, many changes take place in muscle that convert muscle to meat. One of the changes is the contraction and stiffening of muscle known as **rigor mortis**. Muscle is very tender at the time of slaughter. However, as rigor mortis begins, muscle becomes progressively less tender until rigor mortis is complete. In the case of beef, 6 to 12 hours are required for the completion of rigor mortis. Pork requires only 1 to 6 hours. The carcass is chilled immediately after slaughter to prevent spoilage. If the carcass is chilled too rapidly, the result is **cold shortening** and subsequent toughness. Cold shortening occurs when the muscle is chilled to less than 60°F (16°C) before the completion of rigor mortis. If the carcass is frozen before completion of rigor mortis, the result is "thaw rigor" and subsequently extremely tough meat.

Aging of Meat

After the completion of rigor mortis, changes take place in meat that result in beef becoming progressively more tender. This holding of beef in a cooler or beef in the refrigerator is commonly referred to as the **aging** period. The increase in tenderness is due to natural enzymatic changes taking place in the muscle. The increase in beef tenderness continues only for approximately 7 to 10 days after slaughter when the beef is held at approximately 35°F (2°C). Beef held at higher temperatures will tenderize more rapidly, but it also may spoil and develop off-flavors. Lamb and pork are rarely aged. A lack of tenderness usually is not encountered because of lamb's and pork's relatively young age when slaughtered.

Tenderizing

Tenderness, juiciness, and flavor are components of meat palatability. Although juiciness and flavor normally do not vary a great deal, tenderness can vary considerably from one cut to the next. The most causes of variation in tenderness of beef, pork, lamb, and veal include genetics, species and age, feeding, muscle type, suspension of the carcass, **electrical stimulation**, chilling rate, aging, mechanical tenderizing, chemical tenderizing, freezing and thawing, cooking and carving.

Genetics. About 45 percent of the observed variation in tenderness of cooked beef is due to the genetics or parents of the animal from which the beef came. Although many other factors are involved in tenderness, genetics is one of the main reasons such a wide difference in tenderness often exists among identical grades and cuts of meat.

Species and Age. Beef usually is the most variable in tenderness followed by lamb, pork, and veal. The tenderness variation from species to species is due primarily to the chronological age of the animal at time of slaughter. Beef normally is processed at approximately 20 months of age, lamb at 8 months, pork at 5 months, and veal at approximately 2 months of age.

Within a given species such as beef, age of the animal at slaughter also influences tenderness. Beef normally is slaughtered between 9 and 30 months of age. Usually the meat from these animals is fairly tender. However, meat becomes progressively less tender as the animal gets older. The decrease in tenderness with increasing age is due to the changing nature of collagen (gristle), the connective tissue protein found in meat. Pork and lamb from older animals normally are processed into sausage items, so toughness due to age usually is not a problem.

Feeding. What the animal is fed does not directly influence tenderness. In the case of beef, an indirect effect of feeding on tenderness may be observed. Animals that are finished with grain tend to reach a given slaughter weight sooner than animals that are finished to the same slaughter weight on pasture. Thus, grain-fed animals usually are slightly more tender because they are slaughtered at a younger age.

Muscle to Muscle. Within any species, a considerable variation exists in tenderness among muscles. For example, tenderloin is much more tender than the fore shank or heel of round in beef. (See Figure 17-4.) This difference is due to the amount of connective tissue in the various cuts. The tenderloin usually has a small amount of connective tissue compared with the fore shank or heel of round. The amount of connective tissue present is due to the function of the muscles in the live animal.

Suspension of Carcass. Stretching of the muscle during chilling of the carcass affects tenderness. This has different effects on different muscles according to their anatomical location in the carcass. Although most carcasses are hung from the hind leg, a new method of hanging the carcass from the pelvic or hip bone changes the tension applied to some muscles.

Electrical Stimulation. Electrical stimulation of the hot carcass immediately after slaughter increases tenderness. Beef carcasses subjected to approximately one minute of high-voltage electrical current improves tenderness of many cuts of the carcass.

Chilling Rate. If the carcass is chilled too rapidly, the result is cold shortening and subsequent toughness. Cold shortening occurs when the muscle is chilled to less than 60°F (16°C) before the completion of rigor mortis.

Quality Grade. Age of the animal also plays a major role in tenderness as it applies to quality grading in beef (Figures 17-5 and 17-6). The quality grades of beef are USDA Prime, Choice, Select, Standard, Utility, and Commercial. Carcasses from young animals

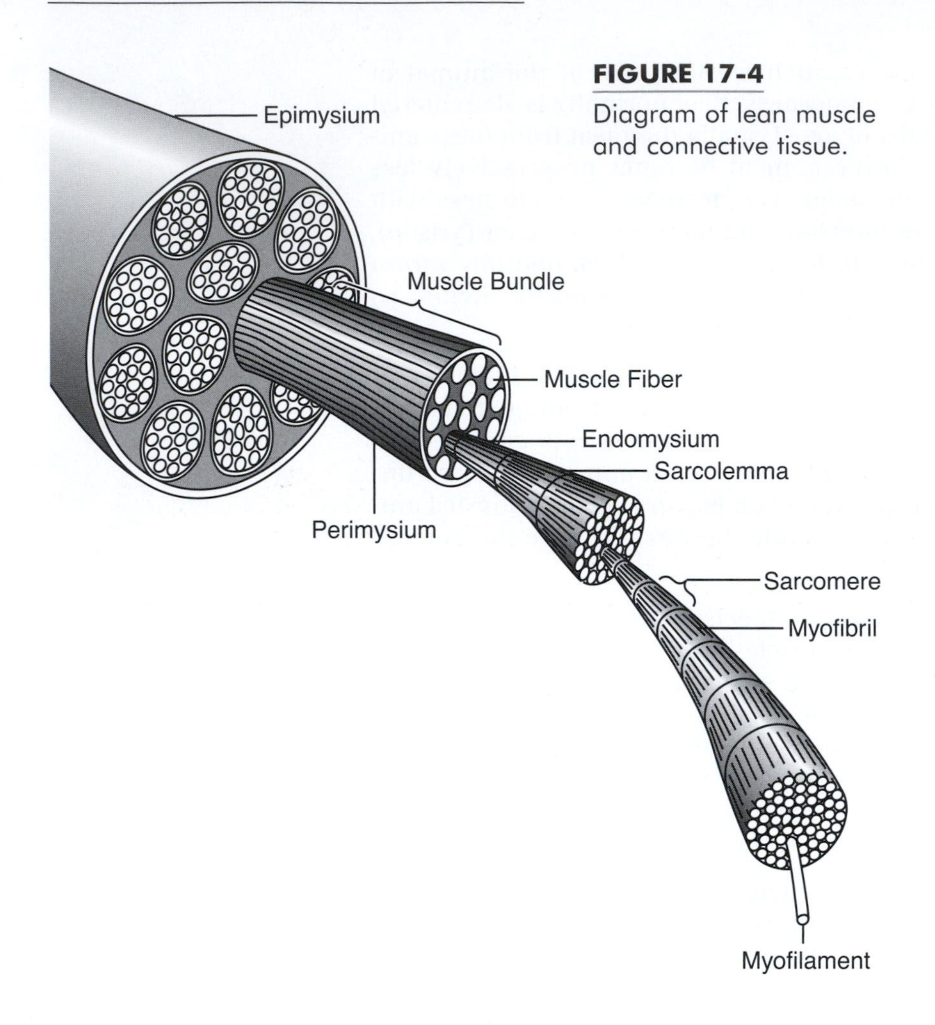

FIGURE 17-4
Diagram of lean muscle and connective tissue.

(up to 40 months of age) are eligible for USDA Prime, Choice, Select, Standard, and Utility grade designations. Carcasses from beef animals older than 40 months are eligible for USDA Commercial and Utility grade designations. Quality grades are not used for pork, but yield grades are given the designation of 1, 2, 3, 4, and 5.

Mechanical. Grinding is a very popular means of increasing tenderness of meat, especially beef (Figure 17-7). The popularity of hamburger and ground beef is due to a texture and tenderness more uniform than that of steaks and roasts. Cubing is another means of mechanically tenderizing meat. The small blades of a cuber simply sever connective tissue in boneless retail cuts so that the connective tissue is broken into smaller pieces.

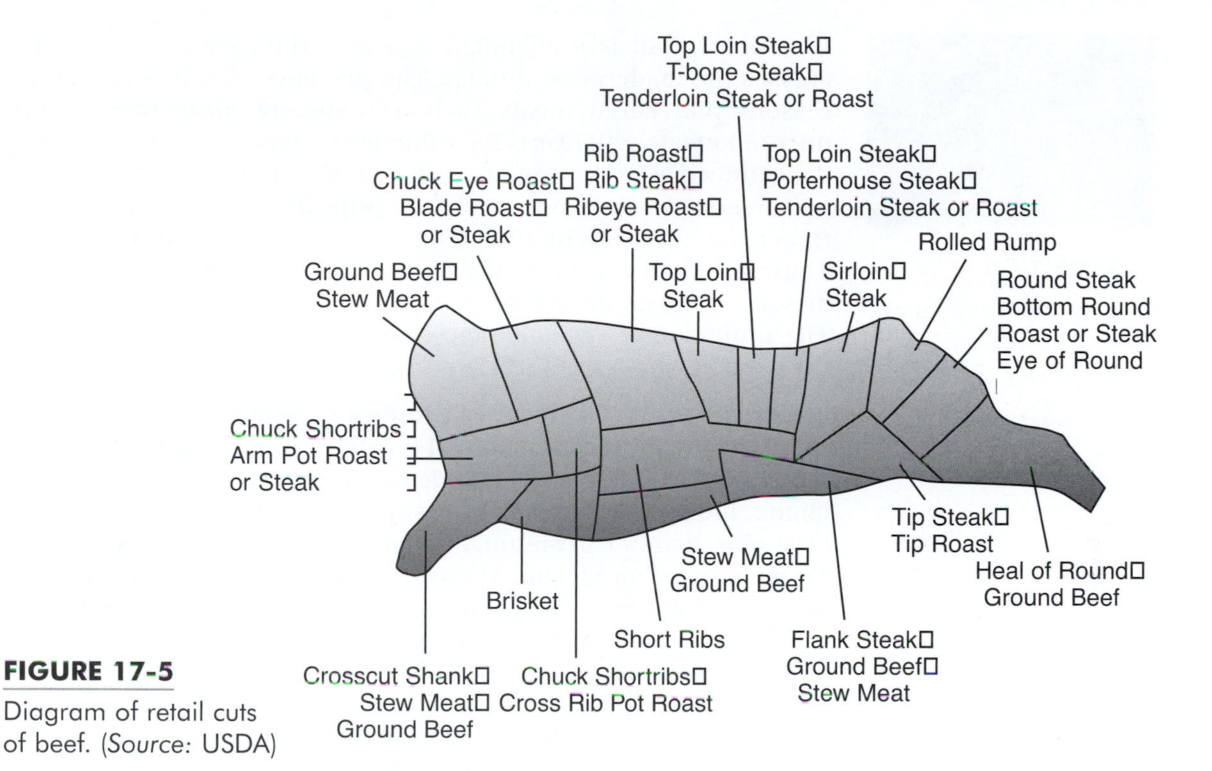

FIGURE 17-5

Diagram of retail cuts of beef. (*Source:* USDA)

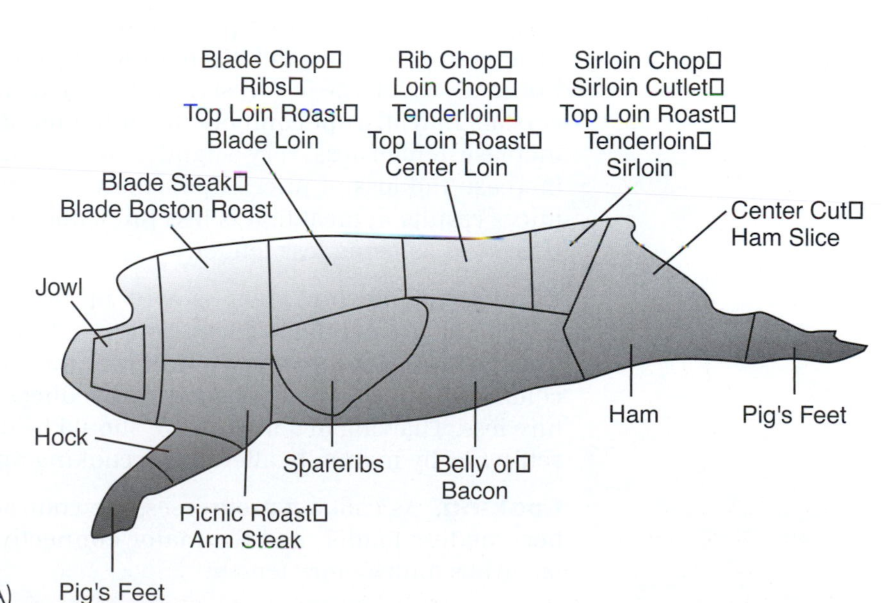

FIGURE 17-6

Diagram of retail cuts of pork. (*Source:* USDA)

FIGURE 17-7

Mixing and grinding of beef. (*Source*: USDA Photography Library)

Chemical. Salt is a chemical that at certain concentrations increases the tenderness of meat. The presence of salt is one of the reasons that cured meats such as ham are more tender than uncured meats. Salt exerts its influence on tenderness by softening the connective tissue protein, collagen, into a more tender form.

Vegetable enzymes such as **papain** (papaya), **bromelin** (pineapple), and **ficin** (fig) tenderize meat (refer to Table 8-2). These are used commercially and in the home. These tenderizers dissolve or degrade the connective tissues, collagen and elastin. The limitation of vegetable enzymes is that their action is sometimes restricted to the surface of meat.

Marinading. Consumers can improve tenderness and add taste variety to the meat component of meals by **marinading**. The basic ingredients of a marinade include salt (or soy sauce), acid (vinegar, lemon, Italian salad dressing, or soy sauce), and enzymes (papain, bromelin, ficin, or fresh ginger root). Some marinade recipes call for addition of an alcohol source (wine or brandy) for flavor. The addition of several tablespoonfuls of olive oil will seal the surfaces from the air and thus result in the meat staying fresher and brighter in color for a longer period of time.

The tenderizing action of marinades occurs through the softening of collagen by the salt, the increased water uptake, and the hydrolysis and breakage of the cross links of the connective tissue by the acids and alcohols.

Freezing. Freezing rate plays a small role in tenderness. When meat is frozen very quickly, small ice crystals form; when meat is frozen slowly, large ice crystals are formed. The formation of large crystals may disrupt components of the muscle fibers in meat and increase tenderness very slightly, but the large ice crystals also increase the loss of juices upon thawing. This increase in loss of juices results in meat that is less juicy upon cooking and usually is perceived as being less tender.

Thawing. Thawing meat slowly in the refrigerator generally results in greater tenderness compared with cooking from the frozen state. Slow thawing minimizes the toughening effect from cold shortening (when present) and reduces the amount of moisture loss. Thawing in a microwave should be done on a lower power setting or by manually alternating cooking and standing times.

Cooking. As cooking progresses, the contractile proteins in meat become less tender, and the major connective tissue protein (collagen) becomes more tender.

Carving. Muscles, muscle bundles, and muscle fibers are all surrounded by connective tissue. When cuts are made from carcasses and wholesale cuts, the normal procedure is to cut at right angles to the length of the muscle. This procedure severs the maximum amount of connective tissue and distributes the bone more evenly among all cuts in that area. Likewise, consumers should carve cooked meat at right angles to the length of the muscle fibers or "against the grain" to achieve maximum tenderness. Cutting with the grain results in "stringiness" and thus less tenderness.

FIGURE 17-8

A variety of cured meats in a Hungarian meat market.

Curing

Curing of meat was used as a preservative method. Now it is used more for flavor and color enhancement (Figure 17-8). Some curing agents include salt, sodium nitrate and nitrite, sugar, and spices. Salt is added to preserve and to add flavor. Sodium nitrate and nitrite fix the red color of meat, act as a preservative, and prevent botulism. Sugar also provides color stability and flavor, and spices of course produce a desired flavor.

Color

The primary color pigment of meat is a protein called **myoglobin**. Its function is to store oxygen in the muscle tissue. When oxygen is present, meat is a bright red color. When oxygen is absent the meat is a purplish color. Myoglobin is denatured by prolonged exposure to air or by cooking. It turns brown.

Smoking

Like many other meat processing practices, **smoking** has been practiced since the beginning of recorded history. The highly smoked products of the past have largely given way to milder smoking methods that have reduced, but not eliminated, the effectiveness of smoke as an inhibitor of microbial growth. As a microbial growth inhibitor, smoke is most effective when used in combination with other preservation techniques. Smoke also protects fat from rancidity, contributes to the characteristic color, and creates unique flavors in processed meats.

Liquid smoke, used widely in industry, not only avoids many of the questionable compounds found in wood smoke but eliminates virtually all the emissions associated with burning wood or sawdust.

Meat Specialties

Dry sausages may or may not be characterized by a bacterial fermentation. When fermented, the intentional encouragement of a lactic acid bacteria growth is useful as a meat preservative as well as producing the typical tangy flavor. The ingredients are mixed with spices and curing materials, stuffed into casings, and put through a carefully controlled, long, continuous air-drying process. Dry sausages include salami or pepperoni.

Semi-dry sausages are usually heated in the smokehouse to fully cook the product and partially dry it. Semi-dry sausages are semi-soft sausages with good keeping qualities due to their lactic acid fermentation. "Summer Sausage" (another word for cervelat) is the general classification for mildly seasoned, smoked, semi-dry sausages like Mortadella and Lebanon bologna.

Most consumers purchase their meat and poultry from retail stores. Some other sources include retailing at a farm, auction markets, direct sales, cooperatives and sometimes door-to-door. Consumers must know important information about the dealer and the company before making a decision they might later regret.

Freezing

Properly wrapped, fresh meat cuts can be frozen and held in frozen storage (0°F [–18°C] or less) for months if the cut is a fatty meat like pork. Beef can be held for years. Storage time for pork and fatty meats is limited because at freezer temperatures the fat gradually oxidizes, producing off-flavors. Once frozen, meat should not be thawed and refrozen. Few cured meats or sausages are frozen because the salt in their formulations increases the rate of the development of rancid flavors. Also, the flavor of the spices used in sausage may change during frozen storage.

Storage

Storage times for frozen meat and refrigerator storage are given in Tables 10-1 and 10-2, and Table 8-4.

Cooking

For cuts that are low in connective tissue–such as steaks and chops from the rib and loin–the recommended method of cooking is dry heat, including pan frying, broiling, roasting, or grilling or barbecuing (Figure 17-9). Dry heat raises the temperature very quickly, and the flavor of meat will develop before the contractile proteins have the opportunity to become significantly less tender. For cuts with a high amount of connective tissue–such as those from the

Roasting, suitable for large tender cuts of beef, veal, pork, and lamb.

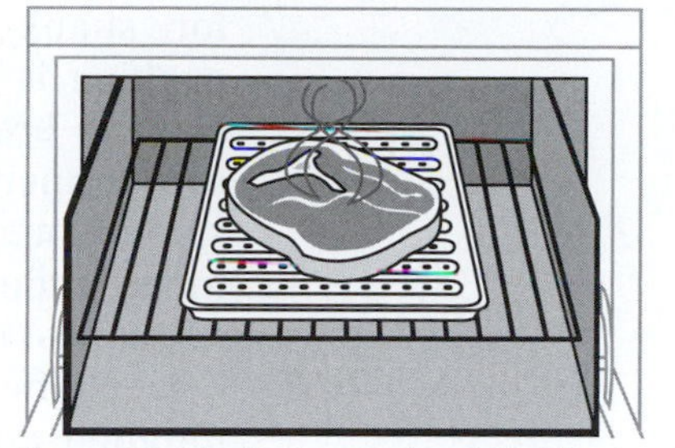

Broiling, suitable for tender beef steaks, lamb chops, pork chops, sliced ham, bacon, and ground beef or lamb.

Panfrying, suitable for comparatively thin pieces of tender meats.

Panbroiling, suitable for tender cuts when cut 1 in. or less thick.

Deep-fat frying, suitable for cooking brains, sweet-breads, liver, croquettes, and leftover meat.

Braising, suitable for less tender cuts of meat.

Cooking in liquid, suitable for large less tender cuts and stews.

FIGURE 17-9 Different methods for cooking meat depending on the cut.

fore shank, heel of round, and chuck–the recommended method of cooking is long and slow at low temperatures using moist heat such as braising. The application of moist heat for a long time at low temperatures (275° to 325°F [135° to 163°C]) results in conversion of tough collagen into tender gelatin and makes this type of cut more tender compared with dry heat cooking of one of the less tender cuts of meat. Degree of doneness significantly affects tenderness.

As the lean is heated, the contractile proteins (muscle proper) toughen and moisture is lost. Both decrease tenderness. Tender cuts of meat cooked to a rare degree of doneness (140°F [60°C]) are more tender than when cooked to medium (155°F [68°C]), and medium in turn is more tender than well-done (170°F [77°C]). Degree of doneness is especially important in the case of beef.

Pork is cooked to approximately 160° or 170°F (71° to 77°C) internal temperature for desirable flavor. Although this temperature range corresponds to well-done in beef, pork still may be slightly pink. Since *Trichinella spiralis* (trichinosis) is destroyed at 137°F (58°C), an internal temperature of 160° to 170°F (71° to 77°C) for pork is definitely safe. More cooking will result in dehydration, loss of juiciness, and unnecessary toughening. Lamb is usually cooked to well-done (approximately 160° to 170°F [71° to 77°C] internal temperature) because the flavor is more desirable compared with lower temperatures.

Precooked meat products such as beef roasts are increasingly being prepared for institutional use, especially in the fast-food industry. The products generally are precooked to a rare state, then chilled and vacuum-packaged. Precooked meat products offer the advantages of closely predictable yield and rapid warming for service.

Cooking time and temperature of roasts, which is critical to microbial safety, is carefully controlled. For example, an internal temperature of 145°F (63°C) must be reached or longer times at lower temperatures are required to control potential pathogenic microorganisms.

Precooked products should be handled carefully to avoid recontamination and incubation of potential pathogenic and spoilage microorganisms. These problems are most serious with repeated warming and chilling. Canning subjects meat products to sufficient heating to control pathogenic and spoilage microorganisms, and sealing prevents recontamination and conditions favorable for the growth of microorganisms.

MEAT SUBSTITUTES

Modern techniques for manufacturing plant protein products are largely the result of two notable advances in processing: (1) a method invented in the 1950s for spinning vegetable proteins into fibers; and (2) the development of the extrusion method. In the spinning technique, the protein is dissolved in an alkaline (basic) solution and then extruded through spinnerets and coagulated to form bundles of fibers. Flavor and color compounds that simulate meat are added to these fibers. A binder, such as egg **albumen** or vegetable gum and such other additives as fats, emulsifiers, and nutrients are also added. The fibers can then be processed into a variety of shapes and textures. In the extrusion method, the vegetable protein, combined with flavor, color, and other ingredients, is formed into a plastic mass in a cooker-extruder. Under pressure this mass is forced through a die to form beeflike strips or other shapes characteristic of meats.

Textured protein products are usually at least 50 percent protein and contain the eight essential amino acids and the vitamins and minerals found in meats. Although soybean protein is most commonly used, other plant proteins–wheat gluten, yeast protein, and most other edible proteins–can be used singly or in combination. The use of textured protein products will probably increase in the future, as populations grow and conventional sources of protein become more scarce.

POULTRY

Meat chicken production is dominated by large **integrated** companies. Typically these companies control hatching egg production, hatching, growing, processing, and marketing of the birds. They often mill their own feed and render the offal and feathers to produce feed ingredients. Any of these steps may be controlled by contract. The company owns all functions except live production. With a production contract, the farmer may provide the growing facility, equipment, litter, brooder, fuel, electricity, and labor. The company usually provides the chicks, feed, medications, bird loading and hauling, and some grow-out supervision. Contract payments are based on a set amount per pound of chicken marketed.

Growing houses are often buildings 40 to 50 feet wide and 400 to 500 feet long (Figure 17-10). Modern facilities control air entering the sides of the building. Exhaust fans located in the building blow air over the birds during hot weather. Overhead fogger lines cool chickens in hot weather. Space allowances range from 0.7 to 1.0 square foot per bird depending on season, house type, and age marketed.

FIGURE 17-10

Broilers in well-ventilated housing where thousands are raised. (*Source:* USDA Photography Library)

Feed is moved to the birds by mechanical conveyers that drop the feed into attached pans. Water is supplied by bird-activated nipples attached to water pipes running the length of the building. Three diets are typically used: starter, grower, and unmedicated finisher or withdrawal feed (last 7 days).

Young chicks require room temperature of about 87°F (31°C) during the first week of life. This is decreased about 5 degrees per week, until ambient temperature is reached. Diseases are controlled by vaccination, medicated feed for coccidiosis control, exclusion of animals that transmit disease (vectors), and sanitation.

A typical broiler production cost per pound produced might be feed $.19, chick $.04, contract payment $.05, other $.02. These costs assume that large operations are more efficient.

Turkey production is also integrated. Most meat production units brood and grow from 50,000 to 75,000 birds three and one half times per year. Many of the larger facilities have a single brooding complex that has the capacity to brood 50,000 to 100,000 poults and that serves two separate grow-out facilities with the same capacity. In this scheme, the brooder facility broods seven times each year and furnishes the poults needed to fill both growing facilities three and one half times per year.

Poults (young turkeys) are brooded with an average density of 1.0 square foot per bird. Toms (males) are placed in grow-out facilities at a density of 3.0 to 4.0 square feet per bird. Hens (females) receive about 2.5 square feet in their grow-out facilities.

Turkeys are no longer produced seasonally. Further processing and the structure of the turkey industry changed turkey production to a year-round activity. Almost all birds are produced on a contractual basis. The producer furnishes the land, facilities, and labor and is paid based on the weight, grade, and feed conversion of the birds delivered to the processing plant. In general, the grower can expect a return of $1.25 to $1.50 per bird.

Processing

Meat chickens may be marketed as broilers, roasters, or game hens (Figure 17-11). Typical eviscerated weights are shown in Table 17-1. Modern commercial meat strains reach an average live weight of 4.0 pounds at 42 days or 4.8 pounds at 49 days of age. The chickens are slaughtered at an appropriate age to get the eviscerated weight desired by the customer.

Turkey hens are marketed between 14 and 16 weeks of age. At this age hens will typically weigh from 14.7 to 17.5 pounds. Toms are often marketed between 17 and 20 weeks of age and will weigh from 26.4 to 32.3 pounds. Market age is determined by the product being produced. About 70 percent of all turkeys grown are further processed. For this market, the industry prefers to grow toms, because they are larger. Hens are also further processed even though the unit cost is higher. About 16 percent of all turkeys are processed for the whole body market.

FIGURE 17-11

Processing chicken parts for wholesale and retail markets. (*Source*: USDA Photography Library)

TABLE 17-1 Weights for Different Types of Meat Chickens

Type	Live Weight (lbs.)	Eviscerated Weight (lbs.)
Cornish game hen	1.2 to 3.1	.75 to 2.0
Broiler/Fryer	4.0 to 6.3	2.8 to 4.4
Roaster	7.4 to10.0	5.0 to 7.0

About 14 percent of all turkeys produced are processed as parts. In the past, parts like wings and drums were often sold at greatly reduced prices. Today, these parts are used extensively in further processing and often end up as part of a further processed product such as ground meat.

Processing Steps

The slaughter and processing of broilers and turkeys is an assembly-line operation conducted under sanitary conditions. Inspecting, classifying, and grading are a part of the processing operation. Although the processing procedures may vary from plant to plant and between broilers and turkeys, the steps usually include the following:

1. Spot inspection of each lot of birds before slaughter (**antemortem** inspection)
2. Suspension and shackling of each bird by the legs

3. Stunning with electrical shock
4. Bleeding
5. Scalding to loosen the feathers
6. Picking off the feathers by machine
7. Removing of the pinfeathers
8. Removing the internal organs or eviscerating
9. Chilling (in ice water)
10. **Postmortem** inspection
11. Grading
12. Packaging

Properties

Meat from chickens and turkeys provides a high-quality protein that is low in fat. The protein is an excellent source of essential amino acids. Poultry meat is also a good source of phosphorus, iron, copper, zinc, and vitamins B_{12} and B_6. Dark meat is higher in fat than white meat, and fat is also high in the skin. A 3-ounce piece of roasted chicken or turkey breast provides about 140 kcal of energy, and 3-ounces of dark turkey meat provides about 160 kcal of energy. Table A-8 provides the composition of many poultry products.

Whether or not a poultry product meets the consumer's expectations depends upon the conditions surrounding various stages in the bird's development from the fertilized egg through production and processing to consumption. Appearance, texture, and flavor are a primary concern in the food industry and to the consumer.

Appearance (Color). Color of cooked or raw poultry meat is important because consumers associate it with the product's freshness. Poultry is unique because it is sold with and without its skin. In addition, poultry has muscles that are dramatic extremes in color–white or breast meat and dark or thigh and leg meat.

Poultry meat color is affected by factors such as bird age, sex, strain, diet, intramuscular fat, meat moisture content, pre-slaughter conditions and processing variables. Color of meat depends upon the presence of the muscle pigments myoglobin and hemoglobin. Discoloration of poultry can be related to the amount of these pigments that are present in the meat, the chemical state of the pigments, or the way in which light is reflected off the meat. The discoloration can occur in an entire muscle, or it can be limited to a specific area.

When an entire muscle is discolored, it is frequently the breast muscle. This occurs because breast muscle accounts for a large portion of the live weight. It is more sensitive to factors that contribute to discoloration, and the already light color makes small changes in color more noticeable. Extreme environmental temperatures or stress due to live handling before processing can cause broiler and turkey breast meat to be discolored. The extent of the discoloration is related to each bird's individual response to the conditions.

Another major cause of poultry meat discoloration is bruising. Approximately 29 percent of all carcasses processed in the United States are downgraded (reduced quality), and the majority of these defects are from bruises. The poultry industry generally tries to identify where (field or plant), how, and when the injuries occur, but this is often difficult to determine. The color of the bruise, the amount of "blood" present, and the extent of the "blood clot" formation in the affected area are good indicators of the age of the injury and may give some clues as to its origin.

ALWAYS ROOM FOR A NEW JELL-O

Gelatin is typically made from the animal protein collagen, which is extracted from skin, bone, and connective tissue of food animals. While competing in the annual "Innovative Uses for Soybeans Contest," Purdue University students created a vegetarian gel dessert. The new vegetarian dessert is made from a gel base made of water, fructose, high-gelling soy protein, and carrageenan (made from seaweed). The students prepared cherry, orange, and lemon flavors of their gelatin for the contest. They call the new product NuSoy Gel.

The new gelatin dessert naturally contains isoflavones, and it is also fortified with calcium and vitamin C. Some studies have suggested that isoflavones could be a part of the reason why Asian cultures, which have diets high in soy-based food products, have a lower incidence of diseases such as breast cancer, prostate cancer, and heart disease.

Besides being a gelatin substitute, the new gel could provide more nutrients for hospital patients who can only have a clear liquid diet while they are in the hospital. The current gelatin dessert served in hospitals does not offer many nutrients. Finally, finding new uses for soybeans is a high priority with soybean farmers.

To learn more about this new product, visit the Web site where it marketed:

<www.welovesoy.com/>

A bruise will vary in appearance from a fresh, "bloody" red color with no clotting minutes after the injury to a normal flesh color 120 hours later. The amount of "blood" present and the extent of clot formation are useful in distinguishing if the injury occurred during catching/transportation or during processing. Injuries that occur in the field are usually magnified by processing plant equipment or handling conditions in the plant.

Texture (Tenderness). After consumers buy a poultry product, they relate the quality of that product to its texture and flavor when they are eating it. Whether or not poultry meat is tender depends upon the rate and extent of the chemical and physical changes occurring in the muscle as it becomes meat (Figure 17-12). When an animal dies, blood stops circulating and muscles receive no new supplies of oxygen or nutrients. Without oxygen and nutrients, muscles run out of energy, and they contract and become stiff. This stiffening is called rigor mortis. Eventually, muscles become soft again, which means that they are tender when cooked.

Anything that interferes with the formation of rigor mortis, or the softening process that follows it, will affect meat tenderness. For example, birds that struggle before or during slaughter cause their muscles to run out of energy quicker, and rigor mortis forms much faster than normal. The texture of these muscles tends to be tough because energy was reduced in the live bird. Exposure to environmental stress (hot or cold temperatures) before slaughter

FIGURE 17-12

Whole chickens being inspected in a processing plant. (*Source:* USDA Photography Library)

creates a similar situation. High pre-slaughter stunning temperatures, high scalding temperatures, longer scalding times, and machine picking can also cause poultry meat to be tough.

Tenderness of boneless or portioned cuts of poultry is influenced by the amount of time between death (postmortem) and the **deboning**. Muscles that are deboned during early postmortem still have energy available for contraction. When these muscles are removed from the carcass, they contract and become tough. To avoid this toughening, meat can be aged for 6 to 24 hours before deboning. This is costly for the processor.

The poultry industry recently started using post-slaughter electrical stimulation immediately after death to hasten rigor development of carcasses and reduce aging time before deboning. When electricity is applied to the dead bird, the treatment acts like a nerve impulse, and causes the muscle to contract, use up energy, and enter rigor mortis at a faster rate. Meat can be boned within 2 hours postmortem instead of the 4 to 6 hours required with normal aging.

Flavor. Flavor is another quality attribute that consumers use to determine the acceptability of poultry meat. Both taste and odor contribute to the flavor of poultry. Generally, distinguishing between the two during consumption is difficult. When poultry is cooked, flavor develops from sugar and amino acid interactions, lipid and thermal (heat) oxidation, and thiamin degradation. These chemical changes are not unique to poultry, but the lipids and fats in poultry are unique and combine with odor to account for the characteristic poultry flavor.

Few factors during production and processing affect poultry meat flavor. Age of the bird at slaughter (young or mature birds) affects the flavor of the meat. Minor effects on meat flavor are related to bird strain, diet, environmental conditions (litter, ventilation, etc.), scalding temperatures, chilling, product packaging, and storage. Overall, these effects are too small for consumers to notice.

The most important aspect of poultry meat is its eating quality –a function of the combined effects of appearance, texture, and flavor. Live production affects poultry meat quality by determining the state of the animal at slaughter. Poultry processing affects meat quality by establishing the chemistry of the muscle constituents and their interactions within the muscle structure. The producer, processor, retailer, and consumer all have specific expectations for the quality attributes of poultry.

Grading

Chickens, turkeys, ducks, geese, guineas, and pigeons are all eligible for grading and certification services provided by the USDA's

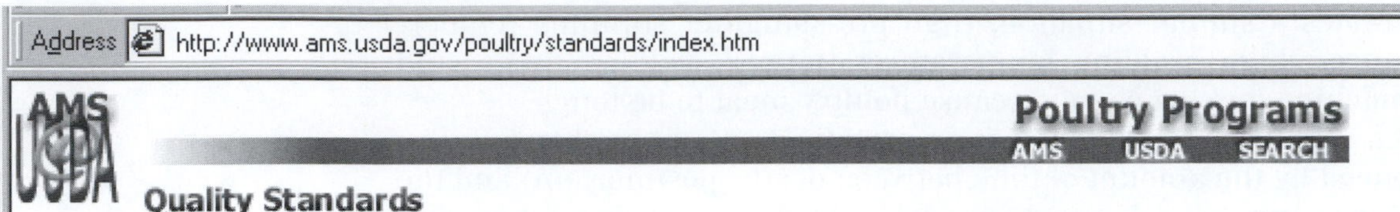

Quality standards have been established by the USDA for most agricultural commodities to facilitate the orderly and efficient marketing of these products among buyers and sellers. Standards allow buyers and sellers to specify the level of quality required or expected for a product without having to see or examine products in advance. U.S. standards are available for free. Buyers and sellers often refer to U.S. standards in their trading specifications and contracts (e.g., "Product must meet U.S Grade A requirements"). The U.S. quality standards for shell eggs, poultry, and rabbits are available below.

At the option of the buyer or seller, products can also be officially graded by USDA for a service fee as meeting the requirements of a U.S. quality standard. Labels for ready-to-cook poultry and rabbit products and shell eggs that include the USDA grade shield indicate that USDA has officially graded that product and that the product met all requirements of the designated quality standard.

◆ Current U.S. Poultry, Shell Egg and Rabbit Standards

FIGURE 17-13

Web site for Agricultural Marketing Service (AMS) of the USDA provides quality standards for poultry products.

Agricultural Marketing Service (AMS) Poultry Program's Grading Branch (Figure 17-13). These services are provided in accordance with federal poultry grading regulations.

Chickens and turkeys are often sold as value-added products. Poultry parts and an increasing number of skinless and/or boneless products are meeting consumer demand for convenient, lower-fat, portion-controlled items. This shift away from whole carcass birds creates special challenges for buyers and sellers, whether they are poultry producers or processors, wholesalers, food manufacturers, food service operators, food retailers, or consumers. All these traders rely on USDA's poultry grading services to ensure that their requirements for quality, weight, condition, and other factors are met.

Grading and USDA Quality Grade Standards. Grading provides a standardized means of describing the marketability of a particular food product. In order for poultry to be eligible for an official USDA grade designation, each carcass or part must be individually graded by a plant grader, then a sample must be certified by a USDA grader. Officially graded poultry that passes this examination and evaluation process is eligible for the grade shield and may be identified as USDA Grade A, B, or C (Figure 17-14). Poultry standards are frequently reviewed, revised, and updated as needed to keep pace with changes in processing and merchandising.

For poultry, the USDA has developed quality grade standards for whole carcasses and parts, as well as boneless and/or skinless parts and products. Depending upon the product, the standards

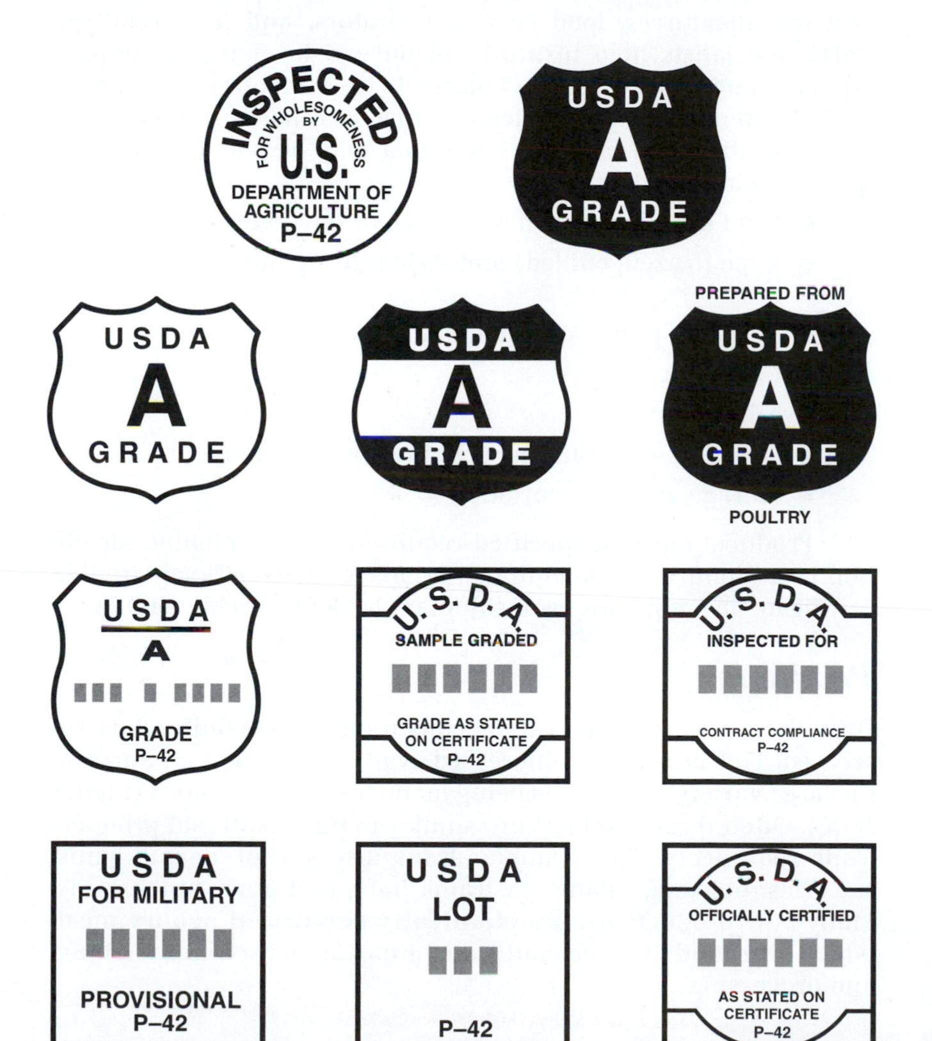

FIGURE 17-14
Various inspection and grade shields that could be found on poultry products. (*Source:* USDA)

define and measure quality in terms of meat yield, fat covering, and freedom from defects such as cuts and tears in the skin, broken bones, and discolorations on the meat and skin. The intensity, aggregate area, location, and number of defects encountered for each quality factor are determined. The final quality rating (A, B, or C) is based on the factor with the lowest rating.

Quality Standards can be found on the Web at <www.ams.usda.gov/poultry/standards/index.htm>.

Contract Acceptance Certification. The contract acceptance service ensures the integrity and quality of poultry and further processed poultry products bought by quantity food buyers such as food manufacturers, food service operators, and food retailers. USDA specialists help institutional buyers develop and prepare explicit poultry specifications tailored to their requirements. USDA graders then provide certification that purchases comply with these specifications. Specific items that may be part of a product specification include:

- Kind and class (species and age of the poultry)
- Type (frozen, chilled) and style (cut-up parts, whole muscle)
- Formula, processing, and fabrication
- Laboratory analysis
- Net weight
- Labeling and marking, packing and packaging
- Storage and transportation

Products meeting specified requirements are eligible for the Contract Compliance identification mark. This official grading certificate accompanies each shipment to the receiving agency.

Products

Over the years, the per capita consumption of poultry has increased. This is due to the increased availability of poultry and also the large variety of products being made from poultry meat (Figure 17-15). Often these products are similar to the traditional products from red meats, for example, frankfurters (hot dogs), hams, sausages, bologna, salami, pastrami, ham, and other lunchmeats. Many new products use **mechanically separated** poultry meat, which is ground to a fine emulsion for curing, seasoning, smoking, and processing.

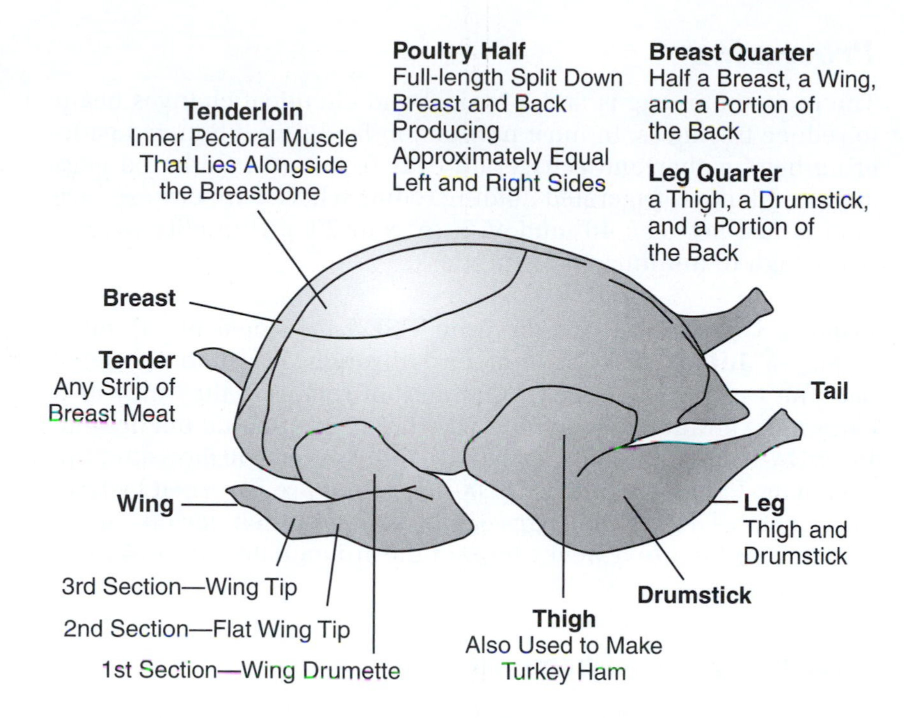

FIGURE 17-15

Diagram showing poultry products obtained from chicken carcass.

EGGS

Maximum production of top-quality eggs starts with a closely controlled breeding program emphasizing favorable genetic factors. The White Leghorn-type hen dominates today's egg industry. This breed reaches maturity early, uses its feed efficiently, has a relatively small body size, adapts well to different climates, and produces a relatively large number of white-shelled eggs, the color preferred by most consumers. In the major egg producing states, flocks of 100,000 laying hens are not unusual, and some flocks number more than 1 million. Each of the 235 million laying birds in the United States produces from 250 to 300 eggs a year.

In today's egg laying facilities, temperature, humidity and light are all controlled and the air is kept circulated. The building is well insulated, is windowless (to aid light control), and is force-ventilated. Most new construction favors the cage system because of its sanitation and efficiency. Because care and feeding of hens, maintenance, sanitation, and egg gathering all take time and money, automation is used whenever possible.

Processing

The moment an egg is laid, physical and chemical changes begin to reduce freshness. In most production facilities automated gathering belts gather and refrigerate eggs frequently. Gathered eggs are moved into refrigerated holding rooms where temperatures are maintained between 40° and 45°F (5° and 7°C). Humidity is relatively high to minimize moisture loss.

Carton Dates. Egg cartons from USDA-inspected plants must display a **Julian date** (a number 1 through 365) indicating the date the eggs were packed. Although not required, they may also carry an expiration date beyond which the eggs should not be sold. In USDA-inspected plants, this date cannot exceed 30 days after the pack date. Plants not under USDA inspection are governed by laws of their states. Fresh shell eggs can be stored in their cartons in the refrigerator for 4 to 5 weeks beyond the Julian date with insignificant quality loss.

Formation and Structure

The structure and characteristics of an egg include its color, shell, white, yolk, air cell, chalaza, germinal disc, and membranes (Figure 17-16).

Color. Egg shell and yolk color may vary, but color has nothing to do with egg quality, flavor, nutritive value, cooking characteristics, or shell thickness.

Shell. The color comes from pigments in the outer layer of the shell and may range in various breeds from white to deep brown. The breed of hen determines the color of the shell.

The egg's outer covering, accounts for about 9 to 12 percent of its total weight depending on egg size (Figure 17-16). The shell is the egg's first line of defense against bacterial contamination. The shell is largely composed of calcium carbonate (about 94 percent) with small amounts of magnesium carbonate, calcium phosphate, and other organic matter including protein.

White. Egg albumen in raw eggs is opalescent and does not appear white until it is beaten or cooked. A yellow or greenish cast in raw white may indicate the presence of riboflavin. Cloudiness of the raw white is due to the presence of carbon dioxide that has not had time to escape through the shell and thus indicates a very fresh egg.

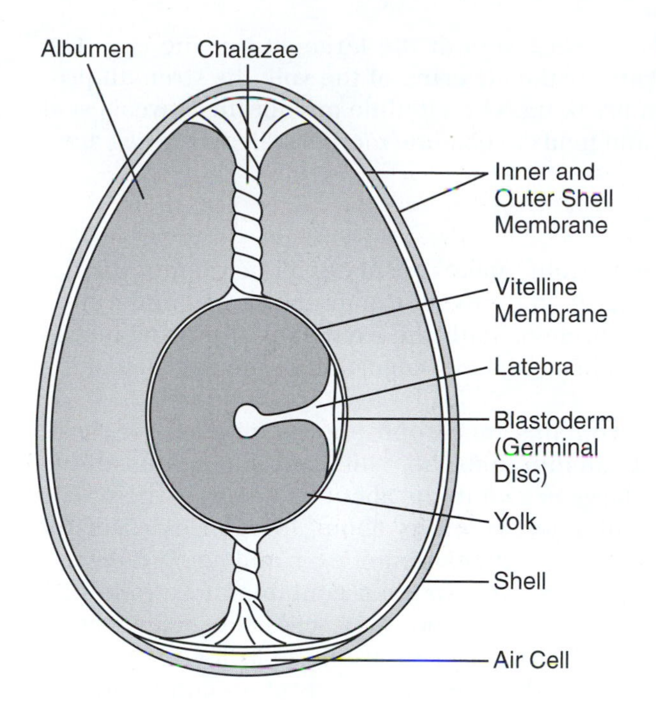

FIGURE 17-16
Diagram of the parts of an egg.

Yolk. Yolk color depends on the diet of the hen. Artificial color additives are not permitted. Gold or lemon-colored yolks are preferred by most buyers in the United States. Yolk pigments are relatively stable and are not lost or changed in cooking.

Air Cell. The air cell (Figure 17-16) is the empty space between the white and shell at the large end of the egg. When an egg is first laid, it is warm. As it cools, the contents contract and the inner shell membrane separates from the outer shell membrane to form the air cell.

Chalaza. Chalaza are ropey strands of egg white that anchor the yolk in place in the center of the thick white. They are neither imperfections nor beginning embryos. The more prominent the chalazae, the fresher the egg.

Germinal Disc. The germinal disc is the channel leading to the center of the yolk. When the egg is fertilized, sperm enter by way of the germinal disc and travel to the center and a chick embryo starts to form.

Membranes. Just inside the shell are two shell membranes, inner and outer. After the egg is laid and it begins to cool, an air cell forms between these two layers at the large end of the egg. The **vitelline membrane** is the covering of the yolk. Its strength protects the yolk from breaking. The vitelline membrane is weakest at the germinal disc and tends to become more fragile as the egg ages.

Composition

The yolk, or yellow portion, makes up about 33 percent of the liquid weight of the egg. It contains all the fat in the egg and a little less than half of the protein. With the exception of riboflavin and niacin, the yolk contains a higher proportion of the egg's vitamins than the white.

The yolk also contains more phosphorus, manganese, iron, iodine, copper, and calcium than the white, and it contains all the zinc. The yolk of a large egg contains about 59 calories; the whole egg including the albumen contains about 75 calories (kcal) of energy. Table A-8 lists the composition of eggs and egg products.

Also known as egg white, albumen contains more than half the egg's total protein, niacin, riboflavin, chlorine, magnesium, potassium, sodium, and sulfur. Protein from the yolk and the albumen provides humans with a high-quality protein containing all the essential amino acids.

Albumen is more opalescent than truly white. The cloudy appearance comes from carbon dioxide. As the egg ages, carbon dioxide escapes, so the albumen of older eggs is more transparent than that of fresher eggs.

Cooking Functions. Although eggs are widely known as breakfast entrees, they also perform in many other ways for the knowledgeable cook. Their cooking properties are varied. They have been called "the cement that holds the castle of cuisine together."

Eggs can bind ingredients as in meat loaves or croquettes. They can also leaven such baked high rises as souffles and sponge cakes. Their thickening talent is seen in custards and sauces. They emulsify mayonnaise, salad dressings, and Hollandaise sauce and are frequently used to coat or glaze breads and cookies. They clarify soups and coffee. In boiled candies and frostings, they retard crystallization. As a finishing touch, they can be hard cooked and used as a garnish.

Grading

Classification is determined by interior and exterior quality and designated by the letters AA, A, or B. In many egg packing plants, the USDA provides a grading service for shell eggs. Its official grade shield certifies that the eggs have been graded under federal supervision according to USDA standards and regulations (Figure 17-17). The grading service is not mandatory. Other eggs are packed under state regulations, which must meet or exceed federal standards.

In the grading process, eggs are examined for both interior and exterior quality and are sorted according to weight (size). Grade quality and size are not related to one another. In descending order of quality, grades are AA, A and B. No difference in nutritive value exists between the different grades.

Because production and marketing methods have become very efficient, eggs move so rapidly from laying house to market that consumers find very little difference in quality between grades AA and A. Although grade B eggs are just as wholesome to eat, they rate lower in appearance when broken out. Almost no grade B's find their way to the retail supermarket. Some go to institutional

FIGURE 17-17

Poster showing how an egg receives the USDA AA grade. (*Source:* USDA)

egg users such as bakeries or food service operations, but most go to egg breakers for use in egg products.

Grade AA. When cracked onto a surface, a grade AA egg will stand up tall. The yolk is firm, and the area covered by the white is small. A large proportion of thick white to thin white exists. The shell approximates the usual shape for an egg. It is generally clean and unbroken. Ridges/rough spots that do not affect the shell strength are permitted.

Grade A. When cracked onto a surface, a grade A egg covers a relatively small area. The yolk is round and upstanding. The thick white is large in proportion to the thin white and stands fairly well around the yolk. The shell approximates the usual shape for an egg. It is generally clean and unbroken. Ridges/rough spots that do not affect the shell strength are permitted.

Both grades AA and A are ideal for any use, but are especially desirable for poaching, frying, and cooking in shell.

Grade B. When cracked onto a surface, a grade B egg spreads out more. The yolk is flattened and there is about as much (or more) thin white as thick white. The shell has an abnormal shape; some slight stained areas are permitted. It is unbroken, and pronounced ridges/thin spots are permitted.

Size. Several factors influence the size of an egg. The major factor is the age of the hen. As the hen ages, her eggs increase in size. The breed of hen from which the egg comes is a second factor. Weight of the bird is another. Environmental factors that lower egg weights are heat, stress, overcrowding, and poor nutrition. All these variables are of great importance to the egg producer. A slight shift in egg weight influences size classification. Size is one of the factors considered when eggs are priced.

Egg sizes are Jumbo, Extra Large, Large, Medium, Small, and Peewee. Medium, Large, and Extra Large are the sizes most commonly available. Sizes are classified according to minimum net weight expressed in ounces per dozen, as shown in Table 17-2.

TABLE 17-2 Weight Classes for Shell Eggs

Size	Weight per Dozen (oz.)
Jumbo	30
Extra large	27
Large	24
Medium	21
Small	18
Peewee	15

Blood Spots. Blood spots are also called meat spots. These are occasionally found on an egg yolk. These tiny spots do not indicate a fertilized egg. Rather, they are caused by the rupture of a blood vessel on the yolk surface during formation of the egg or by a similar accident in the wall of the oviduct. Less than 1 percent of all eggs produced have blood spots.

Storage

Eggs can be stored at 30°F (–1°C) for up to 6 months in the shell. They can be frozen out of the shell for extended storage. The large quantities of eggs required by the food manufacturing industry are preserved by freezing. Eggs may be frozen as the whole egg (minus shell), separately as the white and yolk, or in varying combinations.

Also, after removing from the shell, eggs can be dried (dehydrated) as whole eggs, whites, or yolks. The methods of dehydrating eggs include spray drying, tray drying, foam drying, and freeze-drying.

Salmonella. The inside of the egg had once been considered almost sterile. Recently a bacterial organism, *Salmonella enteritidis,* has been found inside some eggs. Only a very small number of eggs might contain *Salmonella enteritidis.* Even in areas where outbreaks of salmonellosis have occurred, tested flocks show an average of only about 2 to 3 infected eggs out of each 10,000 produced.

The FDA considers these foods "potentially hazardous." The designation is not cause for alarm. It simply means that these foods are perishable and should receive refrigeration, sanitary handling, and adequate cooking. Lack of attention can make any food a hazardous food.

Fertile Eggs

Eggs that can be incubated develop into chicks. Fertile eggs are not more nutritious than nonfertile eggs, do not keep as well as nonfertile eggs, and are more expensive to produce.

Organic Eggs

Organic eggs are from hens fed rations having ingredients that were grown without pesticides, fungicides, herbicides, or commercial fertilizers. No commercial laying hen rations ever contain hormones. Due to higher production costs and lower volume per farm, organic eggs are more expensive than eggs from hens fed conventional feed. The nutrient content of eggs is not affected by whether or not the ration is organic.

Egg Substitutes

With all the attention paid to cholesterol, the level of cholesterol (about 240 mg) in eggs caused consumers to reduce their consumption of eggs. Food manufacturers have taken different approaches to reducing the cholesterol in egg, from physically separating the cholesterol from the yolk to formulating yolks from other products and combining these with the albumen. These products are sold as egg substitutes.

Another approach to reducing the cholesterol and changing the fat content of eggs is to change the genetics of chickens so that they produce the type of egg desired.

Summary

The term "meat" generally refers to the skeletal muscle from the carcasses of animals–beef and veal (cattle), pork (hogs), and lamb (sheep). Inspection takes place at practically every step of the livestock procurement and meatpacking processes. Grading establishes and maintains uniform trading standards and aids in the determination of the value of various cuts of meat. Carcasses are given both a quality and a yield grade. Meat and processed meat products, and other foods of animal origin, provide a complete protein source that contains, in favorable quantities, all the essential animo acids. The most causes of variation in tenderness of beef, pork, lamb, and veal include genetics, species and age, feeding, muscle type, suspension of the carcass, electrical stimulation, chilling rate, aging, mechanical tenderizing, chemical tenderizing, freezing and thawing, cooking and carving.

Poultry includes meat from chickens and turkeys. Meat from chickens and turkeys provides a high-quality protein that is low in fat. The protein is an excellent source of essential amino acids. Appearance, texture, and flavor of poultry meat are a primary concern in the food industry and to the consumer. Poultry meat color is affected by factors such as bird age, sex, strain, diet, intramuscular fat, meat, moisture content, pre-slaughter conditions, and processing variables. Chickens, turkeys, ducks, geese, guineas, and pigeons are all eligible for grading and certification services provided by the USDA's Agricultural Marketing Service (AMS) Poultry Program's Grading Branch.

Although eggs are widely known as breakfast entrees, they also perform in many other ways for the knowledgeable cook. Eggs are an excellent source of amino acids. The structure and characteristics of an egg include its color, shell, white, yolk, air cell, chalaza, germinal disc, and membranes. Classification is determined by interior and exterior quality and designated by the letters AA, A, or B. In many egg packing plants, the USDA provides a grading service for shell eggs. Food manufacturers have taken different approaches to reducing the cholesterol in eggs, from physically separating the cholesterol from the yolk to formulating yolks from other products and combining these with the albumen. Producers are also trying to reduce the cholesterol and change the fat content of eggs by changing the genetics of chickens.

Review Questions

Success in any career requires knowledge. Test your knowledge of this chapter by answering these questions or solving these problems.

1. The general composition of meat is ______ percent water, ______ percent protein, ______ percent fat, and ______ percent ash (mineral).
2. Define what the Contract Acceptance Certificate ensures.
3. Who authorizes meat inspection?
4. Why was the development of the extrusion method important in the meat industry?
5. List three general meat by-products.
6. What are the parts of an integrated meat chicken production company?
7. Explain the difference between a grade AA, a grade A, and a grade B egg.
8. Why are eggs gathered and refrigerated frequently?
9. Beef and veal are from ____________, pork from ____________, and lamb from ____________.
10. Another word for egg white is ____________.
11. As meat is cooked, contractile ____________ become less tender, and ____________ becomes more tender.
12. List the five factors affecting meat tenderness.

Student Activities

1. Visit the Web site of the USDA/AMS and compare the quality grades to the yield grades of beef and pork. Report on your findings.
2. Leave a small piece of meat to set at room temperature for a couple of days. Describe and explain the changes.
3. Make a list of processed meats. Compare your list to other class members' lists to see who has the least duplicated list.
4. Read and research magazine articles/advertisements, television advertisements, and draw some conclusions about

the modern consumers of meat. What do they want in a product? Report on your findings.

5. Display and discuss the colored wall charts of the cuts of beef and pork. These are available from the National Live Stock and Meat Board.
6. Visit a grocery store and make a list of all poultry products. Keep the list separated in terms of canned, frozen, and processed. Report your findings.
7. Visit restaurants and find out how poultry is featured in the menu. Develop a report or presentation on your findings.
8. Visit the Web site for the USDA/AMS Quality Standards at <www.ams.usda.gov/poultry/standards/index.htm>. Develop a short report or presentation that compares the quality grades of poultry to that of beef and pork.
9. Conduct a taste test of some of the wide variety of processed poultry meats such as hams, hot dogs, lunchmeats, sausages, and salami. Compare the taste of these to the same products traditionally prepared from the red meats.
10. Compare the nutrients in one egg to that in a fast food hamburger, candy bar, or other food you eat often. Report your findings. Use Table A-8.

Resources

Bartlett, J. 1996. *The cook's dictionary and culinary reference.* Chicago: Contemporary Books.

Corriher, S. O. 1997. *Cookwise: The hows and whys of successful cooking.* New York: William Morrow and Company, Inc.

Cremer, M. L. 1998. *Quality food in quantity. Management and science.* Berkeley, CA: McCutchan Publishing Corporation.

Ensminger, A. H., M. E. Ensminger, J. E. Konlande, and J. R. Robson. 1994. *Foods and nutrition encyclopedia.* 2 Vols. Boca Raton, FL: CRC Press.

FDA. 2000. *Food irradiation. A safe measure.* Publication No. 00-2329. Washington, DC: USDA.

Horn, J., J. Fletcher, and A. Gooch. 1997. *Cooking a to z. The complete culinary reference source.* Glen Ellen, CA: Cole Publishing Group, Inc.

Mathewson, P. R. 1998. *Enzymes.* St. Paul, MN: Eagan Press.

National Council for Agricultural Education. 1993. *Food science, safety, and nutrition.* Madison, WI: National FFA Foundation.

Vaclavik, V. A., and E. W. Christina. 1999. *Essentials of food science.* Gaithersburg, MD: Aspen Publishers, Inc.

Wagner, S. Ed. 1999. *The recipe encyclopedia: The complete illustrated guide to cooking.* San Diego, CA: Thunder Bay Press.

Internet

Internet sites represent a vast resource of information. The URLs (uniform resource locator) for the World Wide Web sites can change. Using one of the search engines on the Internet such as Yahoo!, HotBot, AltaVista, Excite, Dogpile, About, or Google, find more information by searching for these words or phrases: meat processing, the name of any specific cut of meat, specific meat products names, federal meat inspection, meat/poultry/egg grading, processed meats, beef processing, poultry processing, meat cooking, meat substitutes, turkey production, USDA Quality Grade Standards, Contract Acceptance Certificate, egg production, egg substitutes, organic eggs. Also, Table A-7 provides a listing of some useful Internet sites that can be used as a starting point.

Chapter 18

Fish and Shellfish

Objectives

After reading this chapter, you should be able to:

- Identify three fish and three shellfish used for food
- Describe aquaculture and processing
- Discuss the composition of fish and shellfish
- Identify three spoilage issues associated with fish
- Describe two processes that ensure quality
- List four factors that affect the grading of fish
- List four fish products and by-products
- Describe two methods for preserving fish
- Explain the methods of inspection during processing

Key Terms

aquaculture
breaded
crustaceans
FPC
freezer burn
glazing
HACCP
mollusks
roe
shucked
surimi

Fish and shellfish provide a source of high-quality protein to the diet. Due to the demand and popularity of fish and shellfish, many are commercially cultured and processed. Processed fish and shellfish are checked for quality and graded.

FIGURE 18-1

Catfish ready for processing. (*Source:* USDA Photography Library)

FISH, SHELLFISH, SALT- AND FRESHWATER

Fish (finfish) are classified into saltwater and freshwater varieties. Their flavor depends on the water in which they were grown. Fish are also classified on the basis of their fat content–lean being less than 2 percent fat and fat being more than 5 percent fat. Common species of edible fish include catfish, trout, cod, halibut, haddock, pollock, salmon, tuna, mackerel, herring, shad, tilapia, and eel. The fish are vertebrates (Figure 18-1).

Shellfish include the **mollusks** and the **crustaceans**. Mollusks are soft-bodied and partially or wholly enclosed in a hard shell composed of minerals. Oysters, clams, abalone, scallops, and mussels are examples of mollusks. Crustaceans are covered in a crustlike shell and have segmented bodies (like insects). Common crustaceans used for food include the lobster, crab, shrimp, prawns, and crayfish.

FISHING VERSUS CULTURE

Firms that produce, process, and distribute fish and shellfish are located throughout the United States. American consumers use approximately 8 percent of the total world catch of fish and shellfish. This supply is provided by commercial fishermen, **aquaculture** producers, and imports.

Aquaculture

Many popular fish and shellfish products in the United States are harvested to their full biological capacity in U.S. waters (Figure 18-2). To help meet the demand for some of these products, several varieties of fish and shellfish are grown in both freshwater and marine aquaculture facilities located throughout the United States. Aquaculture facilities cultivate approximately 30 different species of fish and shellfish and grow a variety of aquatic plants. Some of these products include:

- Catfish: Farming concentrated in Mississippi, Arkansas, Alabama, and Louisiana.
- Rainbow Trout: Grown throughout the United States with significant production in Idaho.
- Oysters and Clams: Production centered on the mid-Atlantic coast, Gulf of Mexico, and in Washington State.
- Shrimp and Prawns: Production in southern United States and Hawaii, Southeast Asia, South America, and Central America.
- Salmon: Cultivated in ocean pens in the states of Washington and Maine. Salmon also are partially cultivated in hatcheries on the east and west coasts and are released into the wild.
- Other Products: Other products include baitfish, crayfish (crawfish), hybrid striped bass, tilapia, yellow perch, walleye, bass, sturgeon, alligators, and shrimp.

FIGURE 18-2
Harvesting catfish from a pond in Mississippi. (*Source:* USDA Photography Library)

COMPOSITION, FLAVOR, AND TEXTURE

On average, Americans eat about 15 pounds of fish and shellfish each year. Scientific reports and government guides cite fish and shellfish as low in fat, easily digestible, and a good source of protein, important minerals, and vitamins.

Fish and shellfish contain high-quality protein with all the essential amino acids like red meat and poultry. It is also low in fat–and most of the fat it has is unsaturated. Because many diets now specify unsaturated fat, rather than saturated fat, fish and shellfish make excellent main dishes. Some fish are relatively high in fat such as salmon, mackerel, and catfish. However, the fat is primarily unsaturated.

The cholesterol content of most fish is similar to red meat and poultry–about 20 milligrams per ounce. Some shellfish contain more cholesterol than red meat. Because the fat is mainly polyunsaturated, shellfish may be allowed for some fat- and cholesterol-restricted diets.

Fish is also a good source of B vitamins–B_6, B_{12}, biotin, and niacin. Vitamins D and A are found mainly in fish liver oils, but some high-fat fish are good sources of vitamin A. Fish is a good source of several minerals–especially iodine, phosphorus, potassium, and zinc. Canned fish with edible bones, such as salmon or sardines, are also good sources of calcium. Oysters are a good source of iron and copper. Saltwater fish and shellfish are also excellent sources of iodine. Table A-8 provides the composition of a variety of fish and shellfish products.

Fish, shellfish, and products possess unique flavors and texture. Table 18-1 compares the flavor and texture of a variety of fish, shellfish, and products.

SPOILAGE

Fresh fish held at 61°F (16°C) remains good for only a day or less. At 32°F (0°C) finfish may remain good for 14 to 28 days depending on the species. For some species the time may be less. Fish spoil quicker than other meats because bacteria on the skin and in the digestive tract attack all the tissues once the fish is killed and these bacteria are often adapted to cold temperatures. Fish struggle when they are caught, and they convert all their glycogen to lactic acid before death. Associated with the fat of fish are phospholipids containing trimethylamine. Bacteria and enzymes from the fish split the trimethylamine from the phospholipids, producing the characteristic fishy odor.

Table 18-1 Fish and Shellfish Classified by Flavor and Texture

Mild Flavor	Moderate Flavor	Full Flavor
Delicate Texture	**Delicate Texture**	**Delicate Texture**
Cod	Black Cod	Bluefish
Crabmeat	Buffalo	Mussel
Flounder	Butterfish	Oysters
Haddock	Lake Perch	
Pollock	Lingcod	
Scallops	Whitefish	
Skate	Whiting	
Sole		
Moderate Texture	**Moderate Texture**	**Moderate Texture**
Crawfish	Canned Tuna	Canned Salmon
Lobster	Conch	Canned Sardines
Rockfish	Mullet	Mackerel
Sheepshead	Ocean Perch	Smoked Fish
Shrimp	Shad	
Walleye Pike	Smelt	
	Surimi Products	
	Trout	
Firm Texture	**Firm Texture**	**Firm Texture**
Grouper	Amberjack	Clams
Halibut	Catfish	Marlin
Monkfish	Drum	Salmon
Catfish	Mahi Mahi	Swordfish
Sea Bass	Octopus	Tuna
Snapper	Pompano	
Squid	Shark	
Tautog	Sturgeon	
Tilefish		

Source: U.S. Department of Commerce, *Seafood Inspection Program*

The following guidelines help identify good-quality fish at a fish market (Figure 18-3). The skin of the fish should be shiny, almost metallic, with color that has not faded. As the fish decomposes, its skin markings and colors become less distinctive. Scales of the fish should be brightly colored and tightly attached to the skin. The gills should be red and free from slime. As fish ages, the gills change color, fading gradually to a light pink, then becoming gray and eventually brownish and greenish.

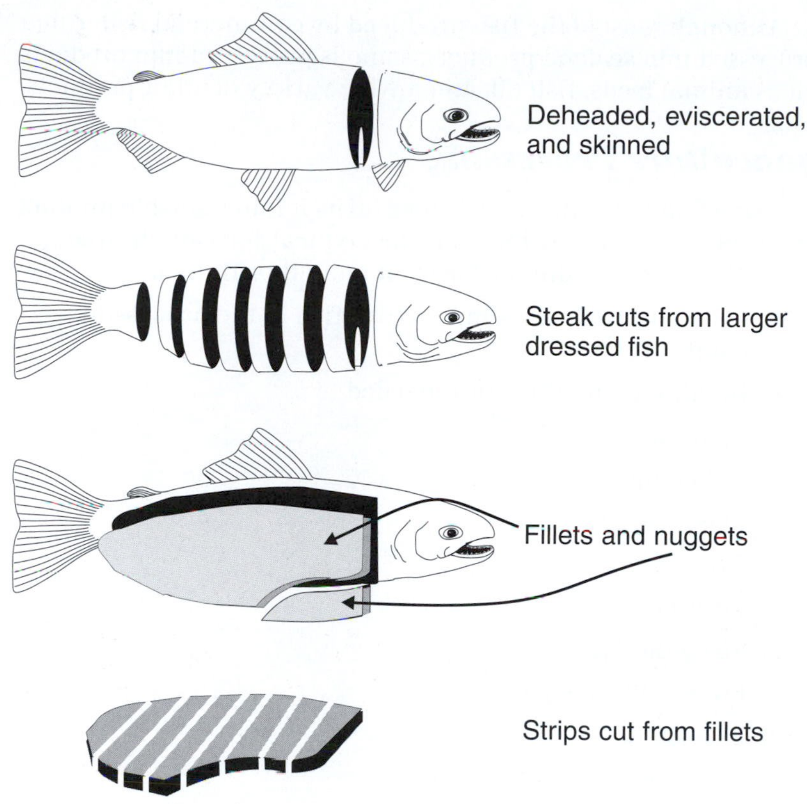

FIGURE 18-3

Catfish cuts available fresh, packed, on ice, or frozen.

If the head is still on the fish, the eyes should be bright, clear, transparent, and full–often protruding. As the fish decomposes, its eyes become cloudy and may turn pink and shrink. The flesh of whole or dressed fish should be firm, elastic, and not separated from the bones. As the fish decomposes, the flesh becomes soft, slimy, and slips away from the bones. Fish fillets should have a fresh cut appearance and color that resembles freshly dressed fish. Last, but not least, the odor should be fresh or mild–not fishy.

PROCESSING

Most domestic and imported fish and shellfish are processed before reaching consumers. The National Marine Fisheries Service estimates that approximately 1,500 plants manufacture fish and shellfish in the United States. Most are small businesses, and many are family owned. A few concentrate on a single species, such as tuna, salmon, and menhaden. Most process several different species to take advantage of the different fisheries in their region.

Although most of the fish produced by commercial fishermen is processed into seafood products, some is processed into products such as animal feeds, fish oil, and a wide variety of other products.

Aquaculture Processing

Processing fish through several steps turns it into a salable product (Figure 18-4). The following steps are typical for catfish processing, but the steps are similar for trout and other finfish.

1. Receiving and weighing the live fish at the processing plant
2. Holding them alive until needed
3. Stunning
4. Deheading
5. Eviscerating
6. Skinning
7. Chilling
8. Size grading
9. Freezing or ice packing
10. Packaging
11. Warehousing
12. Icing
13. Shipping the finished product

FIGURE 18-4
Dressed fish. (*Source:* USDA Photography Library)

Inspection

Unlike the red meat and poultry processing industries, fish processing does not fall under the regulations of the United States Department of Agriculture (USDA). Before beginning operation, fish processors must contact local county health officials to comply with county health regulations and to obtain a health permit. Fish processing operations also must adhere to standards set forth by the Good Manufacturing Practice Code of Federal Regulations, Title 21, Part 110, and are subject to announced and unannounced inspections by the Food and Drug Administration (FDA).

HACCP. Traditionally, industry and regulators have depended on spot-checks of manufacturing conditions and random sampling of final products to ensure safe food. This system tends to be reactive, rather than preventative, and can be less efficient than the new system. The new system is known as Hazard Analysis and Critical Control Point, or **HACCP** (pronounced hassip). Many of its principles already are in place in the FDA-regulated low-acid canned food industry. In December 1995, the FDA issued a final rule establishing HACCP for the seafood industry. Those regulations took effect December 18, 1997.

Quality

As in other industries, the aquaculture industry considers quality a number one priority. Without a quality product, sales of products would quickly decrease.

In order to maintain a quality product and promote consumer confidence, the major commercial fish processors contracted voluntarily with National Marine Fisheries Service (NMFS) to have their plants inspected. NMFS is an agency service of the National Oceanic and Atmospheric Administration (NOAA), an agency of the United States Department of Commerce (USDC). Federal inspectors with the NMFS perform unbiased, official inspections of plants, procedures, and products for firms that pay for these services. The inspectors issue certificates indicating quality and condition of the products.

The NMFS voluntary inspection program provides for the inspection of products and facilities and the grading of products.

Inspection is the examination of fish (seafood) products by a U.S. Department of Commerce inspector or a cross-licensed State or U.S. Department of Agriculture inspector. They determine whether the product is safe, clean, wholesome, and properly labeled. The equipment, facility, and food-handling personnel must also meet established sanitation and hygienic standards.

HACCP EQUALS TWO PLUS FIVE PLUS SEVEN

HACCP stands for Hazard Analysis Critical Control Point; it is pronounced "Hassip." Its concept supports a producer's, manufacturer's, foodservice operator's, retailer's, or distributor's goal to supply safe food—one that will not cause illness or injury to a consumer when used as intended. HACCP is a systematic, two-part process, and it has five preliminary steps and seven principles.

The two-part process:

1. Assess hazards, taking into consideration factors that contribute to most outbreaks and risk evaluation techniques to identify and prioritize hazards.
2. Identify control measures that focus on prevention, establishing controls to reduce, prevent, or eliminate safety risks.

The five primary steps:

1. Assemble the HACCP team.
2. Describe the food and its distribution.
3. Identify intended use and consumers.
4. Develop the flow diagram.
5. Verify the flow diagram.

Finally, the seven principles:

1. Conduct a hazard analysis.
2. Determine the critical control points.
3. Establish critical limits.
4. Establish monitoring procedures.
5. Establish corrective actions.
6. Establish verification procedures.
7. Establish recordkeeping and documentation procedures.

HACCP is as simple as 2, 5, 7.

To learn more about HACCP, search the Web or visit these Web sites:

<vm.cfsan.fda.gov/~lrd/haccp.html>
<www.nal.usda.gov/fnic/foodborne/haccp/index.shtml>
<www.agr.ca/policy/adapt/haccp.html>

Grading

After inspection, grading determines the quality level. Only products that have an established grade standard can be graded. Industry uses the grade standards to buy and sell products. Consumers rely on grading as a guide to purchasing products of high quality. Graded products can bear a U.S. grade mark that shows their quality level. The "U.S. Grade A" mark indicates that the product is of high quality. It is uniform in size, practically free of blemishes and defects, in excellent condition, and has good flavor and odor.

A grading scheme used by trout processors provides an example of how grading works to provide a Grade A mark. In determining the grade of processed fish, each fish is scored for the following factors:

- Appearance–The overall general appearance of the fish, including consistency of flesh, odor, eyes, gills, and skin.
- Discoloration–This refers to any color not characteristic to the species.
- Surface defects–These include the presence of fins; ragged, torn, or loose fins; bruises; and damaged portions of fish muscle.
- Cutting and trimming defects–These include body cavity cuts, improper washing, improper deheading, and evisceration defects.
- Improper boning–For boned styles (fillet) only, this refers to the presence of an unspecified bone or piece of bone.

After inspecting each fish, the number of defects are totaled. Grade A is given when the maximum number of minor defects is three or less and no major defects. Grade B is given to fish with up to five minor defects and one major defect. Grade A fish must also possess good flavor and odor for the species, and grade B must possess reasonably good flavor and odor for the species.

Products

Fresh or frozen fish can be marketed as whole or round, dressed or pan-dressed, fillets, drawn fish, steaks, sticks, or nuggets (Figure 18-5). Whole fish or round fish are just as they come out of the water. Drawn fish have only the entrails removed. Dressed fish are scaled and eviscerated, and they usually have the head, tail, and fins removed. Steaks are crosscut sections of the larger sizes of dressed fish. Fillets are sides of fish cut lengthwise away from the backbone. Sticks are uniform pieces of fish cut lengthwise or

FIGURE 18-5

Fish and shellfish are common even in small grocery outlets.

crosswise from fillets or steaks. Nuggets are like fillets only smaller. Some fish is manufactured into products such as **breaded**, formed, and imitation products. Some fish is cured, and some is canned.

PRESERVATION

Fish are preserved by drying, salting, curing, or smoking, but not all people like fish preserved this way. Some progress is being made on the use of irradiation. Refrigeration, freezing, and canning remain the best methods for preserving the quality of fish.

Large boxes of fish are frozen at temperatures of –22°F (–30°C) or below. When individual fish are frozen, they are sometimes glazed with layers of ice to protect the surface of the fish from oxidation and from **freezer burn** (drying out). **Glazing** is done by dipping the fish in cold water and then freezing a layer before dipping the fish again. Shrimp are also glazed. Even with glazing, frozen fish require packaging in materials that are airtight and moisture-tight. Prebreaded, precooked fish sticks and individual portions are also frozen. High-quality, low-fat fish can be held frozen at –6°F (–21°C) for as long as two years.

Fish with higher fat content like salmon, tuna, and sardines are often canned. Additional fish oil, vegetable oil, or water is often added to the can before sealing it closed. Canned fish products have a shelf life of several years. A typical canning operation involves the following steps:

1. Thaw the partially frozen fish received from a fishing vessel.

2. Eviscerate, clean, and sort.
3. Precook.
4. Cool and separate the meat (usually by hand).
5. Compact meat into shapes to fit cans.
6. Add salt, oil, or water to cans.
7. Vacuum seal cans and sterilize in a retort.

SHELLFISH

Shellfish may be marketed in the shell, **shucked** (removed from the shell), headless (shrimp), and as cooked meat. Shrimp is also sold as peeled, cleaned, and breaded. Shrimp are designated Jumbo, Large, Medium, and Small based on the number per pound. Oysters receive similar designations.

FISH BY-PRODUCTS

Parts of the fish such as the intestines, heads, and gills and less favored fish, are not used for human food. These have been ground up, dried, and converted to fish meal for use as animal feed. This product is called fish meal. Some fish meal is used for fertilizer.

Fish protein concentrate (**FPC**) or fish flour is produced from the dehydrated and defatted fish. It is an excellent source of high-quality protein, which can be used to supplement the breads and cereal products of people in many parts of the world.

Roe

Roe is the mass of eggs and sacs of connective tissue enclosing the thousands of eggs. Some people eat the roe of such fish as the Shad. Fresh roe is usually cooked by parboiling. Caviar is sturgeon roe preserved in brine.

STORAGE

Fish and shellfish must never sit unrefrigerated for long. If necessary, it should be transported in an ice chest. Seafood with bruises or punctures will spoil more rapidly.

As soon as possible, finfish needs to be refrigerated as close to 32°F (0°C) as possible. Fish can be held twice as long at 32°F (0°C) as it can be at 37°F (2.8°C). Fish and shellfish should be cooked within two days of purchase. If not, it can be stored by following a few guidelines. Before storing fresh fish, its package is removed, and it is rinsed under cold water and patted dry. When fish sets in its own juices, the flesh deteriorates more rapidly. To prevent this, cleaned finfish, whole, fillets, or steaks are placed onto a cake rack so that the fish do not overlap. This rack is placed in a shallow pan. Filling the pan with crushed ice allows the fish to keep more that 24 hours. Ice leaches color and flavor from fish and should not come into contact with the fish. The covered pan is placed in a refrigerator, then drained and re-iced as necessary. Each day, the fish, pan, and rack are rinsed and re-iced. Fish with a fishy or ammonia smell after being rinsed should be discarded.

If the fish will not be used within a day or so, it is best to freeze immediately. After rinsing the fish under cold water and patting very dry, the fish should first be wrapped tightly in plastic wrap, squeezing all the air out. Then it is wrapped tightly in aluminum foil and frozen. For best quality of fish frozen at home, it should be used within two weeks.

When frozen fish is thawed, it is always thawed in the refrigerator. Thawing at temperatures higher than 40°F (4.4°C) causes excessive drip loss and adversely affects taste, texture, aroma, and appearance.

Live oysters, clams, and mussels are stored in the refrigerator at a temperature of about 35°F (1.7°C). They should be kept damp, but not placed on ice. Freshwater or an airtight container will kill them.

Freshly shucked oysters, scallops, and clams are stored in their own containers and stored in a refrigerator about 32°F (0°C). Surrounding the containers with ice gives the best results.

Live lobster and crab are stored in the refrigerator in moist packaging (seaweed or damp paper strips), but not in airtight containers, water, or salted water. Lobsters should generally remain alive for about 24 hours.

Just before opening or cooking scallops, mussels, clams, or oysters in the shell, they should be scrubbed under cold water to clean them. Soaking them in water with flour or cornmeal to encourage the creatures to eat to clean out the grit only shortens their life.

Frozen fish and seafood should be stored at 0°F (–18°C) or below. For fish purchased frozen, it is best when used within two months.

NEW PRODUCTS

Machines similar to deboning machines are being used to obtained minced fish flesh from filleting wastes and underutilized fish species. The minced fish flesh is washed to remove solubles

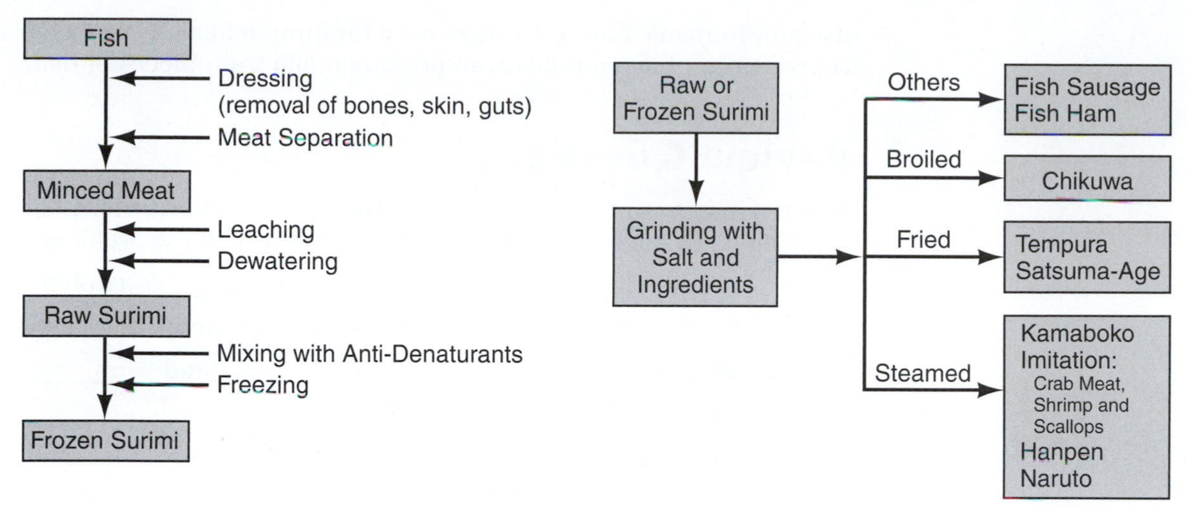

FIGURE 18-6

Diagram showing production of different surimi-based products.

including pigments (color) and flavors. This leaves an odorless, flavorless, high-protein product called **surimi**. This can be combined with other flavors and colors. Then it can be extruded in shapes resembling other products such as crabmeat and lobster (Figure 18-6).

Summary

Fish includes saltwater and freshwater finfish such as catfish, trout, halibut, salmon, tuna, herring, and eel. Shellfish include a group of mollusks and crustaceans like clams, oysters, lobsters, crabs, shrimp, and crayfish. Fish and shellfish are provided by commercial fishing and aquaculture producers. Fish and shellfish provide a high-quality protein and are also a good source of B vitamins, calcium, phosphorus, iodine, and potassium.

Fish spoil easily, so they require strict processing and preservation procedures to maintain quality. Fish processing operations adhere to standards in the Good Manufacturing Practice Code and processors use the HACCP method to monitor quality. After inspection, grading determines the quality. Grade A indicates a product of high quality that is uniform in size, free of blemishes and defects, in excellent condition, with good flavor and odor.

Fish and shellfish are marketed fresh or frozen. Fish can be marketed as whole, dressed, pan-dressed, filleted, steaks, sticks, or nuggets. Fish protein concentrate or fish flour are fish by-products used for humans. Roe is fish eggs used for human food also. Surimi represents a new manufactured product made from pieces of fish.

Review Questions

Success in any career requires knowledge. Test your knowledge of this chapter by answering these questions or solving these problems.

1. Fish are classified on the basis of their _________ content.
2. Describe the difference between fish and shellfish.
3. Fish is a good source of vitamins ________ and ________, _________ fat, and high-quality _________.
4. Discuss three indications of spoilage in fish.
5. List the steps for catfish processing.
6. What does HACCP stand for?
7. Name the four grading factors for fish.
8. What are the head, gills, and intestines of fish used for?
9. Caviar is sturgeon _________ preserved in brine.
10. List four methods of preserving fish.

Student Activities

1. Track your diet for one week, and report on how much fish and what type of fish you eat in a week. Remember to include tuna fish and fast-food fish sandwiches.
2. Compare the protein and energy content of a type of fish and a shellfish to that of steak and chicken breast. Use Table A-8.
3. Name some common lobsters and crabs frequently seen on menus at restaurants.
4. Create a classroom display of color pictures of several types of edible saltwater and freshwater fish, and some of the types of mollusks and crustaceans.
5. Create a sample (taste) test of some of the unique types of fish and shellfish. Use Table 18-1 for ideas.

Resources

Bartlett, J. 1996. *The cook's dictionary and culinary reference.* Chicago: Contemporary Books.

Brody, J. E. 1981. *Jane Brody's nutrition book.* New York: Bantam Books.
Cremer, M. L. 1998. *Quality food in quantity. Management and science.* Berkeley, CA: McCutchan Publishing Corporation.
Ensminger, A. H., M. E. Ensminger, J. E. Konlande, and J. R. Robson. 1994. *Foods and nutrition encyclopedia.* 2 Vols. Boca Raton, FL: CRC Press.
Gardner, J. E., Ed. 1982. *Reader's digest. Eat better, live better.* Pleasantville, NY: Reader's Digest Association, Inc.
Horn, J., J. Fletcher, and A. Gooch. 1997. *Cooking a to z. The complete culinary reference source.* Glen Ellen, CA: Cole Publishing Group, Inc.
Parker, R. 2001. *Aquaculture science.* Albany, NY: Delmar Publishers.

Internet

Internet sites represent a vast resource of information. The URLs (uniform resource locator) for the World Wide Web sites can change. Using one of the search engines on the Internet such as Yahoo!, HotBot, AltaVista, Excite, Dogpile, About, or Google, find more information by searching for these words or phrases: specific types of fish and shellfish, saltwater fish, freshwater fish, aquaculture, HACCP, fish processing, fish grading, fish by-products, fish roe, FPC (fish protein concentrate). Also, Table A-7 provides a listing of some useful Internet sites that can be used as a starting point.

Chapter 19

Cereal Grains, Legumes, and Oilseeds

Objectives

After reading this chapter, you should be able to:

- Diagram the general structure of a grain
- Name three cereal grains
- Describe the general composition of grains, legumes, and oilseeds
- Identify three properties of starch
- List four factors that must be controlled when cooking starch
- Discuss the milling of grain to flour
- Identify five types of wheat flour
- Explain the classes of wheat and grades of flour
- Identify the type of flours other than wheat flour
- List the steps in corn refining
- Name four products derived from corn
- Explain the processes that take place during baking
- List four oilseeds and indicate the use of their products
- Discuss the general use of legumes
- Name four general categories of products from soybean extraction
- Identify five food products of soybean extraction

Key Terms

amylopectin
amylose
bioproducts
bleached
dextrinization
endosperm
extraction
germ
insoluble
middlings
mill starch
milling
refining
retrogradation
shorts
soluble
steeping
straight grade
waxy
weeping

Cereal grains and legumes supply energy (starch) and protein to many people. As a food, they are consumed as seeds, but more often they are consumed in some processed form such as flour, syrup, or vegetable protein extract. Corn refining and soybean extraction provide a wide variety of products for food and technical uses.

CEREAL GRAINS

Many types of grains and seeds are used throughout the world. For example, consumers can purchase bean flour, peanut flour, sunflower flour, buckwheat flour, soy flour, and many others. Cereal markets have expanded the range of uses of these sources due to their consumer appeal and due to manufacturers responding through improved efficiencies and productivity and through advertising, market leaders, lowering prices, and cutting costs. The focus of this chapter is on the grain–wheat, with some discussion of the characteristics of other grains such as corn, rice, oats, and rye.

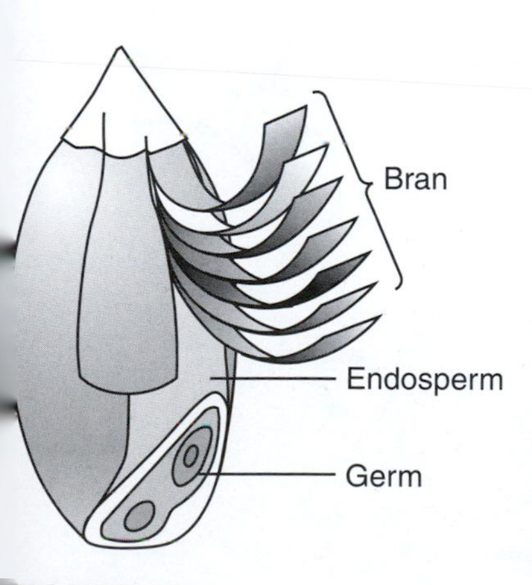

FIGURE 19-1
Diagram showing the parts of a grain seed.

General Structure and Composition

All whole grains have similar structures–outer bran coats, a **germ**, and a starchy **endosperm** portion. Cereal products vary in composition depending on the part or parts of the grain used.

The outer layers of the kernel or the bran constitute about 5 percent of the kernel. The bran is chiefly cellulose with much of the mineral and some of the vitamin of the kernel. As milled, it may also contain some germ and a small amount of the aleurone layer.

Under the bran layers lies the aleurone layer. This layer is rich in proteins, in phosphorus, and thiamine. It also contains starch. About 8 percent of the kernel is aleurone layer (Figure 19-1).

The endosperm is the large central portion of the kernel and contains most of the starch. It also contains most of the protein of the kernel but very little mineral or fiber, and only a trace of fat. The endosperm constitutes about 82 percent of the kernel.

The germ is the small structure at the lower end of the kernel. It is rich in fat, protein, and mineral and contains most of the riboflavin content of the kernel.

The chaffy coat that covers the kernel during growth is eliminated.

Cereals are processed grains that are generally 75 to 80 percent carbohydrates. Fiber is also an important attribute of cereal especially bran cereals, which may contain 10 to 26 grams of fiber per cup. Cereals contain both **soluble** and **insoluble** fiber. Insoluble fiber is good for the digestive tract and helps reduce the risk of certain cancers. Soluble fiber, the type that lowers blood cholesterol, originates in the endosperm of grain and is found in oats, legumes, fruits, and vegetables.

STARCH

Starch is a storage form of carbohydrate deposited as granules or aggregates of granules in the cells of plants. Sizes and shapes of granules differ in starches from various sources but all are microscopic in size (Figure 19-2).

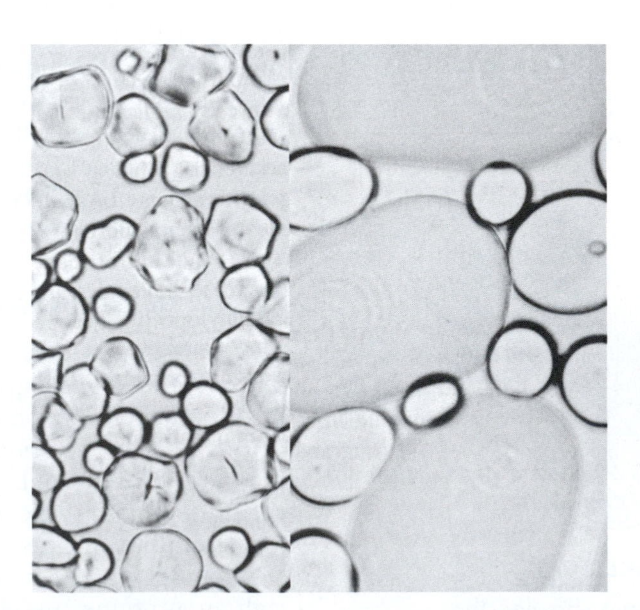

FIGURE 19-2

Representation of a photomicrograph of corn starch granules (left) and potato starch granules (right).

The parts of plants that serve most prominently in the storage of starch are seeds, such as cereals and legumes, and roots and tubers such as parsnips, potatoes, and sweet potatoes. Some starches are derived from the cassava root (marketed as tapioca) and from the pith of a tropical palm (marketed as sago). Starch may be hydrolyzed to form glucose, but several intermediate products, such as dextrin and maltose, are first formed (Chapter 3).

Starch granules are made up of many starch molecules arranged in an organized manner. These molecules are of two types, called fractions, of starch: **amylose** and **amylopectin** (refer to Figure 3-3). The amylose is a polysaccharide of glucose. It contributes gelling characteristics to cooked and cooled starch mixtures. Amylopectin is a highly branched polysaccharide of glucose that provides thickened properties but does not usually contribute to gel formation. Most starches are mixtures of the two fractions. Corn, rice, and wheat starches contain 16 to 24 percent amylose with 84 to 76 percent being amylopectin. Tapioca and potato starches are lower in amylose content than corn, rice, and wheat. Certain strains of corn, rice, grain sorghum, and barley have been developed that are practically devoid of amylose. These are called **waxy** varieties because of the waxy appearance of the cut grain.

Properties of Starch

The starch granule is insoluble in cold water. A nonviscous suspension is formed in which the granules gradually settle to the bottom. When cooked, a colloidal dispersion forms in the resulting starch paste. Some pastes form gels, and some are nongelling. Some are opaque, and some are clear, semiclear, or cloudy in appearance and soft or cohesive in texture. In general, the pastes made with cereal starches, like corn and wheat, are cloudy in appearance, whereas those from root starches, like potato and tapioca, are clearer. Cooked and cooled mixtures of starches containing somewhat larger proportions of amylose, like ordinary cornstarch, tend to become rigid on standing or to gel. Tapioca and potato starch, containing a little less amylose, have fewer tendencies to gel.

Tapioca and potato starch pastes are cohesive and tend to be stringier, partly because of the chain length of the amylose molecules. The "skin" on the surface of cooked starches and cereals is probably due mainly to the amylose's tending to revert to an insoluble state. Waxy varieties of starch, like waxy cornstarch, form thickened viscous pastes that do not gel on cooling. The stringy characteristics of some of these waxy starch pastes may be eliminated if the starch is modified to produce cross-bonding between branches of the amylopectin molecules. High-amylose starches have also been produced. These offer possibilities for the development of edible protective coatings for individual pieces of food like dried fruits, nuts, beans, and candies.

Effect of Dry Heat. When dry heat is applied to starch or starch-containing foods, the color changes to brown, the flavor changes, and the starch becomes more soluble and has reduced thickening power. This process is called **dextrinization**. Brown gravy is usually relatively thin in consistency if the flour is browned in the process of making the gravy. Dry-heat dextrins are formed in the crust of baked flour mixtures, on toast, on fried starchy or starch-coated foods, and on various ready-to-eat cereals.

Effect of Moist Heat. When starches are heated with water the granules swell and the dispersion increases in viscosity until reaching a peak thickness. The dispersion also increases in translucency. The term gelatinization is used to describe these changes. The changes appear to be gradual over a temperature range during gelatinization. The granules are of varying sizes and do not swell at the same rates. The gelatinization temperature ranges also vary from one kind of starch to another.

Potato starch begins to gelatinize at a lower temperature than does cornstarch. Gelatinization is usually complete by 190° to 194°F (88° to 90°C). After maximum swelling has occurred, the granule ruptures. Continued heating under controlled conditions after gelatinization results in decreased thickness. Boiling or cooking starchy sauces and puddings in the home for longer periods of time usually does not produce thinner mixtures because the loss of moisture by evaporation is usually not controlled. The loss of moisture, resulting in increased concentration of the starch, causes increased thickness and offsets the decreasing thickness.

Gel Formation. The presence of amylose encourages the formation of a gel in cooked and cooled starch mixtures. Waxy varieties of starch without amylose do not form gels. The amylose molecules become more soluble as the granules are disrupted and swell during gelatinization. On cooling they tend to **retrogradation**, or gel formation continues as the starch mixture stands. Cornstarch puddings become thicker and more gelled when stored overnight in the refrigerator. Overall, the thickness of a starch paste on heating is not directly related to the strength of gel formed on cooling.

Factors Requiring Control

In order to obtain uniformity in the cooking of starches, five conditions must be standardized and controlled:

1. Temperature of heating
2. Time of heating
3. Intensity of stirring
4. pH of the mixture
5. Addition of other ingredients

Temperature and Time of Heating. Gelatinization temperatures vary for different starches. The larger granules start to swell first and at lower temperatures than smaller sizes, which explains why there is no exact temperature of gelatinization. It is a change that occurs over a range of temperatures. More concentrated mixtures show higher viscosity at lower temperatures than do less concentrated mixtures because of the larger number of granules that swell. Under controlled conditions, starch pastes that are heated rapidly are slightly thicker than similar pastes that are heated slowly.

Intensity of Stirring. Stirring while cooking starch mixtures is desirable in the early stages from a practical standpoint in obtaining a uniform consistency, but it also has value in accelerating gelatinization. However, if stirring is too intense or continued too long, it also accelerates the breakdown or rupturing of the starch and decreases viscosity.

pH of the Mixture. A low pH or acidity causes some fragmentation of starch granules, affecting swelling, and thus decreasing the viscosity of starch pastes. Mixtures with both low and high pH values–pH 2.5 and pH 10.0–gelatinize more rapidly than those with intermediate pH values–pH 4.0 and pH 7.0–and they also break down more rapidly with a decrease in viscosity. In lemon pie filling, cooking the starch paste with lemon juice might be expected to have some effect in decreasing the thickness of the pudding. This degree of acidity does not seem to have a major effect. However, a better natural flavor is obtained when lemon juice is added at the end of the cooking period, reducing the loss of volatile flavor constituents.

Addition of Other Ingredients. Various other ingredients are commonly used with starch in the preparation of food. Some ingredients have a distinct effect on the gelatinization and breakdown of starch and on the gel strength of the cooled mixture. Sugar (disaccharides or monosaccharides) is one ingredient used in many starchy mixtures. If used in a relatively large amount, it interferes with the complete gelatinization of the starch and decreases the thickness of the pastes. It competes with the starch for water. If not enough water is available for the starch granules, they cannot swell sufficiently.

High concentrations of disaccharides (like sucrose) are more effective in inhibiting gelatinization than are equal concentrations of monosaccharides (like glucose). At a concentration of 20 percent or more, all sugars and syrups cause a noticeable decrease in the gel strength of starch pastes.

Egg, fat, salt, and dry milk solids also hinder the gelatinization of starch granules.

Handling of Cooked Starch

Cooked mixtures that are not to be used immediately should be protected against bacterial contamination because they are a good medium for bacterial growth.

Acid mixtures should be cooled quickly to minimize the hydrolyzing effect of acid on the starch. The manner of cooling a starch paste affects its properties. Placing the utensil used for cooking in a refrigerator or cold water and stirring gently cools the mixture in a short time.

Weeping

Starch granules absorb water and swell during cooking. During the early stages of swelling, the water is so loosely held that undercooked starch releases the water easily while cooling and during storage. If the undercooked mixture is refrigerated, **weeping** is accelerated. At the same time, the viscosity of the paste may decrease, sometimes suddenly. This is called collapse of the gel.

MILLING OF GRAINS

Primitive people used stones, wood, and other mechanisms to grind their cereals. Eventually, they converted to the old water driven mills with the large mill stones. Modern **milling** replaces these mill stones with rollers (Figure 19-3).

Although not all grain is milled, most of the wheat used in the United States has been milled to some extent. The aim of millings is to separate the bran covering, germ, and endosperm to the extent

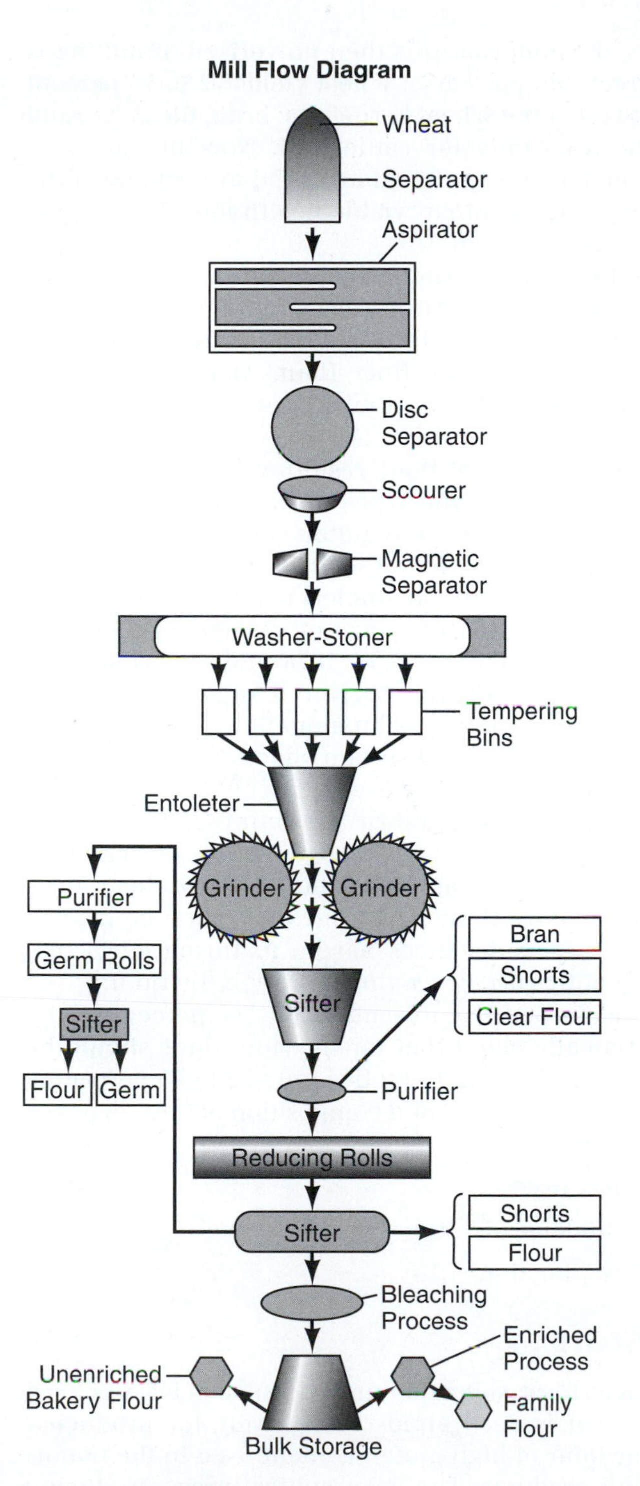

FIGURE 19-3

A flowchart of wheat milling.

desired. Generally, the endosperm is then pulverized. If milling is adequate and correct, 100 pounds of wheat yields 72 to 75 percent straight flour. The rest of the wheat kernel, the bran, the germ, and **shorts** used to be used only for cattle feed. Now nutritionists, health personnel, and a variety of food faddists have expanded the use of these waste products into a "viable health food." A sizeable market exists for bran and germ.

The inner portion of the kernel, which is granulated with difficulty, is known as **middlings**. After its separation from the bran, it is fed through a series of smooth rollers that further reduce the size of the particles and produce finer flour. About six to eight streams of flour are obtained from the rolling and sifting of the purified middlings.

From the many streams of flour resulting from the modern roller process, various grades and types of flours are made. The streams vary in their bran, germ, and gluten content.

A final stage in the production of white flour is often bleaching and/or maturing. Freshly milled, unbleached flour is yellowish in color chiefly because of the presence of carotenoid pigments (carotin). When used for baking bread, it produces a small and fairly coarse-textured loaf. If the flour is stored for several months, the color becomes lighter and the baking qualities improve. Food scientists found that the addition of certain chemical substances to the freshly milled flour will produce similar effects in a much shorter period of time. The Food and Drug Administration permits the use of nitrogen trichloride and nitrogen tetroxide, chlorine dioxide, benzyl peroxide, acetone peroxides, and azodicarbonamide to bleach and mature flour. The flour must then be labeled **bleached**. Most of these substances have a maturing effect that improves baking qualities besides acting to bleach the flour.

Because the endosperm represents about 84 percent of the total kernel, theoretically about that much white flour should be obtained by milling, but in actual practice only 72 to 75 percent is separated as white flour. The kind and composition of flour depends on these characteristics:

1. Class of wheat used
2. Conditions under which the wheat is grown
3. Degree of fractionation

Classes of Wheat

Wheats are classed as hard, soft, and durum (a special class of hard wheat). Durum wheat is used almost exclusively for producing semolina–granular flour of high gluten content used in the manufacture of macaroni products. The geographical areas producing

BREADS WITH SYMBOLIC MEANING

Pretzels and croissants are two popular breads whose origins have been forgotten.

Pretzels are shaped like knots. Traditionally, pretzels are made out of long strips of dough folded over into a loose, trefoil knot before being baked. They have been shaped this way since the seventh century. Thought to bring good luck and prosperity, pretzels have been called the world's oldest snack food.

Medieval monks invented pretzels to carry deep, religious meanings. The folded strips of dough resemble the folded arms of someone who is praying in the usual manner in those days, while the three holes represent the Christian Holy Trinity.

In medieval times, pretzels were given to children as rewards for learning their prayers. Today, they have lost the religious meanings, but pretzels are still among the world's most popular snacks.

The delicate, flaky croissant or crescent roll is a baked pastry that is curved with pointed tips. Although its popular name in English speaking countries is French, the roll itself is of Austrian origin, commemorating a Turkish shape.

According to the most popular story, in 1683 the Ottoman Turks invaded Vienna by trying to tunnel under the city's walls. The Turks were successfully repelled, thanks to the vigilance of the only people who were awake during the nighttime raid: the bakers. In celebration of the victory, the bakers created the croissant, shaping it like the crescent found on the Turkish flag.

Because there are several different stories of this event, the true details may be different. But all sources agree that the croissant's shape is the Turkish crescent, and that it was created in celebration of an Austrian victory over the Turks. The Turkish Siege of Vienna lasted 60 days.

For more about croissants and pretzels, and for recipes, visit these Web sites:

<frenchfood.about.com/home/frenchfood/library/weekly/aa092898.htm>
<www.redstaryeast.net/croiss.htm>
<www.ushistory.org/tour/_pretzel.html>

most of the hard spring wheats are the north central part of the United States and western Canada. Hard winter wheats are grown mainly in the south central and middle central states. Soft winter wheat is grown east of the Mississippi River and in the Pacific Northwest. Because climatic and soil conditions affect the composition of wheat, wide variations may occur within the classes.

Grades of Flour

The miller grades white flours on the basis of the four streams used to make them. **Straight grade** theoretically should contain all the flour streams resulting from the milling process, but actually 2 to 3 percent of the poorest streams is withheld. The poorest streams are those containing most of the outer layers of the endosperm and fine bran particles. Very little flour on the market is straight grade. Patent flours come from the more refined portion of the endosperm and may be made from any class of wheat. They are divided into the following, in order of quality:

- First patent
- Second patent
- First clear
- Second clear
- Red dog

Most patent flours on the market include about 85 percent of the straight flour. Clear grade is made from streams withheld in the making of patent flours.

Types of White Flour

Wheat flour contains a high percentage of starch and 10 to 14 percent protein, most of which is gluten. Gluten is important for binding ingredients together and for its elastic or stretching property. Within limitations, various types of flour may be used interchangeably by altering the proportions of the other ingredients of the mixture. Four common types of white flour include bread, all-purpose, pastry, and cake. White flours produced from wheats grown in some areas are of such poor baking quality that improvers (a gum such as xanthan or guar) are added.

Bread flour has a slightly higher percentage of gluten and a much stronger and more elastic gluten than other types of flour. Most flours are blended; a strong bread flour is made chiefly from hard wheat. Such flour is valuable chiefly for yeast breads but may be used in quick breads.

All-purpose flour (Figure 19-4) has a less strong and elastic gluten than bread flour. It may be a blend of hard and soft wheat flour or may be made entirely from hard or soft winter wheats. Although designated all-purpose flour, it is more popular for some quick breads than for other uses.

Pastry flour is made from soft winter wheat and contains a weaker quality of gluten and a slightly lower percentage of gluten than is found in bread and all-purpose flours. Its chief use is for cakes and pastries, although it is also very useful for all quick breads.

FIGURE 19-4
Many types of flours are available to the consumer.

Cake flours are specially prepared to reduce the gluten content to about 7 percent. They are best made from soft wheat and are finely ground. They are usually highly bleached with chlorine. High starch content and weak quality of gluten make cake flours useful chiefly for cakes.

According to a federal ruling, the terms whole wheat and graham are synonymous terms and refer to products made from the whole-wheat kernel with nothing added or removed. Graham flour was named for Sylvester Graham, an American food reform advocate.

Enriched Flour

Enriched flour is white flour to which specified B vitamins and iron have been added. Optional ingredients are calcium and vitamin D. The enrichment of bakers' white bread and rolls was made compulsory by the federal government in 1941 as a war measure to improve the nutritional status of the people. After the war, enrichment became voluntary. A number of states have passed laws requiring that all white flour sold within their boundaries must be enriched. If enrichment is practiced, it must conform to the Food and Drug Administration standards.

Gluten

The proteins of wheat are so important to the usefulness of flour in baked products that they have been studied for many years. A variety of proteins have been extracted from wheat. About 85 percent of the proteins of white flour are relatively insoluble. These insoluble proteins separate into two fractions called gliadin and glutenin. When flour is moistened with water and thoroughly mixed or kneaded, these insoluble proteins form gluten. Gluten may be extracted from dough by a thorough washing with water to remove the starch. The extracted gluten has elastic and cohesive properties. When gliadin and glutenin are separated from the gluten, the gliadin is a syrupy substance that may bind the mass together and the glutenin exhibits toughness and rubberiness that probably contributes strength.

Other Flours

Cornmeal, a granular product made from either white or yellow corn, is commonly used in several types of quick breads (Figure 19-5). Its chief protein, zein, has none of the properties of the gluten of wheat. To avoid a crumbly product, cornmeal must be combined with some white flour to bind it. Corn flour has the same properties as cornmeal except that it is finer. It is used chiefly in commercial pancake mixes.

Barley flour is rarely used in baked products requiring a gluten structure as there are no gluten-forming proteins. Barley flour may be used in extruded cereals, cakes, cake donuts, cookies, and crackers. The oligosacchrides and pentosans (types of carbohydrates) are useful in these flours. However, it is actually used as a cereal itself, although not currently as popular.

FIGURE 19-5

Wheat, rice, and corn are used in a variety of breads and pastas. (*Source:* USDA ARS Image Gallery)

Although oat flour is not a common flour, it does have some use in extruded cereal products, cakes, cookies, and crackers. The oat flakes have been used in cookies and oat breads. There are different size of flakes themselves. Additionally, some products use coarsely ground groats and dehulled oats. The milling and cleaning of oats is similar to that of wheat.

Rice flour has been used for many products as a substitute flour for those who have an allergy to wheat flour. It cannot be used in products that require gluten. Rice flour is basically rice starch. Although other flours are available, they have little use, either because they lack gluten or some similar constituent, or because they are little known except in certain areas. Potato flour is used in some countries and, like rice flour, is chiefly starch.

Buckwheat is not a seed of the grass family. It is the seed of an herbaceous plant, but because it contains a glutenous substance and is made into flour, it is commonly considered with grain products. Fine buckwheat flour has little of the thick fiber coating included and in that respect is similar to refined white flour. It is prized for its distinctive flavor and is commonly used in the making of pancakes.

Rice

Rice is frequently categorized into short-, medium-, and long-grain. American long-grain white rice is large, fluffy grains with a woody flavor. Glutinous white rice is a short-grained, very sticky, and chewy rice used to make balls or sushi. Long-grain brown (Texmati) rice is a beige, nutty-flavored, fluffy, light rice. Rice bran is ground bran and a soluble fiber. Rice flour is ground from brown or white rice and it is used as a thickener and dusting powder. Rizcous is a cracked rice that may be used interchangeably with bulgur. Short-grain brown rice is unpolished rice with a layer of bran. White rice requires twice as long to cook, giving a nutty, sweet, dense, and chewy food.

CORN REFINING

Corn **refining** is today's leading example of value added agriculture. Over 1.2 billion bushels of corn are used to produce a broad array of food, industrial and feed products for the world market. Corn refiners use shelled corn, which has been stripped from the cob during harvesting. Refiners separate the corn into its components–starch, oil, protein, and fiber–and convert them into higher-value products (Figure 19-6).

Inspection and Cleaning

Refinery staff inspect arriving corn shipments and clean them twice to remove cob, dust, chaff, and foreign materials before **steeping**, the first processing step, begins.

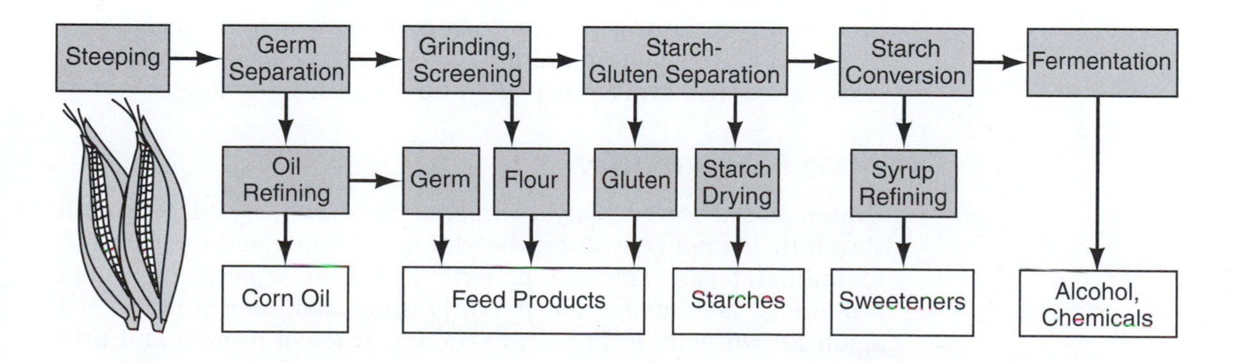

FIGURE 19-6
Flowchart of corn refining.

Steeping

Each stainless steel steep tank holds about 3,000 bushels of corn for 30 to 40 hours of soaking in 50°F (10°C) water. During steeping, the kernels absorb water, increasing their moisture levels from 15 percent to 45 percent and more than doubling in size. The addition of 0.1 percent sulfur dioxide to the water prevents excessive bacterial growth in the warm environment. As the corn swells and softens, the mild acidity of the steepwater begins to loosen the gluten bonds within the corn and release the starch. After steeping, the corn is coarsely ground to break the germ loose from other components. Steepwater is condensed to capture nutrients in the water for use in animal feeds and for a nutrient for later fermentation processes. The ground corn, in a water slurry, flows to the germ separators.

Germ Separation

Cyclone separators spin the low-density corn germ out of the slurry. The germs, containing about 85 percent of the corn's oil, are pumped onto screens and washed repeatedly to remove any starch left in the mixture. A combination of mechanical and solvent processes extracts the oil from the germ. The oil is then refined and filtered into finished corn oil. The germ residue is saved as another useful component of animal feeds.

Fine Grinding and Screening

The corn and water slurry leaves the germ separator for a second, more thorough, grinding in an impact or attrition-impact mill to release the starch and gluten from the fiber in the kernel. The suspension of starch, gluten, and fiber flows over fixed concave screens that catch fiber but allow starch and gluten to pass through. The fiber is collected, slurried, and screened again to reclaim any residual starch or protein, then piped to the feed house as a major ingredient of animal feeds. The starch-gluten suspension, called **mill starch**, is piped to the starch separators.

Starch Separation

Gluten has a low density compared to starch. By passing mill starch through a centrifuge, the gluten is readily spun out for use in animal feeds. The starch, with just 1 or 2 percent protein remaining, is diluted, washed 8 to 14 times, rediluted, and washed again in hydroclones to remove the last trace of protein and produce high-quality starch, typically more than 99.5 percent pure.

Some of the starch is dried and marketed as unmodified cornstarch, some is modified into specialty starches, but most is converted into corn syrups and dextrose.

Syrup Conversion

Starch, suspended in water, is liquified in the presence of acid and/or enzymes that convert the starch to a low-dextrose solution. Treatment with another enzyme continues the conversion process. Throughout the process, refiners can halt acid or enzyme actions at key points to produce the right mixture of sugars like dextrose and maltose for syrups to meet different needs. In some syrups, the conversion of starch to sugars is halted at an early stage to produce low-to-medium sweetness syrups. In others, the conversion is allowed to proceed until the syrup is nearly all dextrose. The syrup is refined in filters, centrifuges, and ion-exchange columns, and excess water is evaporated. Syrups are sold directly, crystallized into pure dextrose, or processed further to create high-fructose corn syrup.

Fermentation

Dextrose is one of the most fermentable of all the sugars. Following conversion of starch to dextrose, many corn refiners pipe dextrose to fermentation facilities where the dextrose is converted to alcohol by traditional yeast fermentation or to amino acids and other **bioproducts** through either yeast or bacterial fermentation. After fermentation, the resulting broth is distilled to recover alcohol or concentrated through membrane separation to produce other bioproducts. Carbon dioxide from fermentation is recaptured for sale, and nutrients remaining after fermentation are used as components of animal feed ingredients.

Bioproducts

The term "bioproducts" designates a wide variety of corn-refining products made from natural, renewable raw materials that replace products made from nonrenewable resources or that are produced by chemical synthesis. The most recognized bioproduct is ethanol–a motor fuel additive fermented from corn.

Fermentation of corn-derived dextrose has created an entirely new group of bioproducts: organic acids, amino acids, vitamins, and food gums.

Citric and lactic acid from corn can be found in hundreds of food and industrial products. They provide tartness to foods and confections, help control pH, and are themselves feedstocks for further products.

Amino acids from corn provide a vital link in animal nutrition systems. Most grain feeds do not have the amount of lysine required by swine and poultry for optimal nutrition. Economical corn-based lysine is now available worldwide to help supplement animal feeds. Threonine and tryptophan for feed supplements also come from corn.

Vitamin C and vitamin E–human nutritional supplements–are now derived from corn, replacing old production systems that relied on chemical synthesis. Even well-known food additives such as monosodium glutamate and xanthan gum are now produced by fermenting a dextrose feedstock.

Corn refiners now have fully commercial products to help deal with the plastic disposal problem and are developing an increasing array of degradable plastic products. Extrusion, the same process used to make snack foods, can alter the physical structure of cornstarch to make totally biodegradable packaging peanuts such as Eco-foam™. Other biodegradable plastics such as Eco-Pla™ are being made by modification of lactic acid.

BREAKFAST CEREALS

Breakfast foods made from cereal grains vary widely in composition, depending on the kind of grain, the part of the grain used, the method of milling, and the method of preparation. Many breakfast foods of the ready-to-serve type are a mixture of several cereals. Breakfast foods may be raw, partially cooked, or completely cooked. Some have added sweeteners, such as sugar, syrup, molasses, or honey. The carbohydrate may be changed by the use of malt or may be browned (dextrinized) by dry heat. Some cereals are reinforced with vitamins (especially the B vitamins) and with iron, calcium, and other minerals.

Raw cereals that are cooked in the home may be in various forms: whole grains, cracked or crushed grains, granular products made from either the whole grain or the endosperm section of the kernel, and rolled or flaked whole grains. The finely cut flaked grains cook in a shorter period of time. Disodium phosphate is sometimes added for quick cooking. It changes the pH of the cereal and causes it to swell faster and cook in a shorter time.

There are many ready-to-eat cereals on the market (Figure 19-7). These include granulated, flaked, shredded, and puffed cereals in many shapes and sizes with various seasonings, sugar, malt, vitamins, and minerals often added.

Table A-8 provides the composition of many cereals, some by brand name.

FIGURE 19-7
Breakfast cereals are grain products popular with children and adults.

PRINCIPLES OF BAKING

The basic foundation of baked products is usually flour and liquid. Fat, sugar, salt, eggs, leavening agents, and flavorings are other common ingredients which may or may not be used in the recipe, depending on the product desired. Each one of these ingredients has its own role and function in baked products. Roles will vary somewhat from one type of batter or dough to another. Liquid in a product may be milk, water, orange juice, or any others. Although these liquids in the baked product may function differently, generally liquid serves as a solvent for salt, sugar, and other solutes, assists in the dispersion of all the colloids and suspensions, assists in the development of gluten, and contributes to both the leavening and gelatinization phenomena during baking.

In addition to sweetening baked products, sugar facilitates air incorporation by shortening, inhibits development of gluten and gelatinization of starch, and elevates the temperature at which egg and flour proteins heat denature. Eggs contribute to the structure of a baked product. They may do this through their contribution of heat denatured proteins, steam for leavening, or moisture for starch gelatinization. Egg yolk is also a rich source of emulsifying agents and thus facilitates the incorporation of air, inhibits starch gelatinization, and contributes to the flavor. The leavening source used in a baked product may serve to produce gas by physical, chemical, or biological methods.

The leavening selected is usually dependent on the balance and kind of ingredients in the formula and the manipulation methods used. Salt and a wide variety of flavorings are used to obtain the type and variety of product wanted. In addition to being used as a flavoring, salt functions to control yeast metabolism in yeast bread.

The roles and functions of the ingredients selected may be maximized or minimized depending on the method of manipulation chosen. The number of different approaches that are used in preparing baked products is almost infinite; however, several basic methods are designated. The individual concerned with preparation of baked products should know the procedures for foods made using the biscuit, pastry, muffin, conventional cake, or straight dough methods of mixing. Each method and its many variations is selected considering both the ingredients used and the characteristics desired in the end products.

A number of factors impact the quality of flour mixtures, including:

- Nature of ingredients in formula, correct proportion, and exact measurements of ingredients
- Proper methods and environment for combining ingredients
- Correct heating temperature, time, and method
- Process melting of fat
- Increased fluidity of batter/dough
- Dissolved ingredients
- Chemical or physical or biological leavening
- Denaturation of protein
- Gelatinization of starch
- Steam leavening
- Maillard reaction

Baking of different flour mixtures brings about a delicate balance between the firming of structure, the leavening action, and the development of optimum flavors and colors. The particular heating process, as with the selection and proportion of ingredients, is specific for each formula.

Pour batters are mechanically leavened. In cream puffs, eggs are an important constituent serving to contribute both leavening with water, emulsifying the high percentage of fat, as well as structure. Both the egg protein and wheat starch are the primary structural components. Popovers also have equal parts of flour and liquid formulating a pour batter. Again, egg and starch are the primary structural components. Another mechanical steam-leavened product is pastry. In mixing a pastry product, the incorporation of fat is critical. Finely mixed fat and flour will make a more mealy pastry. A coarsely mixed flour/fat formula will likely make a more flaky pastry.

Mixing is an important factor in producing any baked product. The blades themselves make a difference. These influence viscosity, degree of dispersion, air incorporation, and other quality characteristics.

General objectives in mixing batters and doughs include the following:

- Uniform distribution of ingredients
- Minimum loss of the leavening agent
- Optimum blending to produce characteristic textures
- Optimum development of gluten for various products

Many different mixing methods and beating utensils exist. Each method serves to prepare a product of particular quality characteristics and/or is adaptable to particular ingredients and/or conditions.

Flour millers and bakers use a mixograph that actually records the changes as the wheat flour and water mass gradually becomes coherent, loses its wet and rough sticky feel, and becomes a smooth homogenous mass.

LEGUMES

Legumes and grains provide protein and energy to much of the world's population. Edible legumes are found almost everywhere in the world. Some of the common legumes and their processing, preparation, and uses are described in Table 19-1.

Nutritional Composition

Table A-8 lists the nutrient composition of many legumes. In general, the seeds of food legumes are good sources of carbohydrates, fats, proteins, minerals, and vitamins. Mixtures of legumes and grains have a protein quality that comes close to that of animal proteins.

TABLE 19-1 Common Legumes

Popular Name(s) ***Scientific Name***	**Processing**	**Preparation**	**Uses**
Adzuki bean *Phaseolus angularis*	Picked when mature; may be left whole or pounded into a meal	Whole beans boiled, then mashed	Vegetable dish; bean flour for cakes and dessert
Alfalfa (Lucerne) *Medicago sativa*	Mature seeds sprouted; flour made from the dried leaves, and protein concentrate from the juice of fresh leaves	Sprouts cooked or raw	Sprouts used in salads, soups, sandwiches; flour added in small amounts to some cereal products; protein concentrate in livestock feed
Bean (Common, French, Kidney, Navy, Pea, Pinto, Snap, Stringless, Green) *Phaseolus vulgaris*	String and mature beans sold fresh, canned, frozen; immature as pods and seeds	Boil, bake, fry	Vegetable dishes, casseroles, soups, stews
Broad bean (Fava bean) *Vicia faba*	Picked when mostly mature then shelled, dried, canned, or frozen	Boiled, cooked, or steamed in other foods; immature pods cooked whole or sliced	Vegetable dishes, soups, stews, casseroles
Chickpea (Garbanzo bean) *Cicer arietinum*	Dehulling then drying	Boil, fry, roast	Snacks, vegetable dishes, soups, salads, stews
Cowpea (Black-eyed Pea) *Vigna sinensis; V. unguiculata*	Mature beans dried, canned, frozen; picked as immature pods	Boil or bake	Vegetable dishes, with or without pork

TABLE 19-1 Common Legumes *(concluded)*

Popular Name(s) ***Scientific Name***	**Processing**	**Preparation**	**Uses**
Field pea *Pisum arvense*	Shelling, drying; whole plant plowed under as green manure or forage peas	Boil or bake; do not become soft like garden peas	Vegetable dishes, livestock feed
Garden pea *Pisum sativum*	Picked immature; sold fresh, frozen, cooked, or canned; mature seeds dried whole or after splitting	Boil or pressure cook	Vegetable dishes, casseroles, soups, stews
Lentil *Lens esculenta*	Picked mature, can be left whole or dehulled, usually dried; sometimes ground into flour	Boil or stew	Vegetable dishes, soups, stews; flour mixed with cereals
Lima bean (Butter bean) *Phaseolus lunatus*	Picked immature or mature; sold fresh, cooked, canned, frozen, dried, or as flour	Boil or bake	Vegetable dishes, casseroles, soups, stews
Mung bean (Golden gram; Green gram) *Phaseolus aureus*	Picked mature	Boil or sprout for eating raw or cooked	Cooked as vegetable dish; sprouts in salad, soups, sandwiches
Peanut (Groundnut) *Archis hypogaea*	Vines harvested, nuts shelled and blanched; oil may be expressed or extracted then cake ground into defatted flour	Boil or roast whole nuts with or without shell	Snack food; flour added to cereal mixtures
Soybean *Glycine max*	Picked mature or immature; immature seeds sold fresh or cooked or canned; oil expressed or extracted; press cake ground into defatted flour; mature seeds roasted or salted	See Figure 19-8 (page 347)	Refer to Table 19-2 (page 346)

Legume Products

Some of the products made from legumes include fermented foods, flours, imitation meats, infant formulas, oils, and sprouts.

Fermented Foods. Soy sauce is produced by fermentation. Soybean products tempeh (mold ripened soybean cake that is fried in deep fat) and tofu (a cheeselike item made from coagulated soybean milk) are now being produced in the United States and are being sold in some supermarkets.

Flours. Soybean flour is used in the United States to make soybean milk and low-gluten baked goods. Similar legume flours have been made from locally grown beans and peas in many countries of the world.

Imitation Meat. Textured vegetable protein (TVP) products are meatlike in taste and texture. Most of the TVP items are made from soy protein that has been extracted from soy flour and spun into fibrous strands. Some of the more popular items are made from mixtures of soy proteins, wheat protein, egg albumen, and various additives.

Infant Formulas. Many infant formulas are made of soy protein concentrates.

Oil. Seeds of legumes such as soybeans and peanuts are extracted to yield oil that is used for cooking. The by-product of this process–a protein meal–is often feed to livestock.

Sprouts. Legume seeds allowed to germinate are called sprouts. They are sold in the fresh produce section of supermarkets. The most popular types of sprouts are made from alfalfa seeds, mung beans, and soybeans. Also, some bean sprouts are canned by the manufacturers of Chinese foods.

SOYBEANS

As soybeans mature in the pod, they ripen into a hard, dry bean. Most soybeans are yellow, but some varieties are brown and black. Whole soybeans can be cooked and used in sauces, stews, and soups. They are an excellent source of protein. Whole soybeans that have been soaked can be roasted for snacks and can be purchased in natural food stores and some supermarkets. But soybeans are more than this.

The soybean is a versatile agricultural product. After preparation and **extraction**, it has edible and technical uses (Table 19-2). Although they can be eaten whole after being boiled or roasted, most soybeans are transformed into a great variety of foods. In addition, a great many foods already found in a kitchen cupboard contain soyfoods, such as soyoil (often called vegetable oil), lecithin, soy protein concentrates, textured soy protein, and many more. Figure 19-8 illustrates the preparation and extraction of soybeans to produce a wide variety of useful products.

Green Vegetable Soybeans

These large soybeans are harvested when the beans are still green and sweet tasting and can be served as a snack or a main vegetable dish, after boiling in slightly salted water for 15 to 20 minutes. They are high in protein and fiber and contain no cholesterol.

Hydrolyzed Vegetable Protein (HVP)

Hydrolyzed vegetable protein (HVP) is a protein obtained from any vegetable, including soybeans. The protein is broken down into amino acids by a chemical process called acid hydrolysis. HVP is a flavor enhancer that can be used in soups, broths, sauces, gravies, flavoring and spice blends, canned and frozen vegetables, meats, and poultry.

Infant Formulas, Soy-Based

Soy-based infant formulas are similar to other infant formulas except that a soy protein isolate powder is used as a base, instead of cow's milk. Carbohydrates and fats are added to achieve a fluid similar to breast milk.

Lecithin

Extracted from soybean oil, lecithin is used in food manufacturing as an emulsifier in products high in fats and oils. It also promotes stabilization, antioxidation, crystallization, and spattering control. Powdered lecithins can be found in natural and health food stores.

Meat Alternatives (Meat Analogs)

Meat alternatives made from soybeans contain soy protein or tofu and other ingredients mixed together to simulate various kinds of meat. These meat alternatives are sold as frozen, canned, or dried foods. Usually, they can be used the same way as the foods they

TABLE 19-2 Soybean Products and Uses

	Oil Products			Soybean Protein Products	
Whole Soybean Products	**Glycerol, Sterols, Fatty Acids**	**Refined Soyoil**	**Soybean Lecithin**	**Soy Flour Concentrates and Isolates**	**Soybean Meal**
Edible Uses Seed Stock feed Soy sprouts Baked soybeans Full-fat soy flour Bread Candy Doughnut mix Frozen desserts Instant milk drinks Pancake flour Pan grease extender Pie crust Sweet goods **Roasted Soybeans** Candies/ Confections Cookie ingredient/ Topping Crackers Dietary items Soynut butter Soy coffee **Soybean Derivates** Miso Soymilk Tempeh Tofu	Oleochemistry Soy diesel Solvents	**Edible Uses** Coffee creamers Cooking oils Filled Milks Margarine Mayonnaise Medicinals Pharmaceuticals Salad dressings Salad oils Sandwich spreads Shortenings **Technical Uses** Anti-corrosion agents Anti-static agents Caulking compounds Composite building material Concrete release agents Core oils Crayons Dust control agent Electrical insulation Epoxies Fungicides Hydraulic fluids Inks—Printing Linoleum backing Lubricants Metal casting/ Working Oiled fabrics Paints Pesticides Plasticizers Protective coatings Putty Soap/Shampoos/ Detergents Vinyl plastics Waterproof cement	**Edible Uses** Emulsifying agent Bakery products Candy/Chocolate coatings Pharmaceuticals Nutritional uses Dietary Medical **Technical Uses** Anti foam agents Alcohol Yeast Anti-spattering agents Margarine Dispersing agents Paint Ink Insecticides Magnetic tape Paper Rubber Stabilizing agent Shortening Wetting agents Calf milk replacers Cosmetics Paint pigments	**Edible Uses** Alimentary pastes Baby food Bakery ingredients Candy products Cereals Diet food products Food drinks Hypoallergenic milk Meat products Noodles Prepared mixes Sausage casings Yeast Beer and ale **Technical Uses** Adhesives Antibiotics Asphalt Emulsions Composite building material Fermentation Aids/ Nutrients Fibers Films for packaging Firefighting foams Inks Leather substitutes Paints—water based Paper coatings Particle boards Plastics Polyesters Pharmaceuticals Pesticides/ Fungicides Textiles	**Feed Uses** Aquaculture Bee foods Calf milk replacer Fish food Fox and mink feed Livestock feeds Poultry feeds Protein concentrates Pet foods **Hulls** Dairy feed

Source: American Soybean Association

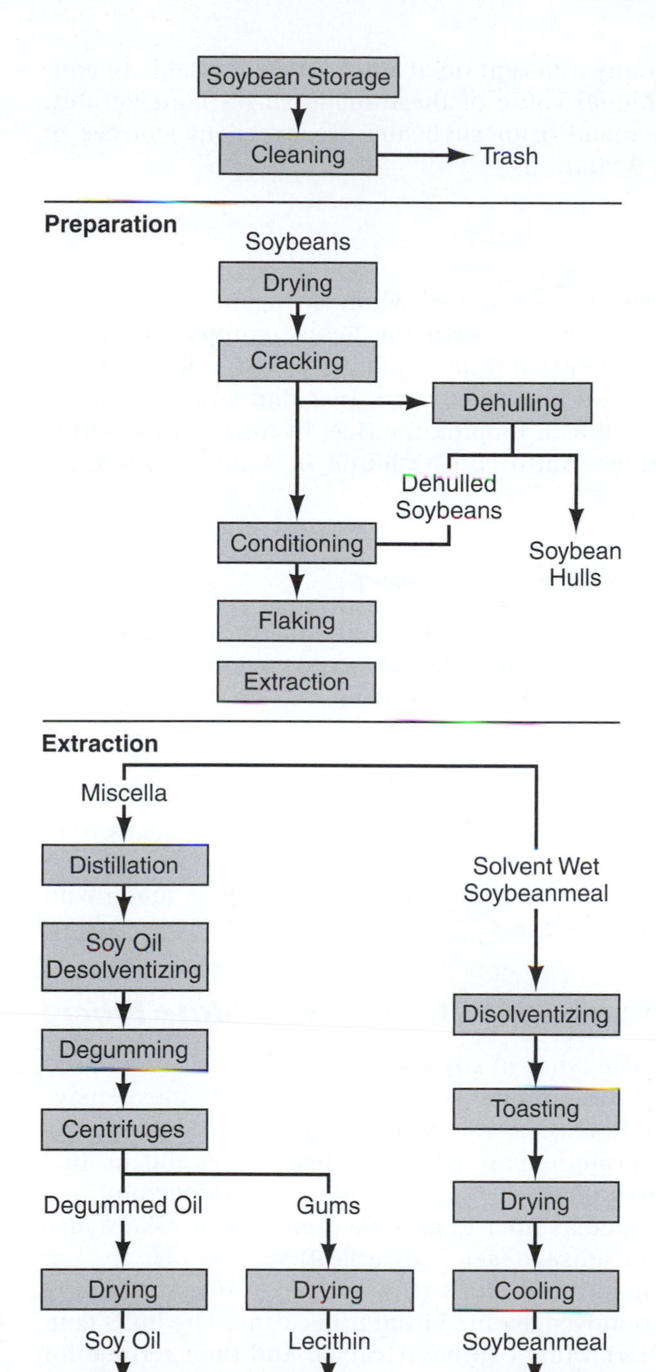

FIGURE 19-8

Flowchart of soybean processing.

replace. With so many different meat alternatives available to consumers, the nutritional value of these foods varies considerably. Meat alternatives made from soybeans are excellent sources of protein, iron, and B vitamins.

Natto

Natto is made of fermented, cooked whole soybeans. Because the fermentation process breaks down the beans' complex proteins, natto is more easily digested than whole soybeans. It has a sticky, viscous coating with a cheesy texture. In Asian countries natto traditionally is served as a topping for rice, in miso soups, and is used with vegetables. Natto can be found in Asian and natural food stores.

Nondairy Soy Frozen Dessert

Nondairy frozen desserts are made from soymilk or soy yogurt. Soy ice cream is one of the most popular desserts made from soybeans and can be found in natural foods stores.

Soy Cheese

Soy cheese is made from soymilk. Its creamy texture makes it an easy substitute for sour cream or cream cheese and can be found in a variety of flavors in natural foods stores. Products made with soy cheese include soy pizza.

Soy Fiber (Okara, Soy Bran, Soy Isolate Fiber)

There are three basic types of soy fiber: Okara, soy bran, and soy isolate fiber. All these products are high-quality, inexpensive sources of dietary fiber. Okara is a pulp fiber by-product of soymilk. It has less protein than whole soybeans, but the protein remaining is of high quality. Okara tastes similar to coconut and can be baked or added as fiber to granola and cookies. Okara also has been made into sausage (see Figure 19-9).

Soy bran is made from hulls (the outer covering of the soybean), which are removed during initial processing. The hulls contain a fibrous material that can be extracted and then refined for use as a food ingredient.

Soy isolate fiber, also known as structured protein fiber (SPF), is soy protein isolate in a fibrous form.

FIGURE 19-9
Soybean products. (*Source:* USDA ARS Image Gallery)

Soy Flour

Soy flour is made from roasted soybeans ground into a fine powder. Three kinds of soy flour are available:

1. Natural or full-fat, containing the natural oils found in the soybean
2. Defatted, having the oils removed during processing
3. Lecithinated, having lecithin added to it

All soy flour gives a protein boost to recipes. However, defatted soy flour is an even more concentrated source of protein than full-fat soy flour. Although used mainly by the food industry, soy flour can be found in natural foods stores and some supermarkets. Soy flour is gluten-free, so yeast-raised breads made with soy flour are more dense in texture.

Soy Grits

Soy grits are similar to soy flour except that the soybeans have been toasted and cracked into coarse pieces, rather than the fine powder of soy flour. Soy grits can be used as a substitute for flour in some recipes. High in protein, soy grits can be added to rice and other grains and cooked together.

Soy Protein Concentrate

Soy protein concentrate comes from defatted soy flakes. It contains about 70 percent protein, while retaining most of the bean's dietary fiber.

Soy Protein Isolates (Isolated Soy Protein)

When protein is removed from defatted flakes, the result is soy protein isolates, the most highly refined soy protein. Containing 92 percent protein, soy protein isolates possess the greatest amount of protein of all soy products. They are a highly digestible source of amino acids.

Soy Protein, Textured

Textured soy protein (TSP) usually refers to products made from textured soy flour, although the term can also be applied to textured soy protein concentrates and spun soy fiber.

Textured soy flour (TSF) is made by running defatted soy flour through an extrusion cooker, which allows for many different forms and sizes. It is widely used as a meat extender. Textured soy flour contains about 70 percent protein and retains most of the bean's dietary fiber. Textured soy flour is sold dried in granular and chunk style.

Soy Sauce

Soy sauce is a dark brown liquid made from soybeans that have been fermented. Soy sauces have a salty taste. Specific types of soy sauce include shoyu, tamari, and teriyaki. Shoyu is a blend of soybeans and wheat. Tamari is made only from soybeans and is a by-product of making miso. Teriyaki sauce can be thicker than other types of soy sauce and includes other ingredients such as sugar, vinegar, and spices.

Soy Yogurt

Soy yogurt is made from soymilk. Its creamy texture makes it an easy substitute for sour cream or cream cheese. Soy yogurt can be found in a variety of flavors in natural foods stores.

Soymilk, Soy Beverages

Soybeans, soaked, ground fine, and strained, produce a fluid called soybean milk, which is a substitute for cow's milk. Plain, unfortified soymilk is an excellent source of high-quality protein, B-vitamins. Soymilk is also sold as a powder, which must be mixed with water.

Soynut Butter

Soynut butter is made from roasted, whole soynuts that are crushed and blended with soyoil and other ingredients. Soynut butter competes with peanut butter.

Soynuts

Roasted soynuts are whole soybeans that have been soaked in water and then baked until browned. Soynuts are similar in texture and flavor to peanuts.

Soyoil and Products

Soyoil is the natural oil extracted from whole soybeans. It is the most widely used oil in the United States. Oil sold in the grocery store under the generic name "Vegetable Oil" is usually 100 percent soyoil or a blend of soyoil and other oils. Soyoil is cholesterol-free and high in polyunsaturated fat. Soyoil also is used to make margarine and shortening (see Table 19-2, page 346).

Sprouts, Soy

Not as popular as mung bean sprouts or alfalfa sprouts, soy sprouts (also called soybean sprouts) are an excellent source of protein and vitamin C. Soy sprouts must be cooked quickly at low heat so they do not get mushy. They can also be used raw in salads or soups, or in stir-fried, sauteed, or baked dishes.

Tempeh

Tempeh, a traditional Indonesian food, is a chunky, tender soybean cake. Whole soybeans, sometimes mixed with another grain such as rice or millet, are fermented into a rich cake of soybeans with a smoky or nutty flavor. Tempeh can be marinated and grilled and added to soups, casseroles, or chili.

Tofu and Tofu Products

Tofu, also known as soybean curd, is a soft cheeselike food made by curdling fresh hot soymilk with a coagulant (refer to Figure 19-9). Tofu is a bland product that easily absorbs the flavors of other ingredients with which it is cooked. Tofu is rich in high-quality protein and B-vitamins and low in sodium. Firm tofu is dense and solid and can be cubed and served in soups, stir-fried, or grilled. Firm tofu is higher in protein, fat, and calcium than other forms of tofu. Silken tofu is a creamy product and can be used as a replacement for sour cream in many dip recipes. Tofu is also available as a powder.

Whipped Toppings, Soy-Based

Soy-based whipped toppings are similar to other nondairy whipped toppings, except that hydrogenated soyoil is used instead of other vegetable oils.

Table 19-2 (page 346) summarizes all of the edible and technical products obtained from the extraction of the legume, soybeans.

Summary

Cereal grains and legume seeds, and products from these, are used as food for people throughout the world. Cereal grains provide mainly starch and some protein. Legumes provide mainly protein, oil, and some starch. Starch has unique properties that are used in foods. Seeds of the grains and legumes are used to produce flour. These flours are used to produce a variety of other food products.

Corn refining and soybean extraction separate the corn seed and soybeans, respectively, into component parts and convert these to high-value products. These processes are some of the best examples of value added agriculture and the application of food science.

Review Questions

Success in any career requires knowledge. Test your knowledge of this chapter by answering these questions or solving these problems.

1. Explain the processes that take place during baking.
2. Soy ____________ is made from roasted soybeans in a fine powder. Soy ____________ are substitutes for cream cheese or sour cream. Soy ____________ is a dark brown liquid made from soybeans that have been fermented. Soy ____________ ____________ comes from defatted soy flakes.
3. List three properties of starch.
4. Name five types of wheat flour.

5. List four general uses of legumes.
6. The __________ is the large central portion of the kernel and contains most of the starch. The __________ is the small structure at the lower end of the kernel. It is rich in fats, proteins, and minerals.
7. Identify five food products of soybean extraction.
8. What is the purpose of the first step of corn refining–steeping?
9. List four factors that must be controlled when cooking starch.
10. Name four products derived from corn.

Student Activities

1. Compare a labeled cross-section of a kernel of grain to that of a legume seed.
2. Collect the labels on five different ready-to-eat breakfast cereals. Report on the grains or grain products used. Do the same for a type of hot breakfast cereal that requires cooking.
3. Visit a corn milling or soybean extraction facility on the Web. Develop a report or presentation on your visit.
4. Develop a bread-making demonstration that explains all the components and the processes taking place. Experiment with different types of flours. Use a bread machine if possible.
5. Use a hand mill or a small electric mill to grind some wheat. Separate and screen to produce flour. Develop a report on your results. Include yield and the evaluation of a product produced with the flour.
6. Create a display of soybean extracted products (Table 19-2, page 346) or products from corn refining.
7. Collect and display the seeds of cereal grains or legumes.

Resources

Fast, R. B., and E. F. Caldwell. 1990. *Breakfast cereals and how they are made.* St. Paul, MN: American Association of Cereal Chemists.

Hoseney, R. C. 1994. *Principles of cereal science and technology.* 2nd Ed. St. Paul, MN: American Association of Cereal Chemists.

Lehner, E., and J. Lehner. 1962. *Folklore and odysseys of food and medicinal plants.* New York: Tudor Publishing Company.
Parker, R. 1998. *Introduction to plant science.* Albany, NY: Delmar.
Posner, E. S., and A. N. Hibbs. 1997. *Wheat flour milling.* St. Paul, MN: American Association of Cereal Chemists.

Internet

Internet sites represent a vast resource of information. The URLs (uniform resource locator) for the World Wide Web sites can change. Using one of the search engines on the Internet such as Yahoo!, HotBot, AltaVista, Excite, Dogpile, About, or Google, find more information by searching for these words or phrases: wheat, specific types of wheat, flour milling, small grains, grain milling, corn milling, cereal grains, breakfast cereals, corn refining. Also, Table A-7 provides a listing of some useful Internet sites that can be used as a starting point.

Chapter 20

Fruits and Vegetables

Objectives

After reading this chapter, you should be able to:

- Identify the parts of a plant considered a vegetable or a fruit
- Describe the nutrient composition of a fresh fruit or vegetable
- Discuss the structure of a plant cell
- Describe the plant tissues and their functions
- Explain climacteric and nonclimacteric with examples
- Name one pigment in fruits or vegetables and describe how it responds to heat or pH
- List four factors affecting the texture of fruits or vegetables
- Name four general compounds that give fruits and vegetables their flavor
- Identify the quality grades for fruits and vegetables
- Describe how quality grade determines the use of a fruit or vegetable
- List five factors considered during storage
- Describe the processing of fruits
- Discuss the processing of vegetables

Key Terms

choice
climacteric
dermal
ethylene
grades
nonclimacteric
standards
storage tissue
tubers
U.S. Fancy
vascular

Fruits, vegetables, and other plant tissues either directly or indirectly supply all foods to humans. An estimated 270,000 plant species exist. The number of crops that fit into humans' dietary picture is probably between 1,000 and 2,000 species. Fruits and vegetables take in a wide variety of edible plant parts. Fruits include apples, pears, peaches, apricots, plums, cherries, bananas, oranges, tangerines, and grapes. The term vegetable includes many different parts of plants, including some fruits.

GENERAL PROPERTIES AND STRUCTURAL FEATURES

Leafy vegetables are generally high in water and low in carbohydrates, protein, and fats. They frequently contain the mechanism for photosynthesis. Some examples include spinach, lettuce, mustard greens, and cabbage.

Seeds vary in water content and are a source of carbohydrates and protein. They may be "fresh" and high in water or "dried" and relatively low in water content. Corn and beans are examples.

Tubers are generally higher in carbohydrates and lower in water content than stem, flower, or leafy vegetables. Tubers are enlarged underground stems. Potatoes are the best example.

The fruit part of the plant can be eaten either as a fruit or as a vegetable. Some examples include cantaloupe, eggplant, squash, and peas (snow peas) (Figure 20-1).

FIGURE 20-1

Fruits are generally the fruit of a plant but vegetables include fruits, leaves, stems, bulbs, flowers, seeds, roots, and tubers of a plant. (*Source*: USDA ARS Image Gallery)

Stems are plant portions generally high in water and fiber. They have relatively little other nutritive value. Asparagus is an example.

Bulbs are generally higher in carbohydrates and lower in water content than stem, flower, or leafy vegetables. Bulbs like onions and garlic are enlargements above the roots.

Broccoli, cauliflower, and artichokes are generally listed as the edible flowers. Flowers are generally high in water and low in carbohydrates.

Roots like carrots are generally higher in carbohydrates and lower in water content than stem, flower, or leafy vegetables. Roots are the part of a plant that grow downward into the soil to anchor the plant and furnish nourishment by absorbing nutrients.

GENERAL COMPOSITION

Fresh as well as canned and frozen vegetables provide a variety of vitamins, minerals, and fiber. They are low in fat.

Fresh fruits and fruit juices contain many vitamins and minerals, they are low in fat (except avocados) and sodium, and they provide dietary fiber. Whole, unpeeled fruit is higher in fiber than peeled fruit or fruit juice. Canned and frozen fruits and fruit juices contain many vitamins and minerals, they are low in fat and sodium.

Table A-8 provides the composition of many fruits and vegetables. Fresh and canned fruits and vegetables contain a high water content. Obviously, the energy content increases when the fruit is dried. The carbohydrate content is high compared to the protein content in fruit. Vegetable nutrient content varies widely. Some fresh and canned vegetables have a high water content with some protein but more carbohydrate. Seeds that are considered vegetables contain high amounts of energy, protein, and carbohydrates.

Fresh Vegetable Labels

Under federal guidelines retailers must provide nutrition information for the 20 most frequently eaten raw vegetables: potatoes, iceberg lettuce, tomatoes, onions, carrots, celery, sweet corn, broccoli, green cabbage, cucumbers, bell peppers, cauliflower, leaf lettuce, sweet potatoes, mushrooms, green onions, green (snap) beans, radishes, summer squash, and asparagus. Information about other vegetables may also be provided. The nutritional information may appear on posters, brochures, leaflets, or stickers near the vegetable display (Figure 20-2). It may include serving size; calories per serving; amount of protein, total carbohydrates, total fat, and sodium per serving; and percent of the U.S. Recommended Daily Allowances for iron, calcium, and vitamins A and C per serving.

FIGURE 20-2

Under federal guidelines, retailers provide nutrition information for raw fruits and vegetables.

Fresh Fruit Labels

Under federal guidelines, retailers must provide nutrition information for the 20 most frequently eaten raw fruits: bananas, apples, watermelons, oranges, cantaloupes, grapes, grapefruit, strawberries, peaches, pears, nectarines, honeydew melons, plums, avocados, lemons, pineapples, tangerines, sweet cherries, kiwifruit, and limes. Information about other fruits may also be provided. The nutritional information may appear on posters, brochures, leaflets, or stickers near the fruit display. It may include serving size; calories per serving; amount of protein, total carbohydrates, total fat, and sodium per serving; and percent of the U.S. Recommended Daily Allowances for iron, calcium, and vitamins A and C per serving.

ACTIVITIES OF LIVING SYSTEMS

Because so many parts of plants are eaten for vegetables or for fruits, an understanding of the plant cell and plant tissues is essential to food scientists.

Plant cell structure depends upon the role and function of the cell (Figure 20-3). The cell wall consists of a primary wall and a secondary wall. The primary walls of two cells are joined together by a common layer called the middle lamella. The cell wall and middle lamella's chief components are cellulose, hemicellulose, and pectic substances.

A vacuole is contained within plant cells. Its size depends on the cell's function. The vacuole is composed of water with soluble

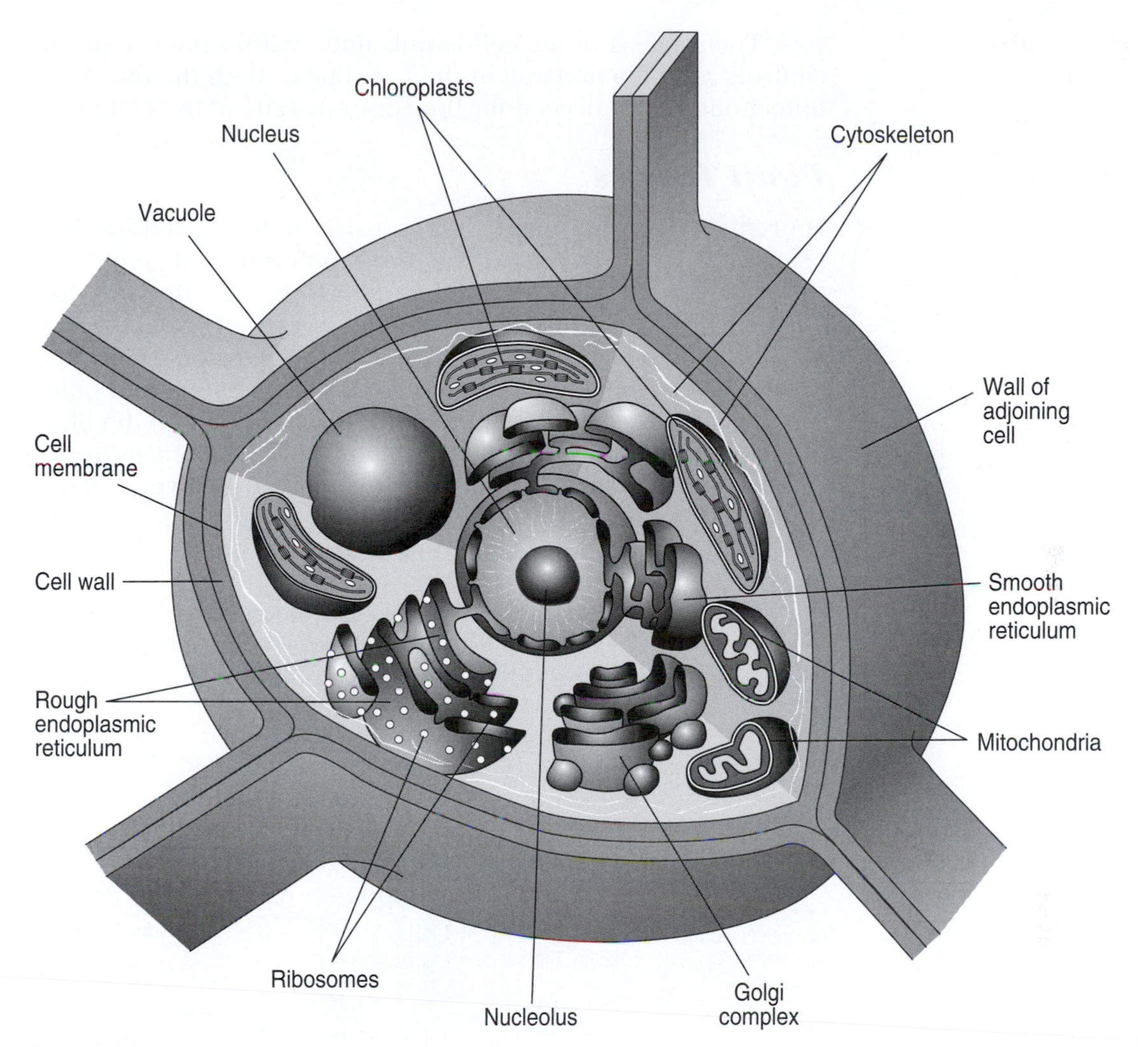

FIGURE 20-3

A plant cell.

substances dissolved within it. These may include sugars, acids, volatile esters, aldehydes, ketones, and water-soluble pigments depending upon the particular fruit or vegetable.

Energy conversion in the cell is carried out by the chloroplasts and mitochondria. The leucoplasts store starch that is used for energy. The mitochondria are small spheres, rods, or filaments that produce energy for the cell through cellular respiration. They contain fats, proteins, and enzymes.

The nucleus of the cell is imbedded within the cytoplasm. It controls reproduction and protein synthesis. Both the nucleus and mitochondria are needed for the continued life of the cell.

Plant Tissues

Four main types of plant tissues exist in fruits and vegetables (Figure 20-4): These are the **dermal** (epidermis and endodermis), **vascular** (xylem and phloera), supporting, and **storage tissue** (cortex). The dermal tissue generally is a layer of protective tissue. It has the stoma that will penetrate to the interior. The vascular system has two clearly defined structures, the xylem and phloem. The xylem is used by the plant to transport water, and the phloem is used to transport food. Supporting tissue varies depending upon the particular plant. Generally, storage or parenchyma plant tissues make up most of the edible portion of fruits and vegetables. The leucoplasts, protein bodies, and other food- and energy-containing bodies are located here.

Storage tissue is located in the cytoplasm in leucoplasts. The leucoplasts are colorless plastids that may accumulate fats and oils (elaioplasts), proteins (aleuroneplasts), and starch (amyloplasts)

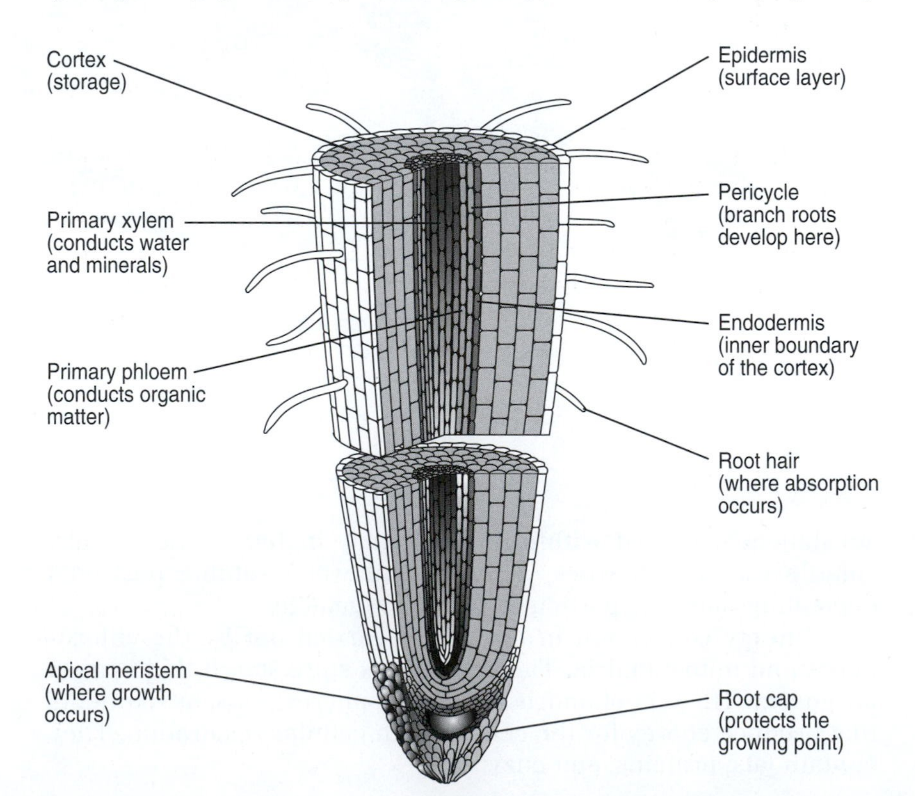

FIGURE 20-4

Plant tissues.

and are the storage structures of the cell. As the cell matures, lipid, protein, or starch content increases. In storage tissue, plastids may dominant the cell. They are particularly dominant in roots, tubers, bulbs, and seeds.

Protective tissue is made up of modified parenchyma cells that contain suberin, may secrete cutin, and grow tightly together to form an epidermal layer for the cell. The thickness and composition of this protective tissue will vary with the part and type of plants. It is the layer that protects against insects, fungi, microorganisms, and small abrasions. The extent of cutin or suberin will affect the quality of a food, particularly in apples. A shiny apple will usually have a higher content of these components in its outer skin.

Quality characteristics are influenced by water transport. This is critical during the growing period and also influences the crispness of the final fruit or vegetable. Water movement occurs because of evaporation and controlled loss of water from stomates in the stems and in the leaves. The more stomata the dermal tissue possesses, the greater the potential for water loss. Dermal tissue is sensitive to light and heat, allowing the stomata to regulate water transport by opening and closing. Supporting tissues are generally cells whose cellulose cell walls have thickened and have become embedded with lignin and/or pectic substances. The cells themselves are frequently elongated, and these portions of the plant have a tough texture.

Although water is lost through the stomata in the stems and particularly in the leaves, the transport and translocation of water and plant sap from the vacuole occurs in vascular tissue or conducting tissue. One vascular category is the xylem, which chiefly transports water and minerals from the roots up the stem and into the leaves. The transport of carbohydrates, amino acids, and other constituents of the cell sap in the vacuole occurs through the second vascular categroy, the phloem. The phloem is essentially the nutrient transport. Some overlap exists between it and the xylem.

HARVESTING

Many factors affect production of the vegetable and fruit to the point at which it will be harvested (Figure 20-5). The soil will vary for the different types of vegetables and fruits. Some of the fruits and vegetables require more water than others, different fertilizers than others, and so forth. The actual days available for the plant to grow and ripen is a critical factor. For that reason, some fruits and vegetables do not have an adequate time at appropriate temperature to grow to maturity.

FIGURE 20-5
Proper harvesting of potatoes is important to maintaining quality.

Harvesting by hand versus machine affects the quality and quantity of fruits and vegetables. Mechanical harvesting machines generally exert more impact and damage to fruits and vegetables. To offset the greater handling force, new cultivars and varieties have been developed. In many cases, these fruits and vegetables are firmer, higher in solids, and of different flavors. For example, horticulturalists and crop scientists have developed just such a firm, high-solid, and high-pH tomato to meet the impacts of mechanical harvesting. The food processor or preparer must recognize these changes that are occurring in fruits and vegetables as they influence both safety and quality. In the case of the tomato, no longer is this vegetable necessarily safe if canned only under hot water. In most instances, pressure canning must be considered.

The season environment and length is important. The producer, and to some extent the consumer, is interested in how long it takes a fruit or vegetable to mature. However, the urban consumer is more interested in when they can purchase selected fruits and vegetables.

Ripening

Many factors and interactions affect ripeness as well as maturation. Temperature, time, or added gases are of real interest and influence ripening. Each fruit appears to function differently.

One aspect of ripening is the production of **ethylene** (C_2H_4), which signals and orchestrates the growth stages of fruits and flowers. Ethylene activates fruit ripening.

Some fruits are **climacteric** and some are **nonclimacteric** fruits. Climacteric fruits produce ethylene gas during ripening, and they are ethylene sensitive (Table 20-1).

TABLE 20-1 Classification of Edible Fruits According to Ripening Patterns

Climacteric	Nonclimacteric
apple	cherry
apricot	cucumber
avocado	fig
banana	grape
cherimoya	grapefruit
feijoa	lemon
mango	melon
papaya	orange
passion fruit	pineapple
papaw	strawberry
peach	
pear	
plum	
sapote	
tomato	

Appearance

Fruits and vegetables get their characteristic color from pigments. Anthocyanins include the purple, blue, and red colors. Anthoxanthins are white to yellow, and betalains are red. Carotenoids are orange to yellow, and chlorophyll is green. These pigments respond differently to the processing environment as shown in Table 20-2.

Texture

Texture of plant foods has been difficult to describe in precise terms. For example, a comparison of crisp versus wilted lettuce, a crisp tender carrot versus a crisp tough carrot, a plump, watery strawberry compared to a plump, pithy strawberry. All these comparisons emphasize the two-sided nature of texture. Ultimately, the texture of fruits and vegetables is mainly caused either by the structural components of the plants themselves or by the process of osmosis and diffusion.

Toughness results from the cell wall components–pectins, hemicelluloses, and cellulose–which change during maturation, storage, and processing. Many factors impacting toughness include tissue conditions, pH, enzymes, and salt concentrations.

The crispness of a vegetable is due to the movement of water in the plant. Crispness of a plant cell is influenced by turgidity factors such as the concentration of osmotically active substances and the permeability of the protoplasm and the elasticity and toughness. Water movement in plants is a combined effect of capillary action, diffusion, transpiration, and osmosis (Figure 20-6).

TABLE 20-2 Affect of Heat, Acid, and Alkali on Pigments

Name of Pigment	Acid	Alkali	Prolonged Heating
Anthocyanins	Red	Purple or Blue	Little effect
Anthoxanthins	White	Yellow	Darkens if excessive*
Betalains	Red	Little effect	Fading if slightly acid
Carotenoids	Less intense	Little effect	May be less intense*
Chlorophylls	Olive-green	Intensifies green	Olive-green

* Heating usually produces little effect

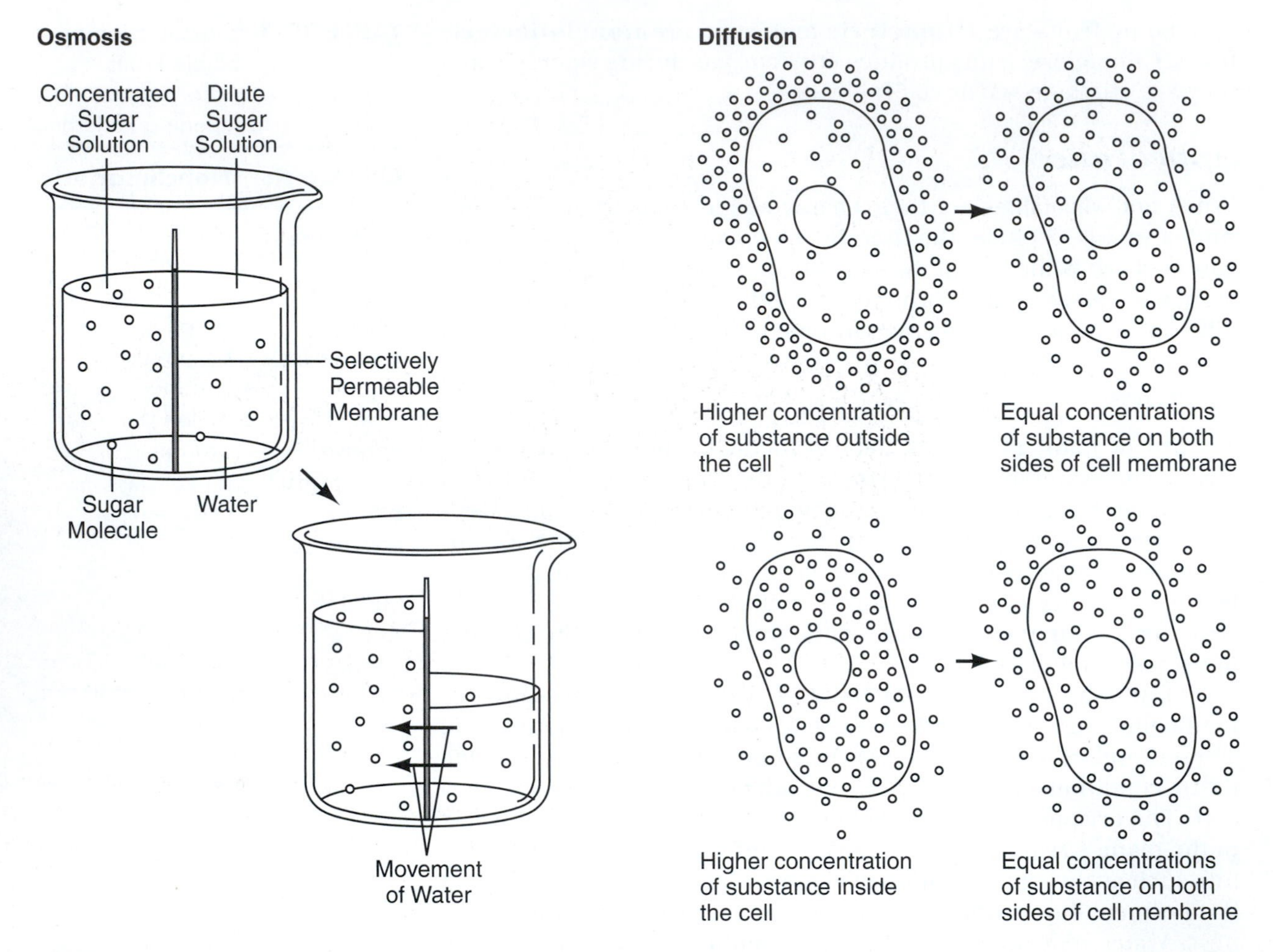

FIGURE 20-6

A diagram of osmosis and diffusion—two important processes in plant cells and tissues.

Flavor

Flavors and aromas in fruits and vegetables are due to a variety of compounds working together to give unique and distinctive characteristics: aldehydes, alcohols, ketones, organic acids, esters, sulfur compounds, and trace amounts of other chemicals.

Astringency in fruits and vegetables is primarily due to the flavonoid pigments classified as tannins or phenolics. These flavor components make the mouth pucker. Some examples include lemons and chokecherries.

Fruity flavor is extremely complex. It can be partially caused by a combination of esters, alcohols, aldehydes, ketones, and minor compounds.

Sweetness of fruits and vegetables has been the one taste perception that is constantly searched for. Some plants such as sugar cane and sugar beet are grown for their sweet component, sucrose. Other foods are consumed for a combination of sugars and other flavor component interactions. Glucose, fructose, maltose, xylose, and less common sugars are also found. The types of sugars in plants vary considerably. Also, "non-sugar" sources of plant sweetness include compounds like glycyrrhizin from licorice.

Acid flavor from fruits and vegetables is formed by many different acids. Although malic and citric acid are the most common acids, a number of others can be found in selected plant foods. For example, grapes have considerable tartaric acid, and oxalic acid (rhubarb) and benzoic acid (plums, cranberries) are found in a number of fruits. These acids, in turn give a range of pH values.

The cabbage and onion family give flavors and odors due to a variety of sulfur compounds. Cooking will develop strong flavors due to hydrogen sulfide and other volatile sulfur compounds. Also, cutting or shredding across the cell wall releases an enzyme that develops the distinctive flavor of onion, garlic, and leek.

Quality Grades for Fresh Vegetables

Some vegetables are labeled with a USDA quality grade. The quality of most fresh vegetables can be judged reasonably well by their external appearance. Vegetables are available year-round from both domestic production and imports from other countries (Figure 20-7).

FIGURE 20-7

Besides variety, consumers expect high-quality fruits and vegetables. (*Source:* USDA ARS Image Gallery)

USDA has established grade **standards** for most fresh vegetables. The standards are used extensively as a basis for trading between growers, shippers, wholesalers, and retailers. They are used to a limited extent in sales from retailers to consumers. Use of U.S. grade standards is voluntary in most cases. Grade designations are most often seen on packages of potatoes and onions. Other vegetables occasionally carry the grade name. Grade designations are Fancy, 1, 2, and 3.

U.S. Fancy. **U.S. Fancy** vegetables are of more uniform shape and have fewer defects than U.S. No. 1.

U.S. No. 1. No. 1 vegetables should be tender and fresh-appearing, have good color, and be relatively free from bruises and decay.

U.S. No. 2 and No. 3. Even though U.S. No. 2 and No. 3 have lower quality requirements than Fancy or No. 1, all **grades** are nutritious. The differences are mainly in appearance, waste, and preference.

Quality Grades for Canned Frozen Vegetables

The grade standards are used extensively by processors, buyers, and others in wholesale trading to establish the value of a product described by the grades.

U.S. Grade A. Grade A vegetables are carefully selected for color, tenderness, and freedom from blemishes. They are the most tender, succulent, and flavorful vegetables produced. The term "fancy" may appear on the label to reflect the Grade A product.

U.S. Grade B. Grade B vegetables are of excellent quality but not quite as well selected for color and tenderness as Grade A. They are usually slightly more mature and therefore have a slightly different taste than the more succulent vegetables in Grade A.

U.S. Grade C. Grade C vegetables are not so uniform in color and flavor as vegetables in the higher grades, and they are usually more mature. They are a thrifty buy when appearance is not too important. These vegetables could be used as an ingredient in a soup, stew, or casserole.

Other names may be used to describe the quality grades of canned or frozen vegetables–Grade A as "Fancy," Grade B as "Extra Standard," and Grade C as "Standard." The brand name of a frozen or canned vegetable may also be an indication of quality. Producers of nationally advertised products spend considerable money and effort to maintain the same quality for their brand labels year after year. Unadvertised brands may also offer an assurance of quality, often at a slightly lower price. Many stores, particularly chain stores, carry two or more qualities under their own name labels (private labels).

Quality Grades for Fresh Fruit

The U.S. Department of Agriculture has established grade standards for most fresh fruits. The grades are used extensively as a basis for trading among growers, shippers, wholesalers, and retailers. Grade standards are used to a limited extent in sales from retailers to consumers. Use of U.S. grade standards is voluntary. Most packers grade their fruits, and some mark consumer packages with the grade. Some state laws and federal marketing programs require grading and grade labeling of certain fruits. Fruits are graded as Fancy, 1, 2, and 3.

U.S. Grade A or Fancy. Grade A or Fancy means premium quality (Figure 20-8). Only a small percentage of fruits are packed in this grade. These fruits have excellent color, uniform size, weight, and shape. They have the proper ripeness and possess few or no blemishes. Fruits of this grade are used for special purposes where appearance and flavor are important.

FIGURE 20-8

Grade A or Fancy apples. (*Source:* USDA ARS Image Gallery)

U.S. No. 1. U.S. No. 1 means good quality and is the most commonly used grade for most fruits.

U.S. No. 2 and U.S. No. 3. U.S. No. 2 is noticeably superior to U.S. No. 3, which is the lowest grade practical to pack under normal commercial conditions.

Quality Grades for Canned and Frozen Fruits

Grading and use of fruits for canning and freezing is slightly different. These fruits are graded U.S. Grade A, B, and C.

U.S. Grade A. Grade A fruits are the very best, with an excellent color and uniform size, weight, and shape. Having the proper ripeness and few or no blemishes, fruits of this grade are excellent to use for special purposes where appearance and flavor are important. This highest grade of fruits is the most flavorful and attractive. Often they are the most expensive. They are excellent to use for special luncheons or dinners, served as dessert, used in fruit plates, or broiled or baked to serve with meat entrees.

U.S. Grade B. Grade B fruits make up much of the fruits that are processed and are of very good quality. Only slightly less perfect than Grade A in color, uniformity, and texture, Grade B fruits have good flavor and are suitable for most uses. Grade B fruits, which are not quite as attractive or tasty as Grade A, are still of good quality. They are used as breakfast fruits; in gelatin molds, fruit cups, or compotes; as topping for ice cream; or as side dishes.

U.S. Grade C. Grade C fruits may contain some broken and uneven pieces. Though flavor may not be as sweet as in higher qualities, these fruits are still good and wholesome. They are useful where color and texture are not of great importance, such as in puddings, jams, and frozen desserts. Grade C fruits vary more in taste and appearance than the higher grades, and they cost less. They are useful in many dishes, especially where appearance is not important–for example, in sauces for meats, in cobblers, tarts, upside-down cakes, frozen desserts, jams, or puddings.

Other names are often used to describe the quality grades of canned and frozen fruits: Grade A as "Fancy," Grade B as "**Choice**," and Grade C as "Standard."

Styles. Both canned and frozen fruit are sold in many forms, shapes, or styles. Larger fruit, such as pears and peaches, may be found in whole, halves, quarters, slices, and diced. Smaller fruits

DEVELOPMENT OF POTATO CHIPS

In terms of consumption, potatoes rank second worldwide behind rice. Many potatoes are consumed as potato chips. The potato chip is more popular in America than in any other part of the world. America's favorite snack food is a direct descendant of another popular potato snack, the French fry.

According to the popular story, a dinner guest was dining at Moon's Lake House in Saratoga Springs, New York, in 1853. He sent his French fries back to the kitchen because they were too thick. The chef, a Native American named George Crum, was annoyed at the guest's complaint, so he responded by slicing the potatoes into extremely thin sections, which he fried in oil and salted. The plan backfired, and soon potato chips began appearing on the menu as Saratoga Chips.

For a time, potato chips were available only in the North. In the 1920s, Herman Lay, a traveling salesman in the South, helped popularize the chip from Atlanta to Tennessee. Lay sold potato chips to Southern grocers out of the trunk of his car. Soon he built a business and a name. Lay's potato chips became the first successfully marketed national brand. Today potato chips have evolved into many forms and varieties, including chips of many flavors, fat-free potato chips cooked in high-tech synthetic chemicals, and even artificially shaped chips pressed from potato pulp and sold in card-board tubes.

For more information about potato chips and other vegetable products, search the Web or visit these Web sites:

<www.ideafinder.com/facts/story/story007.htm>
<www.dmgi.com/chips.html>
<detnews.com/1999/food/0121/panel/panel.htm>

are usually whole, but fruits such as strawberries are sliced and halved. It is best to examine the label for a description of the type or style that will best suit the purpose you have in mind for serving the fruit. The grade, style, and syrup or special flavorings in which processed fruits are prepared all affect the cost of the fruits and how they are used.

Most processed fruits are available in at least two grades. The grade is not often indicated on processed fruits, but individuals can learn to tell differences in quality by trying different brands. Whole fruits, halves, or slices of similar sizes are more expensive than mixed pieces of various sizes and shapes. Some of the most popular fruits, along with the styles in which they are available, are shown in Figure 20-9.

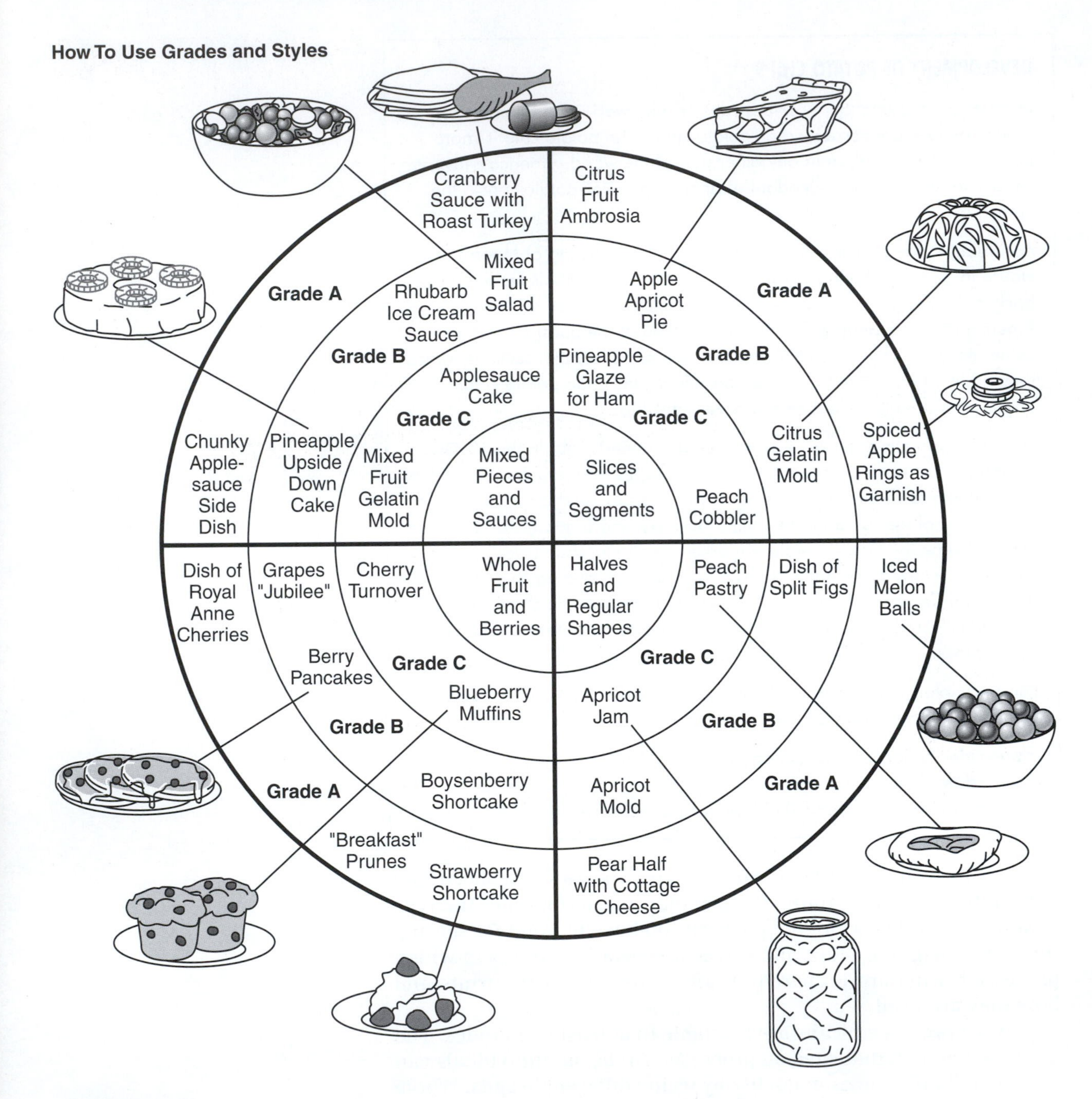

FIGURE 20-9

Fruit grades and their uses. (*Source:* USDA Agricultural Marketing Service)

POST-HARVEST

The quality of the harvested fruit and vegetable can be optimum, but the post-harvest care, or storage, will affect what the processor, food preparer, or consumer receives (Figure 20-10). Post-harvest factors that affect the marketing of quality produce after harvesting include:

- Shipping/storage at optimum temperature
- Control of carbon dioxide (CO_2) and oxygen (O_2)
- Good humidity control
- Minimum exposure to ethylene gas
- Application of appropriate chemicals
- Good sanitation
- Harvest at the correct maturity stage
- Care in handling during harvest, packing, and shipping

Likely the single most important factor influencing quality during storage is temperature reduction and maintenance post-harvest because metabolic activity of the produce and spoilage microorganisms is reduced by low temperature. Each fruit or vegetable has its optimum temperature regime. The generally accepted rule is that for every 50°F (10°C) rise in temperature from optimum, metabolic rate will increase two- to threefold. This increase in sensitivity of fruits and vegetables to temperature and subsequent deterioration make fruit and vegetable transportation and storage by the consumer of critical importance.

Natural or synthetic chemicals are used commercially to maintain quality through prolonged storage life. Three primary gases–water vapor (humidity), carbon dioxide (CO_2), and ethylene –do impact vegetables and fruits. Modified atmospheres (lower O_2, and/or higher CO_2) in the form of gases have been most effective in prolonging storage.

Modified atmosphere is done in the intact or processed fruit or vegetable through such processes as controlled-atmospheric storage procedures (CAS). These conditions attempt to control the level of those gases that influence the respiration of the vegetable–for example, ripening or maturation. Generally, CO_2 evolution occurs with increased maturation. Controlling the CO_2 level is critical.

The interrelationship between ethylene gas and storage life is dynamic. Fruits and vegetables may be classified as either ethylene producers or ethylene sensitive (climacteric versus nonclimacteric). (Refer to Table 20-1, page 363.) Processors and food preparers must be aware of this in their packaging and storage.

FIGURE 20-10

Fresh and processed potato products. (*Source:* USDA ARS Image Gallery)

PROCESSING OF FRUITS

Fruits for canning or freezing are harvested at the proper stage of ripeness so that a good texture and flavor may be preserved. Much of the processing is done by automated equipment, and the fruits are handled little by plant workers. Present-day practices help assure us of wholesome, sanitary products with good flavor and quality.

Table 20-3 lists the common fruits and provides some information about each.

The initial work in preparing canned or frozen fruits is similar. At the processing plant, the fresh fruits are usually sorted into sizes by machine and washed in continuously circulating water or under sprays of water (Figure 20-11). Some fruits, such as apples, pears, and pineapple, are mechanically peeled and cored. Next, they are moved on conveyor belts to plant workers who do any additional peeling or cutting necessary. Pits and seeds are removed by automatic equipment, and the fruits are also prepared in the various styles (halves, slices, or pieces) by machine. Before the fruits are canned or frozen, plant workers remove any undesirable portions.

FIGURE 20-11

Washing and grading apples.

TABLE 20-3 Common Fruits

Fruit	Notes
Apples	The many varieties of apples differ widely in appearance, flesh characteristics, seasonal availability, and suitability for different uses.
Apricots	Apricots develop their flavor and sweetness on the tree, and should be mature but firm at the time that they are picked.
Avocados	Avocados vary greatly in shape, size, and color. Most tend to be pear-shaped, but some are almost spherical. Some have rough or leathery textured skin, while others have smooth skin. The skin color of most varieties is some shade of green, but certain varieties turn maroon, brown, or purplish-black as they ripen.
Bananas	Bananas develop their best eating quality after they are harvested. This allows bananas to be shipped great distances. Bananas are sensitive to cool temperatures and will be injured in temperatures below 55°F (12.8°C), so they should never be kept in the refrigerator. The ideal temperature for ripening bananas is between 60° and 70°F (15.6° and 21.1°C).
Blueberries	Large berries are cultivated varieties, and the smaller berries are wild varieties. A dark blue color with a silvery bloom is the best indication of quality. This silvery bloom is a natural, protective, waxy coating.
Cherries	Good cherries have bright, glossy, plump-looking surfaces and fresh-looking stems. A very dark color is your most important indication of good flavor and maturity in sweet cherries.
Cranberries	Differ considerably in size and color, but are not identified by variety names. Plump, firm berries with a lustrous color provide the best quality. Duller varieties should at least have some red color.
Grapefruit	Several varieties are marketed, but the principal distinction at retail is between those that are "seedless" (having few or no seeds) and the "seeded" type. Another distinction is pink- or red-fleshed fruit and white-fleshed. Grapefruit is picked "tree ripe" and is ready to eat. Thin-skinned fruits have more juice than coarse-skinned ones.
Grapes	European types are firm-fleshed and generally have high sugar content. American-type grapes have softer flesh and are juicier than European types. The outstanding variety for flavor is the Concord, which is blue-black when fully matured. Well-colored, plump grapes that are firmly attached to the stem. White or green grapes are sweetest when the color has a yellowish cast or straw color, with a tinge of amber. Red varieties are better when good red predominates on all or most of the berries. Bunches are more likely to hold together if the stems are predominantly green and pliable.

(continued)

TABLE 20-3 Common Fruits *(continued)*

Fruit	Notes
Kiwifruit	A relatively small, ellipsoid-shaped fruit with a bright green, slightly acid-tasting pulp surrounding many small, black, edible seeds, which in turn surround a pale heart. The exterior of the kiwifruit is unappealing to some, being somewhat "furry" and light to medium brown in color. Kiwifruit contains an enzyme, actinidin, similar to papain in papayas that reacts chemically to break down proteins. Actinidin prevents gelatin from setting.
Lemons	Lemons should have a rich yellow color, reasonably smooth-textured skin with a slight gloss, and should be firm and heavy. A pale or greenish-yellow color means very fresh fruit with slightly higher acidity. Coarse or rough skin texture is a sign of thick skin and not much flesh.
Limes	Limes should have glossy skin and heavy weight for the size.
Cantaloupes	A cantaloupe might be mature, but not ripe. The stem should be gone, leaving a smooth symmetrical, shallow base called a full slip. The netting, or veining, should be thick, coarse, and corky, and should stand out in bold relief over some part of the surface. The skin color (ground color) between the netting should have changed from green to yellowish-buff, yellowish-gray, or pale yellow.
Casaba	This sweet, juicy melon is normally pumpkin-shaped with a very slight tendency to be pointed at the stem end. It is not netted, but has shallow, irregular furrows running from the stem end toward the blossom end. The rind is hard with light green or yellow color. The stem does not separate from the melon, and must be cut in harvesting.
Crenshaw	Its large size and distinctive shape make this melon easy to identify. It is rounded at the blossom end and tends to be pointed at the stem end. The rind is relatively smooth with only very shallow lengthwise furrowing. The flesh is pale orange, juicy, and delicious, and is generally considered outstanding in the melon family.
Honeydew	The outstanding flavor characteristics of honeydews make them highly prized as a dessert fruit. A soft, velvety texture indicates maturity. The stem does not separate from the fruit, and must be cut for harvesting.
Watermelon	Judging the quality of a watermelon is very difficult unless it is cut in half or quartered. Firm, juicy flesh with good red color that is free from white streaks; and seeds that are dark brown or black. Seedless watermelons often contain small white, immature seeds, which are normal for this type.
Nectarines	This fruit combines characteristics of both the peach and the plum. Most varieties have an orange-yellow background color between the red areas, but some varieties have a greenish background color.

TABLE 20-3 Common Fruits *(concluded)*

Fruit	Notes
Oranges	Leading varieties are the Washington Navel and the Valencia, both characterized by a rich orange skin color. The Navel orange has a thicker, somewhat more pebbled skin than the Valencia; the skin is more easily removed by hand, and the segments separate more readily. It is ideally suited for eating as a whole fruit or in segments in salads. The Valencia orange is excellent either for juicing or for slicing in salads. Oranges are required by strict state regulations to be mature before being harvested and shipped out of the producing state. Thus, skin color is not a reliable index of quality, and a greenish cast or green spots do not mean that the orange is immature. Often, fully matured oranges will turn greenish (called "regreening") late in the marketing season. Some oranges are artificially colored to improve their appearance. This practice has no effect on eating quality, but artificially colored fruits must be labeled "color added."
Peaches	Peaches fall into two general types: freestone (flesh readily separates from the pit) and clingstone (flesh clings tightly to the pit). Freestones are usually preferred for eating fresh or for freezing; clingstones are used primarily for canning, although they are sometimes sold fresh.
Pears	The most popular variety of pear is the Bartlett, both for canning and for sale as a fresh fruit.
Pineapple	Pineapples are available all year, but are most abundant from March through June. Present marketing practices, including air shipments, allow pineapples to be harvested as nearly ripe as possible. They should have a bright color, fragrant pineapple aroma, and a very slight separation of the eyes or pips—the berrylike fruitlets patterned in a spiral on the fruit core.
Plums and Prunes	Quality characteristics for both are very similar, and the same buying tips apply to both. Plum varieties differ slightly in appearance and flavor. Only a few varieties of prunes are commonly marketed, and they are all very similar.
Raspberries/ boysenberries, etc.	Blackberries, raspberries, dewberries, loganberries, and youngberries are similar in general structure. They differ from one another in shape or color, but quality factors are about the same for all. Look for berries that are fully ripened, with no attached stem caps.
Strawberries	Berries with a full red color and a bright luster, firm flesh, and the cap stem still attached. The berries should be dry and clean, and usually medium to small strawberries have better eating quality than large ones.
Tangerines	Deep yellow or orange color and a bright luster is your best sign of fresh, mature, good-flavored tangerines. Because of the typically loose nature of tangerine skins, they will frequently not feel firm to the touch.

Canned Fruits

Cans or glass jars are filled with fruit by semiautomatic machines. Next, the containers are moved to machines that fill them with the correct amount of syrup or liquid and then to equipment that automatically seals them. The sealed containers are cooked under carefully controlled conditions of time and temperature to assure that the products will keep without refrigeration. After the containers are cooled, they are stored in cool, dry, well-ventilated warehouses until they are shipped to market.

Frozen Fruits

Frozen fruits are most often packed with dry sugar or syrup. After the initial preparation, packages are filled with fruit by semiautomatic equipment, sugar or syrup is added, and the containers are automatically sealed. The packaged fruit is then quickly frozen in special low-temperature chambers and stored at temperatures of 0°F (−18°C) or lower.

Fruit Juices

Orange juice is probably the most commonly processed juice. Steps in the orange juice process are also common for the manufacture of other juices. The main steps in the production of most juices include:

- Extraction
- Clarification (clearing)
- Deaeration (removal of air)
- Pasteurization
- Concentration
- Essence add-back (flavors)
- Canning or bottling
- Freezing

Not all juices go through all processes. For example, the desired end product determines if the juice will be concentrated and frozen.

Blending of juices is also a popular processing technique. Some blends include mango and apple, apple and cranberry, and mango and orange. Pear juice is used as a base for many juices because its fruit flavor is strong but not necessarily characteristic. Finally, juices low in vitamin C may be fortified with this vitamin.

PROCESSING OF VEGETABLES

Vegetables for canning and freezing are grown especially for that purpose, and the processing preserves much of their nutritional value. Both canning and freezing plants are usually located in the vegetable production areas so that the harvested vegetables can be quickly brought to the plant for processing while fresh and at their peak in quality.

Table 20-4 lists the common vegetables and provides some information about each.

In today's modern processing plants, most of the vegetables are canned or frozen by automated equipment, and the procedures used are similar to those used for fruit. The initial work in preparing canned or frozen vegetables is similar. At the processing plant, the fresh product is usually sorted into sizes by machine and washed in continuously circulating water or sprays of water. Some vegetables, such as carrots, beets, and potatoes, are mechanically peeled. Next, they are moved onto conveyer belts where plant workers do any additional peeling or cutting prior to preparation for the various styles–whole, cut, sliced, and so on (Figure 20-12).

Canned Vegetables

Cans or glass jars are filled with vegetables by semiautomatic or automatic machines. Next, the containers are moved to machines that fill them with the correct amount of brine or liquid and then to machines that preheat them prior to automatically sealing them. The sealed containers are then cooked under carefully controlled conditions of time and temperature to assure that the product will keep without refrigeration. After the containers are cooled, they are stored in cool, dry, well-ventilated warehouses until they are shipped to market.

Vegetables sold in glass jars with screw-on or vacuum-sealed lids are sealed tightly to preserve the contents.

FIGURE 20-12
Packing potatoes.

TABLE 20-4 Common Vegetables

Vegetable	Notes
Artichoke	The globe artichoke is the large, unopened flower bud of a plant belonging to the thistle family. The many leaflike parts making up the bud are called "scales."
Beets	Many beets are sold in bunches with the tops still attached; others are sold with the tops removed. If beets are bunched, you can judge their freshness fairly accurately by the condition of the tops.
Broccoli	A member of the cabbage family, and a close relative of cauliflower. A firm, compact cluster of small flower buds, with none opened enough to show the bright-yellow flower. Bud clusters should be dark green or sage green—or even green with a decidedly purplish cast.
Brussels Sprouts	Another close relative of the cabbage, Brussels sprouts develop as enlarged buds on a tall stem, one sprout appearing where each main leaf is attached. The "sprouts" are cut off and, in most cases, are packed in small consumer containers.
Cabbage	Three major groups of cabbage varieties are available: smooth-leaved green cabbage; crinkly-leaved green Savoy cabbage; and red cabbage. All types are suitable for any use. Cabbage may be sold fresh (called "new" cabbage) or from storage.
Carrots	Freshly harvested carrots are available year round. Most are marketed when relatively young, tender, well colored, and mild-flavored—an ideal stage for use as raw carrot sticks. Larger carrots are packed separately and used primarily for cooking or shredding.
Cauliflower	The white edible portion is called "the curd" and the heavy outer leaf covering is called "the jacket leaves." Cauliflower is generally sold with most of the jacket leaves removed, and is wrapped in plastic film. A slightly granular or "ricey" texture of the curd will not hurt the eating quality if the surface is compact.
Celery	Most celery is of the so-called "Pascal" type, which includes thick-branched, green varieties.
Chinese Cabbage	Primarily a salad vegetable, Chinese cabbage plants are elongated, with some varieties developing a firm head and others an open, leafy form.
Chicory, endive, escarole	Used mainly in salads, are available practically all year round, but primarily in the winter and spring. Chicory or endive has narrow, notched edges, and crinkly leaves resembling the dandelion leaf. Chicory plants often have "blanched" yellowish leaves in the center that are preferred by many people. Escarole leaves are much broader and less crinkly than those of chicory.

TABLE 20-4 Common Vegetables *(continued)*

Vegetable	Notes
Corn	Sweet corn is available practically every month of the year, but is most plentiful from early May until mid-September. Yellow-kernel corn is the most popular, but some white-kernel and mixed-color corn is sold. Sweet corn is produced in a large number of states during the spring and summer. For best quality, corn should be refrigerated immediately after being picked. Corn will retain fairly good quality for a number of days, if it has been kept cold and moist since harvesting.
Cucumber	Produced at various times of the year in many states, and imported during the colder months. The supply is most plentiful in the summer months. Cucumbers should be a good green color, well developed, but not too large in diameter.
Eggplant	Most plentiful during late summer, but is available all year. Although the purple eggplant is more common, white eggplant is occasionally seen in the marketplace.
Greens	A large number of widely differing species of plants are grown for use as "greens." Spinach, kale, collard, turnip, beet, chard, mustard, broccoli leaves, chicory, endive, escarole, dandelion, cress, and sorrel are well-known greens. Many others, some of them wild, are also used to a limited extent as greens.
Lettuce	One of the leading U.S. vegetables, lettuce owes its prominence to the growing popularity of salads in our diets. It is available throughout the year in various seasons from California, Arizona, Florida, New York, New Jersey, and other states. Four types of lettuce are generally sold: iceberg, butterhead, Romaine, and leaf.
Mushrooms	Grown in houses, cellars, or caves, mushrooms are available year-round in varying amounts. They are mainly produced in Pennsylvania, California, New York, Ohio, and other states. They are described as having a cap (the wide portion on top), gills (the numerous rows of paper-thin tissue seen underneath the cap when it opens), and a stem.
Okra	Okra is the immature seedpod of the okra plant, generally grown in southern states.
Onions	The many varieties of onions grown commercially fall into three general classes, distinguished by color: yellow, white, and red. Onions are available year-round, either fresh or from storage. Consumers should avoid onions with thick, hollow, woody centers in the neck or with fresh sprouts.

(continued)

TABLE 20-4 Common Vegetables *(continued)*

Vegetable	Notes
Onions, green (leeks)	Sometimes called scallions, they are similar in appearance, but are somewhat different in nature. Green onions are ordinary onions harvested very young. They have very little or no bulb formation, and their tops are tubular. Leeks have slight bulb formation and broad, flat, dark-green tops.
Parsley	Parsley is generally available year-round. It is used both as a decorative garnish and to add its own unique flavor.
Parsnips	Although available to some extent throughout the year, parsnips are primarily late-winter vegetables because the flavor becomes sweeter and more desirable after long exposure to cold temperatures, below 40°F (4.4°C).
Peppers	Most of the peppers are the sweet green peppers, available in varying amounts throughout the year, but most plentiful during late summer. Fully matured peppers of the same type have a bright red color. A variety of colored peppers are also available, including white, yellow, orange, red, and purple.
Potatoes	Potatoes can be put into three groups: new potatoes, general purpose, and baking. Some overlapping exists. New potatoes frequently describe those potatoes freshly harvested and marketed during the late winter or early spring. The name is also widely used to designate freshly dug potatoes that are not fully matured. The best uses for new potatoes are boiling or creaming. They vary widely in size and shape, depending upon variety, but are likely to be affected by "skinning" or "feathering" of the outer layer of skin. Skinning usually affects only their appearance. General-purpose potatoes include the great majority of supplies, both round and long types, offered for sale in markets. With the aid of air-cooled storage, they are available throughout the year. They are used for boiling, frying, and baking. Potatoes grown specifically for their baking quality also are available. Variety and area where grown affect baking quality. The Russet Burbank is commonly used for baking. Potatoes should be free from blemishes and sunburn (a green discoloration under the skin).
Radishes	Radishes are available year-round, but they are most plentiful from May through July. Medium-size radishes are 3/4 to 1 inch in diameter.
Rhubarb	Rhubarb is a specialized vegetable used like a fruit in sweetened sauces and pies. Very limited supplies are available during most of the year, with best supplies available from January to June. The petiole should be red, tender, and not fibrous.

TABLE 20-4 Common Vegetables *(concluded)*

Vegetable	Notes
Squash (summer)	Summer squash includes those varieties that are harvested while still immature and when the entire squash is tender and edible. They include the yellow Crookneck, the large Straight neck, the greenish-white Patty Pan, and the slender green Zucchini.
Squash (fall and winter)	Winter squash are those varieties that are marketed only when fully mature. Some of the most important varieties are the small-corrugated Acorn (available year-round), Butternut, Buttercup, green and blue Hubbard, green and gold Delicious, and Banana.
Sweet potatoes	Available in varying amounts year-round, moist sweet potatoes, sometimes called yams, are the most common type. They have orange-colored flesh and are very sweet. (The true yam is the root of a tropical vine that is not grown commercially in the United States.) Dry sweet potatoes have pale-colored flesh and are low in moisture.
Tomatoes	Popular and nutritious, tomatoes are in moderate to liberal supply throughout the year. The best flavor usually comes from locally grown tomatoes produced on nearby farms, because the tomato is allowed to ripen completely before being picked. Many areas now ship tomatoes that are picked right after the color has begun to change from green to pink. If tomatoes need further ripening, keep them in a warm place but not in direct sunlight.
Turnips	Popular turnips have white flesh and a purple top (reddish-purple tinting of upper surface). It may be sold "topped" (with leaves removed) or in bunches with tops still on. Rutabagas are distinctly the yellow-fleshed, large-sized relatives of turnips.
Watercress	Watercress is a small, round-leaved plant that grows naturally (or it may be cultivated) along the banks of freshwater streams and ponds. It is prized as an ingredient in mixed green salads and as a garnish, because of its spicy flavor. Watercress is available in limited supply through most of the year.

Frozen Vegetables

After initial preparation, vegetables that are to be frozen are usually blanched, or slightly precooked. This precooking process ensures that the frozen vegetables will retain much of their natural appearance and flavor for long periods of time in storage. Without blanching, the product would prematurely turn brown or oxidize before it could be marketed. The vegetables, after freezing, are packaged in polyethylene bags of varying sizes or may be packaged in retail-size fiber cartons with a labeled overwrap that identifies the product (Figure 20-13).

BY-PRODUCTS

Processing of fruits and vegetables produces many by-products. Often these by-products become feed for livestock. For example, citrus pulp and processed potato wastes, and grape seeds and skins can be fed to cattle depending on location (Figure 20-14).

FIGURE 20-13
Frozen fruit and vegetable products and packaging. (*Source:* USDA ARS Image Gallery)

FIGURE 20-14
Potato by-product being loaded and fed to cattle.

BIOTECHNOLOGY

Biotechnology allows researchers to target the genetics of plants, animals, and microorganisms and to manipulate them to our food production advantage. By doing this, they will be able to design environmentally hardy food-producing plants that are naturally resistant to pests and diseases and capable of growing under extreme conditions of temperature, moisture, and salinity. Or scientists may design an array of fresh fruits and vegetables, with excellent flavor, appealing texture, and optimum nutritional content, that stay fresh for several weeks. Possibly they could custom design plants with defined structural and functional properties for specific food-processing applications. On the control side, scientists might use biotechnology to produce microsensors that accurately measure the physiological state of plants, or temperature-abuse indicators for refrigerated foods, and shelf-life monitors built into food packages. The possibilities are endless.

Summary

Fruits and vegetables include a wide variety of edible parts. They vary widely in carbohydrate and protein content. Many have a high water content. Fruits and vegetables are a good source of many vitamins and minerals. Because many parts of plants are eaten as fruits or vegetables, an understanding of plant tissues is critical to food scientists.

Harvesting of fruits and vegetables can be affected by variety, soil type, water, temperature, and season. Climacteric fruits produce ethylene gas during ripening; nonclimacteric fruits do not. Fruits and vegetables get their characteristic color from numerous pigments. In general, water transport in plant tissues influences the texture of fruits and vegetables; flavors and aromas are due to compounds such as aldehydes, alcohols, ketones, esters, organic acids, and sulfur compounds.

The USDA assigns quality grades to both fruits and vegetables. These quality grades determine the eventual use of the fruit or vegetable. Post-harvest care is critical to maintaining the optimal quality of fruits and vegetables. Fruits and vegetables are sold fresh, canned, and frozen. Most of these processes are automated. By-products from fruit and vegetables processing are often used for livestock feed. Biotechnology offers the promise of providing fruits and vegetables to meet new consumer demands.

Review Questions

Success in any career requires knowledge. Test your knowledge of this chapter by answering these questions or solving these problems.

1. Fruits and vegetables get their characteristic color from ___________.
2. What are the grade designations for fresh fruits and vegetables and canned fruits and vegetables?
3. What is the difference between climacteric and non-climacteric fruits?
4. List four compounds that give fruits or vegetables their flavor.
5. Plant tissues existing in fruits and vegetables are the ___________, a protective tissue layer; the ___________, which contains the xylem and phloem; the ___________, which differs according to the plant; and the ___________ tissue, which is the edible portion of the fruit or vegetable.
6. The crispness of a vegetable is due to the movement of ___________ in the plant.
7. Describe the processing of fruit for canning or freezing.
8. Name the single most important factor influencing quality during storage of fruits and vegetables.
9. Why are frozen vegetables blanched or precooked?
10. List the steps of orange juice processing.

Student Activities

1. Make a list of all vegetables you consume during a week. Determine which part of the plant you are considering as a vegetable–for example, the flower, the stem, the root, the fruit, and so on. Using Table A-8, list the composition of five of these vegetables.
2. Pick a fruit or vegetable and list all the fresh and processed forms for the fruit or vegetable. Compete with other class members to see who found a vegetable or fruit sold in the most forms.
3. Develop a report or presentation on the production, harvesting, and processing of a specific fruit or vegetable listed in Table 20-3, page 373, or Table 20-4, page 378. Use the resources of the World Wide Web to do this.

4. Develop an experiment to extract the pigment from a fruit or vegetable such as carrots, apples, or spinach.
5. Develop an experiment to store a selected vegetable or fruit under different conditions–for example, light or dark, dry or humid, warm or cold, or some combinations. Record what happens to the quality of the fruit or vegetable under each set of conditions.

Resources

Corriher, S. O. 1997. *Cookwise: The hows and whys of successful cooking.* New York: William Morrow and Company, Inc.

Horn, J., J. Fletcher, and A. Gooch. 1997. *Cooking a to z. The complete culinary reference source.* Glen Ellen, CA: Cole Publishing Group, Inc.

Lehner, E., and J. Lehner. 1962. *Folklore and odysseys of food and medicinal plants.* New York: Tudor Publishing Company.

Parker, R. 1998. *Introduction to plant science.* Albany, NY: Delmar.

Shewfelt, R. L., and V. Bruckner. 2000. *Fruit and vegetable quality.* Lancaster, PA: Technomic Publishing Co., Inc.

Smith, D. S., J. N. Cash, W. K. Nip, and Y. H. Hui (Eds.). 2000. *Processing vegetables.* Lancaster, PA: Technomic Publishing Co. Inc.

Somogyi, L. P., D. M. Barrett, H. Ramaswamy, Y. H. and Hui (Eds.). 2000. *Processing fruits.* Lancaster, PA: Technomic Publishing Co. Inc.

Vaclavik, V. A., and E. W. Christina. 1999. *Essentials of food science.* Gaithersburg, MD: Aspen Publishers, Inc.

Vieira, E. R. 1996. *Elementary food science,* 4th ed. New York: Chapman and Hall.

Wagner, S. (Ed.). 1999. *The recipe encyclopedia: The complete illustrated guide to cooking.* San Diego: Thunder Bay Press.

Internet

Internet sites represent a vast resource of information. The URLs (uniform resource locator) for the World Wide Web sites can change. Using one of the search engines on the Internet such as Yahoo!, HotBot, AltaVista, Excite, Dogpile, About, or Google, find more information by searching for these words or phrases: a specific fruit or vegetable; grading of fruits and vegetables such as U.S. Fancy, U.S. No. 1, 2, 3; grades of fruits and vegetables: grade A, B, C; processing of specific fruits or vegetables; harvesting of specific fruits or vegetables. Also, Table A-7 provides a listing of some useful Internet sites that can be used as a starting point.

Chapter 21

Fats and Oils

Objectives

After reading this chapter, you should be able to:

- Explain saturated and unsaturated, cis and trans in terms of fatty acids
- Describe fatty acids
- Discuss melting point and the structure of fatty acids
- Identify six sources of fats and oils
- List eight functions fats and oils serve in foods
- Compare the extraction of fats or oils from animals to that of plants
- Describe the process used on oils after extraction
- List five processes in the refining and modifying of oils or fats after extraction
- Discuss monoglycerides and diglycerides and their uses
- Identify substances that may substitute for fat
- Describe two tests conducted on fats and oils

Key Terms

bleaching
cis
degumming
deordorization
diglycerides
double bonds
fatty acid
glycerol
hydrogenation
iodine value
monoglycerides
peroxide value
rancidity
rendering
saponification value
saturated
trans
triglyceride
unsaturated
winterization

Plants and animals provide sources of fats and oils. Like proteins and carbohydrates, fats and oils are composed of carbon, hydrogen, and oxygen, yet they contain 2.25 times more energy per unit of weight. In foods and food processing, they contribute unique characteristics.

EFFECTS OF COMPOSITION ON FAT PROPERTIES

Fats can be classified into simple lipids, compound lipids, composite lipids, spingolipids, and derived lipids. Food science is concerned most with the simple lipids. These are the **triglyceride** lipids. They make up the major components of fat, butter, shortening, and oil.

A triglyceride molecule of fat is three (tri) fatty acid molecules connected to (or esterified) a **glycerol** molecule. The structure of the **fatty acids** that are esterified to the glycerol determine the properties of fats. For example, the structure determines if they are solid or liquid at room temperature.

Fatty acids are chains of 4 to 28 carbon atoms with the carbons in the chain joined by single or **double bonds**, depending on the number of hydrogen atoms attached. If a fatty acid has all the hydrogens possible attached to the carbons in the chain, it is said to be a **saturated** fatty acid. If some of the carbons in the chain are joined by double bonds, thus reducing the number of hydrogen atoms, the fatty acid is called **unsaturated**. The acid portion of a fatty acid is represented by COOH. For example, the following formulas represent a saturated and an unsaturated fatty acid, respectively:

CH3-CH2-CH2-CH2-CH2-CH2-CH2-CH2-CH2-CH2-CH2-CH2-CH2-CH2-CH2-COOH

CH3-CH2-CH2-CH2-CH2-CH2-CH2-CH==CH-CH2-CH2-CH2-CH2-CH2-CH2-COOH

If a fatty acid contains a double bond, the hydrogen atom or other groups attached to the carbon atoms involved in the double bond may have different orientations. In other words, atoms or other groups may exist as **cis** or **trans** forms around the double bond of the fatty acids. These are called isomers. Cis and trans forms have the same number of carbon, hydrogen, and oxygen atoms but in a different geometrical arrangement. This too gives a fatty acid different chemical and physical properties.

Table 21-1 shows how the number of carbon atoms and saturation influences the melting point of some common fatty acids.

TABLE 21-1 Influence of the Size of a Fatty Acid and Saturation on the Melting Point

Fatty Acid	No. of Carbon Atoms	Melting Point
Saturated Fatty Acids		
Butyric	4	17.8°F (–7.9°C)
Caproic	6	25.0°F (–3.9°C)
Caprylic	8	61.3°F (16.3°C)
Capric	10	88.3°F (31.3°C)
Lauric	12	111.2°F (44.0°C)
Myristic	14	129.9°F (54.4°C)
Palmitic	16	145.0°F (62.8°C)
Stearic	18	157.3°F (69.6°C)
Arachidic	20	167.7°F (75.4°C)
Unsaturated Fatty Acids		
Palmitoleic	16	31.1° to 32.9°F (–0.5° to 0.5°C)
Oleic	18	55.4°F (13.0°C)
Linoleic	18	23.0° to 10.4°F (–5° to –12°C)
Linolenic	18	5.9°F (–14.5°C)
Arachidonic	20	–57.1°F (–49.5°C)

In general, a longer carbon chain increases the melting point; the more double bonds, the lower the melting point; and cis fatty acids have a lower melting point than trans.

In practice, the characteristics of the fatty acids are only a part of the factors influencing the properties of fats. The arrangement of fatty acids upon the glycerol backbone (Figure 21-1) will make a difference. Also, the properties of fats vary because they are made up of a variety of fatty acids and triglycerides.

SOURCES OF FATS AND OILS

Fats and oils come from plant and animal sources, including fish (Figure 21-2). The plant or vegetable fats include cocoa butter, corn

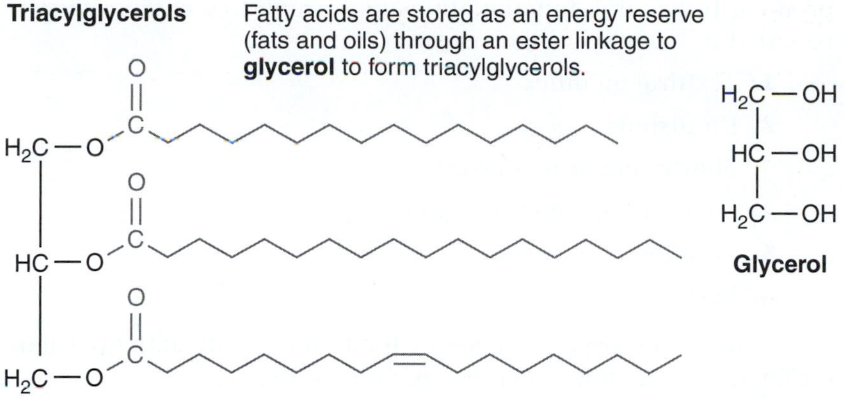

Triacylglycerols Fatty acids are stored as an energy reserve (fats and oils) through an ester linkage to **glycerol** to form triacylglycerols.

H_2C-OH
$HC-OH$
H_2C-OH

Glycerol

FIGURE 21-1
Diagram showing fatty acids attached to a glycerol backbone making a triglyceride.

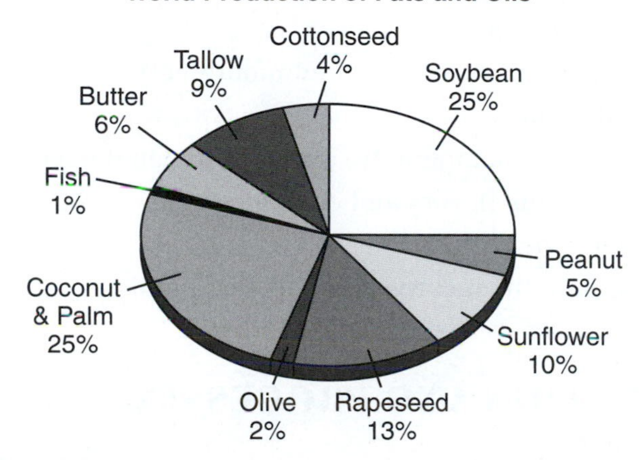

FIGURE 21-2
World production of fats and oils dominated by those produced by plants. (*Source:* USDA)

oil, sunflower oil, soybean oil, cottonseed oil, peanut oil, olive oil, canola oil, and many others. Animal fats include lard from pigs, tallow from beef, and butterfat from milk. Fish oils include cod liver oil, oil from menhaden, and whale oil. (The whale is really a mammal, but whale oil is included as a fish or marine oil.) Fats and oils and their composition are listed in Table A-8.

FUNCTIONAL PROPERTIES OF FATS

From a nutritional perspective, fats and oils have a high caloric value, 2.25 times more energy than carbohydrates or proteins. Each gram of fat contains 9 kcal. Fats and oils also carry the fat-soluble vitamins.

The uses of fats in foods continue to expand as they become more healthy and as the industry learns to modify the natural product. In general, functionalities and properties of fats are represented in these six major uses:

1. Textural qualities
2. Emulsions
3. Shortening or tenderizers
4. Medium for transferring heat
5. Aeration and leavening
6. Spray oils

These uses are influenced by the functionality of the particular fat or oil. The functionalities of fat include:

- Producing satiety (fullness after eating)
- Transfering heat
- Adding flavor
- Providing texture (body and mouthfeel)
- Tenderizing
- Decreasing temperature shock in frozen desserts
- Solubilizing flavors and colors
- Dispersing
- Aiding in the incorporation of air (foaming)

PRODUCTION AND PROCESSING METHODS

Food fats and oils are first extracted from both plant or animal to produce the products butter, margarine, lard, and hydrogenated shortening.

Refined oils include soybean oil, cottonseed oil, sunflower oil, peanut oil, olive oil, corn oil, canola oil, safflower oils, coconut oils, palm oil, and palm-kernel oils.

Fats and oils are extracted by a number of different methods. For example, the adipose tissue (fat) of the pig is heated. This melts the fat and allows it to be further processed. Melting animal fat to extract it is called **rendering**. Rendering can be accomplished by heating meat scraps in steam or water and then skimming or centrifuging to separate the fat. Dry heat and a vacuum can also be used to render fat. The temperature used to render fat can influence its color and flavor.

The extraction of butterfat was covered in Chapter 16, and butter is made by reversing the oil-in-water emulsion of cream from the dairy cow into a water-in-oil emulsion.

Extracting plant fats requires a little more complex method of processing. First, the oil is removed from the plant source by mechanical presses and expellers that squeeze the oil from oilseeds like soybeans, cottonseed, peanuts, and so on. Often the seeds are ground and cooked before the oil is extracted. Another method of extracting oil from seeds is the solvent extraction method. A nontoxic fat solvent such as hexane is percolated through the cracked seeds. The oil is then distilled from the solvent, and the solvent is reused. Often pressing and solvent methods are combined. The oil-free residue from extraction is called meal, and it is used for animal feed.

After removal of the oil from the seed, pod, or grain, it is further refined and modified by **degumming**, refining, **bleaching**, **winterization** (fractionation), **hydrogenation**, **deordorization**, and interesterification (Figure 21-3).

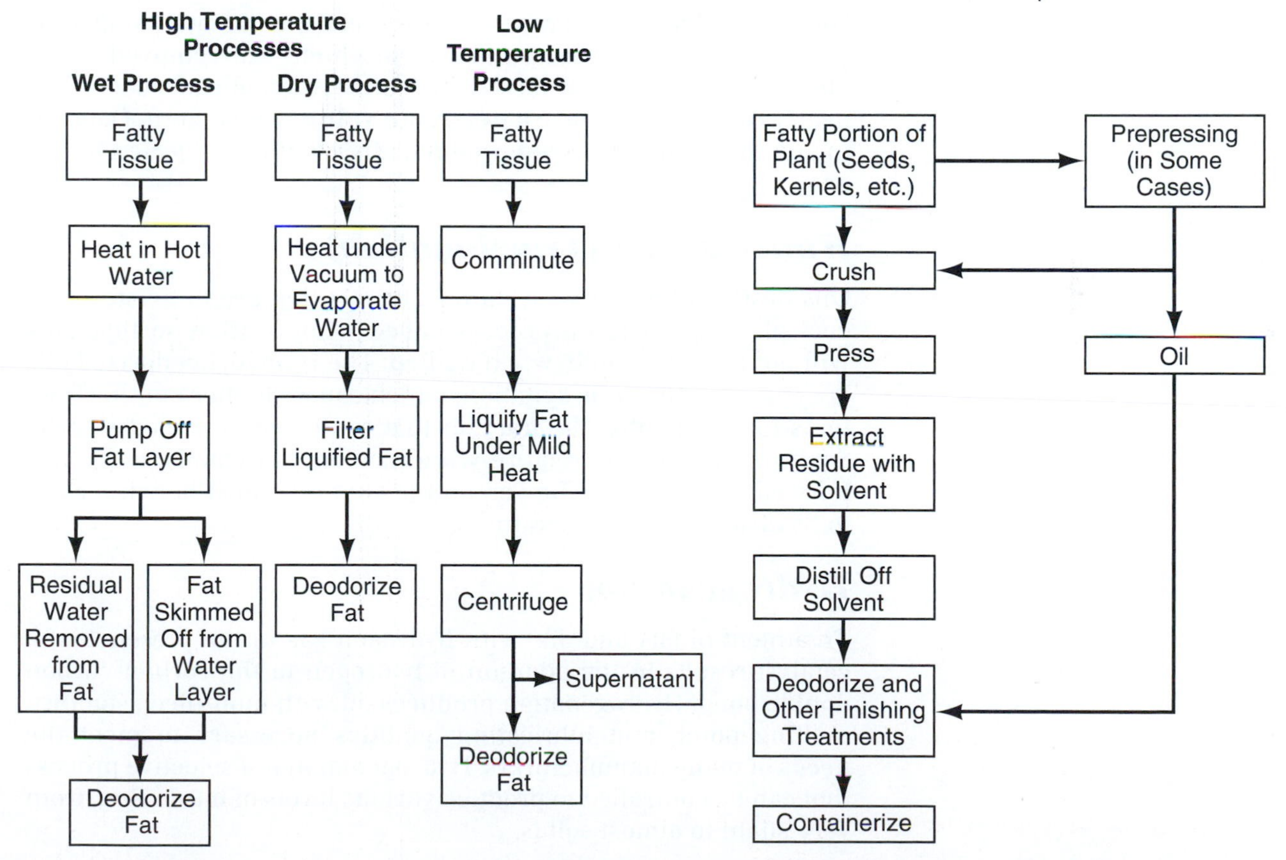

FIGURE 21-3
Flowcharts for the recovery of animal fats and plant oils.

Degumming

The first step in the refining process of many oils is degumming. Oils are degummed by mixing them with water. Degumming may be enhanced by adding phosphoric or citric acid or silica gel. Degumming removes valuable emulsifiers such as lecithin. Cottonseed oils are not degummed, but the process is necessary for such oils as soybean and canola.

Alkali Refining

The degummed oil is then treated with an alkali (base) to remove free fatty acids, glycerol, carbohydrates, resins, metals, phosphatides, and protein meal. The oil and alkali are mixed, allowing free fatty acids and alkali to form a soap. The resulting soapstock is removed through centrifuging. Residual soaps are removed with hot water washings.

Bleaching

During the bleaching process, trace metals, color bodies such as chlorophyll, soaps, and oxidation products are removed using bleaching clays, which adsorb the impurities. Bleached oils are nearly colorless and have a **peroxide value** of near zero. Depending on the desired finished product, oils are then subjected to one or more additional processes.

Winterization (Fractionation)

Oils destined for use as salad oils, or oils that are to be stored in cool places, undergo a process called winterization so that they will not become cloudy when chilled. The refined, deodorized oils are chilled with gentle agitation, which causes higher melting fractions to precipitate. The fraction that settles out is called stearin. Soybean oil does not require winterization, but canola, corn, cottonseed, sunflower, safflower, and peanut oils must be winterized to be clear at cool temperatures.

Hydrogenation

Treatment of fats and oils with hydrogen gas in the presence of a catalyst results in the addition of hydrogen to the carbon-carbon double bond. Hydrogenation produces oil with mouthfeel, stability, melting point, and lubricating qualities necessary to meet the needs of many manufacturers. Hydrogenation is a selective process that can be controlled to produce various levels of hardening, from very slight to almost solids.

Deordorization

Deordorization is a steam distillation process carried out under a vacuum, which removes volatile compounds from the oil. This may be a batch or continuous process. The end product is a bland oil with a low level of free fatty acids and a zero peroxide value. (See Tests on Fats and Oils, later in this chapter.)

This step also removes any residual pesticides or metabolites that might be present, which are more volatile than the triglycerides in the oil. Some manufacturers favor cottonseed oil because it can be deordorized at lower temperatures, which results in more tocopherols (natural antioxidants) being retained. Deordorization produces some of the purest food products available to consumers. Few other products are so thoroughly clean as refined, bleached, and deordorized oil.

GOOD FAT AND BAD FAT

More than anything, too much fat in the diet is bad fat, but some types of fats and oils may have good or bad effects on health. Fats contain both saturated and unsaturated (monounsaturated and polyunsaturated) fatty acids. Saturated fat raises blood cholesterol more than other forms of fat. Reducing saturated fat to less than 10 percent of calories will help lower blood cholesterol level. The fats from meat, milk, and milk products are the main sources of saturated fats in most diets. Many bakery products are also sources of saturated fats. Vegetable oils supply smaller amounts of saturated fat.

Olive and canola oils are particularly high in monounsaturated fats; most other vegetable oils, nuts, and high-fat fish are good sources of polyunsaturated fats. Both kinds of unsaturated fats reduce blood cholesterol when they replace saturated fats in the diet. The fats in most fish are low in saturated fatty acids and contain a certain type of polyunsaturated fatty acid—omega-3. This fatty acid is under study because of a possible association with a decreased risk for heart disease in certain people. Total fat in the diet should be consumed at a moderate level—that is, no more than 30 percent of calories. Mono- and polyunsaturated fat sources should replace saturated fats within this limit.

Partially hydrogenated vegetable oils, such as those used in many margarines and shortenings, contain a particular form of unsaturated fat known as trans-fatty acids that may raise blood cholesterol levels, although not as much as saturated fat.

For more information about fat intake and the types of fats, search these Web sites:

<www.nalusda.gov/fnic/>
<www.ars.usda.gov/>
<www.usda.gov/>
<pufa.co.net/>

Interesterification

Interesterification allows fatty acids to be rearranged or redistributed on the glycerol backbone. This is most often accomplished by catalytic methods at low temperatures. The oil is heated, agitated, and mixed with the catalyst at 194°F (90°C). Also enzymatic systems may be used for interesterification. It does not change the degree of saturation or isomeric state of the fatty acids, but it can improve the functional properties of the oil.

PRODUCTS MADE FROM FATS AND OILS

Butter is, of course, made by churning the butterfat from milk. Margarine is a similar product made from vegetable oils that have been hydrogenated or crystallized for spreading. Some vegetable oils may be mixed with small amounts of animal fat, and like butter, legal margarine contains no less than 80 percent fat. Margarine is produced by agitating a water-in-oil emulsion that is crystallized and platicized (controlled chilling).

The largest portion of the market for edible-oil products is that which includes margarine, spreads, dressings, retail bottled oils, and frying oils. The sources of these lipids have facilitated the selection and development of products that are healthier and more stable. Originally, the use of animal fats and tropical oils (cocoa, palm, and coconut) minimized the problem of stability during food use. These products are not currently widely used. The exception may be the increasing use of butter.

In the United States, soybean, corn, canola, and sunflower oils are now used. Stability and functionality are obtained by their modification through hydrogenation. Stability and healthfulness have been enhanced recently through development of high-oleic-acid sunflower and safflower oils, low-linolenic-acid soybean oil, high-oleic/low-linolenic canola oils. Additionally, high-saturated, low-saturated oils are also being developed. Future food scientists and health professionals need to be constantly aware that the fats and oils they are using may be changing, and as these changes take place, functional properties may change.

MONOGLYCERIDES AND DIGLYCERIDES

Monoglycerides and **diglycerides** are used as emulsifiers in a variety of foods. A monoglyceride is a glycerol molecule with only one fatty acid attached; a diglyceride is a glycerol with two fatty acids attached. Heating a mixture of triglycerides in the presence of a sodium hydroxide catalyst causes some of the fatty acids to disassociate from the glycerol and react with excess glycerol in the mixture. Some groups on the glycerol molecules remain unesterified, producing monoglycerides and diglycerides.

Monoglycerides and diglycerides are hydrophilic (attract water) and hydrophobic (repel water), so they are partially soluble in water and partially soluble in fat, making them excellent emulsifying agents.

FAT SUBSTITUTES

The advantages of decreased cancer and heart disease with low fat and decreased fat in the diet have been extensively studied and reported. The U.S. Surgeon General recommended that fat be reduced to 30 percent of total dietary calories, with saturated fat being less than 10 percent, polyunsaturated less than 10 percent, and monounsaturated 10 to 15 percent. The popular press and the medical community both have investigated, discussed, and reported this over the years. With the placing of fat on the Nutrition Labeling and Education Act (NLEA), it became advantageous to lower fat in various products. However, the FDA allowed a number of terms to evolve indicating those foods with reduced fat (Figure 21-4).

Unfortunately, fat is not just a flavor enhancer but plays many other roles. In most instances, carbohydrates and proteins are modified or used directly, because they contain less calories per gram (9 kcal versus 4 kcal). The fat industry is attempting to modify fat itself.

FIGURE 21-4

Reduced-fat and fat-free products.

Approaches to fat reduction in food generally falls into one of two categories: decreasing fat content, or using fat replacers, substitutes, extenders, mimetics, or synesthetic fat.

The use of fat replacers includes various carbohydrate-based, protein-based, and fat-based replacers for some of the many functions fats perform in food. For example, protein particles can be used in some foods to give a similar mouthfeel to that of fat at about 40 percent of the caloric content of fat. Carbohydrates, such as some forms of cellulose, can be used to increase the viscosity of foods and mimic oil. The many functions of fats and consumer demands suggest a constant demand by consumers for new and improved fat replacers.

One of the newest forms of fat substitutes are the sugar esters. These are chemically similar to fats, but they are not absorbed or metabolized by the body. They do not contribute calories to food. Currently, the most well-known sugar ester is Olestra. It is a chemical derivative of table sugar (sucrose). (For more information about Olestra, visit the Web site at <www.olean.com/>.)

Some fat-replacers by name include:

- Olestra–a sugar ester
- Amalean I–a modified high-amylose corn
- Amalean II–an instant modified high-amylose corn
- Cellulose

TESTS ON FATS AND OILS

Fats and oils are tested to obtain information about their function in foods, to measure any degree of deterioration (stability), and to check for misrepresentation or adulteration. Chemical tests can determine the degree of unsaturation of the fatty acids in a fat. This is expressed as the **iodine value** of the fat. The test is based on the amount of iodine absorbed by a fat on a per 100 grams basis. The higher the iodine value, the greater the degree of unsaturation. Another chemical test yields the peroxide value. This indicates the degree of oxidation that has taken place in a fat or oil. The test is based on the amount of peroxides that form at the site of double bonds. These peroxides release iodine from potassium iodide when it is added to the system.

Hydrolytic **rancidity** refers to the rancidity that occurs under conditions of moisture, high temperature, and natural lipolytic enzymes. The acid value refers to a measure of free fatty acids present in a fat that were released during hydrolysis.

TABLE 21-2 Smoke, Flash, and Fire Points of Oils (°C)

Oil/Fat	Smoke	Flash	Fire
Castor, Refined	200	298	335
Corn, Crude	178	294	346
Corn, Refined	227	326	359
Olive, Virgin	175 to 199	321	361

The average molecular weight of the fatty acids in a fat influences the firmness of the fat, the flavor, and the odor. A **saponification value** indicates the average molecular weight of the fatty acids in a fat. This value represents the number of milligrams of potassium hydroxide needed to saponify (convert to soap) one gram of fat. The saponification value increases and decreases inversely (opposite of) the average molecular weight.

One of the most common physical tests performed on fats is a determination of the melting point. Fats and oils used in frying are subjected to measurements of their smoke point, flash point, and fire point (Table 21-2). Other physical determinations include color and specific gravity.

Summary

Fats and oils contain 2.25 times more energy than carbohydrates or proteins. Food science is concerned most with simple lipids such as the triglycerides that make up fat, butter, shortening, and oil. Some foods such as butter, cooking oils, and shortening are pure fat. Other foods make use of the properties of fats. Chemical characteristics of fats influence their properties. The function and properties of fats in foods are important. For example, fats can be used for textural qualities, emulsions, tenderizers, heat transfer, aeration, flavor, and dispersion.

Fats and oils are extracted from plants and animals. Rendering extracts fats from animal products. Expellers and solvents extract fats from plants. Extracted fats undergo additional processing such as degumming, alkali refining, bleaching, winterization, hydrogenation, deordorization, and interesterification. Chemical tests such as iodine value, rancidity, melting point, smoke point, and saponification provide information about function and stability.

Because fats contain more calories per pound than carbohydrates or proteins, the food industry continues to reduce the fat content of foods or to search for fat substitutes.

Review Questions

Success in any career requires knowledge. Test your knowledge of this chapter by answering these questions or solving these problems.

1. List eight functions fats and oils serve in foods.
2. The number of __________ __________ and __________ influence the melting point of some common fatty acids.
3. Describe fatty acids and how they relate to saturated and unsaturated fats.
4. Name the two methods of extracting oil from plants.
5. List the eight processes used on oils after extraction.
6. __________ and __________ are used as emulsifiers in a variety of foods. A __________ is a glycerol molecule with only one fatty acid attached; a __________ is a glycerol with two fatty acids attached.
7. Explain the difference between cis and trans fatty acids.
8. Name and discuss the two chemical tests on fats and oils.
9. List six sources of fats and oils.
10. What is Olestra?

Student Activities

1. Make a diagram or a three-dimensional model to explain cis and trans.
2. Create a list of foods where the flavor provided by the fat makes the unique "taste" of the food. Compare your list with others.
3. Identify four food items consumed in the last 24 hours. Determine and report the fat content of these foods. Use Table A-8.
4. Find labels that list some substances used to prevent the oxidation of fats. (Refer to Chapter 14.)
5. Demonstrate that some fats are liquid at room temperature and some are solid at room temperature, and discuss the reasons for this. For example, compare lard, shortening, vegetable oil, olive oil, and margarine or butter.
6. Use chemical models to demonstrate the concept of single and double bonds and saturation and unsaturation in fatty acids.
7. Use a fat or oil and demonstrate how to make soap.

Resources

Bartlett, J. 1996. *The cook's dictionary and culinary reference.* Chicago: Contemporary Books.

Corriher, S. O. 1997. *Cookwise: The hows and whys of successful cooking.* New York: William Morrow and Company, Inc.

Cremer, M. L. 1998. *Quality food in quantity. Management and science.* Berkeley, CA: McCutchan Publishing Corporation.

Ensminger, A. H., M. E. Ensminger, J. E. Konlande, and J. R. Robson. 1994. *Foods and nutrition encyclopedia.* 2 Vols. Boca Raton, FL: CRC Press.

Horn, J., J. Fletcher, and A. Gooch. 1997. *Cooking a to z. The complete culinary reference source.* Glen Ellen, CA: Cole Publishing Group, Inc.

Potter, N. N., and J. H. Hotchkiss. 1995. *Food science,* 5th ed. New York: Chapman and Hall.

Smoot, R. C., R. G. Smith, and J. Price. 1990. *Chemistry–A modern course.* Columbus, OH: Merrill Publishing Company.

Vaclavik, V. A., and E. W. Christina. 1999. *Essentials of food science.* Gaithersburg, MD: Aspen Publishers, Inc.

Internet

Internet sites represent a vast resource of information. The URLs (uniform resource locator) for the World Wide Web sites can change. Using one of the search engines on the Internet such as Yahoo!, HotBot, AltaVista, Excite, Dogpile, About, or Google, find more information by searching for these words or phrases: fat substitutes; saturated fats; unsaturated fats; fatty acids; production/processing of fats or oils; names of oils–for example, soybean oil, olive oil, and the like; monoglycerides; diglycerides; fat replacers. Also, Table A-7 provides a listing of some useful Internet sites that can be used as a starting point.

Chapter 22

Candy and Confectionery

Objectives

After reading this chapter, you should be able to:

- Identify three crystalline and three noncrystalline candies
- Describe the relationship between sugar concentration and the boiling point
- Discuss common components of candies and confectioneries
- Identify two ways to produce invert sugar
- Explain caramelization in candymaking
- Name four sugar-based sweeteners developed from cornstarch
- Describe uses of high-fructose corn syrup
- Define cocoa
- Explain conching
- Describe modern candy and confectionery manufacturing
- List four sugar alcohols and four high-intensity sweeteners
- Discuss the labeling information and requirements for candy

Key Terms

caramelization
conching
crystalline
Dutch-processed
enrober
HFCS
hydrolyze
interfering agent
invert sugar
nib
noncrystalline
polymerize
substrate

The word "candy" does not cover only pure-sugar concoctions, but also includes an array of tasty confectioneries combining sugar or similar substances with other compatible ingredients such as fruits, nuts, or chocolate. Some people may not like the idea, but candy is a food. Its basic elements are included in the U.S. Department of Agriculture's Food Pyramid. Some candies such as those containing milk or nuts do offer some beneficial food values.

SUGAR-BASED CONFECTIONERY

Honey, sorghum/molasses, maple syrup, and selected fruit juices and pulps serve as a sweetener substitute for cane sugar and sugar beet sugar. A variety of sweeteners have been developed, including sugar-based sweeteners. Sugar-based sweeteners are those developed from cornstarch.

Confectionery (candy) can be divided into those in which sugar is the main ingredient and those that are based on chocolate. In sugar-based candies, the sugar is manipulated to achieve some desired changes in the texture of the candy. For example, the sugar can be **crystalline**, and these crystals may be large or small. Or the sugar can be **noncrystalline** and glasslike. If the sugar is crystalline or noncrystalline, the sugar structure can be made hard or soft, modified by the amount of moisture, modified by the air that is whipped in, or modified by other ingredients. Crystallinity and moisture in a finished candy are determined by the ingredients, the heat used in cooking, cooling, and stirring.

Candies based on a crystalline sugar include rock candy, fondant, and fudge (Figure 22-1). Noncrystalline sugar candies include hard candies, brittles, chewy candies, and gummy candies.

Composition

Sugars and sugary foods provide a valuable and inexpensive source of energy. Other than energy, some sugary foods provide little other nutrition. Table A-8 provides the nutrient composition for a variety of common candies such as caramels, chocolates, chocolate, fondant, fudge, gum drops, hard candy, and jelly beans. Sugar and honey are also listed in Table A-8.

Ingredients

The principal ingredient of candies, including chocolate, is the sweetener. The most common sweetener used in candies and

FIGURE 22-1
Fudge-making equipment helps small specialty shops provide fresh candy.

chocolates is sucrose, the sugar from cane and sugar beets. At room temperature a concentrated solution of sucrose can be created by adding two parts of sucrose to one part water. If this solution is cooled, it becomes supersaturated. Agitation during cooling causes crystallization, and the addition of a small sugar crystal speeds crystallization.

Heating the water before adding sucrose increases the sucrose concentration and increases the boiling point. The relationship between the sucrose concentration and the boiling point is used in making candy. It determines the percentage of water and sugar in the final product. When a boiling syrup (sugar and water) reaches a specific temperature, it will have the desired sugar concentration, as shown in Table 22-1. The numbers in Table 22-1 do not apply to mixed sugar solutions like sucrose solutions containing corn syrup, glucose, or molasses.

Depending on the type of product being created, the concentration of sugar in a sugar syrup can be measured by a reading on a thermometer or by its appearance when dropped into ice water as indicated in Table 22-2. The temperature and concentration then lead to a predicted behavior of the syrup to form the product.

Candies are made of sugar (sucrose), water or other liquid, and usually some **interfering agent**(s). Interfering agents interfere with the formation of sucrose crystals and provide some secondary properties to the candies. Butter, milk, starch, egg white, gelatin, fats, pectins, gums, cocoa, and corn syrup are commonly used as crystal interfering agents. Also, the secondary properties

TABLE 22-1 Concentration of Sucrose, Water, and the Boiling Point

Percent Sucrose	Percent Water	Boiling Temperature °F/°C
30.0	70.0	212/100
50.0	50.0	216/102
70.0	30.0	223/106
90.0	10.0	253/123
95.0	5.0	284/140
97.0	3.0	304/151
99.5	0.5	331/166
99.6	0.4	340/171

these agents bring to a candy include thickening, chewiness, whipping, flavoring, tenderizing, and lubricating.

Invert Sugar

Heat and acid will **hydrolyze** and invert the sugar into the component monosaccharides–fructose and glucose. This is particularly true if the product is heated. Fructose and glucose production create changes in a product. They are reducing sugars, sucrose is not. This means that they will enhance browning. They are more soluble and more hygroscopic than sucrose. The confectionery industry refers to glucose as dextrose and fructose as levulose. Hydrolysis of sucrose is shown in the following reaction:

$$C_{12}H_{22}O_{11} + H_2O \longrightarrow C_6H_{12}O_6 + C_6H_{12}O_6$$

$$\text{Sucrose} + \text{Water} \longrightarrow \text{Glucose} + \text{Fructose}$$

(***Note:*** Glucose and fructose have the same chemical formula, but the atoms are arranged differently for each.)

Cream of tartar (an acid salt) as an added ingredient in a candy formula serves indirectly to decrease the rate of crystallization as well as crystal size. It does this through its ability to hydrolyze sucrose into its **invert sugar**.

TABLE 22-2 Sugar Concentration, Temperature, and Behavior for Various Candies and Confections

Candy/Confection	Temperature of Syrup at Sea Level	Stage of Concentration Desired	Behavior of Syrup State Desired
Syrup	230° to 234°F (110° to 112°C)	Thread	Spins a 2-inch thread when dropped from fork or spoon
Fondant Fudge Penuchi	234° to 240°F (112° to 115°C)	Soft ball	When dropped into very cold water, forms a soft ball that flattens on removal
Caramels	244° to 248°F (118° to 120°C)	Firm ball	When dropped into very cold water, forms a firm ball that does not flatten on removal
Divinity Marshmallows Nougat Popcorn balls Salt-water taffy	250° to 265°F (121° to 130°C)	Hard ball	When dropped into very cold water, forms a ball that is hard enough to hold its shape, yet plastic
Butterscotch Taffies	270° to 290°F (132° to 143°C)	Soft crack	When dropped into very cold water, separates into threads that are hard but not brittle
Brittle Glace	300° to 310°F (149° to 154°C)	Hard crack	When dropped into very cold water, separates into threads that are hard and brittle
Barley sugar	320°F (160°C)	Clear liquid	Sugar liquifies
Caramel	338°F (170°C)	Brown liquid	Melted sugar becomes brown

The classic example of using invert sugar in foods is chocolate covered cherries. These cherries are made by adding the enzyme, invertase, to fondant. Fondant as a solid crystalline candy is placed around the cherries and coated with chocolate. The cherries are allowed to sit, and the invertase inside hydrolyzes the sucrose into fructose and glucose. The fructose and glucose combination is much more soluble than the sucrose crystals, and so the consumer (eater) perceives a syrup that is very sweet. The reason for the increased sweetness is that fructose is 40 to 70 percent sweeter.

Another category of inversion is the development of the high fructose corn syrup (**HFCS**). HFCS is manufactured from cornstarch. Cornstarch is hydrolyzed by acid or enzyme, and then the resulting glucose is inverted into fructose by an enzyme. The percentage will vary. This is one of the new processing methods in foods, particularly in the sweetener area.

Caramelization

Caramelization is the application of heat to the point that the sugars dehydrate and breakdown and **polymerize**. Although a relatively complex reaction, it can be simply done. Once the melting point is reached, sugars will caramelize. The sugar actually comes apart chemically and forms smaller sugars, some of which join together again to form different smaller sugars. During the caramelizing process new sugars and similar compounds are constantly being formed. By the time a dark caramel is formed, more than 128 different sugars and related compounds have been formed.

CANDY CANE HISTORY

In 1670, the choirmaster at the Cologne Cathedral gave sugar sticks to his young singers to keep them quiet during the long Living Creche ceremony. In honor of the occasion, he had the candies bent into shepherds' crooks. In 1847, a German-Swedish immigrant named August Imgard of Wooster, Ohio, decorated a small pine tree with paper ornaments and candy canes. At the beginning of the 1900s red and white stripes and peppermint flavors became the norm for candy canes. Supposedly, the white and red are symbolic to those of the Christian faith.

In the 1920s, Bob McCormack began making candy canes as Christmas treats for his children, friends, and local shopkeepers in Albany, Georgia. It was a difficult process—pulling, twisting, cutting, and bending the candy by hand. It could only be done for local sales.

In the 1950s, Bob's brother-in-law, Gregory Keller, a Catholic priest, invented a machine to automate candy cane production. Packaging innovations by other members of the McCormack family made it possible to transport the delicate canes. This transformed Bobs Candies, Inc. into the largest producer of candy canes in the world.

Modern technology has made candy canes accessible and plentiful, and a Christmas tradition for many.

For more information on candy canes, search the Web or visit these Web sites:

<www.bobscandies.com/>
<www.candyusa.org/canestor.html>

Each sugar has its own caramelization temperature. Caramel from sucrose forms at 338°F (170°C) or above. Galactose and glucose caramelize at about the same temperature as sucrose. Fructose caramelizes at 230°F (110°C) and maltose at about 356°F (180°C). Caramel is brown. It has a characteristic pungent taste, often bitter, less sweet, and it is noncrystalline.

A good example of caramelization is in the making of peanut brittle. Sugar is slowly and carefully heated in a skillet. This slow heating allows for the uniform "unsaturated polymer formation" to occur. Brown flavorful peanut brittle is a result of this caramelization process. The presence of sugar acids produced during this process is evident when foamy peanut brittle is made. This is produced by taking the heated brown mixture, while still hot, and adding a small amount of baking soda. The reaction of the baking soda with the sugar acids produces carbon dioxide gas, which foams.

Corn Syrups and Other Sweeteners

Sugar-based sweeteners are developed from cornstarch. The development of the various types of corn syrups, maltodextrins, and high-fructose corn syrup from cornstarch sources could be called one of the greatest changes in the sugar and sweetener industry over several centuries. In the late 1800s it was found that cornstarch could be hydrolyzed and a sugar formed. In the 1970s it became a major commercial product bringing about changes in the food industry.

These sweeteners are processed and refined using a series of steeping, separation, and grinding processes. Finally, the products are converted and fermented. From this processing, the five classes of corn sweeteners are formed:

- Corn syrup (glucose syrup)
- Dried corn syrup (dried glucose syrup)
- Maltodextrin
- Dextrose monohydrate
- Dextrose anhydrous

This does not include the corn syrup conversion into high-fructose corn syrup.

Fructose and Fructose Products

Fructose is a monosaccharide that is approximately 75 percent sweeter than sucrose. For this reason, fructose and fructose products are frequently substituted for sucrose. High Fructose corn syrup is often used.

The high-fructose corn syrup story is one of the most revolutionary in food science in the last decade. Consumption has increased since its inception. The products themselves are made up of hydrolyzed cornstarch. The cornstarch is hydrolyzed, and that corn syrup has an invertase that will change glucose into fructose.

High-fructose corn syrups (HFCS) retain moisture and/or prevent drying out, control crystallization, produce an osmotic pressure that is higher than for sucrose or medium invert sugar, and help control microbiological growth or help in penetration of cell membranes. HFCS's provide a ready yeast-fermentable **substrate**. HFCS's provide a controllable substrate for browning and Maillard reaction. They impart a degree of sweetness that is essentially the same as in invert liquid sugar. Less HFCS is required because HFCS is sweeter than liquid sucrose or corn syrup blends. HFCS's blend easily with sweeteners, acids, and flavorings.

These attributes are advantages in many instances. However, these same attributes are a disadvantage as well as an advantage. High Fructose Corn Syrup is extremely soluble and hygroscopic. Generally, baked products made with HFCS will be softer than those made with sucrose. If these products are steamed, they may get gummy. A fast-food hamburger business that precooks and wraps their product may prefer the firmer product.

CHOCOLATE AND COCOA PRODUCTS

Cocoa is finely pulverized, defatted, roasted cacao kernels, to which natural and artificial spices and flavors may be added. It is manufactured by pumping hot chocolate liquor (semiliquid ground cacao kernels) into hydraulic cage presses. Here, under extreme pressure, part of the fat, or cocoa butter, is removed. The fat content of cocoa varies from less than 10 percent to 22 percent. Cocoa may be **Dutch-processed** by mild alkali treatment to change and darken color and improve flavor. Cocoa is the flavoring ingredient in many confectioneries, baked goods, ice creams, puddings, and beverages.

Federal standards define several kinds of chocolate products. Bitter chocolate, or chocolate liquor, is the roasted ground kernel (**nib**) of the cacao bean; it is commonly known as baker's, or baking, chocolate. A minimum of 15 percent liquor mixed with sugar and cocoa butter is sweet chocolate. When the amount of chocolate liquor is greater than 35 percent, the product is bittersweet chocolate. A combination of at least 12 percent dry whole milk solids, sugar, cocoa butter, and at least 10 percent chocolate liquor produces milk chocolate.

Cocoa

Cocoa is made by removing some of the cocoa butter. Eating chocolate is made by adding cocoa butter. This is true of all eating chocolate, whether it is dark, bittersweet, or milk chocolate. Besides enhancing the flavor, the added cocoa butter serves to make the chocolate more fluid.

One example of eating chocolate is sweet chocolate, a combination of unsweetened chocolate, sugar, cocoa butter, and perhaps a little vanilla. Making it involves melting and combining these ingredients in a large mixing machine until the mass has the consistency of dough.

Milk Chocolate

Milk chocolate, the most common form of eating chocolate, goes through essentially the same mixing process, except that it involves using less unsweetened chocolate and adding milk (Figure 22-2). Whatever ingredients are used, the mixture then travels through a series of heavy rollers set one atop the other. During the grinding that takes place here, the mixture is refined to a smooth paste ready for **conching**.

Conching is a flavor development process that puts the chocolate through a "kneading" action and takes its name from the shell-like shape of the containers originally employed. The conches, as the machines are called, are equipped with heavy rollers that plow back and forth through the chocolate mass anywhere from a few hours to several days. Under regulated speeds, these rollers can produce different degrees of agitation and aeration in developing and modifying the chocolate flavors.

In some manufacturing setups, an emulsifying operation either takes the place of conching or supplements it. This operation is carried out by a machine that works like an eggbeater to break up sugar crystals and other particles in the chocolate mixture to give it a fine, velvety smoothness.

After the emulsifying or conching machines, the mixture goes through a tempering interval–heating, cooling, and reheating–and then at last into molds to be formed into the shape of the complete product. Molds take a variety of shapes and sizes, from the popular individual-size bars available to consumers to a ten-pound block used by confectionery manufacturers.

When the molded chocolate reaches the cooling chamber, cooling proceeds at a fixed rate that keeps hard-earned flavor intact. Bars are then removed from the molds and passed along to wrapping machines to be packed for shipment.

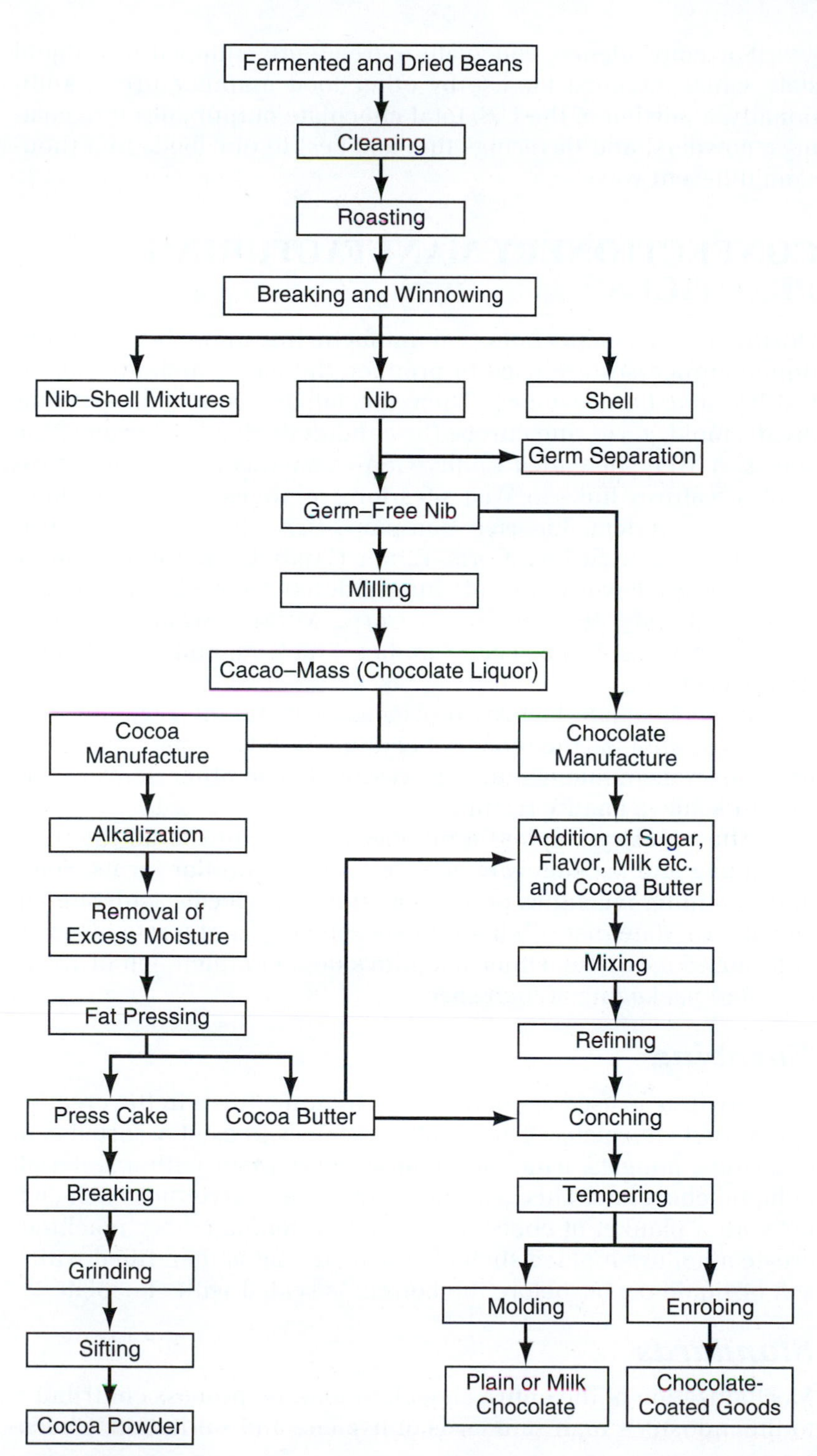

FIGURE 22-2

Flowchart of cocoa and chocolate manufacturing.

For convenience, chocolate is frequently shipped in a liquid state when intended for use by other food manufacturers. Additionally, a portion of the U.S. total chocolate output goes into coatings, powders, and flavorings that add zest to our foods in a thousand different ways.

CONFECTIONERY MANUFACTURING PRACTICES

During modern confectionery manufacturing, many batch or continuous processes are used to produce the basic fondants, taffies, brittles, and hard candies. Then specializing machines extrude, divide, mold, glaze, and enrobe the candies during final production stages. A Web site called Candy USA (<www.candyusa.org/store.html>) features links to Web sites for everyones' favorite candies such as: American Licorice Company, Ben Myerson Candy Co., Clark Bar, America Inc., Certs, Elmer Candy Corporation, Fannie May Candies, Favorite Brands International, Georgia Nut Co. Inc., Gimbal's Candy, Hershey Foods Corp., M&M's Brand Chocolate Candies, Nestles' Foods, See's Candies, Snickers, and Willy Wonka Candy Factory.

In a chocolate factory, precision instruments regulate temperatures, stabilize the moisture content of the air, and control the time intervals of manufacturing operations and other items necessary to achieve quality results.

The industry employs a number of machines to do the work of shaping and packaging chocolate into the familiar forms. Some of the shaping machines perform at amazing speeds, squirting out jets of chocolate that solidify into special shapes at a rate of several hundred a minute. Other machines do a complete job of wrapping and packaging at high speeds.

Enrobing

The **enrober** is used by many candy manufactures in the creation of assorted chocolates. The enrober receives lines of assorted centers (nuts, nougats, fruit, etc.) and showers them with a waterfall of liquid chocolate. This generally covers and surrounds each center with a blanket of chocolate. Yet other confectionery machines create a hollow-molded shell of chocolate that is then filled with a soft or liquid center before the bottom is sealed with chocolate.

Standards

Mechanization of the entire chocolate-making process contributes to the industry's high standards of hygiene and sanitation. Chocolate factories constantly run quality tests, which show whether the process is proceeding within the strict limitations designed for each product. These tests are for the viscosity of chocolate, for the cocoa butter content, for acidity, for the fineness of a product, and for the purity and taste of the desired finished product.

All chocolate manufacturers must meet standards as set forth in the rules and regulations of The Food and Drug Administration. These govern manufacturing formulas, even to the extent of specifying the minimum content of the chocolate liquor and milk used. They also impose strict rules regarding the flavorings and other ingredients that may be used.

FDA standards for cacao products were updated in 1993, and the recent amended regulations were published in May 1993. Those rules are highly technical, down to prescribing analytic techniques and specifying approved processing methods. Specifications for cacao nibs themselves says they may contain "not more than 1.75 percent by weight" of residual shell. The standards provide definitions of intermediate and end products, including chocolate liquor. For example, chocolate liquor "contains not less than 50 percent nor more than 60 percent by weight of cacao fat," among other requirements. The FDA also sets standards for breakfast cocoa, sweet chocolate, semisweet or bittersweet chocolate, milk chocolate, skim milk chocolate, and so on.

TABLE 22-3 Relative Sweetness of Sugar Alcohols Compared to a Sucrose Value of 100

Sugar Alcohol	Sweetness
Xylitol	90
Sorbitol	63
Galactitol	58
Malitol	68
Lactitol	35

SUGAR SUBSTITUTES

Sugar alcohols are made by chemically reducing a sugar to an alcohol. These are not fermentable by bacteria in the mouth, so they do not contribute to tooth decay. But the sugar alcohols are less sweet than sugar, as Table 22-3 shows. Products using the sugar alcohols are often labeled "sugar free," but this does not mean that they are calorie free. Sugar alcohols contain 4 kcal of energy per gram, just like sugar.

High-intensity sweeteners are used in confectioneries to reduce their caloric content. The calories are reduced because less of the sweetener is required or the sweetener is not metabolizable by the body, so they do not contribute calories. Some examples of these sweeteners and their relative sweetness are listed in Table 22-4.

Not all the sweeteners in Table 22-4 are approved for use at this time. Also, high-intensity sweeteners often do not provide the same functional properties as sugar. Additional additives are used to provide bulking, mouthfeel, and other desirable characteristics.

TABLE 22-4 Approximate Sweetness of Sugar Substitutes Compared to a Sucrose Value of 1

Substitute	Sweetness
Acesulfame K	200
Aspartame	180
Cyclamate	30
Dihydrochalcones	300 to 2000
Glycyrrhizin	50 to100
Monellin	1500 to 2000
Saccharin	300
Stevioside	300
Talin	2000 to 3000

LABELING

By law, all packaged foods must bear a label listing ingredients in order of predominance; candy is no exception. Every package of hard candies or chocolate bar must offer such a listing. As part of the new food labeling rules under the 1990 Nutrition Labeling and Education Act, manufacturers must include substantially more nutrition information on labels than in the past.

For example, the label on one popular brand of chocolate-with-peanuts bar, using the new format, includes the information that it provides 280 calories, 8 percent of the daily value for carbohydrate, and a significant 34 percent of the daily value for saturated fat.

A representative label listing of ingredients, which the Food and Drug Administration requires on all packaged foods, reads: sugar, corn syrup, citric acid, artificial flavor, Red No. 40, Yellow No. 5, Yellow No. 6, Blue No. 1. The last four are simply specific FDA-sanctioned food colorings, synthetic additives a few people are allergic to, but otherwise pose no threat to health.

Summary

Candy or confectionery are divided into those in which sugar is the main ingredient and those based on chocolate. Further, candies can be classified as crystalline and noncrystalline. The processes of creating an invert sugar and caramelization produce unique flavors and characteristics. The relationship between sucrose concentration and the boiling point determines the percentage of water and sugar in the final product. Interfering agents interfere with the formation of sucrose crystals and provide secondary properties to candies.

Sugar-based sweeteners developed from cornstarch brought about major changes in the food industry. These sweeteners include corn syrup, maltodextrin, dextrose monohydrate, and dextrose anhydrous. High-fructose corn syrups are produced when cornstarch is hydrolyzed by an invertase. Again, high-fructose corn syrup revolutionized the food industry.

Cocoa is the ground, defatted, roasted cacao kernels. Federal standards define cocoa and other chocolate products including milk chocolate, dark chocolate, and baking chocolate. Much of candy manufacturing, including chocolate manufacturing, relies on automation, and precision instrumentation.

Sugar substitutes and high-intensity sweeteners are often used to reduce the calories in candy and confectionery. Labeling requirements for candy and confectionery are the same as for other foods.

Review Questions

Success in any career requires knowledge. Test your knowledge of this chapter by answering these questions or solving these problems.

1. Define conching.
2. Sugar-based sweeteners are developed from ___________.
3. Candies based on a ___________ sugar include rock candy, fondant, and fudge. ___________ sugar candies include hard candies, brittles, chewy candies, and gummy candies.
4. The ___________ receives lines of assorted centers (nuts, nougats, fruit, etc.) and showers them with a waterfall of liquid chocolate. This generally covers and surrounds each center with a blanket of chocolate.
5. Why are fructose and high-fructose products frequently substituted for sucrose?
6. What is the most common sugar in candies?
7. What labeling information is required on candies and confectioneries?
8. Define cocoa.
9. What substance is added to the ingredients of candy that serves indirectly to decrease the rate of crystallization as well as crystal size?
10. ___________ is the application of heat to the point that sugars dehydrate and break down and polymerize.

Student Activities

1. Visit the Candy USA Web site (<www.candyusa.org/store.html>) and link with the manufacturer of your favorite candy. Report your findings on the manufacturing process.
2. Identify a candy recipe that relies on the formation of an invert sugar from sucrose.
3. Develop a report or presentation on products using some of the sugar alcohols listed in Table 22-3, page 411.
4. Make a list of chocolate products used in cooking.
5. Visit the Hershey Chocolate Web site (<www.hersheys.com/>) and report on what you find at their Web site.

6. Develop a report or presentation on which of the sugar substitutes in Table 22-3 are being used and which have been used.
7. Use a simple recipe for a candy and make it in class. Discuss the changes in the sugar as the final product is reached.
8. Conduct a taste test of a variety of candies and classify them as crystalline or noncrystalline.

Resources

Alexander, R. J. 1998. *Sweeteners: Nutritive.* St. Paul, MN: Eagan Press.

Bartlett, J. 1996. *The cook's dictionary and culinary reference.* Chicago: Contemporary Books.

Corriher, S. O. 1997. *Cookwise: The hows and whys of successful cooking.* New York: William Morrow and Company, Inc.

Horn, J., J. Fletcher, and A. Gooch. 1997. *Cooking a to z. The complete culinary reference source.* Glen Ellen, CA: Cole Publishing Group, Inc.

McGee, H. 1997. *On food and cooking. The science and lore of the kitchen.* New York: Simon and Schuster Inc.

Internet

Internet sites represent a vast resource of information. The URLs (uniform resource locator) for the World Wide Web sites can change. Using one of the search engines on the Internet such as Yahoo!, HotBot, AltaVista, Excite, Dogpile, About, or Google, find more information by searching for these words or phrases: candies; confectionery; specific candy brand names; conching; enrober; types of sugars–sucrose, fructose, corn syrup, and the like; cocoa products; sugar substitutes. Also, Table A-7 provides a listing of some useful Internet sites that can be used as a starting point.

Chapter 23

Beverages

Objectives

After reading this chapter, you should be able to:

- Describe how carbonated nonalcoholic beverages are manufactured
- List the steps in the production of beer
- Compare the production of wine to vinegar
- Indicate how fermentation plays a role in the production of coffee
- Name six ways enzymes are used in the production of beverages
- Discuss how two beverages meet the demand for a healthful drink
- Identify the fastest growing segment of the beverage industry
- Name five herbs used in beverages
- Identify the plants that produce coffee and tea
- Describe how to produce a coffee substitute
- Compare tea to herbal teas

Key Terms

black tea
carbonated
carbonator
green tea
lagering
malt
mashing
nonnutritive
nutritionally enhanced
oolong
steeped
synchrometer
vitamin-fortified
vinification
wort

People drink beverages for their food value, thirst quenching, and stimulating effects. People also drink because consumption is pleasurable, and some drinks are considered healthful.

CARBONATED NONALCOHOLIC BEVERAGES

Carbonated soft drinks are the most popular beverage (Figure 23-1). Carbonated nonalcoholic beverages are usually sweetened, flavored, acidified, colored, artificially carbonated, and chemically preserved. Their origin goes back to Greek and Roman times of naturally occurring mineral waters. But the British chemist Joseph Priestley discovered how to artificially carbonate water in about 1767. Since Priestley's discovery, many flavors and forms of carbonated soft drinks have been developed for today's consumer (Figure 23-2).

Sweeteners

Syrup made from sucrose was commonly used in soft drinks, but now the most common sugar used in soft drinks is high-fructose corn syrup. High-fructose sugars are sweeter than sucrose. The final concentration of sugar in a soft drink is 8 to 14 percent. Besides contributing sweetness, sugar in the soft drink contributes body and mouthfeel.

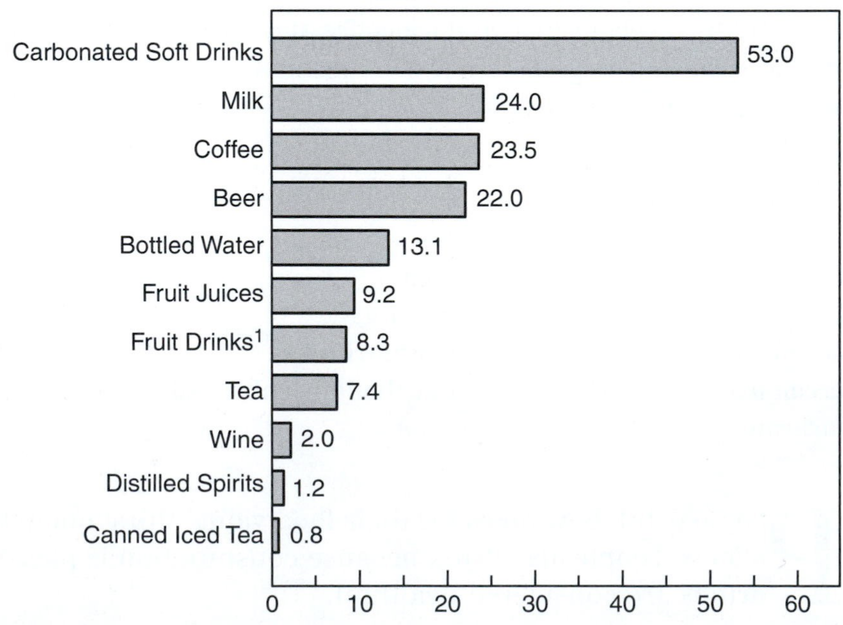

FIGURE 23-1

Per capita beverage consumption in the United States. (*Source:* USDA Economic Research Service)

FIGURE 23-2
Today's consumer can choose from a wide variety of soft drinks in plastic bottles or aluminum cans.

Soft drinks with no calories or reduced calories are popular. **Nonnutritive** sweeteners (no calories) that have been used in soft drinks include saccharin and cyclamate. Currently, reduced-calorie soft drinks use a sweetener called aspartame (trademark Nutra-Sweet®). This sweetener is 150 to 200 times sweeter than sucrose, so very little is needed even though it contains the same number of calories per gram as sugar (4 kcal/g). When these nonnutritive or reduced-calorie sweeteners are used, nonnutritive carbohydrates such as carboxymethyl cellulose or a pectin are sometimes added to give the soft drink a mouthfeel.

Flavors

Flavors used in soft drinks include synthetic flavors (refer to Chapter 14), natural flavor extracts, and fruit juice concentrates. Synthetic flavors are very complex and may easily contain hundreds of distinct compounds. Some like the cola flavors are complex and carefully guarded secrets.

Colors

Important coloring agents for soft drinks are synthetic and are U.S. certified food colors approved by the FDA (refer to Chapter 14). These colors meet strict requirements for purity. Heated sugar produces a caramel color used in the dark beverages. This is a nonsynthetic color, and it is preferred because of its coloring power and color stability. When natural fruit juices or extracts are used in soft drinks, their colors are often supplemented with synthetic colors.

Acid

Carbon dioxide in soft drinks contributes to the acidity, but the main acids used are phosphoric, citric, fumaric, tartaric, and malic acids. Besides acting as preservatives, the acids enhance flavors. Of course, acids in the soft drink lower the pH, as shown in Table 23-1.

Table 23-1 The pH of Common Soft Drinks

Flavor	pH
Colas	2.6
Cherry	3.7
Lemon-lime	3.0
Grape	3.0
Root beer	4.0

Water

Water is the major ingredient of carbonated soft drinks. By volume, it makes up as much as 92 percent. Pure water, or nearly chemically pure, is essential, because traces of impurities may react with the chemicals in the soft drink. Alkalinity, iron, and manganese must be low. Chlorine cannot be present. Obviously the water must be colorless, and it must not have any odor, taste, or organic matter. Bottling plants condition the water they use so that it will have the high standard needed.

Carbon Dioxide

Carbon dioxide (CO_2) gas provides the zest and sparkle of carbonated drinks. Sources of carbon dioxide gas include carbonates, limestone, the burning of fuels, and industrial fermentation. Soft drink manufacturers use carbon dioxide from high-pressure cylinders purchased from manufacturers who produce the gas under strict food purity regulations. The amount of carbon dioxide gas in each beverage is different and is measured out in terms of volumes of gas per volume of liquid. Most beverages are carbonated in the range of 1.5 to 4 volumes of carbon dioxide. The equipment that carbonates a soft drink is called a **carbonator**. Carbonation of the soft drink takes place at a lowered temperature because the solubility of the carbon dioxide is greater at lower temperatures.

Mixing

Operations producing soft drinks include mixing, carbonating, and bottling. Treated, deaerated (removal of air) water, and the flavored syrup are pumped to a **synchrometer**. This metering device measures the syrup and water in fixed proportion to the carbonator. Next, the mix is cooled and sent to the carbonator. After carbonation, the soft drink is placed in cans or bottles and sealed. In restaurants, fast-food establishments, and other businesses, soft drinks are served directly into a container, the syrup from the manufacturer being held in one pressurized tank and carbon dioxide in another. The syrup and carbon dioxide are mixed as the drink is being drawn.

NONCARBONATED HERBAL AND HEALTHFUL BEVERAGES

Not too long ago, carbonated beverages dominated the shelves and cooler space in markets, but now new competition has arrived in the form of fortified, healthful, or natural beverages. The world's first **vitamin-fortified** fruit drinks appeared in 1948. Other herbal and healthful beverages followed, and new ones are being developed and marketed. Four examples of these include Hi-C®, Gatorade®, SoBe®, and Snapple®.

Hi-C®

Hi-C drink was conceived as a result of a weather related shortage of orange juice in the United States. In 50 years, Hi-C has become the world's largest brand of vitamin-fortified fruit drink, and the fifth largest-selling trademark of the Coca-Cola® Company. Hi-C fruit drinks contain 100 percent of the recommended daily intake of vitamin C per serving. In 1987, The MinuteMaid® Company (Coca-Cola) was the first to introduce a calcium-fortified orange juice, and the first and only company to offer drink boxes containing juices and juice drinks that are fortified with calcium and vitamin C. Providing a vitamin-fortified beverage is also important in other countries. For example, Kapo® fruit-flavored beverages fortified with vitamins were originally developed in 1975 under a Chilean government request to develop a low-cost way to provide children with a nutritious drink.

The Web site for Hi-C and other MinuteMaid products is <www.minutemaid.com>.

Gatorade®

In the 1960s, a team of researchers at the University of Florida began a project to develop a product for rapid fluid replacement for the body to help prevent the incidence of severe dehydration and loss of body salts brought about by physical exertion in high temperatures. By 1965, the group developed a formula that was ready for testing. Because football players experience tremendous fluid losses during practices and games, the formula was tested on ten members of the University of Florida football team (the "Gators"). This beverage became known as "Gatorade." The name was later trademarked. The coach's statement that "Gatorade made the difference" was published by *Sports Illustrated,* and this marked the beginning of the Gatorade phenomenon.

In May of 1967, Stokely-Van Camp, then a leading processor and marketer of fruits and vegetables, acquired the rights to produce and sell Gatorade throughout the United States. In 1983, the Quaker Oats Company acquired Stokely-Van Camp, including the Gatorade brand. The Quaker Oats Company marketed the brand nationally.

Gatorade provides the body with fluids, minerals, and energy during exercise. Today it remains the number-one sports drink in America amidst increasing competition. Gatorade sports drink is scientifically formulated to quickly replace the fluids and minerals the body loses during exercise or physical activity. Gatorade is formulated to stimulate rapid fluid absorption, assure rapid rehydration, provide carbohydrate energy to working muscles, and to encourage an individual to drink more. It is composed of water, sucrose, glucose-fructose syrup, citric acid, natural flavors, salt, sodium citrate, and mono-potassium phosphate.

The Web site for Gatorade is <www.gatorade.com/pages/gatorade/index.html>.

SoBe®

SoBe is a registered trademark of South Beach Beverage Company, makers of **nutritionally enhanced** refreshment beverages. The company started in 1995 with the introduction of South Beach brand iced teas and fruit drinks. The four partners who started SoBe were into health and fitness as a way of life. The first product introduced was SoBe **black tea** 3g with ginseng, ginkgo, and guarana. South Beach has since introduced the SoBe brand, a line of energizing wellness beverages targeted at the mass market. The line includes exotic teas, juice blends, elixirs, and "effect" beverages such as Energy, Power, and Wisdom. All these are fortified with herbs, minerals, vitamins, and other nutrient enhancers. SoBe's package designs feature dual lizards representing the yin/yang of life. The bottles are of sculpted glass with the lizard embossed on the neck (Figure 23-3).

More information about SoBe and the claims for their beverages can be found at their Web site at <www.sobebev.com>.

THE OLDEST BRAND OF CARBONATED DRINK

The oldest (continuously) commercially marketed carbonated drink is Moxie, which became available in apothecaries as a medical tonic in 1876. Made by the Nerve Food Company of Lowell, Massachusetts, Moxie was first sold as a beverage in 1884. Although Hires Root Beer was developed in 1876, eight years before Moxie, it was periodically pulled off the market.

By the early twentieth century, the "Nerve Food" was carbonated, brilliantly merchandised, and became a household word. In spite of the claims restrictions placed on Moxie by the Food & Drug Act, many ads from this explosive growth period touted the "healthful" and alleged medicinal benefits of the tonic. An early newspaper ad had Professor Allyn, "Food Authority," giving his esteemed testimonial.

Bottlers were opened all over the country. Frank Archer, who started with the company as a clerk, continued to brilliantly promote Moxie using every promotional gimmick known at the time. In the "glory days," the beverage was strongly associated with amusement parks, dance halls, and East Coast resorts. These were places synonymous with good times and the "vigorous" life that drinking Moxie was supposed to sustain.

Moxie has an odd flavor that has been described as a combination of cola, root beer, and licorice. The first version contained wintergreen as well as a bittersweet herb, root gentian. According to those who like it, it is definitely an acquired taste.

Moxie was also the first carbonated beverage to offer a sugar-free version. The potent brew with its distinctive orange labels is still available in Maine, where it is considered a local classic.

As sugar prices rose after World War II, increased expenses and competition primarily from Coca Cola made the Moxie empire dwindle. Today, it remains as it began, a New England beverage, under the ownership of the Armstrong family of the Monarch Corporation in Atlanta, Georgia.

The word "Moxie" is a proper name that made it to the dictionary as a noun synonymous with having "spunk." It is still common to hear of someone as having "a lot of moxie."

To read more about the Moxie phenomenon, or root beer, visit these Web sites:

<www.whatsgoingon.com/coolest/19980710>
<www.xensei.com/users/iraseski/>
<www.bottlebooks.com/hires.htm>
<pressplus.com/content/cumbo250/rootbeer.html>
<www.icaen.uiowa.edu/~psday/rootbeer.html>

FIGURE 23-3

New products such as SoBe® compete for the market of traditional carbonated beverages.

Snapple®

Snapple beverages claim to be all natural, with no preservatives, no artificial flavoring, and no chemical dyes. Snapple is produced in 50 different flavors, but the top ten flavors are:

1. Lemon Tea
2. Kiwi Strawberry
3. Diet Peach Tea
4. Raspberry Tea
5. Pink Lemonade
6. Mango Madness
7. Diet Raspberry Tea
8. Fruit Punch
9. Diet Cranberry Raspberry
10. Diet Lemon Tea

The Web site for Snapple is <www.snapple.com>.

BOTTLED WATER

Bottled water started out being fashionable, but now it has become a basic staple in the American household. Water sales in the United States rose 9.5 percent in 1998 to $4.3 billion, according to Beverage Marketing Corporation. Bottled water represents the fastest growing segment of the beverage industry, when compared to fruit beverages, other soft drinks, and beer.

This fastest growing segment of the beverage industry can attribute the boom to several factors. Baby boomers are maturing, and their tastes, as well as waistlines, are guiding them toward more natural, less caloric beverages. America's passion with fitness has encouraged consumers to drink beverage alternatives. The deteriorating taste and quality of tap water and fear of unknown contaminants have placed bottled water as the perfect solution.

Companies leading the bottled water business in terms of estimated dollar sales (at wholesale), according to the Beverage Marketing Corporation include Perrier Group of America, Suntory, McKesson Water Products Company, Danone International, Crystal Geyser, and U.S. Filter. All the other companies included in the bottled water industry make up the remaining 45 percent and represent over 900 brands.

The top ten leading brands include:

1. Poland Spring®
2. Arrowhead®
3. Evian®
4. Sparkletts®
5. Hinckley & Schmitt®
6. Zephyrills®
7. Ozarka®
8. Deer Park®
9. Crystal Geyser®
10. Crystal Springs®

These leading brands make up about 40 percent of the market share. All other brands make up the remaining 60 percent of the market share.

Within the bottled water business, two distinct industries and segments are represented. The biggest by volume is the five-gallon or returnable container water business. Companies like Arrowhead, Sparkletts, and Hinckley & Schmitt are leaders in this field. Often associated with the office cooler, bottlers also use 2½- as well as 1-gallon containers for supermarket distribution. This type of bottled water is sold as an alternative to tap water. Premium bottled waters, such as Evian, Vittel®, and Perrier®, are sold as soft drink and alcohol alternatives. Packaging ranges from 6 ounces to 2 liters and from custom glass and PET plastic to aluminum cans (refer to Chapter 15). More and more bottled water producers have switched from glass to polycarbonate because of the increased acceptance of quality with this kind of packaging.

FIGURE 23-4
Water is water, but many brands are available.

In the United States, waters labeled "spring water" may come from a spring source or from a borehole adjacent to a spring. Some famous spring waters in America include Mountain Valley from Arkansas, Belmont Springs from Massachusetts, Saratoga from New York, and Poland Spring from Maine. Artesian water is water from a well that taps a confined underground aquifer and in which the water level stands above the natural water table. Kentwood Springs from Louisiana is a well-known artesian water. Some bottled water is filtered, reprocessed water from municipal sources.

Bottled water is a profitable industry (Figure 23-4). A bottle of water similar in size to a carbonated drink can sell for as much as or more than the carbonated drink. Some current statistics on the bottled water industry are maintained at this Web site: <www.bottledwaterweb.com/statistics.html>.

ALCOHOLIC BEVERAGES

Fermentation of a carbohydrate source such as corn, rye, rice, molasses, agave, wheat, potatoes, and barley creates a variety of alcoholic beverages such as beer, whiskey, sake, vodka, rum, and tequila. Fermentation of a sugar-containing juice such as grape juice or other fruit juices creates wines or brandy. The production of beer and wine is discussed in the sections that follow.

Beer

Beer and ale are produced from **malt**, hops, yeast, malt adjuncts, and water. Malt is prepared from barley that has germinated. It is dried, and the sprouts are removed. Malt adjuncts are starch- or sugar-containing materials such as corn, potato, rice, wheat, and barley.

Malting. Barley grains are soaked (**steeped**) at 50° to 60°F (10° to 15.6°C), germinated at 60° to 70°F (15.6° to 21.1°C) for 5 to 7 days, and then dried. Most of the sprouts are removed (Figure 23-5). The malt serves as a source of amylases (enzymes) that will break down the starch into sugar for the yeast to ferment.

Mashing. **Mashing** involves mixing the ground malt with a previously boiled malt adjunct at a temperature of 65° to 70°F (18.3° to 21.1°C) (Figure 23-6). The enzymes in the malt digest the starch in the adjunct and release sugar. The mix is heated to 75°F (23.9°C) to denature the enzymes and then filtered. The clear filtrate is known as **wort**.

Boiling. Hops are added to the wort and boiled for about 2.5 hours then filtered. Boiling serves to:

- Concentrate the solids
- Kill microorganisms
- Inactivate enzymes
- Coagulate proteins
- Caramelize the sugars

FIGURE 23-5
Barley is sprouted during the malting process.

FIGURE 23-6
Copper kettles for cooking mash and adding hops during beer making.

Fermentation. Wort is inoculated with a beer yeast, *Saccharomyces carlsbergenis.* The temperature is held between 38° and 57°F (3.3° to 14°C). Fermentation is complete in 8 to 14 days. At this time the alcohol content of the wort is about 4.6 percent by volume and the pH is about 4.0. Bacterial growth is minimized as much as is possible. As soon as fermentation is complete, the beer is quickly chilled to 32°F (0°C) and passed through a series of filters to remove yeast and other suspended materials before storage and completion.

Completion. Beer is aged at 32°F (0°C) for from several weeks to several months. This storage time is known as **lagering**. During storage the beer is carbonated to a CO_2 content of 0.45 to 0.52 percent. After storage, the beer is given a final filtration to remove traces of suspended materials and to give the beer a crystal clear appearance.

Beer in cans is pasteurized by heating to about 140°F (60°C) to remove organisms. Draft beer is kept under refrigeration and does not need pasteurization. Supposedly it has a better flavor because it is not pasteurized.

Cold pasteurization or filtration removes leftover yeast and bacteria. This stabilizes the beer without the heat of conventional pasteurization.

Wine

Wine is an alcoholic beverage made from fermented grape juice. It can be made from many fruits and berries, but the grape is the most popular. Growing grapes is a major feature of the economy of many wine-producing countries. Wines may be either red, white, or rose and also dry, medium, or sweet. They fall into three basic categories:

1. Natural, or table wines, with an alcohol content of 8 to 14 percent, generally consumed with meals
2. Sparkling wines, containing carbon dioxide, like champagne
3. Fortified wines, with an alcohol content of 15 to 24 percent with varying sweetness

The various types include port, sherry, and aromatic wines and bitters such as vermouth.

The quality and quantity of grapes depend on geographical, geological, and climatic conditions in the vineyards, and on the grape variety and methods of cultivation. Viniculture is the science and art of growing grapes for wine, and **vinification** is the production of wine from grapes. (Viticulture is the science and art of growing grapes.)

Harvesting. The crop is harvested in the autumn when the grapes contain the optimum balance of sugar and acidity. For the sweet white wines of France and Germany, picking is delayed until the grapes are affected by a beneficial mold, *Botrytis cinerea*, which concentrates the juice by dehydration.

Vinification. For red wine, the grapes are crushed immediately after picking and the stems generally removed. The yeasts present on the skins come into contact with the grape sugars, and fermentation begins naturally. Cultures of the yeast *Saccharomyces ellipoideus* are sometimes added, and the grapes are often treated with sulfur dioxide (SO_2) to control the growth of undesirable microorganisms, especially bacteria. During fermentation the sugars are converted by the yeasts to ethyl alcohol (ethanol) and carbon dioxide (Figure 23-7). The alcohol extracts color from the skins; the longer the vatting period, the deeper the color. Glycerol and some of the esters, aldehydes, and acids that contribute to the character, bouquet (aroma), and taste of the wine are by-products of fermentation. Traditional maturation of red wine takes up to two years in 50-gallon oak casks. During this time the wine is racked–drawn off its lees, or sediment–three or four times into fresh casks to avoid bacterial spoilage. Further aging usually takes place after bottling.

The juice of most grape varieties is colorless. Grapes for white wine are also pressed immediately after picking. Fermentation can proceed until it is completed, which will make a dry white wine. Or it can be stopped to make a sweeter wine. Maturation of white

FIGURE 23-7
Containers for the fermentation of wine.

Burgundy and some California Chardonnays still takes place in oak casks, but vintners tend now to use stainless steel large tanks. Minimum contact with the air retains the freshness of the grapes.

To make rose wines, the fermenting grape juice is left in contact with the skins just long enough for the alcohol to extract the required degree of color. Vinification then proceeds as for white wine.

The best and most expensive sparkling wines are made by the champagne method, in which cultured yeasts and sugar are added to the base wine, inducing a second fermentation in the bottle. The resulting carbon dioxide is retained in the wine. Sparkling wines are also made by carbonation.

The alcohol content of fortified wines is raised by adding alcohol. With port and madeira, brandy, spirit (alcohol) is added during fermentation killing off the yeasts, stopping fermentation, and leaving the desired degree of natural grape sugar in the wine. Sherry is made by adding spirit to the fully fermented wine. Its color, strength, and sweetness are then adjusted to the required style before bottling.

In the United States, the Bureau of Alcohol, Tobacco, and Firearms (BATF) of the U.S. Treasury Department issues regulations governing the labeling and taxing of wines and other alcoholic beverages. The labels must also meet requirements of the Food, Drug, and Cosmetic Act.

COFFEE

Coffee beans or cherries come from a small tree (shrub) of the genus *Coffea.* Three important species include *C. arabica, C. canephora,* and *C. liberica.* After harvesting, the coffee cherries may be dried and the pulp around the beans removed. In wet climates or for particular types of coffee, the harvested cherries may be washed and then pulped to separate the beans. The dry and wet methods of preparation produce distinctive flavors in the beans and, along with the differences between varieties, account for the subtle flavor distinctions between beans from the various growing areas. Removal of pulpy berry from skin is achieved by bacteria that break down pectin (pectinolytic). This is followed by an acid fermentation by lactic acid bacteria. After fermentation, beans are dried, hulled, and roasted.

The flavor of coffee is determined not only by the variety but also by the length of time the green beans are roasted (Figure 23-8). In continuous roasting, hot air that is 400° to 500°F (200° to 260°C) is forced through small quantities of beans for a 5-minute period. In batch roasting, much larger quantities of beans are roasted for a

FIGURE 23-8

Specialty coffees and grinder.

longer time. Dark-roasted coffees (French or espresso roasts) are stronger and mellower than lightly roasted beans.

After roasting, the beans are usually ground and vacuum packed in cans. The flavor of coffee deteriorates rapidly after it is ground, or after a sealed can is opened.

Instant coffee is prepared by forcing an atomized spray of very strong coffee extract through a jet of hot air. This evaporates the water in the extract and leaves dried coffee particles, which are packaged as instant, or soluble, coffee. Another method of producing instant coffee is freeze-drying.

To make decaffeinated coffee, the green bean is processed in a bath of methylene chloride, which removes the caffeine, and steam, to remove the methylene chloride. In a newer method the caffeine is removed using steam only.

COFFEE SUBSTITUTES

An aromatic beverage similar to coffee can be created by roasting certain grains and grain products in combination with other flavor sources. These cereal beverages are commercially available, or they can be homemade. One popular product, Postum®, was developed by C. W. Post, and it helped form the foundation for General Foods Corporation. Another commercial product is called Pero®.

These commercial products are generally prepared from barley, wheat, rye, malt, and bran with additional flavoring from items like molasses, chicory, carob, cassia bark, allspice, and star anise, depending on the manufacturer. Some of these cereal beverages can even be percolated like coffee. Also, a homemade roasted grain drink can be prepared by baking a combination of wheat bran, eggs, cornmeal, and molasses.

TEA

Tea is the beverage made when the processed leaves of the tea plant are infused with boiling water. The tea plant (*Camellia sinesis*) is native to Southeast Asia. Its dark green leaves contain the chemicals caffeine and tannin. Tea ranks first as the most popular beverage in the world.

Types of tea include fermented (black), unfermented (green), and partially fermented (**oolong**). Production of tea is not really a fermentation, but rather the action of enzymes contained within the tea leaves.

Processing

The leaves are hand plucked by experienced workers. Only the smallest, youngest leaves are used to produce tea.

To make black tea, harvested leaves are spread on withering racks to dry. The leaves become soft and pliable and are then roller crushed to break the cell walls and release an enzyme. This process gives the tea its flavor. After rolling, the lumps of tea are broken and spread in a fermentation room to oxidize, which turns the leaves to a copper color. The leaves are finally hot-air dried in a process that stops fermentation and turns the leaves black. After the tea is processed, it is sieved to produce tea leaves of a uniform size and to facilitate blending and packing.

Leaf-grade sizes run from pekoe, the coarsest size, to flowery orange pekoe, the smallest. Tippy golden flowery orange pekoe indicates a tea containing the golden-colored tip, obtained from the bud. The broken teas have leaves that were broken during processing. These include broken orange pekoe, broken pekoe, fannings, pekoe fannings, and dust.

Oolong tea begins like black tea. The aroma develops more quickly. When the leaf is fired or dried, a coppery color forms around the edge of the leaf while the center remains green. The oolong flavor is fruity and pungent.

Green tea is produced much like the others, except that the leaf is heated before rolling in order to destroy the enzyme. The leaf then remains green throughout further manufacture, and the aroma characteristic of black tea does not develop. Green tea is graded by age and style.

Blended and Unblended Varieties

Various blended and unblended teas achieved fame for their characteristic flavors including Assam, Darjeeling, and Keemun also known as English Breakfast tea. Popular blended teas include Irish Breakfast, Russian style, and Earl Grey.

Instant Tea and Bottled Tea

Instant tea is manufactured in a process similar to that of instant coffee. First the tea is extracted from the tea leaves using hot water, 140° to 212°F (60° to 100°C). This extract is concentrated in low-temperature evaporators and is dried using spray driers in a low-temperature vacuum. Just before concentration the aromatics (flavor) are distilled with flavor-recovery equipment. The flavor is added back later. Some teas may be freeze-dried instead of spray dried.

Recently, bottled and canned tea have gained in popularity. The manufacture of these products is similar to carbonated beverages except that they begin with a brewed product. If their pH is above 4.6, retort processing is required.

HERBAL TEA

Tea also includes herb teas. Herb tea is made from many plants, using not just leaves but also flowers, roots, bark, and seeds. Unlike the limited flavor variations of black tea, herb tea exists in a variety of distinctively different flavors, colors, and aromas. Blending the flavors of different herbs results in an infinite variety of taste sensations. Most herb teas contain no caffeine. Drinking herb teas was widespread in Europe long before the arrival of black tea. Some of the lasting favorites include such flavors as chamomile, peppermint, and rose hips.

The parts of the plant picked depend on the type of plant and the intended use. Leaves, flowers, roots, bark, and seed are all potential herb tea ingredients. After harvest, the herbs are dried by either spreading them on large screens or by tying them in bundles and hanging them upside down. This is done indoors or outdoors in the shade, but it must be done quickly to retain the plants' natural oils and color. Oven drying is less effective than natural drying in terms of preserving the natural oils and flavor. After drying, the herbs are bundled into large sacks and wooden chests for shipment to the herb tea maker. The next processes include cleaning, milling, sifting, and blending the herbs into the desired flavor combinations.

The most successful herb tea company is Celestial Seasonings® in Boulder, Colorado. Celestial Seasonings began in 1969 in Aspen, Colorado, where 19-year-old Mo Siegel and a friend, Wyck Hay, gathered wild herbs in the forests and canyons of the Rocky Mountains and made them into herbal teas. Their first tea was produced and packaged with the help of wives and friends and sold in local health food stores.

Now Celestial Seasonings purchases over a hundred different varieties of herbs, spices, and fruits from over 35 different countries. Each herb and spice has its own flavor and taste characteristics–for example, tangy hibiscus, citruslike rose hips, applelike chamomile, soothing peppermint, fruity blackberry leaves, sweet cloves, and exotic cinnamon. These create a variety of teas when blended.

Celestial's herbalists sample and select entire crops for purchase. Once the herbs and spices arrive in Boulder, they are inspected and tested again. The approved herbs are cleaned and inspected once again, and then sent to the milling room to be cut into different degrees of fineness demanded for individual types of teas. Finally, they are sifted to achieve uniformity of texture and purity of content necessary for consistent flavor.

Next, the herbs are blended to match a standard recipe. The batch is compared with previous blends and readjusted until it is absolutely perfect in terms of consistency, quality, and flavor. It is then carefully packaged to protect the flavor.

The most popular herbs for tea are those like hibiscus flowers, lemon grass, and peppermint. These herbal teas are recommended by many physicians and nutrition advocates as beneficial alternatives to such beverages as coffee, black tea, and sweetened, carbonated drinks.

Kraft, Inc. bought Celestial Seasonings in 1984. Kraft's marketing brought Celestial Seasonings to the attention of new consumers, strengthening its lead in the herb tea industry while also introducing a gourmet line of traditional, or black, teas. Then in September 1988, Kraft sold Celestial Seasonings back to its management, returning the tea company to independent ownership. The company remains headquartered in Boulder, in the new corporate facility on Sleepytime Drive. Their Web site is located at <www. celestialseasonings.com>.

Today, the company claims to serve more than 1.2 billion cups of tea per year. It is the largest herb tea manufacturer in North America and is expanding internationally. Some of the herb teas being produced now include the following:

Almond Sunset
Antioxidant Green Tea
Authentic Green Tea
Bengal Spice
Black Raspberry
Caffeine-Free Tea
Ceylon Apricot Ginger
Chamomile
Cinnamon Apple Spice
Country Peach Passion
Cranberry Cove
Detox A.M. Herb Tea
Diet Partner Herb Tea
Echinacea Cold Season
Echinacea Herb Tea
Emerald Gardens Green Tea
Emperor's Choice
English Toffee
Estate Blend
Fast Lane
Fruit Sampler
GingerEase Herb Tea
GinkgoSharp Herb Tea
Ginseng Energy
Grandma's Tummy Mint
Green Tea Sampler
Harvest Chamomile
Heart Health
Honey Lemon Ginseng Green Tea
LaxaTea Herb Tea
Lemon Berry Zinger
Lemon Zinger
Mandarin Orange Spice
Mint Magic
Misty Jasmine Green Tea
Misty Mango Oolong
Mood Mender
Morning Thunder
Nutcracker Sweet
Orange Mango Zinger
Peppermint
Raspberry Zinger
Red Zinger
Roastaroma
Sleepytime
Strawberry Kiwi
Sugar Plum Spice
Sunburst C
Tension Tamer
Throat Soothers Tea
Vanilla Hazelnut
Vanilla Maple
Wild Berry Zinger
Wild Cherry Blackberry

In 1998, the company announced a strategic transition to leverage the Celestial brand. It launched a line of herbal dietary supplements under the Celestial Seasonings Mountain Chai® label. The company also introduced a line of "Wellness" Teas and Organic Teas and reintroduced a line of Green Teas. It also announced the return of the Red Zinger.

Summary

The food industry creates a wide variety of beverages. This includes carbonated, nonalcoholic beverages, with different flavors and with nutritive and nonnutritive sweeteners. Carbonation in beverages provides their zest and sparkle. Recently, the popularity of carbonated beverages has given way to new products created by the food industry such as vitamin-fortified fruit drinks, scientifically formulated sports drinks, nutritionally enhanced beverages, and water. Bottled water represents a profitable, fast-growing segment of the beverage industry.

Fermentation of carbohydrates from corn, rye, rice, molasses, wheat, potatoes, barley, agave, and fruit juices creates alcoholic beverages such as beer, wine, whiskey, vodka, rum, sake, and tequila. Fermentation by yeast changes sugar to alcohol; the taste of the beverage depends on the carbohydrate source and the process.

Coffee comes from roasted beans of a small tree. Coffee flavor is determined by the variety and length of roasting time. Coffee substitutes are produced from roasted grains and grain products in combination with other flavors. Tea comes from the processed leaves of a plant. The type of tea produced depends on the enzymatic action on tea leaves. Both tea and coffee have been produced in an instant form. Tea has been bottled as a beverage.

Herbal teas come from many plants and not just the leaves: flowers, roots, bark, and seeds of herbs all can be used in a blend for herbal tea. Herbal teas are enjoyed because of their wide ranging tastes, and they are often recommended as beneficial alternatives to coffee, tea, and carbonated drinks.

Review Questions

Success in any career requires knowledge. Test your knowledge of this chapter by answering these questions or solving these problems.

1. List the five reasons why hops and wort are boiled in the beer-making process.
2. Name the three categories of wines.
3. What is the difference between viticulture and vinification?
4. How is the flavor of coffee determined?
5. Coffee comes from a ___________; tea is made from a ___________.
6. Bottled ___________ is the fastest growing segment of the beverage industry.
7. List five herbs used in beverages.
8. What is the difference between what plain teas and herbal teas are made from?
9. What are coffee substitutes made from?
10. Name the drink that provides minerals, vitamins, and energy during exercise.

Student Activities

1. List the ingredients on a can or bottle of a beverage and explain the function of each ingredient.
2. Taste test some of the new healthful beverages and make a list of their claims.
3. Develop and conduct a survey to determine the favorite beverage enjoyed by people in your community.
4. Develop a report or presentation on the physiological action of caffeine and beverages containing caffeine.
5. Conduct a taste test (contest) to see if the students can differentiate between the brand name colas and some generic cola.
6. Make one of the coffee substitutes or herbal teas in class and conduct a taste test.

Resources

Alexander, R. J. 1998. *Sweeteners: Nutritive.* St. Paul, MN: Eagan Press.

Bartlett, J. 1996. *The cook's dictionary and culinary reference.* Chicago: Contemporary Books.

Francis. F. J. 1998. *Colorants.* St. Paul, MN: Eagan Press.

Horn, J., J. Fletcher, and A. Gooch. 1997. *Cooking a to z. The complete culinary reference source.* Glen Ellen, CA: Cole Publishing Group, Inc.

Lehner, E., and J. Lehner. 1962. *Folklore and odysseys of food and medicinal plants.* New York: Tudor Publishing Company.

McGee, H. 1997. *On food and cooking. The science and lore of the kitchen.* New York: Simon and Schuster Inc.

Still, J. 1981. *Food selection and preparation.* New York: Macmillan Publishing Co., Inc.

Internet

Internet sites represent a vast resource of information. The URLs (uniform resource locator) for the World Wide Web sites can change. Using one of the search engines on the Internet such as Yahoo!, HotBot, AltaVista, Excite, Dogpile, About, or Google, find more information by searching for these words or phrases: nonalcoholic beverages, teas (specific brand names) and types, colas (specific brand names), wine (specific brand names), beer (specific brand names), malting, mashing, vinegar, cold pasteurization, carbonated soft drinks, diet drinks, carbonation. Also, Table A-7 provides a listing of some useful Internet sites that can be used as a starting point.

SECTION *Four*

Related Issues

Chapter 24

Environmental Concerns and Processing

Objectives

After reading this chapter, you should be able to:

- Describe the properties and the requirements of water used in food processing
- Discuss four methods food-processing industries use to dispose of solid wastes
- Explain how water becomes wastewater during food processing
- Relate the level of solids in wastewater to BOD
- Describe wastewater treatments to lower BOD
- List eight products that could be in wastewater
- Identify four methods of separating water from solid wastes
- Name four ways food processors reduce the amount of solid wastes and water discharge
- List five methods of conserving water during food processing

Key Terms

alkaline
BOD
caustic
flocculation
landfilling
potable
tramp material
turbidity
wastewater
wet scrubber

Solid wastes and wastewater from food processing are environmental concerns. The food-processing industry uses technology to reduce the quantity of high-moisture-content solid wastes generated by washing, cleaning, extracting, and the separation of undesirable solids from fruits and vegetables processing. Research in this area continues to develop ways to reduce the volume and water content of solid wastes, to increase the value of solid wastes sold as animal feed, and to convert the solid wastes into other by-products.

WATER IN FOOD PRODUCTION

The role of water in food production, processing, and preparation to produce a quality food is almost infinite. Water serves as a universal solvent. The ability of water to serve as a solvent has tremendous impact on many of its other roles. It impacts the osmotic pressure of a solution. The most obvious example of structure is the use of water in frozen desserts. There the crystalline water contributes the structure. Water serves as a heat transfer medium. It also serves as a cleansing agent. Managing the use and disposal of water during processing continues to be an important issue for the industry.

PROPERTIES AND REQUIREMENTS OF PROCESSING WATERS

Converting raw products into finished food products requires large quantities of water. Incoming-water quality is important to a food manufacturing operation, and the water quality exiting a food manufacturing operation is important to everyone. All the water is part of the water (hydrologic) cycle (Figure 24-1). Food scientists are always searching for ways to prevent or eliminate contamination of water by food processing.

Food production and processing industries concern themselves with three aspects of water:

1. Microbiological and chemical purity and safety
2. Suitability for processing use
3. Decontamination after use

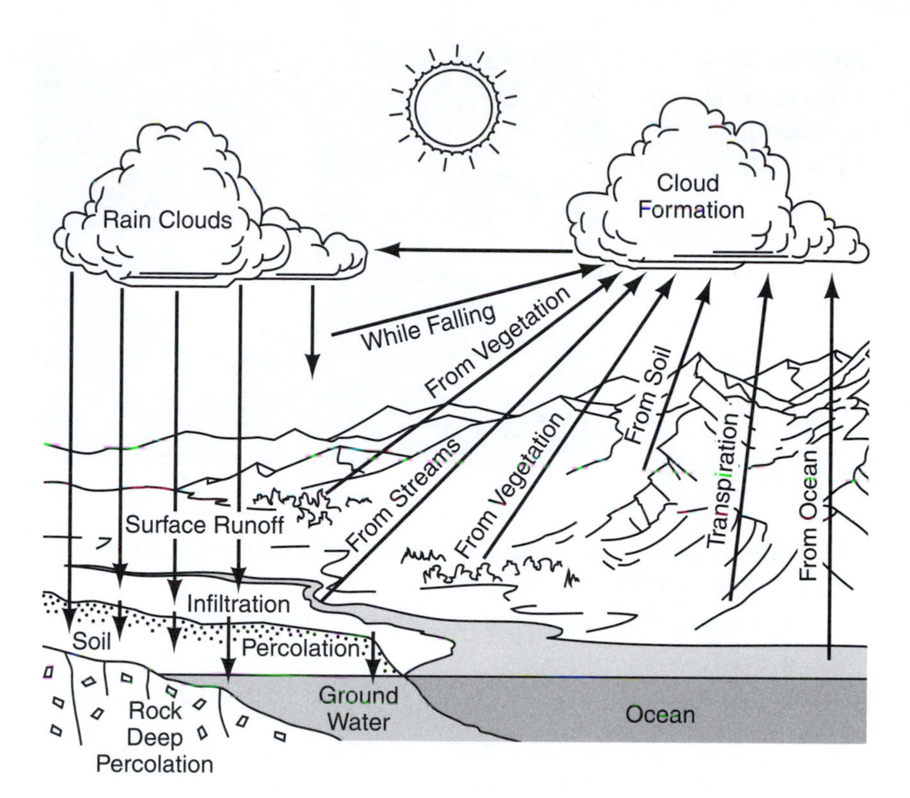

FIGURE 24-1
The water cycle.

Water entering a food processing plant must meet health standards for drinking (**potable**) water. National Primary Drinking Water Regulations are issued by the EPA. These regulations set limits on such primary characteristics as:

- Inorganic chemicals (As, Ba, Cd, Cr, Pb, Hg, NO_3, Se, Ag, F)
- Organic chemicals (Endrin®, Lindane®, Methoxychlor®, Toxaphene®, 2,4-D®, 2,4,5-TP Silvex®)
- **Turbidity** (cloudiness)
- Coliform bacteria

Strict environmental regulations govern the discharge of polluted water from processing plants (Figure 24-2). Contaminated water and the difficulties of disposing of **wastewater** increases the overall cost of manufacturing food.

ENVIRONMENTAL CONCERNS

For many food-processing plants, a large fraction of the solid waste produced at the plant comes from separation of the desired food

FIGURE 24-2
Fresh water storage is important to any community.

components from undesired ones in the early stages of processing. Undesirable components include:

- **Tramp material** (soil and extraneous material)
- Spoiled food stocks
- Fruit and vegetable trimmings, peels, pits, seeds, and pulp

In some food processing plants, **caustic** peeling is used to remove skins from soft fruit and vegetables such as tomatoes. This operation produces a highly **alkaline** or salty solid waste, depending on whether the alkalinity is neutralized. High-moisture solid waste materials can also be generated by water cleanup and reuse operations in which the dissolved or suspended solids are concentrated and separated from wastewater streams.

DISPOSAL OF SOLID WASTES

Currently food-processing solid waste can be disposed of by sewer or **landfilling**, chemical production, as animal feed, and as fuel. The effective recovery of the energy value in the solid waste is often determined by the moisture content and the processing method–gasification, bioconversion, incineration. A final use for solid wastes is composting or land application with limited soil amendment value.

Disposal of solid wastes as animal feed is limited by several factors, including cost of shipping, spoilage during storage and transport, and the presence of undesirable components such as alkaline or salt. For example, distillers' dried grains and solubles have been fed to animals for decades. Water content is a major contributor to shipping costs and to some extent the spoilage rate.

Spoilage reduces the value of solid wastes and limits the animal feeding option to local animal herds.

Incineration and use as a fuel are options in certain cases, but limited to those solid wastes that have relatively low water content and can be further dried with ease. The moisture content of suitable fuels is about 10 percent or less.

Composting (Figure 24-3) is an option for disposal, but odor and leaching of soluble constituents are limiting factors. Composted material is valued as a soil amendment or potting soil, but widespread use and marketability are limited by shipping cost. Composition of the composting materials needs to be controlled to obtain the correct physical mix to allow the natural composting aerobic bioprocesses to proceed. A full range of food-processing wastes can be composted, including fruit and vegetable wastes such as peelings, skins, pumice, cores, leaves, and twigs; fish processing waste such as bones, heads, fins, tails, skin, whole fish, and fish offal; meat processing wastes such as paunch contents, blood, fats, intestines, and manure; and grain processing wastes such as chaff, hulls, pods, stems, and weeds.

Disposal of solid wastes from food processing to domestic sewers is becoming less favorable because of increased sewer rates and the hesitation of municipal sewage treatment plants to accept these waste streams that have high biological oxygen demand (**BOD**) and, in some cases, high salt content.

The practice of landfilling is becoming less favorable due to the generation of foul odors as communities expand and reside next to food-processing plants. Leaching of undesirable constituents (salts, soluble organics) into the soil and groundwater is also an important concern where the groundwater is used by communities or it migrates into nearby streams.

FIGURE 24-3

Composting could possibly solve some food-processing waste issues. (*Source:* USDA Photography Library)

Currently, solid wastes disposed by landfilling or composting are minimally treated using dewatering screens, centrifugal screens or strainers to separate free liquids from the solids. In the case of solids from juice extractors and sorting operations to remove blemished or spoiled fruits and vegetables, the solid waste is not treated before disposal. Similarly, low-moisture-content solid wastes that are potentially suitable as fuel or for incineration undergo minimal processing, such as size reduction and minimal drying.

Solid wastes sent to a sewer undergo size reduction and are mixed with water or other liquid waste streams to produce acceptable flow properties and BOD loading.

Some solid wastes disposed of as animal feeds are not further treated, but are fed to local livestock, particularly dairy and beef

HOH—WATER

Water—the stuff of life—covers three fourths of Earth's surface and represents a major component of the bodies of plants and animals. A 200-lb. (91-kg) human would weigh only about 90 lbs. (41 kg) with all the water removed. This miracle liquid forms from two gases—hydrogen and oxygen. Two atoms of hydrogen (H) and one atom of oxygen (O) combine to form water, making its chemical formula H_2O or HOH. The water molecule is bipolar, having charged poles like a magnet, giving water unique properties.

Depending on the temperature, water exists in three forms. It is a liquid, between 32° and 212°F (0° and 100°C). It is a gas or vapor at the temperatures above 212°F (100°C). Water becomes a solid (ice) at temperatures below 32°F (0°C).

Of all the naturally occurring substances, water has the highest specific heat. This makes it a good coolant in biological systems and makes it resist rapid temperature changes. Specific heat is the amount of heat required to raise the temperature of a substance 1°C.

Water is the universal solvent, dissolving almost everything. It is powerful enough to dissolve rocks yet gentle enough to hold an enzyme in a fragile cell. As a solvent, it acts as a medium for biochemical reactions and carries waste products and nutrients.

This miracle liquid supports life.

To learn more about water and water quality visit these Web sites:

<www.britannica.com/>
<www.epa.gov/OW/>

cattle. For example, solid wastes from orange juice processors and distilleries can be dried and sold as livestock feed.

PROPERTIES OF WASTEWATERS

Some food-processing plants may literally wash profits down the drain. This industry typically uses a large volume of water to process food products and clean plant equipment, yielding large amounts of wastewater that must be treated. Excessive water use and wastewater production adds financial and ecological burdens to the industry and to the environment. However, food processors can take actions that will dramatically reduce water use, wastewater production, and the high costs associated with these problems.

Using water for cleanup in food-processing plants flushes loose meat, blood, soluble protein, inorganic particles, and other food waste to the sewer. Some of these raw materials could be recovered and sold to other industries but instead are lost. Also, most of this waste adds a high level of biological oxygen demand (BOD) to the wastewater.

Wastewater treatment plants (Figure 24-4) use BOD levels to gauge the amount of waste that is present in water–the higher the BOD level, the more treatment the wastewater will require. Sewer plants add surcharges for each pound of BOD that exceeds a set limit. These charges can cost the company hundreds of thousands of dollars each year.

Food-processing companies use current methods and interventions that can assist in effectively managing their water resources. Without knowledge and use of these wastewater management

FIGURE 24-4

Treatment of wastewater.

techniques, companies lose money through water use charges, raw material losses, sewage surcharges, and possible fines from environmental agencies. With the public emphasis on environmental quality, the food industry has further incentive to reduce its water usage and its wastewater production.

WASTEWATER TREATMENT

New dewatering schemes are needed to make food-processing solid wastes more economical to handle for many applications. Common dewatering processes use a variety of mechanical means such as screw presses, belt presses, vacuum filters, and so on. Separations involving other mechanical forces, such as ultrasonic and nonmechanical forces such as electric or magnetic fields, have been used in a few specialized sectors.

Primary treatments such as screening, filtering, centrifuging, skimming, settling, and coagulation or **flocculation** lower the wastewater BOD and the total solids in the wastewater. Secondary treatment often involves the use of trickling filters, activated sludge tanks, and various ponds. Sometimes the secondary treatments are preceded by anaerobic digestion.

Lowering Discharge Volumes

Plant surveys determined where water use occurred and where wastes were generated in food-processing plants. The results showed that over half the waste load resulted from wet cleanup practices. Waste in the form of bits of food were being flushed down the drains.

Where cleanup practices (Figure 24-5) are involved in increasing the wastewater, specialists suggest techniques for dry cleanup that would reduce wastewater production. Dry cleanup uses methods to capture all nonliquid waste and prevent it from entering the wastewater. If most of the waste that comes out of a plant consists of carbohydrates or proteins, dry cleanup allows much of the waste to be reclaimed and put to secondary use–for example, as animal feed. Some of the remaining wastes in an animal-processing facility can be sold to a rendering plant.

A grease trap, solids recovery basin, and an activated sludge system with provisions for pH control are part of an overall wastewater management system. Where odors are also part of the problem, they can be eliminated by passing a building's exhaust air through a **wet scrubber**.

A concept as simple as keeping wastes off the floors and out of the drains will save a company money and reduce the strain on

FIGURE 24-5
Cleaning requires large quantities of water, as for this commercial dishwasher.

the city sewage treatment plant. Most of the changes made to reduce water use and waste cost a company little or nothing. Carelessness can be prevented by focusing employee awareness and management emphasis on the problem.

Common sense approaches to cleanup, such as using trays beneath machines to catch spillage, picking up spillage before hosing down the floors, and placing screens over drains, were used at little cost. Awareness of the serious problems caused by overuse and product waste cost a company only the time needed to educate employees thoroughly. A successful pollution prevention program requires frequent retraining to keep employees focused and careful.

Food-processing plants conserve water by following these conservation guidelines:

- Always treat water as a raw material with a real cost
- Set water conservation goals for the plant
- Make water conservation a management priority
- Install water meters and monitor water use
- Train employees how to use water efficiently
- Use automatic shut-off nozzles on all water hoses
- Use high-pressure, low-volume cleaning systems
- Do not let people use water hoses as brooms
- Reuse water where possible
- Minimize spills of ingredients and of raw and finished product on the floor
- Always clean up the spills before washing

RESPONSIBILITY

Water conservation and waste reduction are important. Water costs and sewer charges are rising. Water quality and availability are threatened by increased consumption and pollution in many areas of the country. Pollution is being attacked aggressively by public agencies and the public at large. Future regulations may require water conservation and elimination of pollutant discharges. A food processor's image can be tarnished and its sales hurt if its plants are perceived as harming the environment. Enforcement actions are becoming more severe. Lawsuits, fines, and even prison terms may face those who are not fully in compliance with environmental laws.

Summary

Food processing produces solid wastes. To remain environmentally friendly, food processors seek ways of disposing of or using solid wastes. Because food processing requires large quantities of potable water and produces large quantities of wastewater, this creates other environmental issues. Food processors must clean up wastewater. They also try to reduce the amount of water used in processing to lower the wastewater production. Responsibility for disposing of solid wastes and for treating and reducing wastewater resides with the food processor to meet environmental standards, laws, and restrictions.

Review Questions

Success in any career requires knowledge. Test your knowledge of this chapter by answering these questions or solving these problems.

1. Water serves as a universal ___________.
2. List five methods of conserving water during food processing.
3. Name the three aspects of water that concern food-production and -processing industries.
4. Define BOD.
5. Explain how water becomes wastewater during food processing.
6. When BOD levels are high, the more ___________ the wastewater will require.
7. List four methods of separating solids from wastes.

8. Name four ways food processors can reduce solid wastes and water discharge.
9. Explain dry cleanup.
10. Name one way dry cleanup can be used.

Student Activities

1. Visit a wastewater treatment facility and develop a report or presentation.
2. Diagram the water (hydrologic) cycle and show how use of water and the disposal of wastewater by food processing fits into the cycle.
3. Collect two current articles that relate to water quality. Summarize these and report on them to the class.
4. Demonstrate the value of clear, clean water. Start with a clear, clean quart bottle of water. Take a drink. Then add a tablespoon of soil, some bread crumbs, a few pieces of hair, a little catsup, some crumbs of hamburger, and a tablespoon of oil. Shake the mixture. This mixture represents wastewater. Devise an experiment to show how this wastewater could be cleaned up.
5. Invite a wastewater treatment specialist to visit the classroom and discuss methods used to treat and clean water.

Resources

Council for Agricultural Science and Technology (CAST). 1995. *Waste management and utilization in food production and processing* (Task force report No. 124). Ames, IA.

Cremer, M. L. 1998. *Quality food in quantity. Management and science.* Berkeley, CA: McCutchan Publishing Corporation.

Kreith, F. (Ed.). 1994. *Handbook of solid waste management.* New York: McGraw-Hill.

Potter, N. N., and J. H. Hotchkiss. 1995. *Food science,* 5th ed. New York: Chapman and Hall.

Tchobanoglous, G. (Ed.). 1991. *Wastewater engineering: Treatment, disposal, and reuse.* New York: McGraw-Hill.

Internet

Internet sites represent a vast resource of information. The URLs (uniform resource locator) for the World Wide Web sites can change. Using one of the search engines on the Internet such as Yahoo!, HotBot, AltaVista, Excite, Dogpile, About, or Google, find more information by searching for these words or phrases: wastewater, wastewater treatment, solid wastes, potable water, processing waters, National Primary Drinking Water Regulations, composting, landfilling, sludge system, water conservation, waste reduction, Environmental Protection Agency, Department of Environmental Quality. Also, Table A-7 provides a listing of some useful Internet sites that can be used as a starting point.

Chapter 25

Food Safety

Objectives

After reading this chapter, you should be able to:

- List three categories of food safety
- Name four factors contributing to the development of a foodborne disease
- List four types of microorganisms that can cause foodborne illness
- List five factors affecting microbial growth
- Identify the microorganisms that provide an index of food sanitation
- Discuss the role of sanitation and cleaning during processing in food safety
- Identify the correct order of sanitizing or cleaning a food contact surface
- Name three types of food soils
- List two types of sanitization
- Identify agencies involved in food safety regulation
- Describe the role of HACCP in food safety

Key Terms

biofilms
COP
cross-contamination
food soil
gastroenteritis
generation time
HVAC
lag time
psychrotrophic
sanitization
SPC
thermotrophic
toxins

Food safety is a very broad topic. Pesticides, herbicides, chemical additives, and spoilage are all of concern, but food scientists, food processors, and consumers focus most on microbiological quality. Microorganisms pose a challenge to the food industry, and most food processes are designed with microbial quality in mind. Microorganisms are too small to be seen with the unaided eye and have the ability to reproduce rapidly. Many of them produce toxins and can cause infections. For these reasons, the microbiological quality of the food is scrutinized closely.

SAFETY, HAZARDS, AND RISKS

Food safety hazards include all microbiological, chemical, and foreign materials that, if consumed, could cause injury or harm. Three conditions result in food safety problems:

1. Illness caused through transmission of disease germs. Pathogenic germs are passed from person to person through soiled objects such as money, doorknobs, railings, common drinking cups, and the like. Food can serve as a mere vehicle of disease transmission. Transmission of animal pathogens to humans by way of food is also possible.
2. Food poisonings and food infections caused by bacteria. The terms "food poisoning" and "food infections" refer to a violent illness of the stomach and intestinal tract (known as **gastroenteritis**) following the consumption of an offending food. The offending food contains, in general, high numbers of bacteria that are capable of producing the gastroenteritis suffered by the victim. Some of the pathogens are able to release **toxins** into the food. These toxins are the direct cause of the illness (an intoxication). Other pathogens do not act until swallowed. Then they cause an infection of the gastrointestinal tract.
3. Food poisoning caused by agents other than microorganisms. The offending food contains poisonous chemicals, or the food is a poisoned plant or animal. Examples would be tuna and mercury, and apples and alar.

Recent developments in diagnosing and tracking reported illnesses have helped the public become more aware that certain types of illness may be related to the food they ate prior to becoming sick.

FOOD-RELATED HAZARDS

Today, foodborne illness is of serious concern. Its frequency is not known because a great majority of cases go unreported. Reporting foodborne diseases to public health authorities is not required in

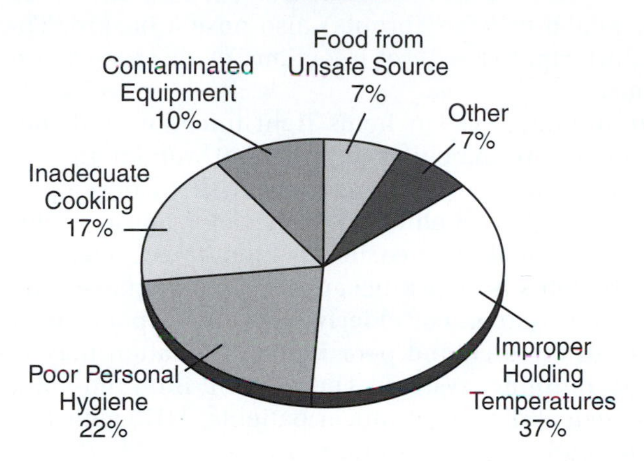

FIGURE 25-1
Sources of foodborne illness. (*Source:* U.S. Department of Health and Human Services, Public Services, 1996)

the United States. Estimates claim as many as 200 million cases in the United States per year. Only a small percentage of these are hospitalized. Most are passed off as traveler's diarrhea, 24-hour flu, or upset stomach.

Most foodborne illness can be avoided if food is handled properly. Statistics from the Centers for Disease Control show that the most commonly reported food preparation practice that contributed to foodborne disease was improper holding temperatures, followed by poor personal hygiene, inadequate cooking, contaminated equipment, and food from an unsafe source, as shown in Figure 25-1.

Cross-Contamination

Cross-contamination is the transportation of harmful substances to food by:

- Hands that touch raw foods, such as chicken, then touch food that will not be cooked, like salad ingredients.
- Surfaces, like cutting boards or cleaning cloths, that touch raw foods, are not cleaned and sanitized, then touch ready-to-eat food.
- Raw or contaminated foods that touch or drip fluids on cooked or ready-to-eat foods.

High-Risk Foods and Individuals

The U.S. Public Health Service classifies moist, high-protein, and/or low-acid foods as potentially hazardous. High-protein foods consist, in whole or in part, of milk or milk products, shell eggs, meats, poultry, fish, shellfish, edible crustacea (shrimp, lobster, crab). Baked or boiled potatoes, tofu and other soy protein foods, plant foods that have been heat-treated, and raw seed sprouts (such as alfalfa or bean sprouts) also pose a hazard. These foods can support rapid growth of infectious or disease-causing microorganisms.

The immune system helps fight infection, but the immune systems of very young children, pregnant women, the elderly, and chronically ill people are at greatest risk to develop foodborne infections. Infants and children in particular produce less acid in their stomachs, making it easier for them to get sick. For pregnant women, the fetus is at risk because it does not have a fully developed immune system. For elderly individuals, poor nutrition, lack of protein in the diet, and poor blood circulation may result in a weakened immune system. Those with immuno-compromised systems, such as diabetics, cancer patients, AIDS patients, and people on antibiotics, are at greater risk.

Microorganisms are everywhere. They hide on the body, in the air, on kitchen counters and utensils, and in food. The main microorganisms are viruses, parasites, fungi, and bacteria.

Table 25-1 describes some of the common foodborne diseases.

MICROORGANISMS

Safety of foods requires an understanding of the basics of food microbiology. Organisms important in food include:

- Bacteria
- Fungi (yeasts and molds)
- Viruses
- Parasites

Although not living organisms, viruses are generally included as biological agents of concern. These materials are combinations of proteins and nucleic acids that can take over cellular functions. Harmful effects of microorganisms include:

- Spoilage of foods
- Foodborne toxins
- Foodborne infections
- Viral borne infections

TABLE 25-1 Microorganisms in Food

Disease and Organism That Causes It	Source of Illness	Symptoms
Botulinum toxin (produced by *Clostridium botulinum* bacteria)	Spores of these bacteria are widespread. But these bacteria produce toxins only in an anaerobic (oxygenless) environment of little acidity. Found in a considerable variety of canned foods, such as corn, green beans, soups, beets, asparagus, mushrooms, tuna, and liver pate. Also in luncheon meats, ham sausage, stuffed eggplant, lobster, and smoked and salted fish.	Onset: Generally 4 to 36 hours after eating. Neurotoxic symptoms, including double vision, inability to swallow, speech difficulty, a progressive paralysis of the respiratory system. **Get medical help immediately. Botulism can be fatal.**
Campylobacteriosis *Campylobacter jejuni*	Bacteria on poultry, cattle, and sheep can contaminate meat and milk of these animals. Chief food sources: raw poultry, meat, and unpasteurized milk.	Onset: Generally 2 to 5 days after eating. Diarrhea, abdominal cramping, fever, and sometimes bloody stools. Lasts 7 to 10 days.
Listerosis *Listeria monocytogens*	Found in soft cheese, unpasteurized milk, imported seafood products, frozen cooked crabmeat, cooked shrimp, and cooked surimi (imitation shellfish). The Listeria bacteria resist heat, salt, nitrite, and acidity better than many other microorganisms. They survive and grow at low temperatures.	Onset: From 7 to 30 days after eating, but most symptoms have been reported 48 to 72 hours after consumption of contaminated food. Fever, headache, nausea, and vomiting. Primarily affects pregnant women and their fetuses, newborns, the elderly, people with cancer, and those with impaired immune systems. Can cause fetal and infant death.
Perfringen food poisoning *Clostridium perfringens*	In most instances, caused by failure to keep food hot. A few organisms are often present after cooking and multiply to toxic levels during cool down and storage of prepared foods. Meats and meat products are the foods most frequently implicated. These organisms grow better than other bacteria between 120° and 130°F (49° and 54°C). So gravies and stuffing must be kept above 140°F (60°C).	Onset: Generally 8 to 12 hours after eating. Abdominal pain and diarrhea, and sometimes nausea and vomiting. Symptoms last a day or less and are usually mild. Can be more serious in older or debilitated people.

(continued)

TABLE 25-1 Microorganisms in Food *(continued)*

Disease and Organism That Causes It	Source of Illness	Symptoms
Salmonellosis *Salmonella enteritis*	Raw meats, poultry, milk and other dairy products, shrimp, frog legs, yeast, coconut, pasta, and chocolate are most frequently involved.	Onset: Generally 6 to 48 hours after eating. Nausea, abdominal cramps, diarrhea, fever, and headache. All age groups are susceptible, but symptoms are most severe for the elderly, infants, and the infirm.
Shigellosis (bacillary dysentery) *Shigella* bacteria	Found in milk and dairy products, poultry, and potato salad. Food becomes contaminated when a human carrier does not wash hands and then handles liquid or moist food that is not cooked thoroughly afterwards. Organisms multiply in food left at room temperature.	Onset: 1 to 7 days after eating. Abdominal cramps, diarrhea, fever, sometimes vomiting, and blood, pus or mucus in stools.
Staphylococcal food poisoning Staphylococcal enterotoxin (produced by *Staphylococcus aureus* bacteria)	Toxins produced when food contaminated with the bacteria is left too long at room temperature. Meats, poultry, egg products, tuna, potato and macaroni salads, and cream-filled pastries are good environments for these bacteria to produce toxins.	Onset: Generally 30 minutes to 8 hours after eating. Diarrhea, vomiting, nausea, abdominal pain, cramps, and prostration. Lasts 24 to 48 hours. Rarely fatal.
Vibrio infection *Vibrio vulnificus*	The bacteria live in coastal waters and can infect humans either through open wounds or through consumption of contaminated seafood. The bacteria are most numerous in warm weather.	Onset: Abrupt. Chills, fever, and/or prostration. At high risk are people with liver conditions, low gastric (stomach) acid, and weakened immune systems.
Amebiasis *Entamoeba histolytics*	A protozoan existing in the intestinal tract of humans and are expelled in feces. Protozoan present in polluted water and vegetables grown in polluted soil spread the infection.	Onset: 3 to 10 days after exposure. Severe crampy pain, tenderness over the colon or liver, loose morning stools, recurrent diarrhea, loss of weight, fatigue, and sometimes anemia.

TABLE 25-1 Microorganisms in Food *(concluded)*

Disease and Organism That Causes It	Source of Illness	Symptoms
Giardiasis *Giardia lamblia*	Most frequently associated with consumption of contaminated water. May be transmitted by uncooked foods that become contaminated while growing or after cooking by infected food handlers. Cool, moist conditions favor organism's survival.	Onset: 1 to 3 days. Sudden onset of explosive watery stools, abdominal cramps, anorexia, nausea, and vomiting. Especially infects hikers, children, travelers, and institutionalized patients.
Hepatitis A virus	Mollusks (oysters, clams, mussels, scallops, and cockles) become carriers when their beds are polluted by untreated sewage. Raw shellfish are especially potent carriers, although cooking does not always kill the virus.	Onset: Begins with malaise, appetite loss, nausea, vomiting, and fever. After 3 to 10 days patient develops jaundice with darkened urine. Severe cases can cause liver damage and death.

Viruses

Viruses are the tiniest, and probably the simplest, form of life. They are not able to reproduce outside a living cell. Once they enter a cell, they force it to make more viruses. Some viruses are extremely resistant to heat and cold. They do not need potentially hazardous food to survive, and once in the food, they do not multiply. The food is mainly a transportation device to get from one host to another.

Parasites

Parasites need to live on or in a host to survive. Examples of parasites that may contaminate food are *Trichinella spiralis* (trichinosis), which affects pork, and *Anisakis roundworm*, which affects fish.

Fungi

Fungi can be microscopic or as big as a giant mushroom. Fungi are found in the air, soil, plants, animals, water, and some food. Molds and yeast are fungi.

Bacteria

Of all the microorganisms, bacteria are the greatest threat to food safety. Bacteria are single-celled, living organisms that can grow quickly at favorable temperatures. Some bacteria are useful. We use them to make foods like cheese, buttermilk, sauerkraut, pickles, and yogurt. Other bacteria are infectious disease-causing agents called pathogens, that use the nutrients found in potentially hazardous foods to multiply.

Some bacteria are not infectious on their own, but when they multiply in potentially hazardous food, they eject toxins that poison humans when the food is eaten.

Factors Affecting Microbial Growth

The effect of organisms on the safety of foods is dependent on the initial numbers of organisms present, processing to eliminate the organisms, control of the environment to prevent growth, and sanitation.

The major factors that influence the growth of microorganisms in food include:

- pH
- Oxygen availability
- Moisture availability
- Nutrient availability
- Storage temperature
- **Lag time**, **generation time**, and numbers

The food industry depends on minimizing microbial populations in the food and/or control of the environment.

pH. Generally, microorganisms, and especially pathogens, cannot grow at pH levels below 4.0. These are termed acid foods and depend upon the low pH to prevent or minimize growth. These include foods like fruit beverages and salad dressings.

Oxygen. Organisms can be classified as aerobic and anaerobic. Aerobic organisms require air (oxygen) for growth and will not grow in the absence of air. These include yeasts and molds and a number of bacteria. A common practice is to hot-fill foods and seal them, so that a slight vacuum is formed in the package. A number of foods will be labeled "Refrigerate after opening"–and this is because yeasts and molds can grow once the product is opened and contains air.

BACTERIUM OR VIRUS—WHAT'S THE DIFFERENCE?

A bacterium is a microscopic single-celled organism, and very different from a virus. The plural of bacterium is bacteria. Bacteria occur everywhere life exists. They possess a tough, rigid outer cell wall through which they absorb their food. Some bacteria have a slimy outer capsule, and some may have whiplike flagella to propel them through liquids. If flagella are positioned all around the bacterium, it is called peritrichous, but if flagella are at each end, it is called lophotrichous. Some bacteria may simply drift in air or water currents.

Bacteria generally reproduce by splitting into two. This is called binary fission, and it may occur once every 15 to 30 minutes. Under favorable conditions, one bacterium could form over 150 trillion bacteria in 24 hours! This usually does not happen. Bacteria are very numerous and very tough. A pinch of soil contains millions, and some bacteria can survive freezing, intense heat, drying, and some disinfectants. To survive adverse conditions, bacteria form spores that can remain active for years.

Bacteria can be classified by their shapes. Bacteria shaped like a sphere are called cocci. Those shaped like rods are called bacilli. Spirillum are in a spiral shape. Vibrio are comma shaped. Mycobacteria are very small rods. Flexibacter form long thin rods.

Many bacteria perform useful functions for humankind. For example, helpful bacteria include those responsible for decay, sewage treatment, cheese and yogurt production, and those responsible for the nitrification process. Some bacteria cause disease. These are called pathogenic. Bacterial infections can be treated with antibiotics or similar drugs, and some can be prevented by vaccination.

A virus is smaller and simpler than a bacterium. In fact, a virus is so small and simple that it is on the borderline between a living organism and an inanimate particle. A virus is so small that it cannot be seen with an ordinary light microscope, but requires the use of an electron microscope.

Viruses live and reproduce inside of other living cells. They depend on living cells to reproduce, but some can live for quite some time outside cells of the body and some can even survive freezing and drying. Because viruses live inside the cells of plants and animals, chemical treatment is often out of the question because this would kill the host cell. Some drugs relieve the symptoms that viruses produce, but the only effective way of controlling a viral infection is to remove the infected individual. The affected individual's own immune system must produce antibodies to counteract the infection. Vaccination provides a means of preventing viral infections.

To learn more about bacteria and viruses, visit these Web sites:

<www.bu.edu/cohis/infxns/common/diarrhea/diarrhea.htm>
<www.microbe.org/microbes/virus_or_bacterium.asp>
<www.cellsalive.com>

Anaerobic organisms only grow in the absence of air. One such organism is *Clostridium botulinum*, which produces a toxin that can be lethal if ingested. Where foods are packed under a vacuum, the protection against *C. botulinum* growth is to heat-treat the food (sterilization) to a time and temperature where any organisms present are destroyed.

Moisture Availability. Organisms need free moisture in order to grow. Drying foods removes the available moisture and prevents growth of bacteria, yeasts, and molds. Control of water activity is a common method of preventing the outgrowth of microorganisms. Water activity is a measure of free (unbound) water available for chemical and biological activity.

Aw is the vapor pressure of a food product at a specified temperature. Aw is a measure of the relative humidity of the food where:

Aw = 1.0 = 100 percent relative humidity

Aw = 0.0 = 0 percent relative humidity

Foods generally have water activities that range between 0.1 for dried foods to 0.96 for fluid foods.

Different organisms require different levels of free water in order to grow, with bacteria requiring more free water than yeasts and molds. A rule of thumb is that bacteria do not grow at an Aw less than 0.85 and that yeasts and molds do not grow at an Aw less than 0.65. But some exceptions exist.

Intermediate moisture foods depend upon the addition of large quantities of sugars to reduce the Aw to below 0.85 and on packaging to prevent the growth of yeasts and molds.

Nutrient Availability. Most foods contain adequate nutrients to support the growth of microorganisms, especially foods that contain both a fermentable carbohydrate and a protein source. Sugar is the most common carbon source. In starch-based foods, the action of amylases will frequently increase the available sugar source.

Storage Temperature. Organisms can be classified on the basis of their ability to grow at different temperatures:

- Psychrophiles–grow best at temperatures <50°F (<10°C)
- Mesophiles–grow best at ambient temperatures 77° to 95°F (25° to 35°C)
- Thermophiles–grow best at temperatures >104°F (>40°C)
- **Psychrotrophic**–tolerate low temperatures and can grow under refrigeration

- **Thermotrophic**–tolerate high temperatures and can grow at 131° to 140°F (55° to 60°C)

Most pathogenic bacteria are mesophilic. A few, such as *Listeria monocytogenes* or *C. botulinum* type E can grow under refrigeration conditions.

Organisms that spoil refrigerated products are generally psychrotrophic, and frequently belong to the genus *Pseudomonas.*

Lag Time, Generation Time, and Numbers. The amount of time required for an organism to reach the log-growth phase is termed the lag time, and the time required to double the population of the organisms is termed the generation time. The amount of time required for an organism to reach a specific number is dependent upon the initial population, the lag time, and the generation time. Under ideal growth conditions, microorganisms can double their number in about 30 minutes. As conditions move away from their optimum, the generation time is decreased–until eventually no growth occurs (Figure 25-2).

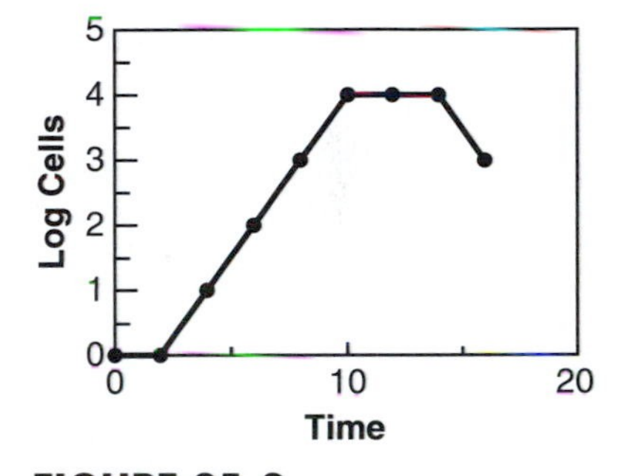

FIGURE 25-2

Logarithmic growth of bacteria over time.

In processed foods, organisms surviving have generally been stressed and commonly may exhibit a lag time of 2 to 4 days. In fermented foods, where cultures are added, the organisms are added in an active-growth phase, and the lag time is 0 or very small.

A rule of thumb for the numbers of organisms required to produce toxins or to produce desired or undesired flavors is one million per gram. For foodborne toxins such as caused by *C. botulinum, Staph. aureous* and *Bacillus cereus*), there has to have been a large number produced. Similarly, the numbers of organisms required for fermented foods (such as yogurt) is also very large.

In the case of organisms causing infectious disease (*E. coli* 0157, *Listeria monocytogenes*, and *Salmonella*), the numbers of organisms that can cause the disease can be very small. FDA generally requires that no organisms can be recovered from 100 g of food after suitable incubation.

Under ideal growth conditions it can take a relatively short period of time to increase from 1 to 1 million per gram (20 generations). At 4 generations/hour, the food can reach more than 1 million in just 5 hours. If the initial load is 1,000 per gram, then the time required to reach 1 million would be 2.5 hours (10 generations).

MICROBIOLOGICAL METHODOLOGY

Testing foods for the presence of pathogenic microorganisms is very important. Although 100 percent of the food cannot be tested, it can be deemed "safe" through proper audit of the food supply. In many instances, the pathogenic microorganisms are present in very small numbers, but for many of these pathogens, small numbers are all that are necessary to transmit disease or illness. For that reason, the presence of other microorganisms is monitored. These microorganisms provide an index of the sanitary quality of the product and may serve as an indicator of potential for the presence of pathogenic species. *Escherichia coli* (*E. coli*) is commonly employed as an indicator microorganism. Because *E. coli* is a coliform bacteria common to the intestinal tract of humans and animals, its relationship to intestinal foodborne pathogens is high.

Total counts of microorganisms are also an indication of the sanitary quality of a food. Referred to as the Standard Plate Count (**SPC**), this total count of viable microbes reflects the handling history, state of decomposition, or degree of freshness of the food. Total counts may be taken to indicate the type of sanitary control exercised in the production, transport, and storage of the food. Most foods have standards or limits for total counts. This is especially true for milk.

In adopting microbiological standards to milk, the first concern is product safety, followed by shelf life. The following bacterial counts are standards for milk as recommended by the U.S. Public Health Service:

- Grade A raw milk for pasteurization: Not to exceed 100,000 bacteria per milliliter (ml) prior to commingling with other produced milk, and not exceeding 300,000 per milliliter as commingled milk prior to pasteurization.
- Grade A pasteurized milk: Not over 20,000 bacteria per milliliter, and not over 10 coliforms per milliliter.

The objective of pasteurization is to reduce the total microbial load, or SPC. In addition, pasteurization must destroy all pathogens that may be carried in the milk from the cow, particularly undulant fever, tuberculosis, Q-fever, and other diseases transmittable to humans. This is accomplished by setting the time and temperature of the heat treatment so that certain heat-resistant pathogens, specifically *Mycobacterium tuberculosis* and *Coxiella burnetii* (causative agents of Q-fever and tuberculosis, respectively), would be destroyed if present. Milk pasteurization temperatures are sufficient to destroy all yeasts, mold, and many of the spoilage bacteria.

A low SPC does not always represent a safe product. It is possible to have a low count of SPC in foods in which toxin-producing organisms have grown. These organisms produce toxins that remain stable under conditions that may not favor the survival of the microbial cell. In adopting microbiological standards, the first concern is product safety, followed by shelf life.

Standard plating methods can take several days. New tests are being developed that will detect specific types of microorganisms in a matter of hours. These new tests are based on being able to detect a specific type of DNA from the microorganisms of interest.

PROCESSING AND HANDLING

Keeping microbial loads at minimal levels is essential to provide safe food of high quality. This requires care in food handling and minimizing microorganisms in the product during processing. Key to this goal is preventing contamination of the food during contact with equipment (or food contact) surface. Cleaning and sanitizing are important steps in the operation of any food plant–and will become more important in the future as the industry deals with new and emerging microorganisms, such as *E. coli* 0157 that first began to be associated with foodborne infections in the 1980s.

Microbial destruction is achieved either by heat or by chemicals. The most common method of making foods safe is to use thermal processing to eliminate pathogens and then to use good sanitation practices to prevent them from re-entering the food after thermal processing. Spores are more resistant than vegetative cells and thus require more heat to kill. Two general heat processes are used:

- Pasteurization–heating to a specific temperature for a specific time to kill the most heat-resistant vegetative pathogen.
- Sterilization–heat to a specific temperature for a specific time to kill the most heat-resistant spore-forming organism.

Neither process kills all the organisms in the food, and nonpathogenic, spoilage organisms can survive to some degree. Because pasteurization does not kill spores, pasteurized products are kept under refrigeration to control the growth of surviving spore formers that grow well at ambient temperature.

RODENTS, BIRDS, AND INSECTS

Rodents carry many diseases and parasites that can be transmitted to humans. These diseases and parasites include leptospirosis, salmonellosis, tapeworms, trichinosis and others. Also, rodents will deposit excreta, urine, and other filth on food products and around food facilities. They will also gnaw on materials in order to build nests. Rodents contaminate much more than they eat.

Some rodents can walk along telephone wires or leap horizontally 18 feet. They can squeeze through gaps the width of a pencil or drop 50 feet without being killed. Their instinct for survival is high. They are extremely prolific.

Birds also carry diseases and parasites potentially hazardous to people. They are capable of flying through any open window, door, or other gaps in a building, and, like rodents, will leave droppings that can contaminate the plant and food products.

Insects seek heat, moisture, and darkness, and can be more elusive than rodents or birds. They leave trails in the dust, and can also be spotted around likely insect hideouts like holes, damp places, behind boxes, and in seams in bags and folds of paper. Some insects like cockroaches have a highly developed survival instinct and they are adaptable. They can develop resistance to poisons within a few insect generations. Insects are even more prolific than rodents. With their hairy legs, they spread dirt, debris, and bacteria.

A reputable pest control company or exterminator is generally the most cost effective method to deal with a rodent, bird, or insect invasion. When professional pest control is used, the owner or manager of a food industry facility should frequently monitor the activities of the pest control representative, know what poisons are being used, and know the location of bait stations and traps. All owners and managers should know an all-purpose pesticide for the food industry does not exist.

Investing in building and grounds maintenance remains the best practice to solve pest problems. Extermination is a poor second choice for pest control and it will cost as much, or more. Following some guidelines will help prevent a rodent, bird, or insect problem at a food industry site, for example:

- Keep grounds surrounding buildings clear of weeds, grass, brush, and standing water
- Windows and doors sealed tightly
- Windows fitted with fine mesh screens
- Holes and cracks in building filled
- Building and equipment cleaned regularly
- No leaks in roof
- Areas of food buildup eliminated
- Eliminate empty space in or around equipment
- Trash, debris, and clutter picked up
- Regular inspection with a checklist
- Garbage covered
- Building humidity low
- Proper building temperature

Preventive and control measures avoid potentially costly, mandated adjustments that might arise when an FDA investigator visits. This also ensures that only quality, safe food products find their way to the consumers.

CLEANING AND SANITIZING

For a food-processing plant, cleaning and sanitizing may be the most important aspects of a sanitation program. Sufficient time should be given to outline proper procedures and parameters. Detailed procedures must be developed for all food-product contact surfaces (equipment, utensils, and so on) as well as for nonproduct surfaces such as nonproduct portions of equipment; overhead structures; shield; walls; ceilings; lighting devices; refrigeration unit; heating, ventilation, and air-conditioning (**HVAC**) systems; and anything else that could impact food safety.

Cleaning frequency must be clearly defined for each process line–for example, daily, after production runs, or more often if necessary. The type of cleaning required must also be identified.

The objective of cleaning and sanitizing food-contact surfaces is to remove food (nutrients) that bacteria need to grow, and to kill those bacteria that are present. The clean, sanitized equipment and surfaces must drain dry and be stored dry so as to prevent bacteria growth. Necessary equipment (brushes, and so on) must also be clean and stored in a clean, sanitary manner.

Cleaning/sanitizing procedures must be evaluated for adequacy through evaluation and inspection procedures. Adherence to prescribed written procedures (inspection, swab testing, direct observation of personnel) should be continuously monitored, and records maintained to evaluate long-term compliance.

The correct order of events for cleaning/sanitizing of food product contact surfaces is rinse, clean, rinse, sanitize.

Cleaning

Cleaning is the complete removal of **food soil** using appropriate detergent chemicals under recommended conditions. People employed in food production need to have a working understanding of the nature of the different types of food soil and the chemistry of its removal.

Equipment can be categorized with regard to the cleaning method as follows:

- Mechanical cleaning. Often referred to as clean-in-place (CIP). Requires no disassembly or partial disassembly.

- Clean-out-of-place (**COP**). Can be partially disassembled and cleaned in specialized COP pressure tanks.
- Manual cleaning. Requires total disassembly for cleaning and inspection.

Food soil is generally defined as unwanted matter on food-contact surfaces. Soil is visible or invisible. The primary source of soil is from the food product being handled. However, minerals from water residue and residues from cleaning compounds contribute to films left on surfaces. Microbiological **biofilms** also contribute to soil buildup on surfaces.

Because soils vary widely in composition, no one detergent is capable of removing all types. Many complex films contain combinations of food components, surface oil or dust, insoluble cleaner components, and insoluble hard-water salts. These films vary in their solubility properties depending upon such factors as heat effect, age, dryness, time, and so on. Soils may be classified as:

- Soluble in water (sugars, some starches, most salts)
- Soluble in acid (limestone and most mineral deposits)
- Soluble in alkali (protein, fat emulsions)
- Soluble in water, alkali, or acid

Sanitation

Appropriate and approved **sanitization** procedures are processes. The duration or time as well as the chemical conditions must be described. The official definition (Association of Official Analytical Chemists) of sanitizing for food product contact surfaces is a process that reduces the contamination level by 99.999 percent (5 logs) in 30 seconds.

Two general types of sanitization include thermal (heat) and chemical. Thermal sanitization involves the use of hot water or steam for a specified temperature and contact time. Chemical sanitization involves the use of an approved chemical sanitizer at a specified concentration and contact time. As with any heat treatment, the effectiveness of thermal sanitizing is dependent upon a number of factors, including initial contamination load, humidity, pH, temperature, and time.

The use of steam as a sanitizing process has limited application. It is generally expensive compared to alternatives, and it is difficult to regulate and monitor contact temperature and time. Further, the by-products of steam condensation can complicate cleaning operations.

Hot-water sanitizing–through immersion (small parts, knives, etc.), spray (dishwashers), or circulating systems–is commonly used. The time required is determined by the temperature of the water (Figure 25-3). Typical regulatory requirements (Food Code 1995) for use of hot water in dishwashing and utensil-sanitizing applications specify immersion for at least 30 seconds at 170°F (77°C) for manual operations; a final rinse temperature of 165°F (74°C) in single tank, single temperature machines and 180°F (82°C) for other machines.

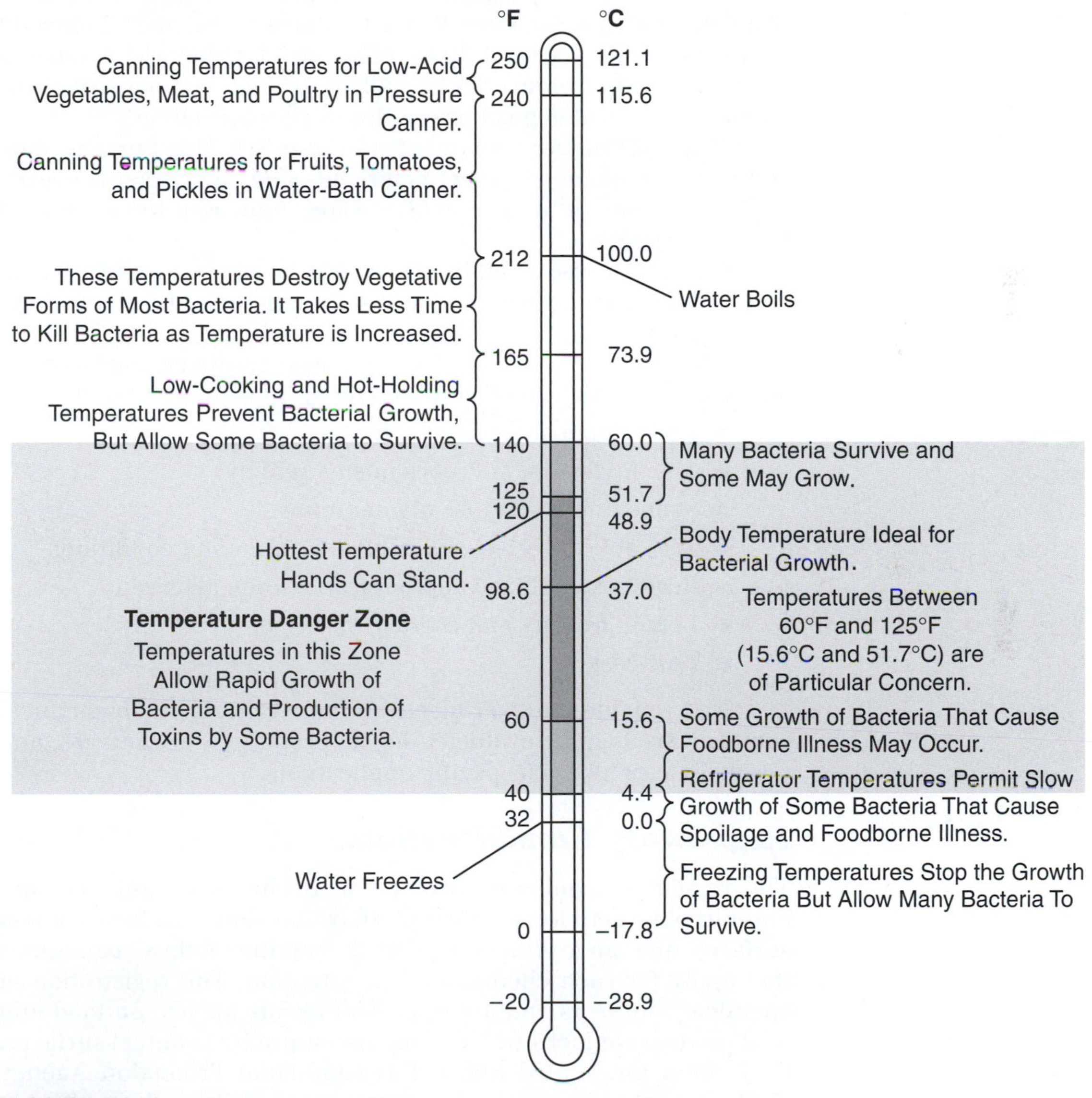

FIGURE 25-3

Relationships between temperature and microbial growth.

Many state regulations require a utensil surface temperature of 160°F (71°C) as measured by an irreversibly registering temperature indicator in utensil-washing machines.

Water comprises approximately 95 to 99 percent of cleaning and sanitizing solutions. Water functions to carry the detergent or the sanitizer to the surface and to carry soils or contamination from the surface.

The impurities in water can drastically alter the effectiveness of a detergent or a sanitizer. Water hardness is the most important chemical property with a direct effect on cleaning and sanitizing efficiency. (Other impurities can affect the food contact surface or may affect the soil deposit properties or film formation.)

Water pH ranges generally from 5.0 to 8.5. This range is of no serious consequence to most detergents and sanitizers. However, highly alkaline or highly acidic water may require additional buffering agents.

Water can also contain significant numbers of microorganisms. Water used for cleaning and sanitizing must be potable and pathogen-free.

Characteristics of the ideal chemical sanitizer approved for food contact surface application include:

- Wide range or scope of activity
- Able to destroy microorganisms rapidly
- Stable under all types of conditions
- Tolerant of a broad range of environmental conditions
- Readily solubilized and possesses some detergency
- Low in toxicity and corrosivity
- Low cost

No available sanitizer meets all these criteria. Each chemical sanitizer needs to be evaluated for its properties, advantages, and disadvantages for each specific application.

Regulatory Considerations

The regulatory concerns involved with chemical sanitizers are antimicrobial activity or efficacy, safety of residues on food-contact surfaces, and environmental safety. Users must follow regulations that apply for each chemical usage situation. The registration of chemical sanitizers and antimicrobial agents for use on food and food product contact surfaces, and on nonproduct contact surfaces, is through the United States Environmental Protection Agency (EPA). Prior to approval and registration, the EPA reviews efficacy and safety data, and product labeling information.

Safe Handling Instructions

This product was inspected for your safety. Some animal products may contain bacteria that could cause illness if the product is mishandled or cooked improperly. For your protection, follow these safe handling instructions.

Keep refrigerated or frozen.
Thaw in refrigerator or microwave.

Keep raw (meats or poultry) separate from other foods. Wash working surfaces (including cutting boards), utensils and hands after touching raw (meat or poultry).

Cook thoroughly.

Refrigerate leftovers within 2 hours.

FIGURE 25-4
Safe handling instructions on many meat and poultry products.

The United States Food and Drug Administration (FDA) is primarily involved in evaluating residues from sanitizer use that may enter the food supply. Thus, any antimicrobial agent and its maximum usage level for direct use on food or on food product contact surfaces must be approved by the FDA. Approved no-rinse food contact sanitizers and nonproduct contact sanitizers, their formulations, and usage levels are listed in the Code of Federal Regulations (21 CFR 178.1010). The United States Department of Agriculture (USDA) also maintains lists of antimicrobial compounds (for example, USDA List of Proprietary Substances and Non Food Product Contact Compounds) that are primarily used in the regulation of meats, poultry, and related products by USDA's Food Safety and Inspection Service (FSIS) (Figure 25-4).

HACCP AND FOOD SAFETY

The Food and Drug Administration is adapting a food safety program developed nearly 30 years ago for astronauts for much of the U.S. food supply. The program for the astronauts focuses on preventing hazards that could cause foodborne illnesses by applying science-based controls, from raw material to finished products. FDA's new system would do the same.

Traditionally, industry and regulators have depended on spot checks of manufacturing conditions and random sampling of final products to ensure safe food. This system, however, tends to be reactive rather than preventative and can be less efficient than the new system.

The new system is known as Hazard Analysis and Critical Control Point, or HACCP (pronounced hassip). Many of its principles already are in place in the FDA-regulated low-acid canned food industry and have been incorporated into FDA's Food Code. The Food Code serves as model legislation for state and territorial agencies that license and inspect food service, retail food stores, and food vending operations in the United States.

The United States Department of Agriculture also has established HACCP for the meat and poultry industry. Larger establishments were required to start using HACCP by January 26, 1998. Smaller companies had until January 25, 1999, and very small plants until January 25, 2000. (USDA regulates meat and poultry; FDA all other foods.)

The FDA is considering developing HACCP regulations as a standard throughout much of the rest of the U.S. food supply. The regulations would cover both domestic and imported foods.

To help determine the degree to which such regulations would be feasible, the agency is now conducting a pilot HACCP program with volunteer food companies that make cheese, frozen dough, breakfast cereals, salad dressing, or other products.

HACCP involves seven steps:

1. Analyze hazards. Potential hazards associated with a food and measures to control those hazards are identified. The hazard could be biological, such as a microbe; chemical, such as a pesticide; or physical, such as ground glass or metal fragments.
2. Identify critical control points. These are points in a food's production–from its raw state through processing and shipping to consumption by the consumer–at which the potential hazard can be controlled or eliminated. Examples are cooking, cooling, packaging, and metal detection.
3. Establish preventative measures with critical limits for each control point. For a cooked food, for example, this might include setting the minimum cooking temperature and time required to ensure the elimination of any microbes.
4. Establish procedures to monitor the critical control points. Such procedures might include determining how

and by whom cooking time and temperature should be monitored.

5. Establish corrective actions to be taken when monitoring shows that a critical limit has not been met–for example, reprocessing or disposing of food if the minimum cooking temperature is not met.

6. Establish procedures to verify that the system is working properly–for example, testing time-and-temperature recording devices to verify that a cooking unit is working properly.

7. Establish effective recordkeeping to document the HACCP system. This would include records of hazards and their control methods, the monitoring of safety requirements and action taken to correct potential problems.

Each of these steps would have to be backed by sound scientific knowledge–for example, published microbiological studies.

New challenges to the U.S. food supply have prompted FDA to consider adopting a HACCP-based food safety system. One of the most important challenges is the increasing number of new food pathogens. For example, between 1973 and 1988, bacteria not previously recognized as important causes of foodborne illness–such as *Escherichia coli* O157:H7 and *Salmonella enteritidis*–became more widespread (Figure 25-5).

Also, public health is increasingly concerned about chemical contamination of food–for example, the effects of lead on the nervous system.

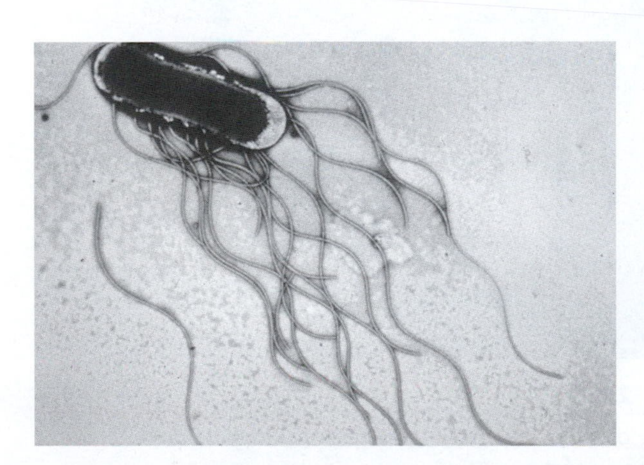

FIGURE 25-5

Microscopic view of the bacterium *Salmonella enteritis.* (*Source:* USDA Photography Library)

Another important factor is that the size of the food industry and the diversity of products and processes have grown tremendously–in the amount of domestic food manufactured and the number and kinds of foods imported. At the same time, FDA and state and local agencies have the same limited level of resources to ensure food safety.

HACCP offers a number of advantages. Some of the most prominent are the following:

- Focus on identifying and preventing hazards from contaminating food (Figures 25-6, 25-7, and 25-8)
- Based on sound science
- Permits more efficient and effective government oversight, primarily because the recordkeeping allows investigators to see how well a firm is complying with food safety laws over a period of time rather than how well it is doing on any given day
- Places responsibility for ensuring food safety appropriately on the food manufacturer or distributor
- Helps food companies compete more effectively in the world market

FIGURE 25-6

Cleaning around the cooking area—an important part of food safety.

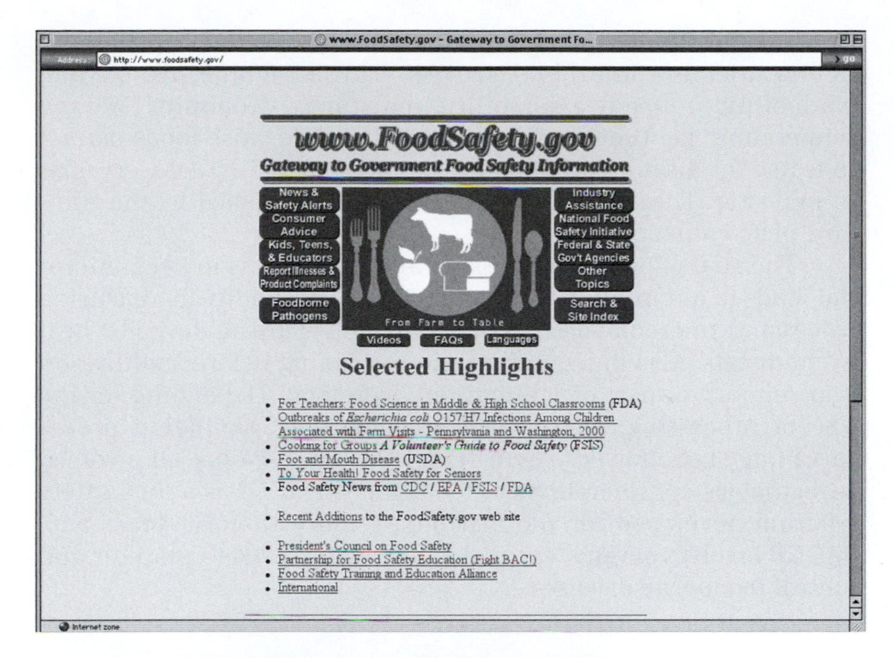

FIGURE 25-7

The U.S. government maintains a Web site promoting food safety.

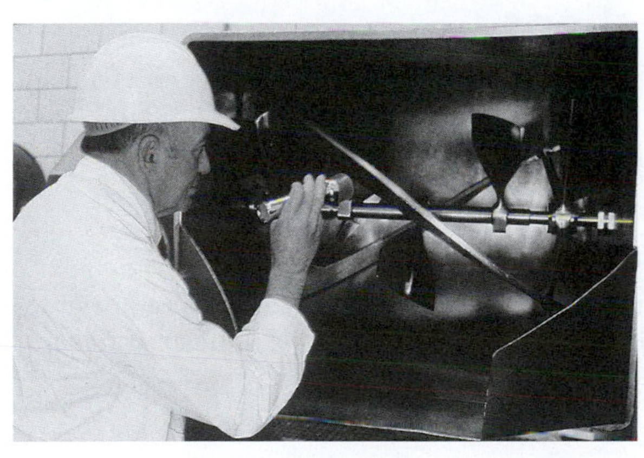

FIGURE 25-8

Food-processing equipment being inspected after cleaning. (*Source:* USDA Photography Library)

Summary

Food safety concerns include pesticides, additives, and spoilage, but most food scientists and consumers focus on microorganisms in food. These microorganisms cause foodborne illnesses that vary in severity. Many people call foodborne illnesses 24-hour flu, upset stomach, or diarrhea. However, for the very young and the old, these foodborne illnesses can be life-threatening. Microorganisms of concern include viruses, bacteria, parasites, and fungi.

To control microorganisms, food scientists must understand factors affecting microbial growth. These include pH, oxygen availability, nutrient availability, moisture availability, storage temperature, lag time, and generation time. Because foods cannot be tested for all microorganisms, the presence of *E. coli* serves as an indicator. The detection of *E. coli* is accomplished by the standard plate count.

The goal of proper processing and handling is to keep microbial loads to a minimum to provide safe, high-quality food. During processing, microbial destruction and control are achieved by heat or chemicals. Also during processing, cleaning before sanitization is an important part of maintaining safe food. Depending on the type of processing, appropriate and approved sanitation procedures must be followed. The USDA, FDA, and EPA are all involved in regulatory considerations of food safety. HACCP is a food safety program developed for the astronauts more than 30 years ago. HACCP involves seven steps and is now being used to monitor and control foodborne diseases.

Review Questions

Success in any career requires knowledge. Test your knowledge of this chapter by answering these questions or solving these problems.

1. List three types of food soils.
2. Briefly list the seven steps involved in HACCP.
3. List four foodborne diseases.
4. A rule of thumb for the numbers of organisms required to produce toxins or to produce desired or undesired flavors is ___________ per gram.
5. Name three ways cross-contamination occurs.
6. The four main microorganisms are ___________, which are the tiniest and simplest forms of life; ___________ which need a host to survive; ___________, which are molds or yeast; and ___________ which, when they multiply, can make food hazardous to eat.
7. Name the two general heat processes for microbial destruction.
8. Name the correct order of events for cleaning/sanitizing of food product contact surfaces.
9. The two types of sanitization include ___________, which involves the use of ___________ or ___________ for a

specified temperature and contact time, and ____________, which involves the use of an approved ____________ sanitizer at a specified concentration and contact time.

10. List the five most commonly reported food preparation practices from the Centers for Disease Control that contribute to foodborne diseases.

Student Activities

1. Choose one of the foodborne diseases in Table 25-1, page 455. Develop a short report or presentation.
2. Collect two current articles that relate to food safety issues. Summarize these and report.
3. Develop a report or presentation on commonly used chemical sanitizers.
4. Invite a guest speaker from an industry to discuss the use of the HACCP process.

Resources

Cremer, M. L. 1998. *Quality food in quantity. Management and science.* Berkeley, CA: McCutchan Publishing Corporation.

National Council for Agricultural Education. 1993. *Food science, safety, and nutrition.* Madison, WI: National FFA Foundation.

Potter, N. N., and J. H. Hotchkiss. 1995. *Food science,* 5th ed. New York: Chapman and Hall.

Seperich, G. J. 1998. *Food science and safety.* Danville, IL: Interstate Printers, Inc.

Vaclavik, V. A., and E. W. Christina. 1999. *Essentials of food science.* Gaithersburg, MD: Aspen Publishers, Inc.

Internet

Internet sites represent a vast resource of information. The URLs (uniform resource locator) for the World Wide Web sites can change. Using one of the search engines on the Internet such as Yahoo!, HotBot, AltaVista, Excite, Dogpile, About, or Google, find more information by searching for these words or phrases: HACCP, food toxins, sanitize, biofilms, psychrophiles, mesophiles, thermophiles, psychrotrophic, thermotrophic, food safety, food poisoning, food hazards, foodborne illness, cross-contamination, U.S. Public Health Service, food microorganisms, foodborne infections, viral borne infections, log-bacteria growth, pasteurization, sanitization, heat sterilization, Environmental Protection Agency (EPA), Food Safety and Inspection Service (FSIS), Food and Drug Administration (FDA). Also, Table A-7 provides a listing of some useful Internet sites that can be used as a starting point.

One of the best sites for food safety information is the Gateway to Government Food Safety Information at <www.foodsafety.gov>.

Chapter 26

Regulation and Labeling

Objectives

After reading this chapter, you should be able to:

- Identify the agencies and laws that regulate foods and labeling
- Describe the functions of a quality assurance department
- Discuss the history of food labels
- List five features of new labels
- Name two general categories of food exempt from food labels
- List six components found on the nutritional panel
- Describe the format of the nutritional panel
- Discuss the use of DRVs
- Identify when these words can be used: free, low, high, less, light, and more
- List two health claim relationships that can be listed on a food package

Key Terms

coliform
compliance
enhancers
Food, Drug, and Cosmetic Act
food labeling
hydrolysates
Infant Health Formula Act
Meat Inspection Act
nutrition labeling
Nutrition Labeling and Education Act
quality assurance (QA)
SPC

The Food and Drug Administration (FDA), the United States Department of Agriculture (USDA), and legislative acts, regulate foods and the labeling of foods and food laws to protect the consumer. Additionally, many states and cities have food laws. Food labeling provides basic information about the ingredients in, and the nutritional value of, food products so that consumers can make informed choices in the marketplace. Recent changes in food labeling provide the consumer with more complete, useful, and accurate information than ever before.

FEDERAL FOOD, DRUG, AND COSMETIC ACT

The U.S. Food and Drug Administration (FDA), operating under the federal **Food, Drug, and Cosmetic Act**, regulates the labeling for all foods other than meat and poultry (Figure 26-1). Meat and poultry products are regulated by the U.S. Department of Agriculture (USDA) under the federal **Meat Inspection Act**.

ADDITIONAL FOOD LAWS

Beside the Food, Drug, and Cosmetic Act, additional federal laws cover specific foods. The Federal Meat Inspection Act of 1906 provides for mandatory inspection of animals, slaughtering conditions, and meat-processing facilities. The Food Safety and Inspection Service (FSIS) of the USDA enforces the act.

The Federal Poultry Products Inspection Act of 1957 is like the Meat Inspection Act, but it obviously applies to poultry and poultry products.

To protect the public and the food industry against false advertising, the Federal Trade Commission Act was amended for food in 1938.

For infant health, the **Infant Health Formula Act** of 1980 provides that manufactured formulas contain the known essential nutrients at the correct levels.

The **Nutrition Labeling and Education Act** of 1990 protects consumers against partial truths, mixed messages, and fraud regarding nutrition information.

Federal grade standards have already been discussed in Chapters 17, 18, 19, and 20. These are standards of quality to help producers, dealers, wholesalers, retailers, and consumers in

FIGURE 26-1
FDA Web site provides information on legal guidelines for food and labels.

marketing and purchasing food. Standards and inspection come under the Agricultural Marketing Service of the Department of Agriculture (<www.ams.usda.gov/standards/index.htm>).

Besides these laws at the federal level, every state and many cities have food laws to further protect the consumer from unsanitary conditions and deception. These laws govern retail food outlets and eating establishments.

LEGAL CATEGORIES OF FOOD SUBSTANCES

In the United States substances that become a part of foods can be legally divided into several categories. Substances added to foods that have a history of being safe based on common usage in food are call "generally recognized as safe" or GRAS. Substances such as common spices, natural seasonings, many flavorings, baking powders, citric acid, malic acid, phosphoric acid, various gums, many emulsifiers, and numerous other substances are included in the list of GRAS.

Food additives are a very specific group of substances that are added intentionally and directly to foods. These are regulated and approved by the FDA. Scientific data must show that the additive is harmless in the intended food application and at the intended level of use. Food additives fall into one of the following categories:

- Preservatives
- Antioxidants
- Flavoring agents
- Sweeteners
- Emulsifiers, stabilizers, and thickeners
- Leavening agents
- Anticaking agents
- Humectants
- Coloring agents
- Bleaches
- Acids, bases, and buffers
- Nutrients

Pesticide residues may become a small part of a food product rather unintentionally through the application of the pesticide to the crop and a minute carryover to the food product. Most toxicologists do not believe that the small residues present a significant risk to human health. Law regulates pesticides. For additional information on additives, refer to Chapter 14.

TESTING FOR SAFETY

Food safety is a very broad topic, and it is covered in Chapter 25. Pesticides, herbicides, chemical additives, and spoilage are all of concern, but food scientists, food processors, and consumers focus most on microbiological quality. Microorganisms pose a challenge to the food industry, and most food processes are designed with microbial quality in mind. Microorganisms are often too small to be seen with the unaided eye and have the ability to reproduce rapidly. Many of them produce toxins and can cause infections. For all these reasons, the microbiological quality of the food we eat is scrutinized closely. *Escherichia coli* (*E. coli*) is commonly employed as an indicator microorganism. Because *E. coli* is a **coliform** bacterium common to the intestinal tract of humans and animals, its relationship to intestinal foodborne pathogens is high.

Total counts of microorganisms are also an indication of the sanitary quality of a food. Referred to as the Standard Plate Count (**SPC**), this total count of viable microbes reflects the handling history, state of decomposition, or degree of freshness of the food. Total counts may be taken to indicate the type of sanitary control exercised in the production, transport, and storage of the food. Most foods have standards or limits for total counts. A low SPC does not always represent a safe product. It is possible to have low-count foods in which toxin-producing organisms have grown. These organisms produce toxins that remain stable under conditions that may not favor the survival of the microbial cell.

QUALITY ASSURANCE

Of all functions in the food industry, **quality assurance** **(QA)** requires many diverse technical and analytical skills. QA personnel continually monitor incoming raw milk and finished milk products to ensure **compliance** with compositional standards, microbiological standards, and various government regulations. A QA manager can halt production, refuse acceptance of raw material, or stop the shipment if specifications for a product or process are not met. This department does not usually have control over the product unless something has gone wrong.

The major functions of the QA department include:

- Compliance with specifications–legal requirements, industry standards, internal company standards, shelf-life tests, customers' specifications.
- Test procedures–testing of raw materials, finished products, and in-process tests.
- Sampling schedules–use a suitable sampling schedule to maximize the probability of detection while minimizing workload.
- Records and reporting–maintain all QA records so that customer complaints and legal problems can be dealt with.
- Troubleshooting–solve various problems caused by poor-quality raw materials, erratic supplies and malfunctioning process equipment, and investigate reasons for poor-quality product to avoid repetition.
- Special problems–customer complaints, production problems, personnel training, short courses, and so on.

A typical QA department may have a chemistry lab, a raw materials inspection lab, a sensory lab, and a microbiology lab (Figure 26-2). All these disciplines work together to assure that the food we consume is of the highest quality. After all, it is quality that will bring a customer back again and again.

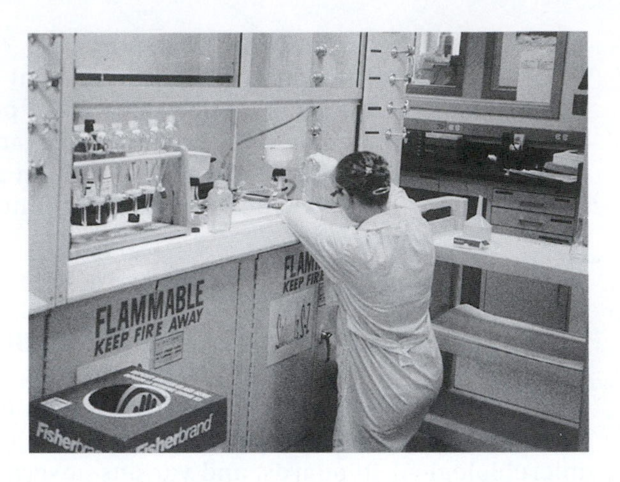

FIGURE 26-2
Modern laboratories and technicians monitor the safety of the food supply.

FOOD LABELING

Food labeling for most of the food products sold in the United States must have the product name, the manufacturer's name and address, the amount of the product in the package, and the product ingredients. The ingredients are listed in descending order, based on their weight. Under the current laws, fresh fruits, vegetables, and meat are exempt from these labeling requirements.

In 1973 the Food and Drug Administration (FDA) established "**nutrition labeling**" or guidelines for labeling the nutrient and caloric content of food products. Nutrition labeling is mandatory only for those foods that have nutrients added or make a nutritional claim. Manufacturers are encouraged, but not required, to provide nutrition labeling of other food products.

The current nutrition labeling regulations from the FDA require a label and have the percentage of the U.S. Recommended Daily Allowances (U.S. RDA). These standards are based on the 1968 edition of the Recommended Dietary Allowances (RDA), but the RDA and the U.S. RDA are not the same!

For each nutrient, the U.S. RDA is the highest RDA for any of the RDA age and sex groups. The U.S. RDA usually apply to people four years of age and older.

FDA nutrition labels must have the serving size; servings per container; calories per serving; grams of protein, carbohydrate, and fat per serving; and the percent of the U.S. RDA for protein, five vitamins, and two minerals.

There is less nutrition information on labels regulated by the USDA. USDA labels list only the serving size; servings per container; calories per serving; and grams of protein, carbohydrate, and fat per serving.

NEW FOOD LABELS

In the 1990 Food Marketing Institute (FMI) survey, over 70 percent of food shoppers identified taste, nutrition, and product safety as being very important factors in making food purchases. In the same survey, 36 percent of shoppers reported that they always read the ingredient and nutrition labels, and another 45 percent said that they sometimes read nutrition labels.

The growing importance of the role of nutrition in promoting health and preventing disease, and consumer demand for clearer and easier-to-understand information, led to the passage of the Nutrition Labeling and Education Act (NLEA) of 1990. Federal regulations, detailing the format and content of food labels, are now in effect.

THE NEW FOOD LABEL

Under new regulations from the Food and Drug Administration of the Department of Health and Human Services and the Food Safety and Inspection Service of the U.S. Department of Agriculture, the food label offers more complete, useful, and accurate nutrition information than ever before.

The purpose of the food label reform was to clear up confusion that has prevailed for years, to help consumers choose more healthful diets, and to offer an incentive to food companies to improve the nutritional qualities of their products. Key features of the new label include:

- Nutrition labeling for almost all foods
- Distinctive, easy-to-read format
- Information on the amount per serving of saturated fat, cholesterol, dietary fiber, and other nutrients
- Nutrient reference values, expressed as Percentage Daily Values
- Uniform definitions for terms that describe a food's nutrient content like light, low-fat, and high-fiber
- Claims about the relationship between a nutrient or food and a disease or health-related condition
- Standardized serving sizes
- Declaration of total percentage of juice in juice drinks
- Voluntary nutrition information for many raw foods

Foods Affected

The regulations, most of which went into effect in 1994, call for nutrition labeling for most foods. In addition, they set up voluntary programs for nutrition information for many raw foods: the 20 most frequently eaten raw fruits, vegetables, and fish each, under FDA's voluntary point-of-purchase nutrition information program, and the 45 best-selling cuts of meat.

Exemptions

Some foods are exempt from nutrition labeling:

- Food served for immediate consumption, such as that served in hospital cafeterias and airplanes, and that sold by food service vendors, such as mall cookie counters, sidewalk vendors, and vending machines
- Ready-to-eat food that is not for immediate consumption but is prepared primarily on-site–for example, bakery, deli, and candy store items
- Food shipped in bulk, as long as it is not for sale in that form to consumers
- Medical foods, such as those used to address the nutritional needs of patients with certain diseases
- Plain coffee and tea, some spices, and other foods that contain no significant amounts of any nutrients
- Food produced by small businesses–based on the number of people a company employs

Although these foods are exempt, they are free to carry nutrition information, when appropriate–as long as it complies with the new regulations. Also, they will lose their exemption if their labels carry a nutrient content or health claim or any other nutrition information.

Nutrition information about game meats–such as deer, bison, rabbit, quail, wild turkey, and ostrich–is not required on individual packages. Instead, it can be given on counter cards, signs, or other point-of-purchase materials. Because few nutrient data exist for these foods, the FDA believes that allowing this option will enable game meat producers to give first priority to collecting appropriate data and make it easier for them to update the information as it becomes available.

Nutrition Panel Title

The new food label features a revamped nutrition panel (Figure 26-3). It has a new title, "Nutrition Facts," which replaces "Nutrition

INGREDIENTS

Water, Flour, Beef (beef, beef broth and salt), **Vegetable Oil, Tomato Paste, Seasoning Blend** (salt, spices),**Seasoning Blend** (spices, salt, powdered onion, granulated garlic), **Hydroxipropyl Methyl-Cellulose** (a stabilizer), **Parsley, Red Pepper.**

Prepared & Packaged by Food Company, Anytown, USA, 55555

For questions or comments call:
1 800 555-1234
1 800 555-7685

Nutrition Facts

Serving Size 1 (126 g)
Servings Per Container 15

Amount per Serving

Calories 240 Calories from Fat 100

	% Daily Value*
Total Fat 12g	**18%**
Saturated Fat 2g	**9%**
Cholesterol 15mg	**5%**
Sodium 710mg	**30%**
Total Carbohydrate 23g	**8%**
Dietary Fiber 3g	**13%**
Sugars 0g	
Protein 11g	**22%**

Vitamin A 8% • Vitamin C 4%
Calcium 4% • Iron 15%

*Percent Daily Values are based on 2,000 calorie diet. Your daily values may be higher or lower based on your calorie needs:

		Calories: 2,000	2,500
Total Fat	Less than	65 g	80 g
Sat Fat	Less than	20 g	25 g
Cholesterol	Less than	300 mg	300 mg
Sodium	Less than	2,400 mg	2,400 mg
Total Carbohydrates		300 g	375 g
Dietary Fiber		25 g	30 g

Calories per gram:
Fat 9 • Carbohydrates 4 • Protein 4

FIGURE 26-3
The nutrition panel provides a variety of information to the consumer.

Information Per Serving." The new title signals that the product has been labeled according to the new regulations. Also, for the first time, there are requirements on type size, style, spacing, and contrast to ensure a more distinctive, easy-to-read label.

Serving Sizes

The serving size remains the basis for reporting each food's nutrient content. However, unlike in the past, when the serving size was up to the discretion of the food manufacturer, serving sizes now are more uniform and reflect the amounts people actually eat. They also must be expressed in both common household and metric measures.

The FDA allows as common household measures the cup, tablespoon, teaspoon, piece, slice, fraction (such as "¼ pizza"), and common household containers used to package food products (such as a jar or tray). Ounces may be used, but only if a common household unit is not applicable and an appropriate visual unit is given–for example, 1 oz. (28 g/about ½ pickle). Grams (g) and milliliters (ml) are the metric units that are used in serving size statements. NLEA defines serving size as the amount of food customarily eaten at one time. The serving sizes that appear on food labels are based on FDA-established lists of "Reference Amounts Customarily Consumed Per Eating Occasion."

The serving size of products that come in discrete units, such as cookies, candy bars, and sliced products, is the number of whole units that most closely approximates the reference amount. Cookies are an example. Under the "bakery products" category, cookies have a reference amount of 30 g. The household measure closest to that amount is the number of cookies that comes closest to weighing 30 g. Thus, the serving size on the label of a package of cookies in which each cookie weighs 13 g would read "2 cookies (26 g)."

Nutrition Information

Dietary components on the nutrition panel include the mandatory (boldface) components and the voluntary components. The following is the order in which they must appear on the label:

- Total calories
- Calories from fat
- Calories from saturated fat
- Total fat
- Saturated fat
- Polyunsaturated fat
- Monounsaturated fat
- Cholesterol
- Sodium
- Potassium
- Total carbohydrate
- Dietary fiber
- Soluble fiber

- Insoluble fiber
- Sugars
- Sugar alcohol (sugar substitutes xylitol, mannitol, and sorbitol)
- Other carbohydrates (the difference between total carbohydrate and the sum of dietary fiber, sugars, and sugar alcohol if declared)
- Protein
- Vitamin A
- Percent of vitamin A present as beta-carotene
- Vitamin C
- Calcium
- Iron
- Other essential vitamins and minerals

If a claim is made about any of the optional components, or if a food is fortified or enriched with any optional component, the nutritional information for these components becomes mandatory.

These mandatory and voluntary components are the only ones allowed on the nutrition panel. The listing of single amino acids, maltodextrin, calories from polyunsaturated fat, and calories from carbohydrates, for example, may not appear as part of the Nutrition Facts on the label.

The required nutrients were selected because they address today's health concerns. The order in which they must appear reflects the priority of current dietary recommendations. Thiamin, riboflavin, and niacin are no longer required in nutrition labeling because deficiencies of each are no longer considered of public health significance. However, they may be listed voluntarily.

Nutrition Panel Format

The format for declaring nutrient content per serving also has been revised. Now, all nutrients must be declared as percentages of the Daily Values–the new label reference values (Figure 26-4). The amount, in grams or milligrams, of macronutrients (such as fat, cholesterol, sodium, carbohydrates, and protein) still must be listed to the immediate right of each of the names of each of these nutrients. But, for the first time, a column headed " Percent Daily Value" appears.

Requiring nutrients to be declared as a percentage of the Daily Values is intended to prevent misinterpretations that arise with quantitative values. For example, a food with 140 milligrams

Nutrition Facts
Serving Size 1/2 cup (114g)
Servings Per Container 4

Amount per Serving
Calories 90 Calories from Fat 30

	% Daily Value*
Total Fat 3g	**5%**
Saturated Fat 0g	**0%**
Cholesterol 0mg	**0%**
Sodium 300mg	**13%**
Total Carbohydrate 13g	**4%**
Dietary Fiber 3g	**12%**
Sugars 3g	
Protein 3g	

Vitamin A 80% • Vitamin C 60%
Calcium 4% • Iron 4%

*Percent Daily Values are based on a 2,000 calorie diet. Your daily values may be higher or lower based on your calorie needs:

	Calories:	2,000	2,500
Total Fat	Less than	65 g	80 g
Sat Fat	Less than	20 g	25 g
Cholesterol	Less than	300 mg	300 mg
Sodium	Less than	2,400 mg	2,400 mg
Total Carbohydrates		300 g	375 g
Dietary Fiber		25 g	30 g

Calories per gram:
Fat 9 • Carbohydrates 4 • Protein 4

FIGURE 26-4
The ideal nutrition label. (*Source:* FDA)

(mg) of sodium could be mistaken for a high-sodium food because 140 is a relatively large number. Actually, that amount represents less than 6 percent of the Daily Value for sodium, which is 2,400 mg.

On the other hand, a food with 5 g of saturated fat could be construed as being low in that nutrient. In fact, that food would provide 25 percent of the total Daily Value because 20 g is the Daily Value for saturated fat based on a 2,000-calorie diet.

Nutrition Panel Footnote. The Percent (%) Daily Value listing carries a footnote saying that the percentages are based on a 2,000-

SNAKE OIL

White Eagle Rattle Snake Oil was sold as a cure-all for any kind of pain. Of course it did not, and today the term snake oil is used to describe a liquid concoction of questionable medical value sold as an all-purpose curative, especially by traveling hucksters. Snake oil also is used to describe any type of questionable cure-all. The term "nostrum" is also used to describe a medicine sold with false or exaggerated claims—quack medicine.

Nostrums permeated American society by the late nineteenth century. These products appealed to exotica, the medical knowledge of Native Americans, death, religion, patriotism, mythology, and especially new developments in science. There was nothing to stop patent medicine makers from claiming anything and putting anything in their products. For example, Lydia E. Pinkham's Vegetable Compound was first marketed in 1875 as the "female complaint" nostrum. It was widely advertised in the backs of newspapers and women's magazines. After many years and successful sales, people discovered that Lydia E. Pinkham's Vegetable Compound contained 15 to 20 percent alcohol! This is typical of many of the early nostrums. They contained varying levels of alcohol and sometimes morphine.

The rise of advertising in America paralleled the rise of nostrums. At the same time, the biomedical sciences in this country were still in their infancy, and medicine was ill equipped to deal with most diseases. Enterprising individuals were prepared to step in and alleviate the suffering with such products as:

- Ayer's Sarsaparilla: a "blood purifier"
- Dr. Morse's Indian Root Pills: cured everything from tapeworm to skin eruptions

calorie diet. Some nutrition labels–at least those on larger packages–have these additional footnotes:

- A sentence noting that a person's individual nutrient goals are based on his or her calorie needs
- Lists of the daily values for selected nutrients for a 2,000- and a 2,500-calorie diet

An optional footnote for packages of any size is the number of calories per gram of fat (9), and carbohydrate and protein (4).

Format Modifications. In limited circumstances, variations in the format of the nutrition panel are allowed. Some are mandatory.

- Pond's Extract: relieved pain of every kind inside and outside the body
- Rumford Yeast Powder: restored the phosphorus lost by using flour
- Fat Off: a cream that cured obesity
- Dr. Bonker's Egyptian Oil: cured pains and ills in adults, children, and horses
- Dr. Lindley's Epilepsy Remedy: cured epilepsy, fits, spasms, and convulsions
- Eckman's Alterative: cured all throat and lung diseases including tuberculosis
- Rite Wate Vegetable Compound: reduced body fat
- Liquozone: killed disease germs (first bottle free)
- Kickapoo Sagwa: claimed to be a "renovator"
- Anti-Morbific Liver and Kidney Medicine: prevented all diseases
- Dr. Shreves' Anti-Gallstone: treated gallstones and kidney stones
- Mixer's Cancer and Scrofula Syrup: cured cancer, piles, ulcers, tumors, abscesses, and all blood diseases

As ridiculous as these claims seem, some similar products are still sold today. Many people appear quite gullible and look for an easy solution to complex problems.

For more information about nostrums of the past, visit these Web sites or search the FDA Web site for "snake oil."

<www.fda.gov/cder/>
<www.fda.gov/cder/about/history/gallery/Gallery1.htm>
<www.mc.vanderbilt.edu/biolib/hc/nostrums/>
<www.healthcentral.com/drdean/drdean.cfm>
<www.fda.gov/>

For example, the labels of foods for children under 2 (except infant formula, which has special labeling rules under the Infant Formula Act of 1980) may not carry information about saturated fat, polyunsaturated fat, monounsaturated fat, cholesterol, calories from fat, or calories from saturated fat.

The reason is to prevent parents from wrongly assuming that infants and toddlers should restrict their fat intake, when, in fact, they should not. Fat is important during these years to ensure adequate growth and development.

Some format exceptions exist for small and medium-size packages. Packages with less than 12 square inches of available labeling space (about the size of a package of chewing gum) do not have to carry nutrition information unless a nutrient content or health claim is made for the product. However, they must provide an address or telephone number for consumers to obtain the required nutrition information.

Daily Values–DRVs

The new label reference value, Daily Value, comprises two sets of dietary standards: Daily Reference Values (DRVs) and Reference Daily Intakes (RDIs). Only the Daily Value term appears on the label, though, to make label reading less confusing (Figure 26-5).

DRVs have been established for macronutrients that are sources of energy: fat, carbohydrate (including fiber), and protein; and for cholesterol, sodium, and potassium, which do not contribute calories.

DRVs for the energy-producing nutrients are based on the number of calories consumed per day. A daily intake of 2,000 calories has been established as the reference. This level was chosen, in part, because it approximates the caloric requirements for postmenopausal women. This group has the highest risk for excessive intake of calories and fat. DRVs for the energy-producing nutrients are calculated as follows:

- Fat based on 30 percent of calories
- Saturated fat based on 10 percent of calories
- Carbohydrate based on 60 percent of calories
- Protein based on 10 percent of calories (The DRV for protein applies only to adults and children over 4.)
- Fiber based on 11.5 g of fiber per 1,000 calories

Because of current public health recommendations, DRVs for some nutrients represent the uppermost limit that is considered desirable. The DRVs for fats and sodium are:

Reference Values for Nutrition Labeling[1] (Based on a 2,000 Calorie Intake; for Adults and Children 4 or More Years of Age)		
Nutrient[2]	**Unit of Measure**	**Daily Values**
Total fat	grams (g)	65
Saturated fatty acids	grams (g)	20
Cholesterol	milligrams (mg)	300
Sodium	milligrams (mg)	2,400
Potassium	milligrams (mg)	3,500
Total carbohydrate	grams (g)	300
Fiber	grams (g)	25
Protein	grams (g)	50
Vitamin A	International Unit (IU)	5,000
Vitamin C	milligrams (mg)	60
Calcium	milligrams (mg)	1,000
Iron	milligrams (mg)	18
Vitamin D	International Unit (IU)	400
Vitamin E	International Unit (IU)	30
Vitamin K	micrograms (µg)	80
Thiamin	milligrams (mg)	1.5
Riboflavin	milligrams (mg)	1.7
Niacin	milligrams (mg)	20
Vitamin B_6	milligrams (mg)	2.0
Folate	micrograms (µg)	400
Vitamin B_{12}	micrograms (µg)	6.0
Biotin	micrograms (µg)	300
Pantothenic acid	milligrams (mg)	10
Phosphorus	milligrams (mg)	1,000
Iodine	micrograms (µg)	150
Magnesium	milligrams (mg)	400
Zinc	milligrams (mg)	15
Selenium	micrograms (µg)	70
Copper	milligrams (mg)	2.0
Manganese	milligrams (mg)	2.0
Chromium	micrograms (µg)	120
Molybdenum	micrograms (µg)	75
Chloride	milligrams (mg)	3,400

[1] *Source:* U.S. Food and Drug Administration (FDA), Center for Food Safety and Applied Nutrition. Revised January 30, 1998.

[2] Nutrients in this table are listed in the order in which they are required to appear on a label. This list includes only those nutrients for which a Daily Reference Value (DRV) has been established or a Reference Daily Intake (RDI).

FIGURE 26-5

Table of Reference Values for nutrition labeling. (*Source:* FDA)

- Total fat: less than 65 g
- Saturated fat: less than 20 g
- Cholesterol: less than 300 mg
- Sodium: less than 2,400 mg

Nutrient Content Descriptions

The regulations also spell out what terms may be used to describe the level of a nutrient in a food and how they can be used. These include free, low, lean and extra lean, high, good source, reduced, less, light, and more.

Free. Free means that a product contains no amount of, or only trivial or "physiologically inconsequential" amounts of, one or more of these components: fat, saturated fat, cholesterol, sodium, sugars, and calories. Synonyms for "free" include "without," "no" and "zero."

Low. The term "low" can be used on foods that can be eaten frequently without exceeding dietary guidelines for one or more of these components: fat, saturated fat, cholesterol, sodium, and calories.

low-fat: 3 g or less per serving

low-saturated fat: 1 g or less per serving

low-sodium: 140 mg or less per serving

very low sodium: 35 mg or less per serving

low-cholesterol: 20 mg or less and 2 g or less of saturated fat per serving

low-calorie: 40 calories or less per serving

Synonyms for low include "little," "few," and "low source of."

Lean and Extra Lean. Lean and extra lean can be used to describe the fat content of meat, poultry, seafood, and game meats.

lean: less than 10 g fat, 4.5 g or less saturated fat, and less than 95 mg cholesterol per serving and per 100 g

extra lean: less than 5 g fat, less than 2 g saturated fat, and less than 95 mg cholesterol per serving and per 100 g

High. High can be used if the food contains 20 percent or more of the Daily Value for a particular nutrient in a serving.

Good Source. Good source means that one serving of a food contains 10 to 19 percent of the Daily Value for a particular nutrient.

Reduced. Reduced means that a nutritionally altered product contains at least 25 percent less of a nutrient or of calories than the regular, or reference, product. However, a reduced claim cannot be made on a product if its reference food already meets the requirement for a "low" claim.

Less. Less means that a food, whether altered or not, contains 25 percent less of a nutrient or of calories than the reference food. For example, pretzels that have 25 percent less fat than potato chips could carry a "less" claim. "Fewer" is an acceptable synonym.

Light. Light can mean two things. First, a nutritionally altered product contains one-third fewer calories or half the fat of the reference food. If the food derives 50 percent or more of its calories from fat, the reduction must be 50 percent of the fat. Second, the sodium content of a low-calorie, low-fat food has been reduced by 50 percent. In addition, "light in sodium" may be used on food in which the sodium content has been reduced by at least 50 percent.

The term "light" still can be used to describe such properties as texture and color, as long as the label explains the intent–for example, "light brown sugar" and "light and fluffy."

More. More means that a serving of food, whether altered or not, contains a nutrient that is at least 10 percent of the Daily Value more than the reference food (Figure 26-6). The 10 percent of Daily Value also applies to "fortified," "enriched," and "added" claims, but in those cases, the food must be altered.

Country Home

Old Fashioned
Enriched White Bread

Ingredients Wheat Flour, (Enriched with Niacin, Iron, Thiamine Mononitrate, Riboflavin, and Folic Acid), Water, Sugar, Salt, Potato Flour, All Vegetable Shortening (Partially Hydrogenated Soybean and Cottonseed Oils), Yeast, Gluten, Cultured Whey, Vinegar, Lecithin.

Net Wt 1.5 Lb (680 g)

Nutrition Facts
Serving Size 1 slice (40g)
Servings Per Container 17

Amount per Serving
Calories 100 Calories from Fat 10

	% Daily Valve*
Total Fat 1g	2%
Saturated Fat 0g	0%
Cholesterol 0mg	0%
Sodium 230mg	10%
Total Carbohydrate 21g	7%
Dietary Fiber 0g	0%
Sugars 3g	
Protein 2g	

Vitamin A 0% • Vitamin C 0%
Calcium 0% • Iron 4%

*Percent Daily Values are based on a 2,000 calorie diet. Your daily values may be higher or lower based on your calorie needs:

		Calories: 2,000	2,500
Total Fat	Less than	65g	80g
Sat Fat	Less than	20g	25g
Cholesterol	Less than	300mg	300mg
Sodium	Less than	2,400mg	2,400mg
Total Carbohydrates		300g	375g
Dietary Fiber		25g	30g

Calories per gram:
Fat 9 • Carbohydrates 4 • Protein 4

The Bread Shop Inc.

FIGURE 26-6
Label from enriched or fortified products.

Other Definitions

The label regulations also address other claims such as percent fat free, implied, healthy, and fresh.

- Percent fat free: A product bearing this claim must be a low-fat or a fat-free product. In addition, the claim must accurately reflect the amount of fat present in 100 grams of the food. Thus, if a food contains 2.5 grams fat per 50 grams, the claim must be "95 percent fat free."
- Implied: These types of claims are prohibited when they wrongfully imply that a food contains or does not contain a meaningful level of a nutrient. For example, a product claiming to be made with an ingredient known to be a source of fiber (such as "made with oat bran") is not allowed unless the product contains enough of that ingredient (for example, oat bran) to meet the definition for "good source" of fiber.

Healthy. A "healthy" food must be low in fat and saturated fat and contain limited amounts of cholesterol and sodium. In addition, if it is a single-item food, it must provide at least 10 percent of one or more of vitamins A or C, iron, calcium, protein, or fiber. If it is a meal-type product, such as frozen entrees and multicourse frozen dinners, it must provide 10 percent of two or three of these vitamins or minerals or of protein or fiber, in addition to meeting the other criteria.

Fresh. Although not mandated by NLEA, the FDA issued a regulation for the term "fresh." The regulation was issued because of concern over the term's possible misuse on some food labels.

The regulation defines the term "fresh" when it is used to suggest that a food is raw or unprocessed. "Fresh" can be used only on a food that is raw, has never been frozen or heated, and contains no preservatives, but irradiation at low levels is allowed. "Fresh frozen," "frozen fresh," and "freshly frozen" can be used for foods that are quickly frozen while still fresh. Blanching (brief scalding before freezing to prevent nutrient breakdown) is allowed. Other uses of the term "fresh," such as in "fresh milk" or "freshly baked bread," are not affected.

USDA's Meat Grading Program

USDA has quality grades for beef, veal, lamb, yearling mutton, and mutton. It also has yield grades for beef, pork, and lamb. Although there are USDA quality grades for pork, these do not carry through to the retail level as do the grades for other kinds of meat.

USDA meat grades are based on nationally uniform Federal standards of quality. They are applied by experienced USDA graders, who are routinely checked by supervisors who travel throughout the country to make sure that all graders are interpreting and applying the standards in a uniform manner. A USDA Choice rib roast, for example, must have met the same grade criteria no matter where or when you buy it.

When meat is graded, a shield-shaped purple mark is stamped on the carcass. With today's close trimming at the retail level, however, you may not see the USDA grade shield on meat cuts at the store. Instead, retailers put stickers with the USDA grade shield on individual packages of meat. In addition, grade shields and inspection legends may appear on bags containing larger wholesale cuts.

Health Claims

Claims for eight relationships between a nutrient or a food and the risk of a disease or health-related condition are now allowed (Figure 26-7). They can be made in several ways: through third-party references, such as the National Cancer Institute; statements; symbols, such as a heart; and vignettes or descriptions. Whatever the case, the claim must meet the requirements for authorized health claims; for example, they cannot state the degree of risk reduction and can only use "may" or "might" in discussing the nutrient or food-disease relationship. And they must state that other factors play a role in that disease.

The claims also must be phrased so that consumers can understand the relationship between the nutrient and the disease and the nutrient's importance in relationship to a daily diet. An example of an appropriate claim is: "While many factors affect

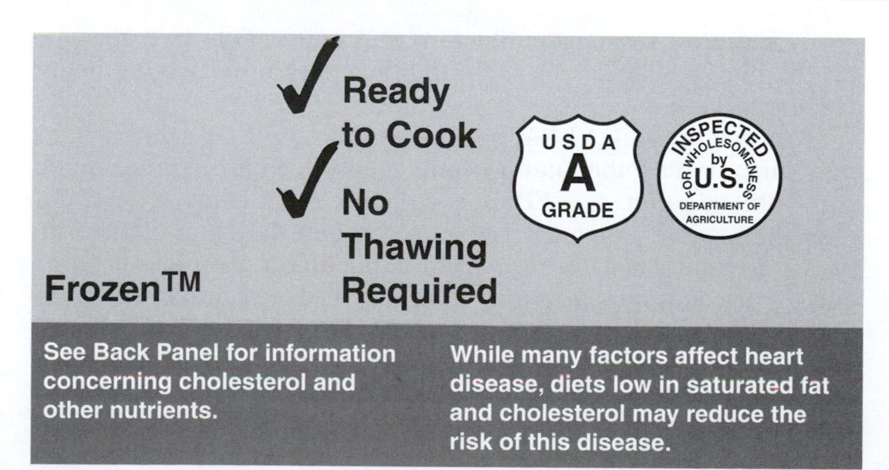

FIGURE 26-7
Label making a health claim.

heart disease, diets low in saturated fat and cholesterol may reduce the risk of this disease."

The allowed nutrient-disease relationship claims include:

- Calcium and osteoporosis
- Fat and cancer
- Saturated fat and cholesterol and coronary heart disease (CHD)
- Fiber-containing grain products, fruits, and vegetables and cancer
- Fruits, vegetables, and grain products that contain fiber and risk of CHD
- Sodium and hypertension (high blood pressure)
- Fruits and vegetables and cancer
- Folic acid and neural tube defects

Ingredient Labeling

The list of ingredients has undergone some changes, too. Chief among them is a requirement for full ingredient labeling on "standardized foods," which previously were exempt. Ingredient declaration is now required on all foods that have more than one ingredient. Also, the ingredient list includes, when appropriate:

- FDA-certified color additives, such as FD&C Blue No. 1, by name
- Sources of protein **hydrolysates**, which are used in many foods as flavors and flavor **enhancers**
- Declaration of caseinate as a milk derivative in the ingredient list of foods that claim to be nondairy, such as coffee whiteners

The main reason for these new requirements is that some people may be allergic to such additives and now may be better able to avoid them.

As required by NLEA, beverages that claim to contain juice now must declare the total percentage of juice on the information panel. In addition, the FDA's regulation establishes criteria for naming juice beverages. For example, when the label of a multi-juice beverage states one or more–but not all–of the juices present, and the predominantly named juice is present in minor amounts, the product's name must state that the beverage is flavored with that juice or declare the amount of the juice in a 5 percent range–for example, "raspberry-flavored juice blend" or "juice blend, 2 to 7 percent raspberry juice."

Summary

The FDA, operating under the federal Food, Drug, and Cosmetic Act, regulates the labeling for all foods except meat and poultry. The USDA under the federal Meat Inspection Act regulates meat and poultry products. Additional federal acts cover specific foods.

Recent changes in the food label provide consumers with more accurate and more useful nutritional information. This information is provided on the nutrition panel. Some of the components of the food label are mandatory and some are voluntary. DRVs establish dietary standards for labels. Guidelines for labels promote the uniform definitions for words such as free, low, lean, high, reduced, and less. Additionally, label regulations address how and when certain health claims can be made for foods. Finally, food-labeling requirements also address ingredient labeling.

Review Questions

Success in any career requires knowledge. Test your knowledge of this chapter by answering these questions or solving these problems.

1. What are DRVs?
2. What is the new requirement on ingredient labels?
3. Name three agencies that regulate foods and labeling.
4. List three foods exempt from food labels.
5. Name four functions of quality assurance programs.
6. What type of claim would this be on a nutrition label–"While many factors affect heart disease, diets low in saturated fat and cholesterol may reduce the risk of this disease"?
7. List six components found on the nutritional panel.
8. ____________ means that a nutritionally altered product contains at least 25 percent less of a nutrient or of calories than the regular, or reference, product.
9. Name the two terms used to describe the fat content of meat, poultry, seafood, and game meats.
10. What are the DRVs for fats and sodium?

Student Activities

1. Visit the FDA's Web site and select one of the food guidance (good manufacturing) documents and develop a short report on what is available on one of the topics:

chemical and pesticides, food additives, food labeling, food processing, low acid and acidified foods, milk sanitation, produce and sanitation. The address for the Web site is <vm.cfsan.fda.gov/~dms/guidance.html>.

2. Find an article in a newspaper, a magazine, or on the Internet that promotes the healthy effects of some food–for example, an article discussing foods low in cholesterol or foods with the ability to reduce cholesterol or to reduce the risk of cancer or the risk of heart disease (CHD). Summarize this article and report to the class. As an additional part of the assignment, find the food described in the article and see if the label on the food makes the same claim.
3. Contribute five different labels from foods around your house to a classroom bulletin board displaying labels of all kinds. When the bulletin board display is complete, make a list of any claims of "low," "free," "light," "reduced," "good source," or "high," and any health claims made on any of the labels.
4. Cut the nutrition label off of a familiar food. Make a transparency or handout from it and then go over each part of the label point-by-point. Lead students in discovering what they can learn about the food from its label.

Resources

Potter, N. N., and J. H. Hotchkiss. 1995. *Food science*, 5th ed. New York: Chapman and Hall.

Vaclavik, V. A., and E. W. Christina. 1999. *Essentials of food science*. Gaithersburg, MD: Aspen Publishers, Inc.

Vieira, E. R. 1996. *Elementary food science*, 4th ed. New York: Chapman and Hall.

Internet

Internet sites represent a vast resource of information. The URLs (uniform resource locator) for the World Wide Web sites can change. Using one of the search engines on the Internet such as Yahoo!, HotBot, AltaVista, Excite, Dogpile, About, or Google, find more information by searching for these words or phrases: food labeling, nutrition panel title, serving size, nutrition information, Infant Health Formula Act, Nutrition Labeling and Eduction Act, quality assurance (QA), food laws, food substances, Daily Reference Values, nutrient panel descriptions, percentage daily values. Also, Table A-7 provides a listing of some useful Internet sites that can be used as a starting point.

Chapter 27

World Food Needs

Objectives

After reading this chapter, you should be able to:

- Discuss the effects of hunger and malnutrition
- Describe the impact of hunger worldwide
- Discuss possible causes of world hunger
- List seven steps identified by the United Nations for eliminating hunger
- Explain the role of technology in eliminating hunger
- Identify agencies and organizations involved in preventing and eliminating hunger
- Discuss the Plan of Action developed at the World Food Summit
- Recognize agencies and organizations concerned with eliminating hunger

Key Terms

famine
food security
foreign aid
hunger
integrated pest management (IPM)
malnutrition
stunting
sustainable
undernutrition
underweight
wasting

World hunger is a serious problem with no simple solution. Awareness, many times, is the first step in solving a problem. About 800 million people are hungry. In fact, if all the world's undernourished people were gathered in one place, their population would be greater than every continent except Asia. Each year, people die from hunger or problems caused by hunger. Many of these are children under 5 years of age.

WORLD FOOD HUNGER AND MALNUTRITION

What is **hunger**? Hunger means different things to different people. World hunger usually means **malnutrition**, **undernutrition**, or **famine**. Undernutrition means a person does not get enough food to have a healthy life. People can go short periods of time without eating and not suffer any permanent damage. Over long periods of time, hunger can slow physical and mental development in children. Because hunger weakens the body, many times diseases can cause death.

Malnutrition implies that a person eats but does not receive the amounts of nutrients needed to keep the body healthy. Because of this, a malnourished person does not always feel hunger. These people just do not get enough of the right things to eat.

Hunger (malnutrition and undernutrition) does not only kill. It can cause serious physical injury to the body–injuries like brain damage and other physical defects. Children suffer the most. The effects of hunger on children are classified as **stunting**, **underweight**, and **wasting** (Figure 27-1). A high proportion of children in the developing world suffer from undernutrition resulting from a combination of inadequate food intake and diseases such as diarrhea that prevent the proper digestion of food.

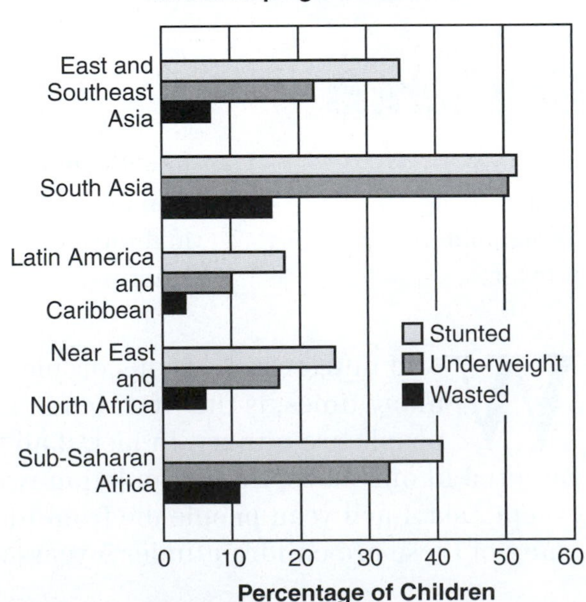

FIGURE 27-1

Undernutrition affects children in many areas of the world. (*Source:* FAO)

Hunger can be a famine (a great shortage of food), but more often it is a very limited diet. Often undernutrition occurs on a seasonal basis. In developing countries, the time before harvest is difficult. The people have used all their money and food. They are counting on the new harvest to bring them food and money. If one of the many natural disasters occurs (drought, insects, freeze, and others), the people will be without food or the money to buy any food. If they are lucky and they get to harvest their crops, they will have food. Each year this problem happens during the harvest time.

Many of the people who suffer from hunger are people who live in the rural areas in developing countries. Strangely, the people in the country who grow the food are hungry. However, when there is a shortage of food or a crisis, these people are the hardest to get to with food (Figure 27-2).

Another group of people who suffer from hunger are those who live in slums. A slum is an area, mostly in large cities, that is usually overcrowded. They have unhealthy living conditions; there is high poverty, and most people are without jobs (Figure 27-3).

FIGURE 27-2

Rural hunger—a problem in many developing countries. (*Source:* FAO)

FIGURE 27-3

Slums of large cities—many hungry people. (*Source:* FAO)

Supplying food for a city of 10 million people is a massive undertaking that stretches the resources of a region. A city of this size requires at least 6,000 tons (12 million pounds) of food to be imported each day!

Worldwide more people are living in large cities (Figure 27-4), and these large cities have slums associated with them. More people live in slums than ever before, and the availability of food is limited (Figure 27-5).

Most of the world's malnourished people live in Southeast and South Asia, Latin America, the Caribbean, the Near East and North Africa, and Sub-Saharan Africa (Figure 27-6). Many of these people live in poverty.

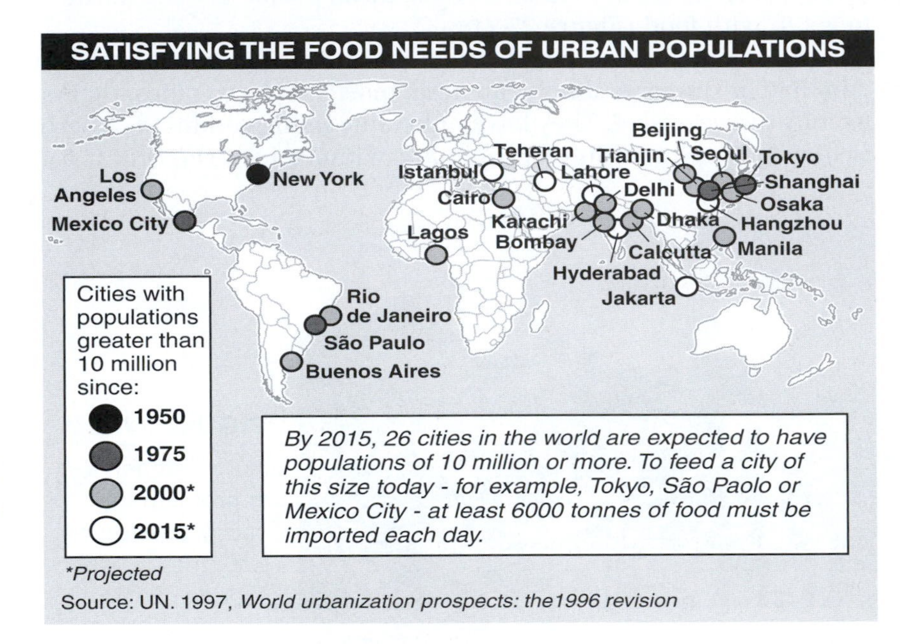

FIGURE 27-4

Location of current and projected cities of more than 10 million people. (*Source:* United Nations)

FIGURE 27-5

Garbage dumps outside large cities become a source of food for some. (*Source:* FAO)

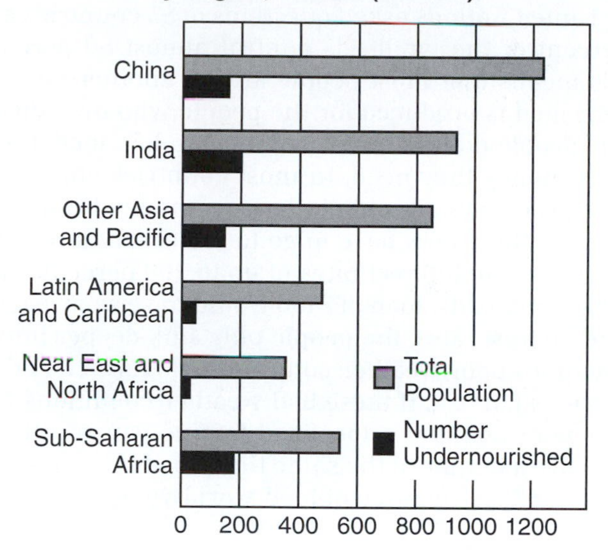

FIGURE 27-6

Proportion of undernourished people in various regions of the world. (*Source:* FAO)

Causes of Hunger

Four common misconceptions about the causes of hunger include the following:

1. Not enough food is available to feed everyone
2. The population is too large
3. Governments cause hunger
4. **Foreign aid** helps eliminate hunger

Not Enough Food Available to Feed Everyone. Enough food to feed everyone in the world is produced. Other conditions affect how food is produced or distributed. If the food produced in the world could be divided so that everyone would get the same amount, about 10 percent of the food produced would be left over. Enough grain is produced in the world to give every man, woman, and child 2 pounds each day. This does not include all the beans, potatoes, fruits, and vegetables that are produced. Two pounds of grain can provide 3,000 calories. The average recommended daily minimum is 2,300 calories.

Most every country in the world has the resources necessary for its people to rid the country of hunger. In many areas where food is produced, after harvesting the food is sent to other areas where the people can pay more. Food production is not the issue, poverty is. If the people had more money to pay for the harvested food, they would stay in or move to an area.

The United Nations asked questions of 83 countries and found that 3 percent of the landlords control almost 80 percent of the land. This means that most people are on another person's land. Usually the food is produced for the people who own the land.

If the people do not have the money to buy food, they cannot borrow the money they need. In most countries only 5 to 20 percent of the producers are able to borrow money from institutions like banks. All the others have to go to either landlords or moneylenders who charge interest rates of up to 200 percent. An interest rate of 200 percent on a loan of $100 would have a pay back of $300. With large interest rates, the people only sink deeper into poverty.

Sometimes poor weather conditions (Figure 27-7) destroy an entire season's harvest. If these bad weather conditions happen in a small or poor country, a localized famine can result. If several countries have a famine at the same time, food becomes hard to get and can even effect the price of food worldwide.

The Population Is Too Large. Total world population is not as important as the location of the people. People in poor countries tend to have many children. Many children can drain a country's food supply. But children are a very important part of countries that are still developing. In many areas, poor people have a lot of children because of poor health care and nutrition, and many of

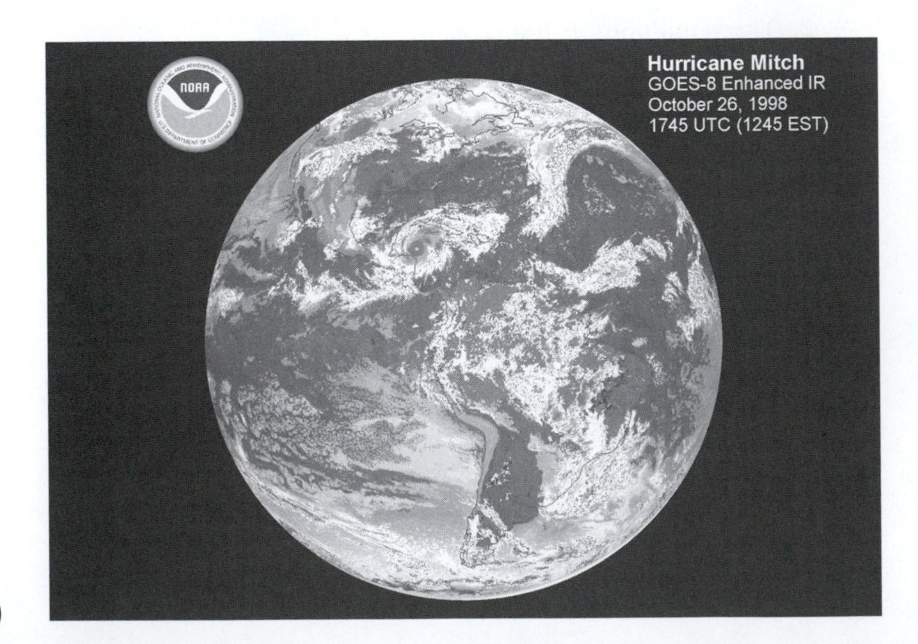

FIGURE 27-7

Satellite technology allows weather forecasting worldwide, but not weather control. (*Source:* NOAA National Data Centers)

FIGURE 27-8

Nutritional assessment helps determine the extent that children in underdeveloped countries suffer from undernourishment. (*Source:* FAO)

the children will not survive (Figure 27-8). In developing countries, children help with the work. They bring in extra money to the family. Most of the time these people depend on their children to take care of them when they get old. In countries like the United States and Canada, many people have been able to save money for when they retire. This is not always true in developing countries. Children may use a lot of the resources of a country, but they are also necessary to the economy and well-being of the country.

Still, a dense population does not always mean that the people suffer from hunger. Surveys have shown that countries that have more people per acre (population density–the average number of people per space unit) are not always the most hungry. For example, Taiwan feeds twice as many people per acre as Bangladesh.

Governments Cause Hunger. The government is just one part of who controls the country. Other agencies or organizations that control part of the country are multinational corporations (businesses that are in more than one country), international agencies, and other governments. Together they form a group that lives differently than the rural people of the country. Though a country's government may add to the hunger of its people, it is seldom if ever the only cause.

Foreign Aid Helps Eliminate Hunger. The United States and other countries send large quantities of food to hungry people in many countries. Much of the food that is supposed to get to hungry people never reaches them. Even if it did, it would only be treating the symptoms and not the cause. It would provide a temporary relief. It also puts more food on the foreign market. This causes the price of food to drop. Because the price of food has gone down, the rural farmer gets less money for what is produced.

Because the food is free and available, people start to depend on it. When the food is gone, the people are in as bad if not worse shape than when the food first arrived. The people now depend on the free food and do not try to get food on their own. Foreign aid needs to help fix the problems that cause hunger. If the problem of hunger is to be solved, the most important thing should be growing and preserving more food in a region. This approach has been called the Green Revolution. It teaches countries to produce enough food for their own needs.

FIGHTING THE PROBLEM

No easy answers exist to fix the problem of world hunger. Sending food to countries where there is an emergency can help. This is not a permanent solution. According to the United Nations' publication, *Ending Hunger: An Idea Whose Time Has Come*, in the last 88 years, 75 countries have done away with hunger. Forty-one of these 75 countries have done this since 1960.

The Food and Agriculture Organization (FAO) of the United Nations (<www.fao.org>) created a seven-step plan against hunger.

Step 1: More Self-Sufficiency

Many countries that are suffering from hunger purchase a lot of the items that they need from other countries. The countries that suffer from hunger need to learn to do more by themselves. Then these countries will not need to purchase as much from other countries.

Step 2: Check Farming Regulations

Many developing countries have rules and regulations for the farmers. These rules need to be checked to make sure that farmers will still want to produce food. Regulations should also make sure that the farmers get a fair amount of money for their food. For the farmers to do their part, they need training to use their land and water wisely.

MOST IMPORTANT GRAIN

For more than 60 percent of the world's population, rice is the main food. Although the production of wheat is greater in absolute tonnage, rice directly supports more of the world's people than any other crop. In China, the most populous country in the world, each person eats about a pound of rice, on average, everyday.

Rice is the source of a wide variety of products, including beer, wine, straw, and paper, as well as edible grain. It is a tropical grass originating in Southeast Asia, and was first cultivated at least 10,000 years ago. Today, it is grown in every tropical area of the world, and is also successfully cultivated in areas where the climate is far cooler, such as on mountain terraces in Nepal.

In most rice-growing areas, young rice seedlings are planted in standing water, in fields called paddies. The water remains until the seed heads emerge, after which the paddies are drained while the crop ripens.

For more information about rice and culture, visit these Web sites:

<www.mpiz-koeln.mpg.de/~rsaedler/schau/OryzasativaL./Rice.html>
<gnome.agrenv.mcgill.ca/breeding/students/max/rice/rice.htm>
<lux.ucs.indiana.edu/~japan/digest6.html>

Step 3: Proper Storage

After the food is harvested, it has to be properly stored until it is needed. When the food is needed, it has to be taken where and when it is needed most. Both the storage and transportation of food need improvement.

Step 4: Check Food Aid

Food aid has to be checked ensuring it gets to the hungry people. Then the country's production needs to be checked. The country must not be allowed to produce less food because of the food aid.

Step 5: Work Together

Developed and developing countries need to work together more. This makes trade easier, and can stabilize food prices.

Step 6: Prevent Waste

Countries need to be kept from using too much food and from wasting food.

Step 7: Pay Off Debt

Many developing countries owe a lot of money. These countries need to pay off their debts. This can be done by paying back money they make on exports (things that they sell to other countries). Because these countries owe a lot of money, it makes it difficult for them to borrow money.

ROLES OF TECHNOLOGY

To alleviate hunger, technology research in the following areas should receive high priority:

- Improving the application of technology to natural resource management
- Protection of crops without heavy reliance on pesticides
- Genetic improvement of key crops
- Global action to advance scientific knowledge and its application

Resource Management

Using the technology available, the agricultural resource base needs to be characterized and evaluated–both in relation to existing cultivated areas and to land currently under forest or pasture. This should be done using such modern tools and techniques as remote sensing, aerial photography, sonic devices, and geographic information systems.

Even with an evaluation of the resource base, three factors determine the future success of resource management:

1. The extent to which cultivars and cropping systems can be adapted to fit the resource rather than trying to modify the resource itself.
2. The degree to which the local community can be involved in the planning and management of the resource and feel ownership of it.
3. The willingness of governments to adjust their policies to encourage efficient and **sustainable** resource management.

Using this information, a country could manage soil fertility for higher productivity by improving soil and water management in rain-fed farming and irrigated land. Techniques such as drainage and canal lining could reduce salinity and water logging.

Protection of Crops

Important progress is being made toward the protection of crops. Less toxic chemicals and more efficient methods of application have been developed. Plant breeding allows the introduction of host-plant resistance. This is being accelerated by gene mapping and genetic engineering techniques to identify sources of genetic resistance in crops and their wild relatives. Also, biological control through the use of predators and parasites is used on some crops.

Integrated pest management (IPM) combines biological and crop management practices to develop cost-effective control with reduced dependence on pesticides. This method could be used on a wider scale worldwide, but it requires highly-skilled farmers. IPM may require community cooperation and crop consulting to help producers monitor pests. Because pests and diseases do not respect national boundaries, crop protection merits high priority for international cooperation to better understand and monitor its status, improve diagnosis, maintain databases, and identify natural enemies and sources of resistance.

Genetic Improvement

Plant breeding is the cornerstone of yield-increasing technology, and it also plays a key role in preventing yield losses from disease, weather, and insects (Figure 27-9). Raising yields in regions where they are already high–for example, wheat in Western Europe and irrigated rice in East Asia–may not be realistic. Plant breeding must also improve nutritional quality of the food, especially of nutrients like vitamins. Plant breeding may also help crops overcome stresses such as drought, extreme temperatures, soil acidity, and other nutrient problems that keep yields low in many other areas.

Biotechnology offers hope of finding solutions to these problems. It already contributes to genetic improvement of cereals, root crops, vegetables, industrial crops, and to animal health. Biotechnology will also improve diagnostic and asexual propagation techniques, in raising the nutritional density of crops, and in increasing resistance to disease, weather, and insects. The evolution of new human-made species to complement existing cereals and improvement of tolerance to stresses will eventually be a reality.

FIGURE 27-9

Research in genetic improvement of plants and animals offers hope for feeding more hungry people. (*Source:* USDA ARS Photo Gallery)

Global Action

Whether the food needs of the world in 2020 can be met both in amount and quality likely will depend on the successful enlistment of scientific resources for research and on the improvement of farmers' skills to manage their resources. Sustainability will require knowledge distribution, training, and accessibility of inputs (seed, fertilizer, money, and so on). Four hopeful signs in this area are:

1. Strong international support for seed and tissue (germplasm) collection, conservation, and evaluation. In many countries, an adequate supply of quality seed is still a problem at the farm level.
2. Increasing international cooperation in resource assessment and monitoring.
3. Development and application of the potential for expanding knowledge created by modern information networks, videos, computers, and related technology.
4. Expansion of regional research and technology transfer networks among developing countries.

WORLD FOOD SUMMIT

The Committee on World Food Security at the 22nd Session on 31 October 1996 submitted to the World Food Summit a Declaration on World Food Security. The first part of the Declaration reads as follows:

> *We, the Heads of State and Government, or our representatives, gathered at the World Food Summit at the invitation of the Food and Agriculture Organization of the United Nations, reaffirm the right of everyone to have access to safe and nutritious food, consistent with the right to adequate food and the fundamental right of everyone to be free from hunger.*
>
> *We pledge our political will and our common and national commitment to achieving* ***food security*** [emphasis added] *for all and to an on-going effort to eradicate hunger in all countries, with an immediate view to reducing the number of undernourished people to half their present level no later than 2015.*
>
> *We consider it intolerable that more than 800 million people throughout the world, and particularly in developing countries, do not have enough food to meet their basic nutritional needs. This situation is unacceptable. Food supplies have increased substantially, but constraints on access to food and continuing inadequacy of household and national incomes to purchase food, instability of supply and demand, as well as natural and man-made dis-*

asters, prevent basic food needs from being fulfilled. The problems of hunger and food insecurity have global dimensions and are likely to persist, and even increase dramatically in some regions, unless urgent, determined and concerted action is taken, given the anticipated increase in the world's population and the stress on natural resources.

We reaffirm that a peaceful, stable and enabling political, social and economic environment is the essential foundation, which will enable States to give adequate priority to food security and poverty eradication. Democracy, promotion and protection of all human rights and fundamental freedoms, including the right to development, and the full and equal participation of men and women are essential for achieving sustainable food security for all.

Poverty is a major cause of food insecurity and sustainable progress in poverty eradication is critical to improve access to food. Conflict, terrorism, corruption and environmental degradation also contribute significantly to food insecurity. Increased food production, including staple food, must be undertaken. This should happen within the framework of sustainable management of natural resources, elimination of unsustainable patterns of consumption and production, particularly in industrialized countries, and early stabilization of the world population. We acknowledge the fundamental contribution to food security by women, particularly in rural areas of developing countries, and the need to ensure equality between men and women. Revitalization of rural areas must also be a priority to enhance social stability and help redress the excessive rate of rural-urban migration confronting many countries.

The entire Declaration and proceedings can be read at the Web site for the World Food Summit (Figure 27-10).

World Food Summit Plan of Action

Food security exists when all people, at all times, have physical and economic access to sufficient, safe, and nutritious food to meet their dietary needs and food preferences for an active and healthy life.

Eradication of poverty is essential to improve the access to food. Most of those who are undernourished cannot produce or cannot afford to buy enough food. They have inadequate access to

FIGURE 27-10
World Food Summit continues to fight against hunger.

means of production such as land, water, inputs, improved seeds and plants, appropriate technologies, and farm credit. In addition, wars, civil strife, natural disasters, climate-related ecological changes, and environmental degradation have adversely affected millions of people. Although food assistance may be provided to ease their plight, it is not a long-term solution to the underlying causes of food insecurity.

A peaceful and stable environment in every country is a fundamental condition for the realization of sustainable food security. Governments should create an environment that facilitates private and group initiatives to allocate their skills, efforts, and resources, and in particular investment, towards the common goal of food for all. Farmers, fishers, foresters, and other food producers and providers, have critical roles in achieving food security, and their full involvement and ability are crucial for success.

The goal of enough food for all can be reached. The 5.8 billion people in the world today have, on average, 15 percent more food per person than the global population of 4 billion people had 20 years ago. (*Source:* World Food Summit.) Large increases in world food production, through the sustainable management of natural resources, are required to feed a growing population and to achieve improved diets. Increased production, including traditional crops and their products, in combination with food imports, reserves, and international trade can strengthen food security and address regional disparities.

The follow-up to the World Food Summit includes actions at the national, intergovernmental, and interagency levels. The international community, and the United Nations (UN) system, including the Food and Agriculture Organization (FAO), as well as other agencies, has important contributions to the implementation of the World Food Summit Plan of Action. The FAO Committee on World Food Security (CFS) will have responsibility to monitor the implementation of the Plan of Action.

Plan of Action to achieve food security should allow people to enjoy their human rights, including the right to development. Also the Plan should recognize various religious and ethical values, cultural backgrounds, and philosophical convictions of individuals and their communities.

HUNGER AGENCIES AND ORGANIZATIONS

Table 27-1 lists some of the agencies and organizations involved in hunger-related projects. These organizations can serve as useful sources of information on fighting hunger, and they provide educational materials about hunger.

Summary

One word explains hunger–poverty. Poverty is the state of one who lacks a usual or socially acceptable amount of money or material possessions. If someone has so few things or so little money that their society says they are poor, then they live in poverty. Enough food is produced. Countries have the resources to eliminate hunger. But often people do not have the money needed to buy the food that is produced.

Poverty, hunger, and malnutrition are some of the principal causes of accelerated migration from rural to urban areas in developing countries. The largest population shift of all times is now underway.

Harmful seasonal and midyear instability of food supplies can be reduced. Food security progress should include minimizing the vulnerability to, and impact of, climate fluctuations and pests and diseases.

Unless national governments and the international community address the many causes of food insecurity, the number of hungry and malnourished people will remain very high in developing countries, particularly in Africa, south of the Sahara. Sustainable food security will not be achieved. The resources required for investment in food security will be generated mostly from domestic, private, and public organizations and groups.

TABLE 27-1 Hunger Organizations

Name of Organization and Web Site	Address
Agricultural Development Council www.chesco.org/agri.html	1290 Avenue of the Americas New York, NY 10019
Bread for the World www.bread.org	1100 Wayne Ave., Suite 1000 Silver Spring, MD 20910
Brookings Institute www.brook.edu	1775 Massachusetts Ave., NW Washington, DC 20036
CARE International www.care.org	151 Ellis Street, NE Atlanta, GA 30303
Carnegie Endowment for International Peace www.ceip.org	1779 Massachusetts Ave., NW Washington, DC 20036
Center of Concern www.igc.org/coc/home.htm	1225 Otis Street, NE Washington, DC 20017
Center for Science in the Public Interest www.cspinet.org	1875 Connecticut Ave., NW, Suite 300 Washington, DC 20009
Committee for Economic Developoment www.ced.org	477 Madison Ave. New York, NY 10022
Committee for UNICEF www.unicef.org	UNICEF House 3 United Nations Plaza New York, NY 10017
The Conference Board www.conference-board.org	845 Third Ave. New York, NY 10022
Church World Service www.churchworldservice.org	475 Riverside Drive New York, NY 10115
Food and Agriculture Organization of the United Nations www.fao.org	116 F Street, NW Washington, DC 20437
Free from Hunger www.freefromhunger.org	1644 DaVinci Court Davis, CA 95616
Friends Committee on National Legislation www.fcnl.org	245 2nd Street, NW Washington, DC 20002
The Hunger Project www.thp.org	15 East 26th Street New York, NY 10010
Institute for Food and Development Policy www.foodfirst.org	398 60th Street Oakland, CA 94618
Institute for Policy Studies www.ips-dc.org	733 15th Street, NW, Suite 1020 Washington, DC 20005
International Bank for Reconstruction and Development (IBRD) (World Bank) www.worldbank.org	The World Bank (Headquarters) 1818 H Street, NW Washington, DC 20433

TABLE 27-1 Hunger Organizations *(concluded)*

Name of Organization and Web Site	Address
International Food Policy Research Institute www.cgiar.org/IFPRI/2index.htm	2003 K Street, NW Washington, DC 20006
International Institute for Environment and Development www.iied.org	3 Endsleigh Street London WC1H 0DD England
National Academy of Sciences National Academy of Engineering National Research Council Institute of Medicine www.nas.edu	2101 Constitution Ave., NW Washington, DC 20418
National Conference of Catholic Bishops United States Catholic Conference www.nccbuscc.org	3211 4th Street, NE Washington, DC 20017
Overseas Development Council www.odc.org	1875 Connecticut Ave., NW, Suite 1012 Washington, DC 20009
Oxfam America www.oxfamamerica.org	Oxfam America (Main Office) 733 15th Street, NW, Suite 340 Washington, DC 20005
Pan American Health Organization www.paho.org	525 23rd Street, NW Washington, DC 20037
Resources for the Future www.rff.org	1616 P Street, NW Washington, DC 20036
United Nations Systems www.unsystems.org	No address, just a listing of all UN organizations
United Nations Development Program (UNDP) www.undp.org	1 United Nations Plaza New York, NY 10017
World Food Program www.wfp.org	2175 K Street, Suite 300 Washington, DC 20437
World Health Organization (WHO) www.who.int/home	Regional Office 525 23rd Street, NW Washington, DC 20037
World Hunger Education Service www.worldhunger.org	PO Box 29056 Washington, DC 20017
World Hunger Year (WHY) www.worldhungeryear.org	505 Eighth Ave., 21st Floor New York, NY 10018
Worldwatch Institute www.worldwatch.org	1776 Massachusetts Ave., NW Washington, DC 20036

Review Questions

Success in any career requires knowledge. Test your knowledge of this chapter by answering these questions or solving these problems.

1. What are the three effects of hunger on children?
2. List five agencies involved in world food hunger.
3. Many of the people who suffer from hunger are people who live in the _______ areas in developing countries.
4. Name the four ways that technology research can help with hunger.
5. What three things does world hunger usually mean?
6. List the seven steps identified by the United Nations for eliminating hunger.
7. Define food security.
8. List the four common misconceptions about the causes of hunger.
9. ____________ eradication is essential to improve access to food.
10. Hunger can be a ____________, but more often it is a very limited ____________.

Student Activities

1. Develop a report or presentation about hunger and malnutrition in an area of the world.
2. Visit the Web site for the Food and Agriculture Organization (FAO) and develop a report on the type of information they provide.
3. Assuming that a city of 10 million people requires at least 6,000 tons of food each day, find some different ways to express this. For example, how many pounds per year is this or how many semitruck loads are required each week or month?
4. Contact one of the agencies listed in Table 27-1, page 514, and report on the information they provide.
5. Develop a report or presentation on a new food, a new food product, or a new food process that has the potential to help eliminate hunger in one area of the world.

Resources

Brunner, B. (Ed.). 1999. *Time almanac 2000.* Boston: Information Please.
Food and Agriculture Organization (FAO). 2000. *The state of food insecurity in the world 2000.* (Available at the FAO Web site: <www.fao.org>).
World Food Summit Fact Sheets: <www.fao.org/wfs/fs/e/fshm-e.htm>.
Worldwatch Institute. 2000. *State of the world 2000.* New York: W. W. Norton and Company, Inc.

Internet

Internet sites represent a vast resource of information. The URLs (uniform resource locator) for the World Wide Web sites can change. Using one of the search engines on the Internet such as Yahoo!, HotBot, AltaVista, Excite, Dogpile, About, or Google, find more information by searching for these words or phrases: hunger organizations, famine, malnutrition, food security, and foreign aid. Refer to Table 27-1 for a listing of hunger organizations, their addresses, and Web sites. Also, Table A-7 provides a listing of some useful Internet sites that can be used as a starting point.

Chapter 28

Careers in Food Science

Objectives

After reading this chapter, you should be able to:

- List the basic skills and knowledge needed for successful employment and job advancement
- Describe the thinking skills needed for the workplace of today
- Identify the traits of an entrepreneur
- List six occupational areas of the food industry
- Identify the careers that require a science background
- Describe the general duties of the occupations in six areas of the food industry
- Describe the education and experience needed to enter six areas of the food industry
- List six general competencies needed in the workplace
- List eight guidelines for choosing a job
- List ten guidelines for filling out an application form
- Describe a letter of inquiry or application
- List the elements of a résumé or data sheet
- Describe ten reasons an interview may fail
- Discuss what research studies indicate about basic skills and thinking skills for the workplace

Key Terms

competencies
creative thinking
cultural diversity
data sheet
demographic
entrepreneur
follow-up letter
letter of application
letter of inquiry
résumé

In terms of value of shipments, food processing is the largest manufacturing industry in the United States. The major technological support of the food processing industry comes from food scientists, technicians, and other industry employees who use their training and experience to convert raw foods into quality products quickly, efficiently, and with a minimum of waste. They are directly concerned with the industry's high standards of quality, new manufacturing methods, new preservation techniques, and new packaging materials. A knowledge of chemistry, microbiology, engineering, and other basic and applied sciences plays an important part in maintaining the flavor, color, texture, nutritional value, and safety of our food.

GENERAL SKILLS AND KNOWLEDGE

Over the past few years, research study after research study indicated that potential employees never receive some very basic skills and knowledge. Without these basic skills and knowledge, the specific skills and knowledge for employment in the food industry is of little value. Also, the new workplace demands a better prepared individual than in the past. Finally, those individuals working for themselves must develop a trait called entrepreneurship. This may also be a good trait for any employee.

Basic Skills

Success in the workplace requires that individuals possess skills in reading, writing, mathematics, listening, and speaking, at levels identified by employers nationwide (Figure 28-1).

Reading. An individual ready for the workplace of today and the future demonstrates reading with the following **competencies**:

- Locates, understands, and interprets written information, including manuals, graphs, and schedules to perform job tasks
- Learns from text by determining the main idea or essential message
- Identifies relevant details, facts, and specifications
- Infers or locates the meaning of unknown or technical vocabulary

FIGURE 28-1
Using computers and interpreting data—an important component for jobs in the food industry. (*Source:* USDA ARS Image Gallery)

- Judges the accuracy, appropriateness, style, and plausibility of reports, proposals, or theories of other writers

Reading skills in the food industry are necessary to keep up with new information and to read directions and instructions for the position.

Writing. An individual ready for the workplace of today and the future demonstrates writing abilities with the following competencies:

- Communicates thoughts, ideas, information, and messages
- Records information completely and accurately
- Composes and creates documents such as letters, directions, manuals, reports, proposals, graphs, and flowcharts with the appropriate language, style, organization, and format
- Checks, edits, and revises for correct information, emphasis, form, grammar, spelling, and punctuation.

In the food industry, writing skills are necessary for such tasks as keeping records, making reports, and communicating with coworkers, as well as others inside and outside the industry.

Mathematics. The workplace of today and the future requires individuals with competencies in mathematics. Mathematics is the science of computing with numbers by the operations of addition, subtraction, multiplication, and division. These important competencies are:

- Performing basic computations
- Using numerical concepts such as whole numbers, fractions, and percentages in practical situations
- Making reasonable estimates of mathematic results without a calculator
- Using tables, graphs, diagrams, and charts to obtain or convey information
- Approaching practical problems by choosing from a variety of mathematical techniques
- Using quantitative data to construct logical explanations of real-world situations
- Expressing mathematical ideas and concepts verbally and in writing
- Understanding the role of chance in the occurrence and prediction of events

Mathematics is used in the food industry to figure conversion, ratios, or composition.

Listening. Individuals working today and in the future must demonstrate an ability to really listen. This means to receive, attend to, and interpret verbal messages and other cues such as body language. Real listening means that the individual comprehends, learns, evaluates, appreciates, or supports the speaker, without interruption.

Speaking. Finally, an individual successful in the workplace of today and the future must demonstrate these speaking competencies:

- Organizes ideas and communicates oral messages appropriate to listeners and situations
- Participates in conversation, discussion, and group presentations
- Uses verbal language, body language, style, tone, and level of complexity appropriate for audience and occasion
- Speaks clearly and communicates the message
- Understands and responds to listener feedback
- Asks and answers questions when needed

Thinking Skills

Contrary to the old workplace, many research studies indicate that employers in the new workplace want workers who can think. Employers search for individuals showing competencies in these areas: **creative thinking**, decision making, problem solving, mental visualization, knowing how to learn, and reasoning (Figure 28-2).

Creative Thinking. Creative thinkers generate new ideas by making nonlinear or unusual connections or by changing or reshaping goals to imagine new possibilities. These individuals use imagination, freely combining ideas and information in new ways.

Decision Making. Individuals who use thinking skills to make decisions are able to specify goals and limitations to a problem. Next, they generate alternatives and consider the risks before choosing the best alternative.

Problem Solving. As silly as it sounds, the first step to problem solving is recognizing that a problem exists. After this, individuals with problem-solving skills identify possible reasons for the problem and then devise and begin a plan of action to resolve it. As the problem is being solved, problem solvers monitor the progress and fine-tune the plan. Being able to recognize the need for a new product and looking for solutions are good examples of problem solving in food science.

FIGURE 28-2

The modern workplace encourages employees to use thinking skills.

Mental Visualization. This thinking skill requires an individual to see things in the mind's eye by organizing and processing symbols, pictures, graphs, objects, or other information.

Knowing How to Learn. Perhaps of all the thinking skills, this is most important with the rapid changes in available technology. This type of individual recognizes and can use learning techniques to apply and adjust existing and new knowledge and skills in familiar and changing situations. Knowing how to learn means awareness of personal learning styles–formal and informal learning strategies.

Reasoning. The individual who uses reasoning discovers the rule or principle connecting two or more objects and applies this to solving a problem. For example, chemistry teaches the theory of pH measurements, but the reasoning individual is able to use this information in understanding pH of food chemistry.

General Workplace Competencies

Besides the basic skills and the thinking skills, the workplace of today and the future demands general competencies in the use of resources, interpersonal skills, information use, systems, and technology.

Resources. Resources of a business include time, money, materials, facilities, and people. Individuals in the workplace must know how to manage–

- Time by the use of goals, priorities, and schedules
- Money with budgets and forecasts
- Material and facility resources such as parts, equipment, space, and products
- Human resources by determining knowledge, skills, and performance levels

Interpersonal Skills. More than ever, people cannot act in a vacuum. Most people are members of a team where they contribute to the group (Figure 28-3). They teach others in their workplace when new knowledge or skills are needed. More than ever and at all levels, individuals must remember to serve customers and satisfy their expectations. Through teams, individuals frequently exercise leadership to communicate, justify, encourage, persuade, or motivate individuals or groups. As part of employment teams, individuals negotiate resources or interests to arrive

FIGURE 28-3
Working in teams—an important skill for employees. (*Source:* USDA ARS Image Gallery)

at a decision. Finally, all interpersonal skills require individuals to work with and use **cultural diversity**.

Information. The information age is here. Individuals in the workplace must cope with and use information. Successful individuals will identify the need for information and evaluate the information as it relates to a specific job. With the computer, individuals in the workplace must gather, organize, and process information in a systematic way (Figure 28-4). Also, with all this information available, individuals must interpret and communicate information to others using oral, written, or graphic methods. To manage production information, computer skills are a key.

Systems. No longer can any aspect of a business or industry be viewed as a part that stands alone. Every part is part of a system, and individuals now seek to understand systems whether these are social, organizational, technological, or biological. With an understanding of the systems in a business, trends can be determined and predictions can be made. Individuals then modify the system to improve the product or service. For example, successful development and marketing of a new product requires understanding of the entire food system.

FIGURE 28-4

Colleges and universities provide training in computer skills.

Technology. Technology makes life easier only for those who know how to select it, use it, maintain it, and troubleshoot it. Technology is complicated. Successful individuals learn to apply appropriate new technology through all the basic skills, the thinking skills, and general workplace competencies.

Personal Qualities

After all the training in basic skills, thinking skills, and general workplace competencies, individuals still fail for lack of some personal qualities. These include responsibility, self-esteem, sociability, self-management, and integrity or honesty. These qualities together describe the term "work ethics."

Responsible individuals work hard at tasks even when the task is unpleasant. Responsibility shows in high standards of attendance, punctuality, enthusiasm, vitality, and optimism in starting and finishing tasks.

Those possessing self-esteem believe in themselves and maintain a positive view of themselves. These individuals know their skills, abilities, and emotional capacity. They feel good about themselves.

Successful individuals demonstrate understanding, friendliness, adaptability, empathy, and politeness to other people. These skills are demonstrated in familiar and unfamiliar social situations. The best examples are sincere individuals who take an interest in what others say and do.

Along with self-esteem is self-management. Individuals successful in business accurately assess their own knowledge, skills, and abilities while setting well-defined and realistic personal goals. Then, once goals are set, those who manage themselves monitor their progress and motivate themselves to achieve these goals. Self-management also implies a person who exhibits self-control and responds to feedback unemotionally and nondefensively.

Finally, to be successful in the food industry, an employee or **entrepreneur** requires good old-fashioned honesty and integrity. Good work ethics are still a part of good business.

ENTREPRENEURSHIP

The most common view of an entrepreneur is one who takes risks and starts a new business (Figure 28-5). Although this may be true for some in the food industry, some traits of entrepreneurship are desirable at many levels of employment. Within any organization, an entrepreneur may–

- Find a better or higher use for resources
- Apply technology in a new way
- Develop a new market for an existing product

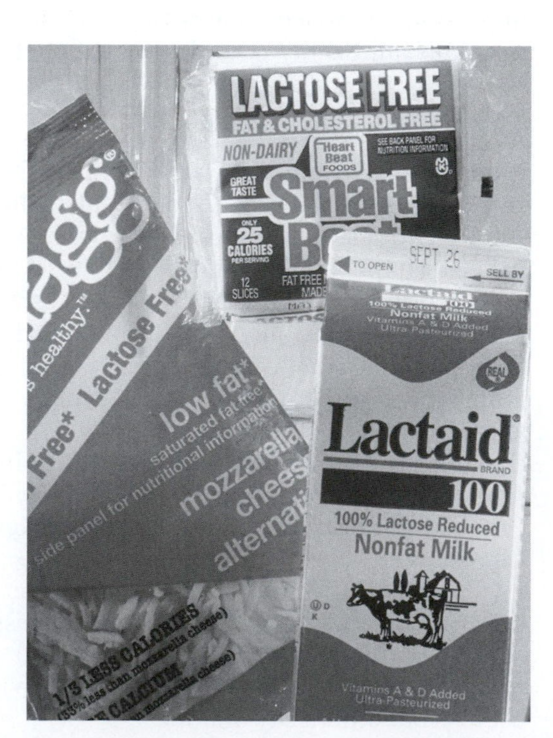

FIGURE 28-5

Success of new products and businesses requires entrepreneurship. (*Source:* USDA ARS Image Gallery)

- Use technology to develop a new approach to serving an existing market
- Develop a new idea that creates a new business or diversifies an existing business

Anyone can be an entrepreneur. It is more of an attitude–yet, an attitude that incorporates many desired traits. The attitude of an entrepreneur includes–

- Risk-taking with clear expectations of the odds
- Focusing on opportunities and not problems
- Seeking constant improvement
- Being impressed with productivity and not appearances
- Recognizing the importance of example
- Keeping things simple
- Providing open door and personal contact leadership
- Focusing on the customer
- Encouraging flexibility
- Being purposeful and communicating a vision

Entrepreneurs are ready for the unexpected, differences, new needs, change, **demographic** shifts, changes in perception, and new knowledge. Entrepreneurs are good employees and good employers. Entrepreneurs keep the food industry growing.

JOBS AND COURSES IN THE FOOD INDUSTRY

Opportunities open to graduates in the food or allied industries include research, development, and production work, technical sales within the food industry or in closely related areas such as the container and equipment manufacturing fields, extension work, research work in experimental stations or in other branches of government, food consulting, and promotional work with public or private utilities.

Food industry skills and knowledge allow an individual to–

- Enjoy a career in a dynamic, multibillion dollar business
- Develop new food products and food-processing technologies
- Improve nutritional quality of foods and insure that food is safe and wholesome
- Manage food-processing companies
- Pursue an advanced graduate or professional degree

Food Science is a very broad scientific field. Processing of food is the largest manufacturing industry in the world. Food scientists are employed by the food industry, government, and universities. Employment opportunities with an excellent pay scale are plentiful. The following is a general list of opportunities in food-related positions:

- Production Manager or Supervisor: Manages food production and processing facilities.
- Product Development Technologist: Assists in designing, researching, and developing new food products.
- Food Engineer: Involved in design and manufacture of machinery necessary for production of processed food that is safe and nutritious.
- Food Microbiologist: Involved with microbiological safety of foods. Uses microorganisms to produce new kinds of foods or improve existing foods.
- Quality Control Scientist: Inspects and determines the quality of food in one or more stages of manufacturing.
- Research Technician: Helps government and university scientists in performing food-related research.
- Technical Representative: Provides technical support to salespeople in food manufacturing or equipment manufacturing plants.

EDUCATION AND EXPERIENCE

Education and experience vary widely for jobs and careers in the food industry. Some require only a high school education with on-the-job training provided (Figure 28-6). Others require college training in a technical area; and still others require bachelors, masters, and doctorate degrees. Some of the careers have their own certification programs. Many of the careers require experience before advancing through the career choices. In the sections that follow, the requirements, education, and experience for jobs and careers are suggested.

IDENTIFYING A JOB

Some of the more specific careers or jobs in the food industry include the areas of food safety and inspection, food service industry, food retail and wholesale industry, research and development, food scientist, and marketing and communication. In many areas

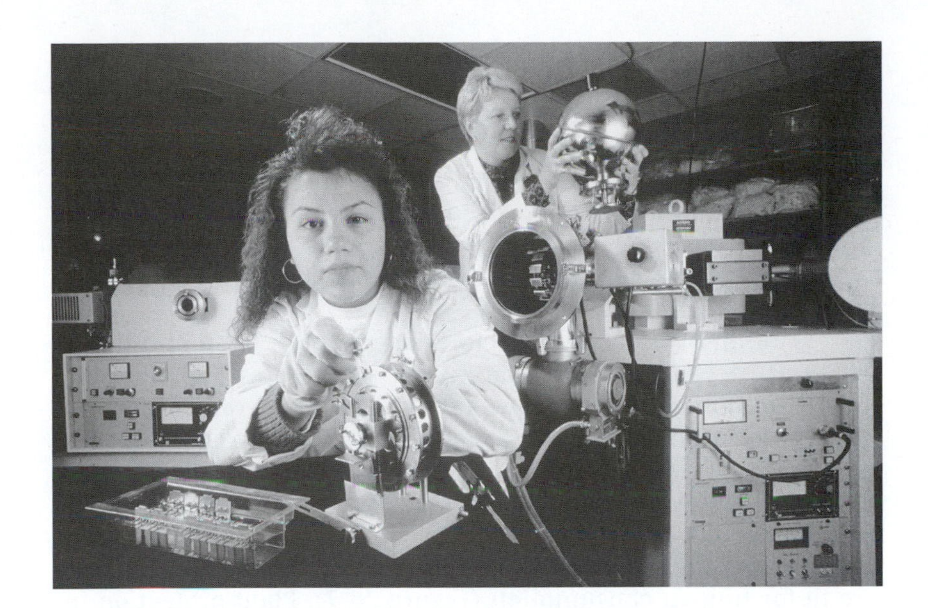

FIGURE 28-6

Most jobs require on-the-job training. (*Source*: USDA ARS Image Gallery)

that follow, specific jobs are described and the educational requirements are listed.

Food Inspection and Safety

With the ongoing demand for food quality and safety, high standards must be maintained. Individuals working in this field monitor the safety and quality of foods, verify the quantity and composition of food products, and verify the accuracy of labels. They also help businesses develop inspection systems that focus on preventing contamination in the entire food-manufacturing process. Some of the job titles in this area are:

- Crop Certification Inspector
- Dairy Products Inspector
- Fish and Fish Products Inspector
- Fruit and Vegetable Inspector
- Grain Inspector
- Livestock Inspector
- Plant Protection Inspector
- Poultry Inspector
- Public Health and Restaurant Inspector

Education and Experience. A bachelor's degree or college diploma in agriculture, biology, or food processing technology is required with several years of experience in agricultural production or fish processing. Frequently, individuals will be required to complete in-house training courses.

Job Description. In general, food inspectors check out agricultural and fish products to ensure that they conform to prescribed standards for production, storage, and transportation. Depending on the area of specialization, inspectors look for a variety of different things.

Fruit and vegetable inspectors inspect both fresh and frozen fruit and vegetables and prepare reports on crop production and market conditions. Grain inspectors inspect and grade all classes of grain at terminal elevators, monitor the fumigation of infested grain and of storage, handling, and transportation and equipment to ensure that sanitary procedures are followed. Meat inspectors monitor the operations and sanitary conditions of slaughtering or meat-processing plants and inspect carcasses to ensure that they are fit for human consumption (Figure 28-7). Plant protection inspectors certify seed crops, oversee the quarantine, treatment, or destruction of plants and plant products, and the fumigation of plants and plant product imports and exports. Many of these individuals work for various levels of government, food processing plants, and slaughter houses.

Public health inspectors inspect workplaces to ensure that equipment, materials, and production processes do not present a health or safety hazard to either employees or the general public. Restaurant inspectors inspect the sanitary conditions of restaurants, hotels, schools, hospitals, and public facilities or institutions and investigate the outbreaks of diseases and poisonings resulting from spoiled food supplies. These individuals often work in municipal, state, or federal government positions.

FIGURE 28-7

Large processing plants employ many inspectors. (*Source:* USDA Photography Library)

Food Service Industry

Individuals in the food service industry area prepare quick meals for the lunchtime crowd, or bake breads and pastries at a local or industrial bakery. Careers in the food service industry include bakers, butchers and meat cutters, chefs, cooks, food service counter attendants, food preparers or kitchen helpers, food service supervisors, and restaurant and food service managers.

Bakers. Some of the job titles associated with bakers include baker, baker apprentice, bakery supervisor, and head baker .

Education. Generally bakers require a high school diploma, a three- or four-year apprenticeship, or a college or other program for bakers. Sometimes on-the-job training is possibly provided.

Job Description. Bakers prepare pies, breads, rolls, muffins, cookies, cakes, icings and frostings, and many other foods, depending on where they work. In local bakeries and the hospitality industries, bakers will–

- Experiment with new recipes and ingredients
- Prepare their own creations and special customer orders
- Draw up production schedules to determine the types and quantities of goods to produce
- Order and/or purchase baking supplies
- Market and sell baked goods

In commercial bakeries or supermarkets, bakers will–

- Prepare dough for pies, bread and rolls, and sweet goods, and prepare batters for muffins, cookies, cakes, icings and frostings according to recipes
- Frost and decorate cakes or other baked goods
- Hire and train baking personnel

Most bakers will work in local and commercial bakeries, supermarkets, hotels/resorts, restaurants, clubs, ships, or their own bakeries.

Butchers and Meat Cutters. Job titles as butchers and meat cutters could include butcher, butcher apprentice, head butcher, meat cutter, or supermarket meat cutter.

Education. A high school diploma may be required. College or other program in meat cutting may be required. On-the-job training in food stores is usually provided. Frequently, trade certification is available after working in the industry.

FIGURE 28-8

Meat cutter.

Job Description. Retail and wholesale butchers and meat cutters prepare those standard cuts of meat, poultry, fish, or shellfish that you buy at a wholesaler, in the supermarket, or at a specialty shop in your town or city. They also–

- Cut, trim, and otherwise prepare the food for sale at self-serve counters or according to customers' special orders
- Grind meats and slice cooked meats using powered grinders and slicing machines
- Shape, lace, and tie roasts and other meats, poultry, or fish
- Prepare special displays
- Supervise other butchers or meat cutters

Butchers and meat cutters work in supermarkets (Figure 28-8), grocery stores, butcher shops, or fish stores, or they own their own businesses.

Chefs. Job titles associated with career choices surrounding chefs include chef, chef de cuisine, chef de partie, corporate chef, executive chef, executive sous-chef, garde manager chef, head chef, master chef, pastry chef, saucier, and specialist chef.

Education. A high school diploma and three-year cook's apprenticeship program is required, or some individuals take formal training abroad. Some executive chefs usually require several years of experience in commercial food preparation, including two years in a supervisory capacity and experience as a sous-chef, specialist chef, or chef. Sous-chefs, specialist chefs, and chefs usually require several years of experience in commercial food preparation. The American Chefs' Federation (ACF) provides certification.

Job Description. Chefs who prepare delicious meals are revered around the world. There are even Culinary Olympics to show off chefs' talents. Anyone interested in becoming a chef can even specialize because quite a variety of chefs exists.

Executive chefs often plan and direct the food preparation and cooking activities of several restaurants in a hotel, restaurant chain, hospital, or other establishment with food services. They plan menus and ensure that the food meets quality standards; estimate food requirements and may estimate food and labor costs; supervise the work of sous-chefs, specialist chefs, chefs, and cooks and may even prepare and cook food on a regular basis, or for special guests and functions.

Sous-chefs usually supervise the activities of specialist chefs, chefs, cooks, and other kitchen workers. They may demonstrate new cooking techniques and new equipment to cooking staff and plan menus and requisition food and kitchen supplies. They may also prepare and cook meals or specialty foods.

Chefs and specialist chefs prepare and cook complete meals, banquets, or specialty foods, such as pastries, sauces, soups, salads, vegetables, and meat, poultry, and fish dishes. They may also create decorative food displays such as fish carved out of ice. Chefs and specialist chefs often instruct cooks in preparation, cooking, garnishing, and presenting food; plan menus; and requisition food and kitchen supplies.

Most chefs of all types work in restaurants, hotels/resorts, hospitals, and other health care institutions, central food commissaries, clubs and similar establishments, and on ships (Figure 28-9). Executive chefs may progress to managerial positions in food preparation businesses.

Cooks. Apprentice cook, cook, dietary cook, grill cook, hospital cook, institutional cook, journeyman/woman cook, licensed cook, second cook, and short order cook are all titles for the career choice of a cook.

Education. A high school diploma is usually required with an apprenticeship program for cooks. Some individuals attend a college or other program in cooking, or they gain several years of commercial cooking experience. Frequently, trade certification is available for work experience.

FIGURE 28-9

Chef and chef's assistant begin daily meal production.

Job Description. Cooks perform some or all of the following duties:

- Prepare and cook complete meals or individual dishes and foods
- Prepare and cook special meals for patients as instructed by a dietitian or chef
- Plan menus, determine the size of food portions, estimate food requirements and costs
- Monitor and order supplies
- Supervise kitchen helpers in the handling of food

Cooks work in restaurants, hotels/resorts, hospitals and other health care institutions, central food commissaries, educational institutions, construction or logging camp sites, and even ships.

Food Service Counter Attendants and Food Preparers/ Kitchen Helpers. Food service titles include cafeteria counter attendant, fast-food preparer, food preparer, ice cream counter attendant, salad bar attendant, or sandwich maker.

Education. Some high school may be required, but most learn their jobs by on-the-job training provided by the organization or company. Some college programs are available.

Job Description. Food service counter attendants and food preparers take customer orders. They prepare food such as sandwiches, hamburgers, salads, milkshakes, and ice cream dishes and take the money for the food purchased. Some may also serve customers at counters or buffet tables.

With additional training and experience, an individual can move into other occupations in food preparation and service, such as cook or waiter. Most of these people work in cafeterias, fast-food outlets, restaurants, hotels, hospitals, and other establishments.

Food Service Supervisors. Job titles for food service supervisors include cafeteria supervisor, catering supervisor, canteen supervisor, and food service supervisor.

Education. A high school diploma is often required. Some community colleges provide programs in food service administration, hotel and restaurant management, or a related discipline. Also, several years of experience in food preparation or service can lead to a job in food service supervision.

Job Description. A good supervisor has to have the ability to deal effectively with people. Food service supervisors coordinate and schedule the activities of staff who prepare and portion food. They may estimate and order ingredients and supplies and prepare food order summaries for chefs according to requests from dietitians, patients in hospitals, or other customers. They supervise and check the assembly of regular and special diet trays and delivery of food trolleys in a hospital and ensure that the food and service meet quality standards.

They work in hospitals and other health care establishments, cafeterias, catering companies, and other food service establishments (Figure 28-10).

FIGURE 28-10
Food service workers in a school lunch program.

Restaurant and Food Service Managers. Titles for restaurant and food service managers can include assistant manager/manager, banquet manager, bar manager, cafeteria manager, catering service manager, dining room manager, food services manager, hotel food and beverage service, and restaurateur.

Education. Most of these jobs require college or other program related to hospitality or food and beverage management. Also, several years of experience in the food service sector, including supervisory experience, are necessary.

Job Description. These individuals manage a restaurant or other food service establishment. They plan, organize, direct, and control the operations of the restaurant, bar, cafeteria, or other food or beverage service, and often they interact with the customers. Restaurant and food services managers also perform some or all of the following duties:

- Determine the type of services to be offered and implement operational procedures
- Recruit staff and oversee staff training
- Set staff work schedules and monitor staff performance
- Control the costs and inventories, monitor revenues, and modify procedures and prices
- Negotiate arrangements with suppliers for food and other supplies
- Negotiate arrangements with clients for catering or the use of facilities for banquets or receptions
- Resolve customer complaints
- Ensure that health and safety regulations are followed

Restaurant and food service managers work in food and beverage service establishments or they may be self-employed.

Food Retail and Wholesale Industry

Much of the food industry relies on buyers and sales representatives. These individuals work for organizations that buy and sell the food products for customers. The general categories of careers include retail and wholesale buyers, and sales representatives.

Job possibilities as retail and wholesale buyers could include food buyer, beverage taster and buyer, and produce buyer.

Education. Jobs as retail and wholesale buyers require a high school diploma. A university degree or college diploma in business, marketing, or a related program is usually required. Also, experience as a sales supervisor or sales representative in an occupation related to the product usually is required. Supervisors and senior buyers require experience.

Job Description. Retail and wholesale buyers decide on the merchandise grocery stores and local cooperatives will carry. Experienced buyers specialize in particular product lines. They study market reports, trade periodicals and promotion materials, and visit trade shows and factories. Once they have seen what is available, they select the products that best fit their company's needs, interview suppliers, and negotiate prices, discounts, credit terms, and transportation arrangements. They oversee the distribution of the products to different outlets and maintain adequate stock levels. More and more these careers will require the use of computers linked to suppliers to place orders.

Retail and wholesale buyers work for food wholesalers, supermarket chains, and other retail outlets.

Sales Representative. For wholesale trade sales representatives possible job titles include: account executive, food products sales rep (representative), and liquor sales rep (Figure 28-11).

Education. High school diploma and a university degree or college diploma in business administration, marketing, or a related program are usually required. Experience as a sales supervisor or sales representative in an occupation related to the product is usually required. Supervisors and senior buyers require experience.

FIGURE 28-11
Maintaining product displays—a sales position in many stores.

Many universities and colleges offer business administration and marketing programs.

Job Description. Sales representatives in the wholesale trade sell products and services to retail, wholesale, commercial, industrial, and professional customers. They look for new customers and take good care of existing ones. Sales reps have to convince customers of the benefits of their products, and provide cost estimates, credit terms, warranties, and delivery dates. They prepare or oversee the preparation of sales contracts. They consult with customers after the sale and provide ongoing support, and review information regarding product innovations, competitors, and market conditions. All this means almost constant contact with people. Sales reps may also travel a lot.

Sales representatives work for companies that produce or provide products and services including food, beverage, and tobacco products.

Research and Development

Research and development provides a career for those interested in developing the next food craze, or breaking down the bacteria behind all the animal and plant diseases that cost farmers millions of dollars every year. This type of career in the food industry is steeped in science–biology, chemistry, or biotechnology. The possibilities for scientists in the food industry include applied chemical technician, applied chemical technologist, food bacteriological technician, food scientist, and food chemist.

Applied Chemical Technician. Applied chemical technician could be a formulation technician, laboratory technician, or food-processing quality control technician (Figure 28-12).

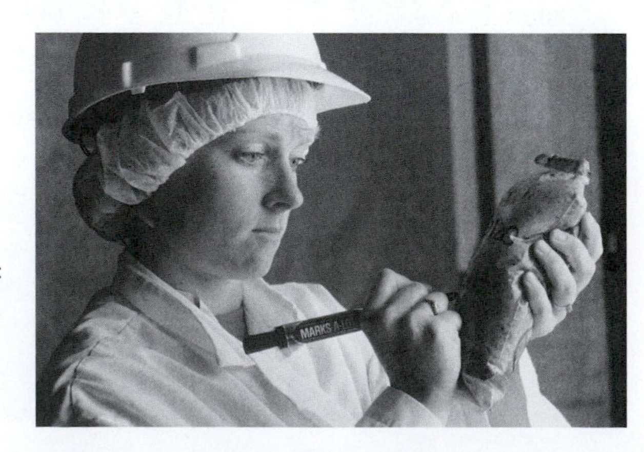

FIGURE 28-12

Technicians work in many aspects of the food industry. (*Source:* USDA Photography Library)

Education. These careers require the completion of a one- or two-year college program in chemical or biochemical technology. Some chemical technicians and technologists are university graduates.

Job Description. Generally, applied chemical technicians work under chemists and technologists, helping them with their research. Applied chemical technicians will–

- Assist in setting up and conducting chemical experiments
- Operate and maintain laboratory equipment and prepare solutions, formulations, and so on
- Compile records for analytical studies
- Carry out a limited range of other technical functions
- Assist in the design and fabrication of experimental apparatus

Applied chemical technicians work in laboratories, food-processing industries; health, education, and government establishments; or they may be self-employed.

Applied Chemical Technologist. An applied chemical technologist can be called a food technologist. This career requires the completion of a two- or three-year college program in chemical, biochemical, or closely related discipline. Many colleges offer chemistry programs. Certification programs are available and may be required by an employer.

Applied chemical technologists or food technologists will conduct chemical experiments, tests, and analyses. They operate and maintain laboratory equipment and prepare solutions, reagents, and sample formulations, and they compile records and interpret results.

Applied chemical technologists work in laboratories, food-processing industries; health, education, and government establishments; or they may be self-employed.

Food Bacteriological Technician. A career as a food bacteriological technician requires the completion of a one- to two-year college program in a related field. Many types of technicians rely on a certification program.

Generally, bacteriological technicians provide technical support to scientists, engineers, and other professionals working in the fields of agriculture, plant and animal biology, microbiology, and cell and molecular biology. Biological technicians assist in conducting biological, microbiological, and biochemical tests and laboratory analyses. They perform a limited range of technical functions in support of agriculture, plant breeding, animal husbandry, and biology. Technicians assist in conducting field research and surveys to collect data and samples of water, soil, plant and animal populations, and they assist in analysis of data and preparation of reports.

Technicians work in laboratory and field settings for governments; manufacturers of food products and pharmaceuticals; biotechnology companies; health, research, and educational institutions; and environmental consulting companies; or they may be self-employed.

Food Scientist and Related Scientists

Food Science is the discipline in which biology, physical sciences, and engineering are used to study the nature of foods, the causes of their deterioration, and the principles underlying food processing.

Education. A food science career requires a bachelor's degree in a related discipline. A master's or doctorate degree is necessary for employment as a research scientist. Postdoctoral research experience is usually required before employment in academic departments or research institutions.

Job Description. Individuals can specialize in the particular discipline that most interests them. Generally, scientists conduct basic and applied research to develop new practices and products related to food and agriculture.

Food scientists work in laboratory (Figure 28-13) and field settings for governments, pharmaceutical and biotechnology companies, as well as for health and educational institutions.

Food Chemist. Chemists play an important role in the development of new foods and nonfood uses such as the development of cosmetics made from milk ingredients. A bachelor's degree in chemistry, biochemistry, or a related discipline is necessary. A master's or doctorate degree is usually required to work as a research chemist.

Chemists analyze, synthesize, purify, modify, and characterize chemical or biochemical compounds. They conduct research to develop new chemical formulations and processes, and research the synthesis and properties of chemical compounds and the mechanisms of chemical reactions. Chemists participate in interdisciplinary research and development projects working with biologists, microbiologists, agronomists, or other professionals, and they act as technical consultants in a particular field of expertise.

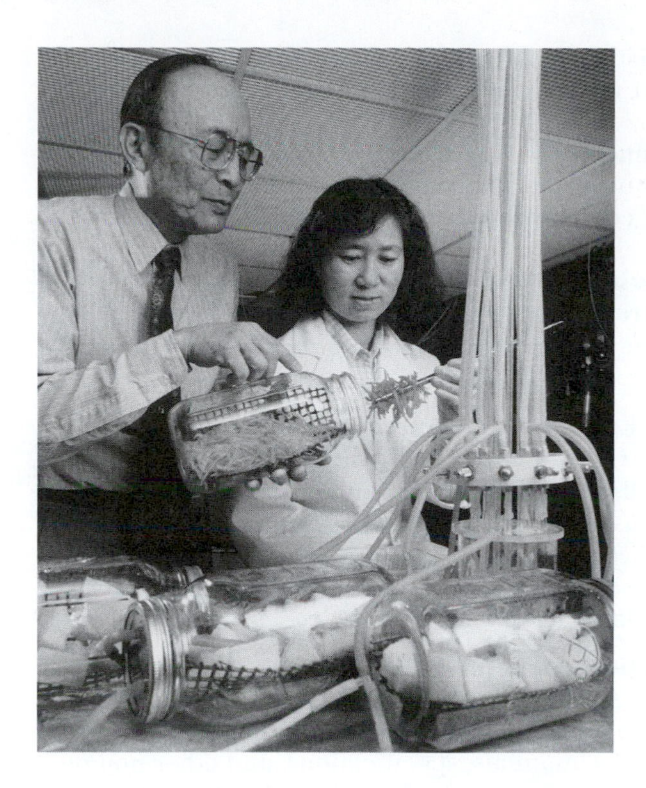

FIGURE 28-13

Food scientists constantly research new products and new processing methods. (*Source:* USDA ARS Image Gallery)

Chemists work in research, development, and quality control laboratories; food processing and biotechnology companies; government and educational institutions.

Marketing and Communications

Marketing products locally or worldwide and communicating globally provide another career area for individuals in the food industry. Successful individuals could end up preparing big-time marketing campaigns for the giants of the food industry. Advertising and marketing specialists and writing and public relations professionals are titles for these careers.

Education. Advertising and marketing specialists or managers require a university degree or college diploma in business administration with sales or marketing specialization, public relations, communications, marketing, journalism, or a related field. Managers also require several years' experience as advertising, public relations, or communications officers.

Job Description. Advertising and marketing specialists analyze advertising needs and current marketing strategies and provide advice on advertising and marketing strategies. They may also plan, develop, and implement advertising campaigns for print or electronic media.

Advertising and marketing specialists work for commercial, industrial, and wholesale organizations; marketing and public relations consulting companies; and government departments.

Writing and Public Relations Professionals. Writing and public relations professionals are called writers, editors, or journalists, or they can be specialists in public relations, communications, and lobbying.

Education. Educational requirements differ depending on the specialization. Usually a university degree or college diploma in English, French, journalism, marketing, communications, or another discipline is required. Most writers require talent and ability, as demonstrated by a work portfolio, in order to be hired. Editors must have several years' experience in journalism, writing, publishing, or a related field.

Job Description. Writers and public relations professionals are usually responsible for writing the copy and preparing the events that help promote the food industries. Writers research and write books, speeches, manuals, specifications, and other nonjournalistic articles. Copywriters study and determine the selling features of products and services and write text for advertisements and commercials. Writers may specialize in a particular subject or type of writing.

Editors review, evaluate, and edit manuscripts, articles, news reports, and other material for publication and broadcast, and coordinate the activities of staff. Journalists research and write for newspapers, television, radio, and other media. Specialists in public relations and communications develop and implement communications strategies and information programs, publicize activities and events, and maintain media relations on behalf of clients. They may also work as lobbyists on behalf of the food industry or associations.

Writers and public relations professionals work for governments, nongovernment associations, private industry, or they are self-employed.

Others

A group of other occupational areas may also have food industry-related jobs. Some of these include the following:

- Environmental agriculture

- Exporting
- Livestock and/or crop production
- Horticulture
- Policy development and analysis
- Veterinary and animal health

As individuals learn their abilities, capabilities, and areas of interest, they can find food industry careers in these occupational areas.

FOOD INDUSTRY SUPERVISED AGRICULTURAL EXPERIENCE

A Supervised Agricultural Experience (SAE) is designed to provide students the opportunity to gain experience in agricultural areas based on their interests. An SAE represents the actual, planned application of concepts and principles learned in agricultural education. Students experience and apply what is learned in the classroom to real-life situations. Students are supervised by agriculture teachers in cooperation with parents/guardians, employers, and other adults who assist them in the development and achievement of their educational goals. The purpose is to help students develop skills and abilities leading toward a career.

Planning and conducting an SAE for food science could include areas of interest such as food processing, food chemistry, nutrition, food packaging, food commodities, and food regulations. Students should work with their instructors to–

- Identify an appropriate SAE opportunity in the community
- Ensure that the SAE represents meaningful learning activities benefiting the student, the agriculture education program, and the community
- Obtain classroom and individual instruction on SAE
- Adopt a suitable record keeping system
- Plan the SAE and acquire needed resources
- Coordinate release time and visits to SAE
- Sign a training agreement along with the employer, teacher, and parent/gardian
- Report on and evaluate the SAE and records resulting from it

Additional help and ideas for planning and conducting an SAE can be found through the National FFA Web site (<www. ffa.org>).

GETTING A JOB

After some job possibilities are identified, the work begins. Getting the job is difficult and requires some preparation. Again whole books, videos, and seminars teach how to get a job. A few tips follow.

After a job is identified, do a little research on the company and the job before applying. Know these things about the job and the company:

- Name of the company
- Name of the personnel manager
- Company address and phone number
- Position available
- Requirements for the position
- Geographic scope of the company–local, county, state, regional, national
- Company's product(s)
- Recent company developments
- Responsibilities of the position
- Demand for the company's product(s)

Before you get too far along in the application process, be certain that the job is what you want to pursue. Money is not everything in a job. Compare the requirements and demands of the occupation with the characteristics you possess.

Application Forms

If the company requires an application form, remember that you are trying to sell yourself by the information given. Review the entire application form before you begin. Pay particular attention to any special instructions to print or write in your own handwriting. When answering ads that require potential employees to apply in person, be prepared to complete an application form on the spot. Take an ink pen. Prepare a list of information you will need to complete the application form: your social security number; the addresses of schools you have attended; names, phone numbers, and addresses of previous employers and supervisors; names, phone numbers, and addresses of references. The following guidelines will provide you with some direction when completing application forms:

1. Follow all instructions carefully and exactly.
2. If handwritten, rather than typed, write neatly and legibly.

Handwritten answers should be printed unless otherwise directed.

3. Application forms should be written in ink unless otherwise requested. If you make a mistake, mark through it with one neat line.
4. Be honest and realistic.
5. Give all the facts for each question.
6. Keep answers brief.
7. Fill in all the blanks. If the question does not pertain to you, write "not applicable" or "N/A." If there is no answer, write "none" or draw a short line through the blank.
8. Many application forms ask what salary you expect. If you are not sure what is appropriate, write "negotiable," "open," or "scale" in the blank. Before applying, try to find out what the going rate for similar work is at other locations. Give a salary range rather than exact figure.

Letters of Inquiry and Application

The purpose of a **letter of inquiry** is to obtain information about possible job vacancies. The purpose of a **letter of application** is to apply for a specific position that has been publicly advertised. Both letters indicate your interest in working for a particular company, acquaint employers with your qualifications, and encourage the employer to invite you for a job interview.

Letters of inquiry and application represent you. They should be accurate, informative, and attractive. Your written communications should present a strong, positive, professional image both as a job seeker and future employee. The following list should be used as a guide when writing letters of inquiry and application:

1. Short and specific, one or two pages (details left to **résumé**). Use 8½ × 11-inch white typing paper, not personal or fancy paper.
2. Neatly typed and error free.
3. Attractive form, free from smudges.
4. Write to a specific person. Use "To Whom It May Concern" if answering a blind ad.
5. Logical organized paragraphs that are to the point.
6. Carefully constructed sentences, free from spelling or grammatical errors.

7. Positive in tone.
8. Ideas expressed in a clear, concise direct manner.
9. Avoid slang words and expressions.
10. Avoid excessive use of the word "I."
11. Avoid mentioning salary and fringe benefits.
12. Write a first draft, then make revisions.
13. Proofread final letter yourself, and also have someone else proofread.
14. Address and sign correctly. Type envelope addresses.

This information should be included in a letter of inquiry:

1. Specify the reasons why you are interested in working for the company and ask if there are any positions available now or expected in the near future.
2. Express your interest in being considered a candidate for a position when one becomes available.
3. Because you are not applying for a particular position, you cannot relate your qualifications directly to job requirements. (You can explain how your personal qualifications and work experience would help meet the needs of the company.)
4. Mention and include your résumé.
5. State your willingness to meet with a company representative to discuss your background and qualifications. (Include your address and phone number where you can be reached.)
6. Address letters of inquiry to the personnel manager unless you know his or her name.

A letter of application should include:

1. Your source of the job lead.
2. The particular job you are applying for and the reason for your interest in the position and the company.
3. How your personal qualifications meet the needs of the employer.
4. How your work experience relates to job requirements.
5. Your résumé.
6. A request for an interview and a statement of your willingness. (Include your address and phone number where you can be reached.)

Résumé or Data Sheet

Some jobs require a résumé (Figure 28-14) or **data sheet**. The following information should be considered when writing a résumé or data sheet:

- Name, address, and phone number
- Brief, specific statement of career objective
- Educational background–names of schools, dates, major field of study, degrees or diplomas–listed in reverse chronological order
- Leadership activities, honors, and accomplishments
- Work experience, listed in reverse chronological order
- Special technical skills and interests related to job
- References
- One page if possible
- Neatly typed and error free
- Logically organized
- Honestly listed qualifications and experiences

Employers look for a quick overview of who you are and how you fit into their business. On the first reading, the employer will spend 10 to 15 seconds reading a résumé. Be sure to present relevant information clearly and concisely in an eye-catching format.

The Interview

The next step in the job-hunting process is the interview. While many do's and don'ts of an interview are available, perhaps the best advice comes from the interviewer's side of the desk. This list provides common reasons interviewers give for not being able to place applicants in a job.

1. Poor attitude
2. Unstable work record
3. Bad references
4. Lack of self-selling ability
5. Lack of skill and experience
6. Not really anxious to work
7. "Bad mouthing" former employers
8. Too demanding (wanting too much money or to work only under certain conditions)
9. Unable to be available for interviews or canceling out

RÉSUMÉ
Susan Smith

CURRENT ADDRESS
PO Box 1238
Anywhere, ID 00000

Telephone: 000/888-8888
E-mail address: sasmith3@micron.com

EDUCATION
- Local High School, Anywhere, ID: Graduated 1999.
- College of Southern Idaho, Twin Falls, ID, 1999–2001: A.S. Food Science.

CAREER OBJECTIVE
Obtain satisfying job in the food science industry that provides advancement opportunities during my career.

ACTIVITIES AND HONORS
- Active member of 4-H Club for three years. Learned food preparation.
- Member FFA for four years and was elected president during senior year.
- Member Postsecondary Agricultural Student (PAS) organization 2000–2001.
- Advisor to local 4-H Club 2000 to present.

EMPLOYMENT AND WORK EXPERIENCES
- **January 1999 to Present:** Sunrise Bakery, Arco, ID; general help; prepare breads and donuts, work some in sales.
- **July 1997 to December 1999:** ABC Grocery, McCall, ID; restocked shelves; boxed groceries; worked into checker position.

REFERENCES
Available on request.

FIGURE 28-14
Neat, complete résumés—a necessary component for career development.

10. Poor appearance
11. Lack of manners and personal courtesy
12. Chewing gum, smoking, fidgeting
13. No attempt to establish rapport; not looking the interviewer in the eye
14. Being interested only in the salary and benefits of the job
15. Lack of confidence; being evasive
16. Poor grammar, use of slang
17. Not having any direction or goals

Follow-up Letters

Follow-up letters are sent immediately after an interview. The follow-up letter demonstrates your knowledge of business etiquette and protocol. Always send a follow-up letter regardless of whether or not you had a good interviewing experience and regardless of whether you are interested in the position. When employers do not receive follow-up letters from job candidates, they often assume that the candidate is not aware of the professional courtesy and protocol they will need to demonstrate on the job.

The major purpose of a follow-up letter is to thank those individuals who participated in your interview. In addition, a follow-up letter reinforces your name, application, and qualifications to the employer and indicates whether you are still interested in the job position.

OCCUPATIONAL SAFETY

Employees in the food industry should expect a safe and healthful workplace. Still, individuals may encounter such hazards as:

- Toxic chemicals in cleaning products
- Slippery floors
- Hot cooking equipment
- Sharp objects
- Heavy lifting
- Stress
- Harassment
- Poor workstation designs

To prevent or minimize exposure to occupational hazards, employers are expected to provide employees with safety and health training, including providing information on chemicals that could be harmful to an individual's health. If an employee is injured or becomes ill because of a job, many employers pay for medical care and lost wages are sometimes provided.

Not only are food industry employers responsible for creating and maintaining a safe workplace. Employees must do their part, including:

- Follow all safety rules and instructions
- Use safety equipment and protective clothing when needed
- Look out for coworkers
- Keep work areas clean and neat
- Know what to do in an emergency
- Report any health and safety hazard to the supervisor

A JOB IS MORE THAN MONEY

Before taking a job, be certain that it is what you want. Although the salary or the wage is important, job satisfaction is something quite different and very important. Jobs quickly become routine and mundane. For example, a job with little fulfillment and challenge for some people can easily become a chore just to go to. Before taking a job or even while looking for a job, answer these questions:

1. Does the job description fit your interests?
2. Is this the level of occupation in which you wish to engage?
3. Does this type of work appeal to your interests?
4. Are the working conditions suitable to you?
5. Will you be satisfied with the salaries and benefits offered?
6. Can you advance in this occupation as rapidly as you would like? What are the advancement opportunities?
7. Does the future outlook satisfy you?
8. Is the occupation in demand now and in the foreseeable future?
9. Do you have or can you get the education needed for the occupation?
10. What type of training is available after taking the job?
11. Can you get the finances needed to get into the occupation?
12. Can you meet the health and physical requirements?
13. Will you be able to meet the entry requirements?
14. Do you know of any other reasons you might not be able to enter this occupation?
15. Is the occupation available locally or are you willing to move to a part of the country where it is available?

For additional information about personal and occupational safety practices in the workplace contact OSHA (Occupational Safety and Health Administration) on the Web at <www.osha.gov>, the National Institute for Occupational Safety and Health (NIOSH) on the Web at <www.cdc.gov/niosh/homepage.html>, or the Department of Labor at <www.dol.gov>.

Summary

The goal of education and training is primarily to become employable and stay employable–to get and keep a job, career, or to run a successful business. The world of work requires people who can read, write, do math, and communicate. Rapidly changing technology has made this even more critical. Also, the modern workplace

---

Also, before taking a job or looking for a job, do a little personality inventory of yourself. Consider the following:

1. Do I like to be alone or with people?
2. Am I mechanical or artistic?
3. Would I rather work independently or work under supervision?
4. Would I like to think or be active?
5. Could I take authority and responsibility for others?
6. Must I have freedom to express creativity?
7. What things do I like to do? Make a list.
8. At what time of day can I work best?
9. Can I work under pressure or stress?
10. Make a list of your strong points. Consider skills, hobbies, and leisure time activities you can offer an employer.

Do your research and your job will be more rewarding and you will feel better about yourself.

For more information about building a career, search the Web or visit these Web sites:

<www.careermotiv8.com/career.html>
<career-planning.com/>
<www.careerperfect.com/CareerPerfect/careerplan.htm>
<www.careersnet.com/>
<www.hawk.igs.net/~jobs/career-counseling/>
<www.asktheemployer.com/>

---

now looks for people who possess thinking skills. Even with a solid set of basic skills, future employees also need to relate to other people; must be able to use information; need to understand the concept of systems; and know how to use technology. Old-fashioned ideas like responsibility, self-esteem, sociability, self-management, and integrity are not out of date.

Jobs or careers in the food industry range from those very closely tied to the industry to those that support the food industry. In general, potential job or career areas include production management, product development, food engineer, microbiology, quality control, research, technical representative, sales, and service. Education and training in the food industry vary from on-the-job training to high school and college degrees and certificates.

After training and education, finding and getting the right job or career may still be a challenge. Good resources exist for locating a job. Still, one of the best resources is personal contact. Well-written letters of inquiry, application forms, a clear eye-catching résumé, and being prepared for the job interview will help secure a job.

Review Questions

Success in any career requires knowledge. Test your knowledge of this chapter by answering these questions or solving these problems.

1. What are the six basic skills required in the workplace?
2. Define "entrepreneur."
3. List the parts of a résumé.
4. A letter of ___________ is to obtain information about possible job vacancies. A letter of ___________ is to apply for a specific position that has been publicly advertised.
5. Name ten reasons why an interview may fail.
6. ___________ processing is the largest manufacturing industry in the world.
7. List the five thinking skills needed in the new workplace.
8. Research and development in any area of the food industry requires a ___________ background.
9. What education is required for a chef certification?
10. Identify four jobs in the food safety and inspection area of the food industry.

Student Activities

1. Develop your own résumé or data sheet.
2. Collect position announcements and classified ads for jobs in the food industry. Write a letter of job inquiry and a letter of job application for a selected job using this information.
3. Develop a list of questions frequently asked during an interview. Use the questions in role-playing job interviews and videotape the interviews.
4. Organize a field trip to a public or private placement office. Following the field trip, discuss the office's policies and how they affect job searchers and employers. Alternatively, invite a representative from a state employment agency to explain how employment agencies can help students gain employment.
5. Hold a food industry career field day. Invite individuals currently employed in the food industry to present a panel discussion on career opportunities. For example, invite representatives from research, education, and government.
6. Select one career in the food industry of interest and prepare a research paper on the career using a computer and word processing software. The paper should identify the knowledge and skills required and the employment opportunities.
7. Collect pictures or photographs of people engaged in various careers with food. Use them to prepare a bulletin board or collage.
8. Invite a resource person such as a business owner or personnel manager to discuss what he or she looks for in résumés, application letters, and forms and during interviews.
9. Invite a panel of local businesspeople to discuss the importance of employee work habits, basic skills, and attitudes and how they affect the entire business.

Resources

Aslett, D. 1993. *Everything I needed to know about business I learned in the barnyard.* Pocatello, ID: Marsh Creek Press.

Bolles, R. N. 1995. *What color is your parachute? A practical manual for job-hunters and career-changers.* Berkeley, CA: Ten Speed Press.

Business Council for Effective Literacy. (BCEL) Bulletin. (1987). *Job-related basic skills: A guide for planners of employee programs.* New York.

Ricketts, C. 1997. *Leadership: Personal development and career success.* Albany, NY: Delmar.

U.S. Department of Education. (1991). *America 2000: An education strategy.* Sourcebook. Washington, D.C.: United States Department of Education.

Ziglar, Z. 1993. *See you at the top.* Gretna, LA: Pelican Publishing Company, Inc.

Internet

Internet sites represent a vast resource of information. The URLs (uniform resource locator) for the World Wide Web sites can change. Using one of the search engines on the Internet such as Yahoo!, HotBot, AltaVista, Excite, Dogpile, About, or Google, find more information by searching for these words or phrases: résumé writing, data sheet, food science careers, entrepreneur, competencies (reading, writing, listening, speaking), letter of inquiry, letter of application, job interview, any specific food science career listing. Also, Table A-7 provides a listing of some useful Internet sites that can be used as a starting point.

Appendix A

Due to its location in a book and because of the implications of its name, an appendix is often ignored by the reader. But an appendix contains valuable information that can enhance a reader's understanding and learning. Moreover, the information in an appendix is quick and easy to find.

The information in this appendix includes a variety of useful conversions, conversion factors, measurement standards, and common measures. This appendix contains a table listing some Internet sites (URLs) that lead to many sources of data and information. Finally, the last table in this appendix is a food composition table that lists many common foods.

By making full use of this appendix, the reader can understand more, plan better, do more, and learn more.

TABLE A-1 Conversion Tables for Common Weights and Measures

Metric Conversions	Equaled Amount
1 pound	454 grams
2.2 pounds	1 kilogram
1 quart	1 liter
1 gram	15.43 grains
1 metric ton	2,205 pounds
1 inch	2.54 centimeters
1 centimeter	10 millimeters or .39 inch
1 meter	39.37 inches
1 acre	.406 hectare

TABLE A-2 Weight Conversions

Measurements	Equaled Amount
8 tablespoons	1/4 pound
3 teaspoons	1 tablespoon
1 pint	1 pound
2 pints	1 quart
4 quarts	1 gallon or 8 pounds
2,000 pounds	1 ton
16 ounces	1 pound
27 cubic feet	1 cubic yard
1 peck	8 quarts
1 bushel	4 pecks
Other Conversions	
1%	.01
1%	10,000 ppm
1 Megacalorie (M-cal)	1,000 calories
1 calorie (big calorie)	1,000 calories (small calorie)
1 M-cal	1 therm

TABLE A-3 Common Measures and Approximate Equivalents

1 liquid teaspoon =	5 milliliters (ml)
3 liquid teaspoons =	1 liquid tablespoon = 15 ml
2 liquid tablespoons =	1 liquid ounce = 30 ml
8 liquid ounces =	1 liquid cup = 0.24 liter
2 liquid cups =	1 liquid pint = 0.47 liter
2 liquid pints =	1 liquid quart = 0.9463 liter
4 liquid quarts =	1 liquid gallon (U.S.) = 3.7854 liters

Table A-4 Fahrenheit to Centigrade Temperature Conversions

°F	°C	°F	°C	°F	°C
100	37.8	77	25.0	54	12.2
99	37.2	76	24.4	53	11.7
98	36.7	75	23.9	52	11.1
97	36.1	74	23.3	51	10.6
96	35.6	73	22.8	50	10.0
95	35.0	72	22.2	49	9.4
94	34.4	71	21.7	48	8.9
93	33.9	70	21.1	47	8.3
92	33.3	69	20.6	46	7.8
91	32.8	68	20.0	45	7.2
90	32.2	67	19.4	44	6.7
89	31.7	66	18.9	43	6.1
88	31.1	65	18.3	42	5.6
87	30.6	64	17.8	41	5.0
86	30.0	63	17.2	40	4.4
85	29.4	62	16.7	39	3.9
84	28.9	61	16.1	38	3.3
83	28.3	60	15.6	37	2.8
82	27.8	59	15.0	36	2.2
81	27.2	58	14.4	35	1.7
80	26.7	57	13.9	34	1.1
79	26.1	56	13.3	33	0.6
78	25.6	55	12.8	32	0.0

Formulas used: °C = (°F – 32) × 5/9 or °F = (°C × 9/5) + 32

TABLE A-5 Conversion Factors for English and Metric Measurements

To Convert the English	To the Metric Multiply by	To Convert Metric	Multiply by	To get English
acres	0.4047	hectares	2.47	acres
acres	4047	m^2	0.000247	acres
BTU	1055	joules	0.000948	BTU
BTU	0.0002928	kwh	3415.301	BTU
BTU/hr.	0.2931	watts	3.411805	BTU/hr.
bu.	0.03524	m^3	28.37684	bu.
bu.	35.24	L	0.028377	bu.
$ft.^3$	0.02832	m^3	35.31073	$ft.^3$
$ft.^3$	28.32	L	0.035311	$ft.^3$
$in.^3$	16.39	cm^3	0.061013	$in.^3$
$in.^3$	1.639×10^{-5}	m^3	61012.81	$in.^3$
$in.^3$	0.01639	L	61.01281	$in.^3$
$yd.^3$	0.7646	m^3	1.307873	$yd.^3$
$yd.^3$	764.6	L	0.001308	$yd.^3$
ft.	30.48	cm	0.032808	ft.
ft.	0.3048	m	3.28084	ft.
ft./min.	0.508	cm/sec.	1.968504	ft./min.
ft./sec.	30.48	cm/sec.	0.032808	ft./sec.
gal.	3785	cm^3	0.000264	gal.
gal.	0.003785	m^3	264.2008	gal.
gal.	3.785	L	0.264201	gal.
gal./min.	0.06308	L/sec.	15.85289	gal./min.
in.	2.54	cm	0.393701	in.
in.	0.0254	m	39.37008	in.
mi.	1.609	km	0.621504	mi.
mph	26.82	m/min.	0.037286	mph

(continued)

TABLE A-5 Conversion Factors for English and Metric Measurements

To Convert the English	To the Metric Multiply by	To Convert Metric	Multiply by	To get English
oz.	28.349	gm	0.035275	oz.
fl. oz.	0.02947	L	33.93281	fl. oz.
liq. pt.	0.4732	L	2.113271	liq. pt.
lb.	453.59	gm	0.002205	lb.
qt.	0.9463	L	1.056747	qt.
$ft.^2$	0.0929	m^2	10.76426	$ft.^2$
$yd.^2$	0.8361	m^2	1.196029	$yd.^2$
tons	0.9078	tonnes	1.101564	tons
yd.	0.0009144	km	1093.613	yd.
yd.	0.9144	m	1.093613	yd.

Table A-6 More Conversion Factors for Metric and English Units

Length
1 mile = 1.609 kilometers; 1 kilometer = 0.621 mile
1 yard = 0.914 meter; 1 meter = 1.094 yards
1 inch = 2.54 centimeters; 1 centimeter = 0.394 inch

Area
1 square mile = 2.59 square kilometers; 1 square kilometer = 0.386 square mile
1 acre = 0.00405 square kilometer; 1 square kilometer = 247.1 acres
1 acre = 0.405 hectare; 1 hectare = 2.471 acres

Volume
1 acre/inch = 102.8 cubic meters; 1 cubic meter = 0.00973 acre/inch
1 quart = 0.946 liter; 1 liter = 1.057 quarts
1 bushel = 0.352 hectoliter; 1 hectoliter = 2.838 bushels

Weight
1 pound = 0.454 kilogram; 1 kilogram = 2.205 pounds
1 pound = 0.00454 quintal; 1 quintal = 220.5 pounds
1 ton = 0.9072 metric ton; 1 metric ton = 1.102 tons

Yield or Rate
1 pound/acre = 1.121 kilograms/acre; 1 kilogram/acre = 0.892 pound/acre
1 ton/acre = 2.242 tons/hectare; 1 ton/hectare = 0.446 ton/acre
1 bushel/acre = 1.121 quintals/hectare; 1 quintal/hectare = 0.892 bushel/acre
1 bushel/acre = (60#) = 0.6726 quintal/hectare; 1 quintal/hectare = 1.487 bushel/acre (60#)
1 bushel/acre = (56#) = 0.6278 quintal/acre; 1 quintal/acre = 1.597 bushels/acre (56#)

Temperature
To convert Fahrenheit (F) to Celsius (C): $0.555 \times (F - 32)$
To convert Celsius (C) to Fahrenheit (F): $1.8 \times (C + 32)$

TABLE A-7 Food Science Resources on the Internet[1]

Name/Topic	URL
Agriculture and Agri-Food Canada (careers)	<www.cfa-fca.ca/careers/index1.html>
American Association of Cereal Chemists	<www.scisoc.org/aacc/>
American Culinary Federation	<www.acfchefs.org/>
American Dairy Science Association	<www.adsa.org/>
American Dietetic Association	<www.eatright.org/adafansa.html>
American Egg Board	<www.aeb.org/>
American Institute of Baking	<www.aibonline.org/>
American Meat Institute	<www.meatami.org/>
American Soybean Association	<www.asa-europe.org/>
Bakery Online	<www.bakeryonline.com/content/homepage/>
Beef.Org	<www.beef.org>
Burger King	<www.burgerking.com>
California Agricultural Technology Institute	<www.atinet.org/CATI/webs/>
Canning Basics for Food Preservation	<www.thevision.net/DMS/canning.htm>
Center for Food Safety and Applied Nutrition	<vm.cfsan.fda.gov/list.html>
Coca-Cola	<www.coca-cola.com/gateway.html>
Cornell Food Science and Technology	<www.nysaes.cornell.edu/fst/>
Council for Agricultural Science and Technology	<www.cast-science.org/>

[1] This table is not a complete listing of food, food science, food processing, or food industry resources on the Internet. It is merely a starting point. Sources in this table will lead to other sources. Many more resources can be found by using one of the search engines available such as Yahoo!, HotBot, Northernlight, Dogpile, About, or Google.

TABLE A-7 Food Science Resources on the Internet *(continued)*

Name/Topic	URL
Dairy Network.com	<www.dairynetwork.com/content/homepage/>
FDA's Bad Bug Book	<vm.cfsan.fda.gov/~mow/intro.html>
Fermentation	<www.uwrf.edu/biotech/workshop/activity/act1/act1.htm>
Food and Agriculture Organization of the United Nations	<www.fao.org/>
Food and Drug Administration (FDA)	<www.fda.gov/>
Food and health (University of Minnesota Extension Service)	<www.extension.umn.edu/nutrition/>
Food and Nutrition Information Center	<www.nalusda.gov/fnic/etext/fnic.html>
FoodCom	<www.foodcom.com/foodcom/>
Food Design Net	<www.foodesignet.com/>
Food Industry Resource	<www.foodfront.com/index.htm>
FoodNet	<foodnet.fic.ca/>
Food preservation (KSU)	<www.oznet.ksu.edu/ext_f&n/foodpreservation/drying.htm>
Food Marketing Institute	<www.fmi.org/>
Food Safety and Inspection Service (USDA)	<www.fsis.usda.gov/index.htm>
Food safety and irradiation (Iowa State)	<www.exnet.iastate.edu/foodsafety/rad/irradhome.html>
Foster Farms	<www.fosterfarms.com>
FruitNet	<www.fruitnet.com/>
General Mills	<www.genmills.com/>
Government Food Safety Information	<www.foodsafety.gov/>
Grocery Manufacturers of America	<www.gmabrands.com/>
Horizon Organic Dairy	<www.horizonorganic.com/>

(continued)

TABLE A-7 Food Science Resources on the Internet *(continued)*

Name/Topic	URL
Hormel Foods	<www.hormel.com/>
IBP, Inc.	<www.ibpinc.com/>
Institute of Food Technologists	<www.ift.org/>
International Food Information Council	<www.ificinfo.health.org/>
International Organization for Standardization (ISO)	<www.iso.ch/welcome.html>
Just Food.com	<just-food.com/>
KFC	<www.kfc.com>
Kraft	<www.kraftfoods.com/index.cgi>
Kroger	<www.kroger.com/>
Loders and Croklaan	<www.croklaan.com/>
McDonald's	<www.mcdonalds.com>
Meat and Poultry Online	<www.meatandpoultryonline.com/content/homepage/>
Meat News.com	<www.meatnews.com/default.cfm>
MinuteMaid	<www.minutemaid.com/>
Nasco	<www.nascofa.com/>
National Food Safety Database (USDA)	<www.foodsafety.ufl.edu/index.html>
National Meat Association	<www.nmaonline.org/>
National Pork Producers Council	<www.nppc.org/>
National Food Processors Association	<www.nfpa-food.org/>
National Food Service Management Institute	<www.nfsmi.org/>
Net Food Directory	<www.foodwine.com/digest/Food-Bev/index.html>
Nutrition and Food Curriculum Guide	<www.uen.org/utahlink/lp_res/nutri375.html>

TABLE A-7 Food Science Resources on the Internet *(concluded)*

Name/Topic	URL
Nutrient Data Laboratory (USDA)	<www.nal.usda.gov/fnic/foodcomp/Data/index.html>
PastryWiz Food Resource Center	<www.sweettechnology.com/>
Pepsi	<www.pepsi.com>
Perdue Farms	<www.perdue.com/>
Performance Food Group	<www.pfgc.com/frame.htm>
Pizza Hut	<www.pizzahut.com>
Postharvest series	<www.bae.ncsu.edu/programs/extension/publicat/postharv/>
Produce Marketing Association	<www.pma.com/>
Safefood.org	<www.safefood.org/>
Seafood education	<www.vims.edu/adv/seafood/>
Seafood Network Information Center	<seafood.ucdavis.edu/>
Smithfield Foods	<www.smithfield.com>
Soy Protein Council	<www.spcouncil.org/>
Taco Bell	<www.tacobell.com>
The Kitchen Link	<www.kitchenlink.com/cgi/public_frames?page=search3>
Tyson Foods, Inc.	<www.tyson.com/>
U.S. Meat Animal Research Center	<www.marc.usda.gov/>
U.S. Poultry and Egg Association	<www.poultryegg.org/>
U.S. Soyfoods Directory	<www.soyfoods.com/>
Wendy's	<www.wendys.com>
World food issues (Iowa State University)	<www.ag.iastate.edu/grants/SALV96/salv97.cp.html>

TABLE A-8 Food Composition Table

No.	Description of Food		Wt. (g)	Water (%)	Energy (kcal)	Protein (g)	Fat (g)	Sat (g)	Mono (g)	Poly (g)	Cholest (mg)	Carb (g)	Ca (mg)	P (mg)	Fe (mg)	K (mg)	Na (mg)	Vit A (IU)	Vit A (RE)	Thmn (mg)	Ribofl (mg)	Niacin (mg)	Vit C (mg)
Beverages																							
10	Beer; Regular	12 Fl Oz	360	92	150	1	0	0	0	0	0	13	14	50	0.1	115	18	0	0	0.02	0.09	1.8	0
20	Beer; Light	12 Fl Oz	355	95	95	1	0	0	0	0	0	5	14	43	0.1	64	11	0	0	0.03	0.11	1.4	0
30	Gin; Rum; Vodka; Whiskey 80-Proof	1.5 F Oz	42	67	95	0	0	0	0	0	0	0	0	0	0	1	0	0	0	0	0	0	0
40	Gin; Rum; Vodka; Whiskey 86-Proof	1.5 F Oz	42	64	105	0	0	0	0	0	0	0	0	0	0	1	0	0	0	0	0	0	0
50	Gin; Rum; Vodka; Whiskey 90-Proof	1.5 F Oz	42	62	110	0	0	0	0	0	0	0	0	0	0	1	0	0	0	0	0	0	0
60	Wine; Dessert	3.5 F Oz	103	77	140	0	0	0	0	0	0	8	8	9	0.2	95	9	0	0	0.01	0.02	0.2	0
70	Wine; Table; Red	3.5 F Oz	102	88	75	0	0	0	0	0	0	3	8	18	0.4	113	5	0	0	0	0.03	0.1	0
80	Wine; Table; White	3.5 F Oz	102	87	80	0	0	0	0	0	0	3	9	14	0.3	83	5	0	0	0	0.01	0.1	0
90	Club Soda	12 Fl Oz	355	100	0	0	0	0	0	0	0	0	18	0	0	0	78	0	0	0	0	0	0
100	Cola; Regular	12 Fl Oz	369	89	160	0	0	0	0	0	0	41	11	52	0.2	7	18	0	0	0	0	0	0
110	Cola; Diet; Aspartame + Saccharin	12 Fl Oz	355	100	0	0	0	0	0	0	0	0	14	39	0.2	7	32	0	0	0	0	0	0
111	Cola; Diet; Saccharin Only	12 Fl Oz	355	100	0	0	0	0	0	0	0	0	14	39	0.2	7	75	0	0	0	0	0	0
112	Cola; Diet; Aspartame Only	12 Fl Oz	355	100	0	0	0	0	0	0	0	0	14	39	0.2	7	23	0	0	0	0	0	0
120	Ginger Ale	12 Fl Oz	366	91	125	0	0	0	0	0	0	32	11	0	0.1	4	29	0	0	0	0	0	0
130	Grape Soda	12 Fl Oz	372	88	180	0	0	0	0	0	0	46	15	0	0.4	4	48	0	0	0	0	0	0
140	Lemon-Lime Soda	12 Fl Oz	372	89	155	0	0	0	0	0	0	39	7	0	0.4	4	33	0	0	0	0	0	0
150	Orange Soda	12 Fl Oz	372	88	180	0	0	0	0	0	0	46	15	4	0.3	7	52	0	0	0	0	0	0
160	Pepper-Type Soda	12 Fl Oz	369	89	160	0	0	0	0	0	0	41	11	41	0.1	4	37	0	0	0	0	0	0
170	Root Beer	12 Fl Oz	370	89	165	0	0	0	0	0	0	42	15	0	0.2	4	48	0	0	0	0	0	0
180	Coffee; Brewed	6 Fl Oz	180	100	0	0	0	0	0	0	0	0	4	2	0	124	2	0	0	0	0.02	0.4	0
190	Coffee; Instant; Prepared	6 Fl Oz	182	99	0	0	0	0	0	0	0	1	2	6	0.1	71	0	0	0	0	0.03	0.6	0
200	Fruit Punch Drink; Canned	6 Fl Oz	190	88	85	0	0	0	0	0	0	22	15	2	0.4	48	15	20	2	0.03	0.04	0	61
210	Grape Drink; Canned	6 Fl Oz	187	86	100	0	0	0	0	0	0	26	2	2	0.3	9	11	0	0	0.01	0.01	0	64
220	Pineapple-Grapefruit Juice Drink	6 Fl Oz	187	87	90	0	0	0	0	0	0	23	13	7	0.9	97	24	60	6	0.06	0.04	0.5	110
230	Lemonade; Concentrate; Frz; Undil	6 Fl Oz	219	49	425	0	0	0	0	0	0	112	9	13	0.4	153	4	40	4	0.04	0.07	0.7	66
240	Lemonade; Concen;Frzen; Diluted	6 Fl Oz	185	89	80	0	0	0	0	0	0	21	2	2	0.1	30	1	10	1	0.01	0.02	0.2	13
250	Limeade; Concentrate; Frzn; Undil	6 Fl Oz	218	50	410	0	0	0	0	0	0	108	11	13	0.2	129	0	0	0	0.02	0.02	0.2	26
260	Limeade; Concen; Frozen; Diluted	6 Fl Oz	185	89	75	0	0	0	0	0	0	20	2	2	0	24	0	0	0	0	0	0	4
270	Tea; Brewed	8 Fl Oz	240	100	0	0	0	0	0	0	0	0	0	2	0	36	1	0	0	0	0.03	0	0
280	Tea; Instant; Preprd; Unsweetend	8 Fl Oz	241	100	0	0	0	0	0	0	0	1	1	4	0	61	1	0	0	0	0.02	0.1	0
290	Tea; Instant; Prepard; Sweetened	8 Fl Oz	262	91	85	0	0	0	0	0	0	22	1	3	0	49	0	0	0	0	0.04	0.1	0

TABLE A-8 Food Composition Table *(continued)*

No.	Description of Food		Wt. (g)	Water (%)	Energy (kcal)	Protein (g)	Fat (g)	Sat (g)	Mono (g)	Poly (g)	Cholest (mg)	Carb (g)	Ca (mg)	P (mg)	Fe (mg)	K (mg)	Na (mg)	Vit A (IU)	Vit A (RE)	Thmn (mg)	Ribofl (mg)	Niacin (mg)	Vit C (mg)
Cheeses																							
300	Blue Cheese	1 Oz	28.4	42	100	6	8	5.3	2.2	0.2	21	1	150	110	0.1	73	396	200	65	0.01	0.11	0.3	0
310	Camembert Cheese	1 Wedge	38	52	115	8	9	5.8	2.7	0.3	27	0	147	132	0.1	71	320	350	96	0.01	0.19	0.2	0
320	Cheddar Cheese	1 Oz	28.4	37	115	7	9	6	2.7	0.3	30	0	204	145	0.2	28	176	300	86	0.01	0.11	0	0
330	Cheddar Cheese	1 Cu In	17	37	70	4	6	3.6	1.6	0.2	18	0	123	87	0.1	17	105	180	52	0	0.06	0	0
340	Chedddar Cheese; Shredded	1 Cup	113	37	455	28	37	23.8	10.6	1.1	119	1	815	579	0.8	111	701	1200	342	0.03	0.42	0.1	0
350	Cottage Cheese; Cremd; Lrge Curd	1 Cup	225	79	235	28	10	6.4	2.9	0.3	34	6	135	297	0.3	190	911	370	108	0.05	0.37	0.3	0
360	Cottage Cheese; Cremd; Smll Curd	1 Cup	210	79	215	26	9	6	2.7	0.3	31	6	126	277	0.3	177	850	340	101	0.04	0.34	0.3	0
370	Cottage Cheese; Cremd; W/Fruit	1 Cup	226	72	280	22	8	4.9	2.2	0.2	25	30	108	236	0.2	151	915	280	81	0.04	0.29	0.2	0
380	Cottage Cheese; Lowfat 2%	1 Cup	226	79	205	31	4	2.8	1.2	0.1	19	8	155	340	0.4	217	918	160	45	0.05	0.42	0.3	0
390	Cottage Cheese; Uncreamed	1 Cup	145	80	125	25	1	0.4	0.2	0	10	3	46	151	0.3	47	19	40	12	0.04	0.21	0.2	0
400	Cream Cheese	1 Oz	28.4	54	100	2	10	6.2	2.8	0.4	31	1	23	30	0.3	34	84	400	124	0	0.06	0	0
410	Feta Cheese	1 Oz	28.4	55	75	4	6	4.2	1.3	0.2	25	1	140	96	0.2	18	316	130	36	0.04	0.24	0.3	0
420	Mozzarella Cheese; Whole Milk	1 Oz	28.4	54	80	6	6	3.7	1.9	0.2	22	1	147	105	0.1	19	106	220	68	0	0.07	0	0
430	Mozzarella Chese; Skim; Lomoist	1 Oz	28.4	49	80	8	5	3.1	1.4	0.1	15	1	207	149	0.1	27	150	180	54	0.01	0.1	0	0
440	Muenster Cheese	1 Oz	28.4	42	105	7	9	5.4	2.5	0.2	27	0	203	133	0.1	38	178	320	90	0	0.09	0	0
450	Parmesan Cheese; Grated	1 Cup	100	18	455	42	30	19.1	8.7	0.7	79	4	1376	807	1	107	1861	700	173	0.05	0.39	0.3	0
460	Parmesan Cheese; Grated	1 Tbsp	5	18	25	2	2	1	0.4	0	4	0	69	40	0	5	93	40	9	0	0.02	0	0
470	Parmesan Cheese; Grated	1 Oz	28.4	18	130	12	9	5.4	2.5	0.2	22	1	390	229	0.3	30	528	200	49	0.01	0.11	0.1	0
480	Provolone Cheese	1 Oz	28.4	41	100	7	8	4.8	2.1	0.2	20	1	214	141	0.1	39	248	230	75	0.01	0.09	0	0
490	Ricotta Cheese; Whole Milk	1 Cup	246	72	430	28	32	20.4	8.9	0.9	124	7	509	389	0.9	257	207	1210	330	0.03	0.48	0.3	0
500	Ricotta Cheese; Part Skim Milk	1 Cup	246	74	340	28	19	12.1	5.7	0.6	76	13	669	449	1.1	307	307	1060	278	0.05	0.46	0.2	0
510	Swiss Cheese	1 Oz	28.4	37	105	8	8	5	2.1	0.3	26	1	272	171	0	31	74	240	72	0.01	0.1	0	0
520	Pasterzd Proces Cheese; American	1 Oz	28.4	39	105	6	9	5.6	2.5	0.3	27	0	174	211	0.1	46	406	340	82	0.01	0.1	0	0
530	Pasterzd Proces Cheese; Swiss	1 Oz	28.4	42	95	7	7	4.5	2	0.2	24	1	219	216	0.2	61	388	230	65	0	0.08	0	0
540	Pasterzd Proces Cheese Food; Amr	1 Oz	28.4	43	95	6	7	4.4	2	0.2	18	2	163	130	0.2	79	337	260	62	0.01	0.13	0	0
550	Pasterzd Proces Cheese Spred; Amr	1 Oz	28.4	48	80	5	6	3.8	1.8	0.2	16	2	159	202	0.1	69	381	220	54	0.01	0.12	0	0
Milk and Milk Products																							
560	Half and Half; Cream	1 Cup	242	81	315	7	28	17.3	8	1	89	10	254	230	0.2	314	98	1050	259	0.08	0.36	0.2	2
570	Half and Half; Cream	1 Tbsp	15	81	20	0	2	1.1	0.5	0.1	6	1	16	14	0	19	6	70	16	0.01	0.02	0	0
580	Light; Coffee or Table Cream	1 Cup	240	74	470	6	46	28.8	13.4	1.7	159	9	231	192	0.1	292	95	1730	437	0.08	0.36	0.1	2
590	Light; Coffee or Table Cream	1 Tbsp	15	74	30	0	3	1.8	0.8	0.1	10	1	14	12	0	18	6	110	27	0	0.02	0	0

TABLE A-8 Food Composition Table *(continued)*

No.	Description of Food		Wt. (g)	Water (%)	Energy (kcal)	Protein (g)	Fat (g)	Sat (g)	Mono (g)	Poly (g)	Cholest (mg)	Carb (g)	Ca (mg)	P (mg)	Fe (mg)	K (mg)	Na (mg)	Vit A (IU)	Vit A (RE)	Thmn (mg)	Ribofl (mg)	Niacin (mg)	Vit C (mg)
Milk and Milk Products *(continued)*																							
600	Whipping Cream; Unwhiped;Light	1 Cup	239	64	700	5	74	46.2	21.7	2.1	265	7	166	146	0.1	231	82	2690	705	0.06	0.3	0.1	1
610	Whipping Cream; Unwhiped;Light	1 Tbsp	15	64	45	0	5	2.9	1.4	0.1	17	0	10	9	0	15	5	170	44	0	0.02	0	0
620	Whipping Cream; Unwhiped;Heavy	1 Cup	238	58	820	5	88	54.8	25.4	3.3	326	7	154	149	0.1	179	89	3500	1002	0.05	0.26	0.1	1
630	Whipping Cream; Unwhiped;Heavy	1 Tbsp	15	58	50	0	6	3.5	1.6	0.2	21	0	10	9	0	11	6	220	63	0	0.02	0	0
640	Whipped Topping; Pressurized	1 Cup	60	61	155	2	13	8.3	3.9	0.5	46	7	61	54	0	88	78	550	124	0.02	0.04	0	0
650	Whipped Topping; Pressurized	1 Tbsp	3	61	10	0	1	0.4	0.2	0	2	0	3	3	0	4	4	30	6	0	0	0	0
660	Sour Cream	1 Cup	230	71	495	7	48	30	13.9	1.8	102	10	268	195	0.1	331	123	1820	448	0.08	0.34	0.2	2
670	Sour Cream	1 Tbsp	12	71	25	0	3	1.6	0.7	0.1	5	1	14	10	0	17	6	90	23	0	0.02	0	0
680	Imitation Creamers; Liquid Frz	1 Tbsp	15	77	20	0	1	1.4	0	0	0	2	1	10	0	29	12	10	1	0	0	0	0
690	Imitation Creamers; Powdered	1 Tsp	2	2	10	0	1	0.7	0	0	0	1	0	8	0	16	4	0	0	0	0	0	0
700	Imitation Whipped Topping;Frzn	1 Cup	75	50	240	1	19	16.3	1.2	0.4	0	17	5	6	0.1	14	19	650	65	0	0	0	0
710	Imitation Whipped Topping;Frzn	1 Tbsp	4	50	15	0	1	0.9	0.1	0	0	1	0	0	0	1	1	30	3	0	0	0	0
720	Imitation Whipd Toping;Pwdrd;Prp	1 Cup	80	67	150	3	10	8.5	0.7	0.2	8	13	72	69	0	121	53	290	39	0.02	0.09	0	1
730	Imitation Whipd Toping;Pwdrd;Prp	1 Tbsp	4	67	10	0	0	0.4	0	0	0	1	4	3	0	6	3	10	2	0	0	0	0
740	Imitation Whipd Toping;Pressrzd	1 Cup	70	60	185	1	16	13.2	1.3	0.2	0	11	4	13	0	13	43	330	33	0	0	0	0
750	Imitation Whipd Toping;Pressrzd	1 Tbsp	4	60	10	0	1	0.8	0.1	0	0	1	0	1	0	1	2	20	2	0	0	0	0
760	Imitation Sour Dressing	1 Cup	235	75	415	8	39	31.2	4.6	1.1	13	11	266	205	0.1	380	113	20	5	0.09	0.38	0.2	2
770	Imitation Sour Dressing	1 Tbsp	12	75	20	0	2	1.6	0.2	0.1	1	1	14	10	0	19	6	0	0	0	0.02	0	0
780	Milk; Whole; 3.3% Fat	1 Cup	244	88	150	8	8	5.1	2.4	0.3	33	11	291	228	0.1	370	120	310	76	0.09	0.4	0.2	2
790	Milk; Lowfat; 2%; No Addedsolid	1 Cup	244	89	120	8	5	2.9	1.4	0.2	18	12	297	232	0.1	377	122	500	139	0.1	0.4	0.2	2
800	Milk; Lowfat; 2%; Added Solids	1 Cup	245	89	125	9	5	2.9	1.4	0.2	18	12	313	245	0.1	397	128	500	140	0.1	0.42	0.2	2
810	Milk; Lowfat; 1%; No Addedsolid	1 Cup	244	90	100	8	3	1.6	0.7	0.1	10	12	300	235	0.1	381	123	500	144	0.1	0.41	0.2	2
820	Milk; Lowfat; 1%; Added Solids	1 Cup	245	90	105	9	2	1.5	0.7	0.1	10	12	313	245	0.1	397	128	500	145	0.1	0.42	0.2	2
830	Milk; Skim; No Added Milksolid	1 Cup	245	91	85	8	0	0.3	0.1	0	4	12	302	247	0.1	406	126	500	149	0.09	0.34	0.2	2
840	Milk; Skim; Added Milk Solids	1 Cup	245	90	90	9	1	0.4	0.2	0	5	12	316	255	0.1	418	130	500	149	0.1	0.43	0.2	2
850	Buttermilk; Fluid	1 Cup	245	90	100	8	2	1.3	0.6	0.1	9	12	285	219	0.1	371	257	80	20	0.08	0.38	0.1	2
860	Sweetened Condensed Milk Cnnd	1 Cup	306	27	980	24	27	16.8	7.4	1	104	166	868	775	0.6	1136	389	1000	248	0.28	1.27	0.6	8
870	Evaporated Milk; Whole; Canned	1 Cup	252	74	340	17	19	11.6	5.9	0.6	74	25	657	510	0.5	764	267	610	136	0.12	0.8	0.5	5
880	Evaporated Milk; Skim; Canned	1 Cup	255	79	200	19	1	0.3	0.2	0	9	29	738	497	0.7	845	293	1000	298	0.11	0.79	0.4	3
890	Buttermilk; Dried	1 Cup	120	3	465	41	7	4.3	2	0.3	83	59	1421	1119	0.4	1910	621	260	65	0.47	1.89	1.1	7
900	Nonfat Dry Milk; Instantized	1 Envlpe	91	4	325	32	1	0.4	0.2	0	17	47	1120	896	0.3	1552	499	2160	646	0.38	1.59	0.8	5

TABLE A-8 Food Composition Table (continued)

No.	Description of Food		Wt. (g)	Water (%)	Energy (kcal)	Protein (g)	Fat (g)	Sat (g)	Mono (g)	Poly (g)	Cholest (mg)	Carb (g)	Ca (mg)	P (mg)	Fe (mg)	K (mg)	Na (mg)	Vit A (IU)	Vit A (RE)	Thmn (mg)	Ribofl (mg)	Niacin (mg)	Vit C (mg)
Milk and Milk Products *(continued)*																							
910	Nonfat Dry Milk; Instantized	1 Cup	68	4	245	24	0	0.3	0.1	0	12	35	837	670	0.2	1160	373	1610	483	0.28	1.19	0.6	4
920	Chocolate Milk; Regular	1 Cup	250	82	210	8	8	5.3	2.5	0.3	31	26	280	251	0.6	417	149	300	73	0.09	0.41	0.3	2
930	Chocolate Milk; Lowfat 2%	1 Cup	250	84	180	8	5	3.1	1.5	0.2	17	26	284	254	0.6	422	151	500	143	0.09	0.41	0.3	2
940	Chocolate Milk; Lowfat 1%	1 Cup	250	85	160	8	3	1.5	0.8	0.1	7	26	287	256	0.6	425	152	500	148	0.1	0.42	0.3	2
950	Cocoa Pwdr with Nonfat Drymilk	1 Oz	28.4	1	100	3	1	0.6	0.3	0	1	22	90	88	0.3	223	139	0	0	0.03	0.17	0.2	0
960	Cocoa Pwdr W/Nofat Drmlk;Prpd	1 Servng	206	86	100	3	1	0.6	0.3	0	1	22	90	88	0.3	223	139	0	0	0.03	0.17	0.2	0
970	Coca Pwdr W/O Nonfat Dry Milk	3/4 Oz	21	1	75	1	1	0.3	0.2	0	0	19	7	26	0.7	136	56	0	0	0	0.03	0.1	0
980	Coca Pwdr W/O Nofat Drymlk;Prd	1 Servng	265	81	225	9	9	5.4	2.5	0.3	33	30	298	254	0.9	508	176	310	76	0.1	0.43	0.3	3
990	Eggnog	1 Cup	254	74	340	10	19	11.3	5.7	0.9	149	34	330	278	0.5	420	138	890	203	0.09	0.48	0.3	4
1000	Malted Milk; Chocolate; Powder	3/4 Oz	21	2	85	1	1	0.5	0.3	0.1	1	18	13	37	0.4	130	49	20	5	0.04	0.04	0.4	0
1010	Malted Milk;Chocolate; Pwdrppd	1 Servng	265	81	235	9	9	5.5	2.7	0.4	34	29	304	265	0.5	500	168	330	80	0.14	0.43	0.7	2
1020	Malted Milk;Natural; Powder	3/4 Oz	21	3	85	3	2	0.9	0.5	0.3	4	15	56	79	0.2	159	96	70	17	0.11	0.14	1.1	0
1030	Malted Milk;Natural; Pwdr Pprd	1 Servng	265	81	235	11	10	6	2.9	0.6	37	27	347	307	0.3	529	215	380	93	0.2	0.54	1.3	2
1040	Shakes; Thick; Chocolate	10 Oz	283	72	335	9	8	4.8	2.2	0.3	30	60	374	357	0.9	634	314	240	59	0.13	0.63	0.4	0
1050	Shakes; Thick; Vanilla	10 Oz	283	74	315	11	9	5.3	2.5	0.3	33	50	413	326	0.3	517	270	320	79	0.08	0.55	0.4	0
1060	Ice Cream; Vanilla; Regular 11%	1/2 Gal	1064	61	2155	38	115	71.3	33.1	4.3	476	254	1406	1075	1	2052	929	4340	1064	0.42	2.63	1.1	6
1070	Ice Cream; Vanilla; Regular 11%	1 Cup	133	61	270	5	14	8.9	4.1	0.5	59	32	176	134	0.1	257	116	540	133	0.05	0.33	0.1	1
1080	Ice Cream; Vanilla; Regular 11%	3 Fl Oz	50	61	100	2	5	3.4	1.6	0.2	22	12	66	51	0	96	44	200	50	0.02	0.12	0.1	0
1090	Ice Cream; Vanilla; Soft Serve	1 Cup	173	60	375	7	23	13.5	6.7	1	153	38	236	199	0.4	338	153	790	199	0.08	0.45	0.2	1
1100	Ice Cream; Vanilla; Rich 16% Ft	1/2 Gal	1188	59	2805	33	190	118.3	54.9	7.1	703	256	1213	927	0.8	1771	868	7200	1758	0.36	2.27	0.9	5
1110	Ice Cream; Vanilla; Rich 16% Ft	1 Cup	148	59	350	4	24	14.7	6.8	0.9	88	32	151	115	0.1	221	108	900	219	0.04	0.28	0.1	1
1120	Ice Milk; Vanilla; 4% Fat	1/2 Gal	1048	69	1470	41	45	28.1	13	1.7	146	232	1409	1035	1.5	2117	836	1710	419	0.61	2.78	0.9	6
1130	Ice Milk; Vanilla; 4% Fat	1 Cup	131	69	185	5	6	3.5	1.6	0.2	18	29	176	129	0.2	265	105	210	52	0.08	0.35	0.1	1
1140	Ice Milk; Vanilla;Softserv 3%	1 Cup	175	70	225	8	5	2.9	1.3	0.2	13	38	274	202	0.3	412	163	175	44	0.12	0.54	0.2	1
1150	Sherbet; 2% Fat	1/2 Gal	1542	66	2160	17	31	19	8.8	1.1	113	469	827	594	2.5	1585	706	1480	308	0.26	0.71	1	31
1160	Sherbet; 2% Fat	1 Cup	193	66	270	2	4	2.4	1.1	0.1	14	59	103	74	0.3	198	88	190	39	0.03	0.09	0.1	4
1170	Yogurt; W/Lowfat Milk;Fruitflv	8 Oz	227	74	230	10	2	1.6	0.7	0.1	10	43	345	271	0.2	442	133	100	25	0.08	0.4	0.2	1
1180	Yogurt; W/Lowfat Milk; Plain	8 Oz	227	85	145	12	4	2.3	1	0.1	14	16	415	326	0.2	531	159	150	36	0.1	0.49	0.3	2
1190	Yogurt; W/Nonfat Milk	8 Oz	227	85	125	13	0	0.3	0.1	0	4	17	452	355	0.2	579	174	20	5	0.11	0.53	0.3	2
1200	Yogurt; W/Whole Milk	8 Oz	227	88	140	8	7	4.8	2	0.2	29	11	274	215	0.1	351	105	280	68	0.07	0.32	0.2	1

TABLE A-8 Food Composition Table *(continued)*

No.	Description of Food		Wt. (g)	Water (%)	Energy (kcal)	Protein (g)	Fat (g)	Sat (g)	Mono (g)	Poly (g)	Cholest (mg)	Carb (g)	Ca (mg)	P (mg)	Fe (mg)	K (mg)	Na (mg)	Vit A (IU)	Vit A (RE)	Thmn (mg)	Ribofl (mg)	Niacin (mg)	Vit C (mg)
Eggs																							
1210	Eggs; Raw; Whole	1 Egg	50	75	75	6	5	1.6	1.9	0.7	213	1	25	89	0.7	60	63	320	95	0.03	0.25	0	0
1220	Eggs; Raw; White	1 White	33	88	15	4	0	0	0	0	0	0	2	4	0	48	55	0	0	0	0.15	0	0
1230	Eggs; Raw; Yolk	1 Yolk	17	49	60	3	5	1.6	1.9	0.7	213	0	23	81	0.6	16	7	320	97	0.03	0.11	0	0
1240	Eggs; Cooked; Fried	1 Egg	46	69	90	6	7	1.9	2.7	1.3	211	1	25	89	0.7	61	162	390	114	0.03	0.24	0	0
1250	Eggs; Cooked; Hard-Cooked	1 Egg	50	75	75	6	5	1.6	2	0.7	213	1	25	86	0.6	63	62	280	84	0.03	0.26	0	0
1260	Eggs; Cooked; Poached	1 Egg	50	75	75	6	5	1.5	1.9	0.7	212	1	25	89	0.7	60	140	320	95	0.02	0.22	0	0
1270	Eggs; Cooked; Scrambled/Omelet	1 Egg	61	73	100	7	7	2.2	2.9	1.3	215	1	44	104	0.7	84	171	420	119	0.03	0.27	0	0
Fats and Oils																							
1280	Butter; Salted	1/2 Cup	113	16	810	1	92	57.1	26.4	3.4	247	0	27	26	0.2	29	933	3460	852	0.01	0.04	0	0
1281	Butter; Unsalted	1/2 Cup	113	16	810	1	92	57.1	26.4	3.4	247	0	27	26	0.2	29	12	3460	852	0.01	0.04	0	0
1290	Butter; Salted	1 Tbsp	14	16	100	0	11	7.1	3.3	0.4	31	0	3	3	0	4	116	430	106	0	0	0	0
1291	Butter; Unsalted	1 Tbsp	14	16	100	0	11	7.1	3.3	0.4	31	0	3	3	0	4	2	430	106	0	0	0	0
1300	Butter; Salted	1 Pat	5	16	35	0	4	2.5	1.2	0.2	11	0	1	1	0	1	41	150	38	0	0	0	0
1301	Butter; Unsalted	1 Pat	5	16	35	0	4	2.5	1.2	0.2	11	0	1	1	0	1	1	150	38	0	0	0	0
1310	Fats; Cooking/Vegetable Shortening	1 Cup	205	0	1810	0	205	51.3	91.2	53.5	0	0	0	0	0	0	0	0	0	0	0	0	0
1320	Fats; Cooking/Vegetable Shortening	1 Tbsp	13	0	115	0	13	3.3	5.8	3.4	0	0	0	0	0	0	0	0	0	0	0	0	0
1330	Lard	1 Cup	205	0	1850	0	205	80.4	92.5	23	195	0	0	0	0	0	0	0	0	0	0	0	0
1340	Lard	1 Tbsp	13	0	115	0	13	5.1	5.9	1.5	12	0	0	0	0	0	0	0	0	0	0	0	0
1350	Margarine; Imitation 40% Fat	8 Oz	227	58	785	1	88	17.5	35.6	31.3	0	1	40	31	0	57	2178	7510	2254	0.01	0.05	0	0
1360	Margarine; Imitation 40% Fat	1 Tbsp	14	58	50	0	5	1.1	2.2	1.9	0	0	2	2	0	4	134	460	139	0	0	0	0
1370	Margarine; Regular; Hard; 80% Fat	1/2 Cup	113	16	810	1	91	17.9	40.5	28.7	0	1	34	26	0.1	48	1066	3740	1122	0.01	0.04	0	0
1380	Margarine; Regular; Hard; 80% Fat	1 Tbsp	14	16	100	0	11	2.2	5	3.6	0	0	4	3	0	6	132	460	139	0	0.01	0	0
1390	Margarine; Regular; Hard; 80% Fat	1 Pat	5	16	35	0	4	0.8	1.8	1.3	0	0	1	1	0	2	47	170	50	0	0	0	0
1400	Margarine; Regular; Soft; 80% Fat	8 Oz	227	16	1625	2	183	31.3	64.7	78.5	0	1	60	46	0	86	2449	7510	2254	0.02	0.07	0	0
1410	Margarine; Regular; Soft; 80% Fat	1 Tbsp	14	16	100	0	11	1.9	4	4.8	0	0	4	3	0	5	151	460	139	0	0	0	0
1420	Margarine; Spread; Hard; 60% Fat	1/2 Cup	113	37	610	1	69	15.9	29.4	20.5	0	0	24	18	0	34	1123	3740	1122	0.01	0.03	0	0
1430	Margarine; Spread; Hard; 60% Fat	1 Tbsp	14	37	75	0	9	2	3.6	2.5	0	0	3	2	0	4	139	460	139	0	0	0	0
1440	Margarine; Spread; Hard; 60% Fat	1 Pat	5	37	25	0	3	0.7	1.3	0.9	0	0	1	1	0	1	50	170	50	0	0	0	0
1450	Margarine; Spread;Soft; 60% Fat	8 Oz	227	37	1225	1	138	29.1	71.5	31.3	0	0	47	37	0	68	2256	7510	2254	0.02	0.06	0	0
1460	Margarine; Spread;Soft; 60% Fat	1 Tbsp	14	37	75	0	9	1.8	4.4	1.9	0	0	3	2	0	4	139	460	139	0	0	0	0
1470	Corn Oil	1 Cup	218	0	1925	0	218	27.7	52.8	128	0	0	0	0	0	0	0	0	0	0	0	0	0

TABLE A-8 Food Composition Table *(continued)*

No.	Description of Food		Wt. (g)	Water (%)	Energy (kcal)	Protein (g)	Fat (g)	Sat (g)	Mono (g)	Poly (g)	Cholest (mg)	Carb (g)	Ca (mg)	P (mg)	Fe (mg)	K (mg)	Na (mg)	Vit A (IU)	Vit A (RE)	Thmn (mg)	Ribofl (mg)	Niacin (mg)	Vit C (mg)
Fats and Oils *(continued)*																							
1480	Corn Oil	1 Tbsp	14	0	125	0	14	1.8	3.4	8.2	0	0	0	0	0	0	0	0	0	0	0	0	0
1490	Olive Oil	1 Cup	216	0	1910	0	216	29.2	159.2	18.1	0	0	0	0	0	0	0	0	0	0	0	0	0
1500	Olive Oil	1 Tbsp	14	0	125	0	14	1.9	10.3	1.2	0	0	0	0	0	0	0	0	0	0	0	0	0
1510	Peanut Oil	1 Cup	216	0	1910	0	216	36.5	99.8	69.1	0	0	0	0	0	0	0	0	0	0	0	0	0
1520	Peanut Oil	1 Tbsp	14	0	125	0	14	2.4	6.5	4.5	0	0	0	0	0	0	0	0	0	0	0	0	0
1530	Safflower Oil	1 Cup	218	0	1925	0	218	19.8	26.4	162.4	0	0	0	0	0	0	0	0	0	0	0	0	0
1540	Safflower Oil	1 Tbsp	14	0	125	0	14	1.3	1.7	10.4	0	0	0	0	0	0	0	0	0	0	0	0	0
1550	Soybean Oil; Hydrogenated	1 Cup	218	0	1925	0	218	32.5	93.7	82	0	0	0	0	0	0	0	0	0	0	0	0	0
1560	Soybean Oil; Hydrogenated	1 Tbsp	14	0	125	0	14	2.1	6	5.3	0	0	0	0	0	0	0	0	0	0	0	0	0
1570	Soybean-Cottonseed Oil; Hydrgn	1 Cup	218	0	1925	0	218	39.2	64.3	104.9	0	0	0	0	0	0	0	0	0	0	0	0	0
1580	Soybean-Cottonseed Oil; Hydrgn	1 Tbsp	14	0	125	0	14	2.5	4.1	6.7	0	0	0	0	0	0	0	0	0	0	0	0	0
1590	Sunflower Oil	1 Cup	218	0	1925	0	218	22.5	42.5	143.2	0	0	0	0	0	0	0	0	0	0	0	0	0
1600	Sunflower Oil	1 Tbsp	14	0	125	0	14	1.4	2.7	9.2	0	0	0	0	0	0	0	0	0	0	0	0	0
Dressings																							
1610	Blue Cheese Salad Dressing	1 Tbsp	15	32	75	1	8	1.5	1.8	4.2	3	1	12	11	0	6	164	30	10	0	0.02	0	0
1620	French Salad Dressing; Regular	1 Tbsp	16	35	85	0	9	1.4	4	3.5	0	1	2	1	0	2	188	0	0	0	0	0	0
1630	French Salad Dressing; Localor	1 Tbsp	16	75	25	0	2	0.2	0.3	1	0	2	6	5	0	3	306	0	0	0	0	0	0
1640	Italian Salad Dressing; Regular	1 Tbsp	15	34	80	0	9	1.3	3.7	3.2	0	1	1	1	0	5	162	30	3	0	0	0	0
1650	Italian Salad Dressing; Localor	1 Tbsp	15	86	5	0	0	0	0	0	0	2	1	1	0	4	136	0	0	0	0	0	0
1660	Mayonnaise; Regular	1 Tbsp	14	15	100	0	11	1.7	3.2	5.8	8	0	3	4	0.1	5	80	40	12	0	0	0	0
1670	Mayonnaise; Imitation	1 Tbsp	15	63	35	0	3	0.5	0.7	1.6	4	2	0	0	0	2	75	0	0	0	0	0	0
1680	Mayonnaise Type Salad Dressing	1 Tbsp	15	40	60	0	5	0.7	1.4	2.7	4	4	2	4	0	1	107	30	13	0	0	0	0
1690	Tartar Sauce	1 Tbsp	14	34	75	0	8	1.2	2.6	3.9	4	1	3	4	0.1	11	182	30	9	0	0	0	0
1700	1000 Island; Salad Drsng;Reglr	1 Tbsp	16	46	60	0	6	1	1.3	3.2	4	2	2	3	0.1	18	112	50	15	0	0	0	0
1710	1000 Island; Salad Drsng;Local	1 Tbsp	15	69	25	0	2	0.2	0.4	0.9	2	2	2	3	0.1	17	150	50	14	0	0	0	0
1720	Cooked Salad Drssing; Home Rcp	1 Tbsp	16	69	25	1	2	0.5	0.6	0.3	9	2	13	14	0.1	19	117	70	20	0.01	0.02	0	0
1730	Vinegar And Oil Salad Dressing	1 Tbsp	16	47	70	0	8	1.5	2.4	3.9	0	0	0	0	0	1	0	0	0	0	0	0	0
Fish and Seafood																							
1740	Clams; Raw	3 Oz	85	82	65	11	1	0.3	0.3	0.3	43	2	59	138	2.6	154	102	90	26	0.09	0.15	1.1	9
1750	Clams; Canned; Drained	3 Oz	85	77	85	13	2	0.5	0.5	0.4	54	2	47	116	3.5	119	102	90	26	0.01	0.09	0.9	3
1760	Crabmeat; Canned	1 Cup	135	77	135	23	3	0.5	0.8	1.4	135	1	61	246	1.1	149	1350	50	14	0.11	0.11	2.6	0

TABLE A-8 Food Composition Table *(continued)*

No.	Description of Food		Wt. (g)	Water (%)	Energy (kcal)	Protein (g)	Fat (g)	Sat (g)	Mono (g)	Poly (g)	Cholest (mg)	Carb (g)	Ca (mg)	P (mg)	Fe (mg)	K (mg)	Na (mg)	Vit A (IU)	Vit A (RE)	Thmn (mg)	Ribofl (mg)	Niacin (mg)	Vit C (mg)
Fish and Seafood *(continued)*																							
1770	Fish Sticks; Frozen; Reheated	1 Stick	28	52	70	6	3	0.8	1.4	0.8	26	4	11	58	0.3	94	53	20	5	0.03	0.05	0.6	0
1780	Flounder or Sole; Baked; Buttr	3 Oz	85	73	120	16	6	3.2	1.5	0.5	68	0	13	187	0.3	272	145	210	54	0.05	0.08	1.6	1
1790	Flounder or Sole; Baked; Margrn	3 Oz	85	73	120	16	6	1.2	2.3	1.9	55	0	14	187	0.3	273	151	230	69	0.05	0.08	1.6	1
1800	Flounder or Sole; Baked; W/Ofat	3 Oz	85	78	80	17	1	0.3	0.2	0.4	59	0	13	197	0.3	286	101	30	10	0.05	0.08	1.7	1
1810	Haddock; Breaded; Fried	3 Oz	85	61	175	17	9	2.4	3.9	2.4	75	7	34	183	1	270	123	70	20	0.06	0.1	2.9	0
1820	Halibut; Broiled; Butter; Lemju	3 Oz	85	67	140	20	6	3.3	1.6	0.7	62	0	14	206	0.7	441	103	610	174	0.06	0.07	7.7	1
1830	Herring; Pickled	3 Oz	85	59	190	17	13	4.3	4.6	3.1	85	0	29	128	0.9	85	850	110	33	0.04	0.18	2.8	0
1840	Ocean Perch; Breaded; Fried	1 Fillet	85	59	185	16	11	2.6	4.6	2.8	66	7	31	191	1.2	241	138	70	20	0.1	0.11	2	0
1850	Oysters; Raw	1 Cup	240	85	160	20	4	1.4	0.5	1.4	120	8	226	343	15.6	290	175	740	223	0.34	0.43	6	24
1860	Oysters; Breaded; Fried	1 Oyster	45	65	90	5	5	1.4	2.1	1.4	35	5	49	73	3	64	70	150	44	0.07	0.1	1.3	4
1870	Salmon; Canned; Pink; W/Bones	3 Oz	85	71	120	17	5	0.9	1.5	2.1	34	0	167	243	0.7	307	443	60	18	0.03	0.15	6.8	0
1880	Salmon; Baked; Red	3 Oz	85	67	140	21	5	1.2	2.4	1.4	60	0	26	269	0.5	305	55	290	87	0.18	0.14	5.5	0
1890	Salmon; Smoked	3 Oz	85	59	150	18	8	2.6	3.9	0.7	51	0	12	208	0.8	327	1700	260	77	0.17	0.17	6.8	0
1900	Sardines; Atlntc; Cnned; Oil;Drn	3 Oz	85	62	175	20	9	2.1	3.7	2.9	85	0	371	424	2.6	349	425	190	56	0.03	0.17	4.6	0
1910	Scallops; Breaded; Frzn; Reheat	6 Scallops	90	59	195	15	10	2.5	4.1	2.5	70	10	39	203	2	369	298	70	21	0.11	0.11	1.6	0
1920	Shrimp; Canned; Drained	3 Oz	85	70	100	21	1	0.2	0.2	0.4	128	1	98	224	1.4	104	1955	50	15	0.01	0.03	1.5	0
1930	Shrimp; French Fried	3 Oz	85	55	200	16	10	2.5	4.1	2.6	168	11	61	154	2	189	384	90	26	0.06	0.09	2.8	0
1940	Trout; Broiled; W/Butter;Lemj	3 Oz	85	63	175	21	9	4.1	2.9	1.6	71	0	26	259	1	297	122	230	60	0.07	0.07	2.3	1
1950	Tuna; Cannd; Drnd; Oil; Chk;Lght	3 Oz	85	61	165	24	7	1.4	1.9	3.1	55	0	7	199	1.6	298	303	70	20	0.04	0.09	10.1	0
1960	Tuna; Cannd; Drnd; Water; White	3 Oz	85	63	135	30	1	0.3	0.2	0.3	48	0	17	202	0.6	255	468	110	32	0.03	0.1	13.4	0
1970	Tuna Salad	1 Cup	205	63	375	33	19	3.3	4.9	9.2	80	19	31	281	2.5	531	877	230	53	0.06	0.14	13.3	6
Fruits and Fruit Products																							
1980	Apples; Raw; Unpeeled; 3 Per Lb	1 Apple	138	84	80	0	0	0.1	0	0.1	0	21	10	10	0.2	159	0	70	7	0.02	0.02	0.1	8
1990	Apples; Raw; Unpeeled; 2 Per Lb	1 Apple	212	84	125	0	1	0.1	0	0.2	0	32	15	15	0.4	244	0	110	11	0.04	0.03	0.2	12
2000	Apples; Raw; Peeled; Sliced	1 Cup	110	84	65	0	0	0.1	0	0.1	0	16	4	8	0.1	124	0	50	5	0.02	0.01	0.1	4
2010	Apples; Dried; Sulfured	10 Rings	64	32	155	1	0	0	0	0.1	0	42	9	24	0.9	288	56	0	0	0	0.1	0.6	2
2020	Apple Juice; Canned	1 Cup	248	88	115	0	0	0	0	0.1	0	29	17	17	0.9	295	7	0	0	0.05	0.04	0.2	2
2030	Applesauce; Canned; Sweetened	1 Cup	255	80	195	0	0	0.1	0	0.1	0	51	10	18	0.9	156	8	30	3	0.03	0.07	0.5	4
2040	Applesauce; Canned; Unsweetnd	1 Cup	244	88	105	0	0	0	0	0	0	28	7	17	0.3	183	5	70	7	0.03	0.06	0.5	3
2050	Apricots; Raw	3 Aprcot	106	86	50	1	0	0	0.2	0.1	0	12	15	20	0.6	314	1	2770	277	0.03	0.04	0.6	11
2060	Apricots; Canned; Heavy Syrup	1 Cup	258	78	215	1	0	0	0.1	0	0	55	23	31	0.8	361	10	3170	317	0.05	0.06	1	8

TABLE A-8 Food Composition Table *(continued)*

No.	Description of Food		Wt. (g)	Water (%)	Energy (kcal)	Protein (g)	Fat (g)	Sat (g)	Mono (g)	Poly (g)	Cholest (mg)	Carb (g)	Ca (mg)	P (mg)	Fe (mg)	K (mg)	Na (mg)	Vit A (IU)	Vit A (RE)	Thmn (mg)	Ribofl (mg)	Niacin (mg)	Vit C (mg)
Fruits and Fruit Products *(continued)*																							
2070	Apricots; Canned; Heavy Syrup	3 Halves	85	78	70	0	0	0	0	0	0	18	8	10	0.3	119	3	1050	105	0.02	0.02	0.3	3
2080	Apricots; Canned; Juice Pack	1 Cup	248	87	120	2	0	0	0	0	0	31	30	50	0.7	409	10	4190	419	0.04	0.05	0.9	12
2090	Apricots; Canned; Juice Pack	3 Halves	84	87	40	1	0	0	0	0	0	10	10	17	0.3	139	3	1420	142	0.02	0.02	0.3	4
2100	Apricots; Dried; Uncooked	1 Cup	130	31	310	5	1	0	0.3	0.1	0	80	59	152	6.1	1791	13	9410	941	0.01	0.2	3.9	3
2110	Apricots; Dried; Cooked;Unsweetnd	1 Cup	250	76	210	3	0	0	0.2	0.1	0	55	40	103	4.2	1222	8	5910	591	0.02	0.08	2.4	4
2120	Apricot Nectar; No Added Vit C	1 Cup	251	85	140	1	0	0	0.1	0	0	36	18	23	1	286	8	3300	330	0.02	0.04	0.7	2
2130	Avocados; California	1 Avocdo	173	73	305	4	30	4.5	19.4	3.5	0	12	19	73	2	1097	21	1060	106	0.19	0.21	3.3	14
2140	Avocados; Florida	1 Avocdo	304	80	340	5	27	5.3	14.8	4.5	0	27	33	119	1.6	1484	15	1860	186	0.33	0.37	5.8	24
2150	Bananas	1 Banana	114	74	105	1	1	0.2	0	0.1	0	27	7	23	0.4	451	1	90	9	0.05	0.11	0.6	10
2160	Bananas; Sliced	1 Cup	150	74	140	2	1	0.3	0.1	0.1	0	35	9	30	0.5	594	2	120	12	0.07	0.15	0.8	14
2170	Blackberries; Raw	1 Cup	144	86	75	1	1	0.2	0.1	0.1	0	18	46	30	0.8	282	0	240	24	0.04	0.06	0.6	30
2180	Blueberries; Raw	1 Cup	145	85	80	1	1	0	0.1	0.3	0	20	9	15	0.2	129	9	150	15	0.07	0.07	0.5	19
2190	Blueberries; Frozen; Sweetened	10 Oz	284	77	230	1	0	0	0.1	0.2	0	62	17	20	1.1	170	3	120	12	0.06	0.15	0.7	3
2200	Blueberries; Frozen; Sweetened	1 Cup	230	77	185	1	0	0	0	0.1	0	50	14	16	0.9	138	2	100	10	0.05	0.12	0.6	2
2210	Cherries; Sour; Red; Cannd; Water	1 Cup	244	90	90	2	0	0.1	0.1	0.1	0	22	27	24	3.3	239	17	1840	184	0.04	0.1	0.4	5
2220	Cherries; Sweet; Raw	10 Chery	68	81	50	1	1	0.1	0.2	0.2	0	11	10	13	0.3	152	0	150	15	0.03	0.04	0.3	5
2230	Cranberry Juice Cocktal W/Vitc	1 Cup	253	85	145	0	0	0	0	0.1	0	38	8	3	0.4	61	10	10	1	0.01	0.04	0.1	108
2240	Cranberry Sauce; Canned; Swtnd	1 Cup	277	61	420	1	0	0	0.1	0.2	0	108	11	17	0.6	72	80	60	6	0.04	0.06	0.3	6
2250	Dates	10 Dates	83	23	230	2	0	0.1	0.1	0	0	61	27	33	1	541	2	40	4	0.07	0.08	1.8	0
2260	Dates; Chopped	1 Cup	178	23	490	4	1	0.3	0.2	0	0	131	57	71	2	1161	5	90	9	0.16	0.18	3.9	0
2270	Figs; Dried	10 Figs	187	28	475	6	2	0.4	0.5	1	0	122	269	127	4.2	1331	21	250	25	0.13	0.16	1.3	1
2280	Fruit Cocktail; Cnnd; Heavy Syrup	1 Cup	255	80	185	1	0	0	0	0.1	0	48	15	28	0.7	224	15	520	52	0.05	0.05	1	5
2290	Fruit Cocktail; Cnnd; Juice Pack	1 Cup	248	87	115	1	0	0	0	0	0	29	20	35	0.5	236	10	760	76	0.03	0.04	1	7
2300	Grapefruit; Raw; White	1/2 Frut	120	91	40	1	0	0	0	0	0	10	14	10	0.1	167	0	10	1	0.04	0.02	0.3	41
2301	Grapefruit; Raw; Pink	1/2 Frut	120	91	40	1	0	0	0	0	0	10	14	10	0.1	167	0	310	31	0.04	0.02	0.3	41
2310	Grapefruit; Canned; Syrup Pack	1 Cup	254	84	150	1	0	0	0	0.1	0	39	36	25	1	328	5	0	0	0.1	0.05	0.6	54
2320	Grapefruit Juice; Raw	1 Cup	247	90	95	1	0	0	0	0.1	0	23	22	37	0.5	400	2	20	2	0.1	0.05	0.5	94
2330	Grapefruit Juice; Canned;Unswt	1 Cup	247	90	95	1	0	0	0	0.1	0	22	17	27	0.5	378	2	20	2	0.1	0.05	0.6	72
2340	Grapefruit Juice; Canned;Swtnd	1 Cup	250	87	115	1	0	0	0	0.1	0	28	20	28	0.9	405	5	20	2	0.1	0.06	0.8	67
2350	Grapefrt Juice; Frzn;Cncn;Unswten	6 Fl Oz	207	62	300	4	1	0.1	0.1	0.2	0	72	56	101	1	1002	6	60	6	0.3	0.16	1.6	248
2360	Grapefrt Juice; Frzn;Dltd;Unswten	1 Cup	247	89	100	1	0	0	0	0.1	0	24	20	35	0.3	336	2	20	2	0.1	0.05	0.5	83

TABLE A-8 Food Composition Table *(continued)*

No.	Description of Food		Wt. (g)	Water (%)	Energy (kcal)	Protein (g)	Fat (g)	Sat (g)	Mono (g)	Poly (g)	Cholest (mg)	Carb (g)	Ca (mg)	P (mg)	Fe (mg)	K (mg)	Na (mg)	Vit A (IU)	Vit A (RE)	Thmn (mg)	Ribofl (mg)	Niacin (mg)	Vit C (mg)
Fruits and Fruit Products *(continued)*																							
2370	Grapes; European; Raw; Thompson	10 Grape	50	81	35	0	0	0.1	0	0.1	0	9	6	7	0.1	93	1	40	4	0.05	0.03	0.2	5
2380	Grapes; European; Raw; Tokay	10 Grape	57	81	40	0	0	0.1	0	0.1	0	10	6	7	0.1	105	1	40	4	0.05	0.03	0.2	6
2390	Grape Juice; Canned	1 Cup	253	84	155	1	0	0.1	0	0.1	0	38	23	28	0.6	334	8	20	2	0.07	0.09	0.7	0
2400	Grapejce;Frzn;Concen;Swtnd;W/C	6 Fl Oz	216	54	385	1	1	0.2	0	0.2	0	96	28	32	0.8	160	15	60	6	0.11	0.2	0.9	179
2410	Grapejce;Frzn;Dilutd;Swtnd;W/C	1 Cup	250	87	125	0	0	0.1	0	0.1	0	32	10	10	0.3	53	5	20	2	0.04	0.07	0.3	60
2420	Kiwifruit; Raw	1 Kiwi	76	83	45	1	0	0	0.1	0.1	0	11	20	30	0.3	252	4	130	13	0.02	0.04	0.4	74
2430	Lemons; Raw	1 Lemon	58	89	15	1	0	0	0	0.1	0	5	15	9	0.3	80	1	20	2	0.02	0.01	0.1	31
2440	Lemon Juice; Raw	1 Cup	244	91	60	1	0	0	0	0	0	21	17	15	0.1	303	2	50	5	0.07	0.02	0.2	112
2450	Lemon Juice; Canned	1 Cup	244	92	50	1	1	0.1	0	0.2	0	16	27	22	0.3	249	51	40	4	0.1	0.02	0.5	61
2460	Lemon Juice; Canned	1 Tbsp	15	92	5	0	0	0	0	0	0	1	2	1	0	15	3	0	0	0.01	0	0	4
2470	Lemon Juice; Frzn; Single-Strngh	6 Fl Oz	244	92	55	1	1	0.1	0	0.2	0	16	20	20	0.3	217	2	30	3	0.14	0.03	0.3	77
2480	Lime Juice; Raw	1 Cup	246	90	65	1	0	0	0	0.1	0	22	22	17	0.1	268	2	20	2	0.05	0.02	0.2	72
2490	Lime Juice; Canned	1 Cup	246	93	50	1	1	0.1	0.1	0.2	0	16	30	25	0.6	185	39	40	4	0.08	0.01	0.4	16
2500	Mangos; Raw	1 Mango	207	82	135	1	1	0.1	0.2	0.1	0	35	21	23	0.3	323	4	8060	806	0.12	0.12	1.2	57
2510	Cantaloup; Raw	1/2 Meln	267	90	95	2	1	0.1	0.1	0.3	0	22	29	45	0.6	825	24	8610	861	0.1	0.06	1.5	113
2520	Honeydew Melon; Raw	1/10 Mel	129	90	45	1	0	0	0	0.1	0	12	8	13	0.1	350	13	50	5	0.1	0.02	0.8	32
2530	Nectarines; Raw	1 Nectrn	136	86	65	1	1	0.1	0.2	0.3	0	16	7	22	0.2	288	0	1000	100	0.02	0.06	1.3	7
2540	Oranges; Raw	1 Orange	131	87	60	1	0	0	0	0	0	15	52	18	0.1	237	0	270	27	0.11	0.05	0.4	70
2550	Oranges; Raw; Sections	1 Cup	180	87	85	2	0	0	0	0	0	21	72	25	0.2	326	0	370	37	0.16	0.07	0.5	96
2560	Orange Juice; Raw	1 Cup	248	88	110	2	0	0.1	0.1	0.1	0	26	27	42	0.5	496	2	500	50	0.22	0.07	1	124
2570	Orange Juice; Canned	1 Cup	249	89	105	1	0	0	0.1	0.1	0	25	20	35	1.1	436	5	440	44	0.15	0.07	0.8	86
2580	Orange Juice; Chilled	1 Cup	249	88	110	2	1	0.1	0.1	0.2	0	25	25	27	0.4	473	2	190	19	0.28	0.05	0.7	82
2590	Orange Juice; Frozen Concentrte	6 Fl Oz	213	58	340	5	0	0.1	0.1	0.1	0	81	68	121	0.7	1436	6	590	59	0.6	0.14	1.5	294
2600	Orange Juice; Frzn;Cncn;Diluted	1 Cup	249	88	110	2	0	0	0	0	0	27	22	40	0.2	473	2	190	19	0.2	0.04	0.5	97
2610	Orange+Grapefruit Juice; Cannd	1 Cup	247	89	105	1	0	0	0	0	0	25	20	35	1.1	390	7	290	29	0.14	0.07	0.8	72
2620	Papayas; Raw	1 Cup	140	86	65	1	0	0.1	0.1	0	0	17	35	12	0.3	247	9	400	40	0.04	0.04	0.5	92
2630	Peaches; Raw	1 Peach	87	88	35	1	0	0	0	0	0	10	4	10	0.1	171	0	470	47	0.01	0.04	0.9	6
2640	Peaches; Raw; Sliced	1 Cup	170	88	75	1	0	0	0.1	0.1	0	19	9	20	0.2	335	0	910	91	0.03	0.07	1.7	11
2650	Peaches; Canned; Heavy Syrup	1 Cup	256	79	190	1	0	0	0.1	0.1	0	51	8	28	0.7	236	15	850	85	0.03	0.06	1.6	7
2660	Peaches; Canned; Heavy Syrup	1 Half	81	79	60	0	0	0	0	0	0	16	2	9	0.2	75	5	270	27	0.01	0.02	0.5	2
2670	Peaches; Canned; Juice Pack	1 Cup	248	87	110	2	0	0	0	0	0	29	15	42	0.7	317	10	940	94	0.02	0.04	1.4	9

TABLE A-8 Food Composition Table *(continued)*

No.	Description of Food		Wt. (g)	Water (%)	Energy (kcal)	Protein (g)	Fat (g)	Sat (g)	Mono (g)	Poly (g)	Cholest (mg)	Carb (g)	Ca (mg)	P (mg)	Fe (mg)	K (mg)	Na (mg)	Vit A (IU)	Vit A (RE)	Thmn (mg)	Ribofl (mg)	Niacin (mg)	Vit C (mg)
Fruits and Fruit Products *(continued)*																							
2680	Peaches; Canned; Juice Pack	1 Half	77	87	35	0	0	0	0	0	0	9	5	13	0.2	99	3	290	29	0.01	0.01	0.4	3
2690	Peaches; Dried	1 Cup	160	32	380	6	1	0.1	0.4	0.6	0	98	45	190	6.5	1594	11	3460	346	0	0.34	7	8
2700	Peaches; Dried; Cooked;Unswetnd	1 Cup	258	78	200	3	1	0.1	0.2	0.3	0	51	23	98	3.4	826	5	510	51	0.01	0.05	3.9	10
2710	Peaches; Frozen; Swetned;W/Vit C	10 Oz	284	75	265	2	0	0	0.1	0.2	0	68	9	31	1.1	369	17	810	81	0.04	0.1	1.9	268
2720	Peaches; Frozen; Swetned;W/Vit C	1 Cup	250	75	235	2	0	0	0.1	0.2	0	60	8	28	0.9	325	15	710	71	0.03	0.09	1.6	236
2730	Pears; Raw; Bartlett	1 Pear	166	84	100	1	1	0	0.1	0.2	0	25	18	18	0.4	208	0	30	3	0.03	0.07	0.2	7
2740	Pears; Raw; Bosc	1 Pear	141	84	85	1	1	0	0.1	0.1	0	21	16	16	0.4	176	0	30	3	0.03	0.06	0.1	6
2750	Pears; Raw; D'anjou	1 Pear	200	84	120	1	1	0	0.2	0.2	0	30	22	22	0.5	250	0	40	4	0.04	0.08	0.2	8
2760	Pears; Canned; Heavy Syrup	1 Cup	255	80	190	1	0	0	0.1	0.1	0	49	13	18	0.6	166	13	10	1	0.03	0.06	0.6	3
2770	Pears; Canned; Heavy Syrup	1 Half	79	80	60	0	0	0	0	0	0	15	4	6	0.2	51	4	0	0	0.01	0.02	0.2	1
2780	Pears; Canned; Juice Pack	1 Cup	248	86	125	1	0	0	0	0	0	32	22	30	0.7	238	10	10	1	0.03	0.03	0.5	4
2790	Pears; Canned; Juice Pack	1 Half	77	86	40	0	0	0	0	0	0	10	7	9	0.2	74	3	0	0	0.01	0.01	0.2	1
2800	Pineapple; Raw; Diced	1 Cup	155	87	75	1	1	0	0.1	0.2	0	19	11	11	0.6	175	2	40	4	0.14	0.06	0.7	24
2810	Pineapple; Canned; Heavy Syrup	1 Cup	255	79	200	1	0	0	0	0.1	0	52	36	18	1	265	3	40	4	0.23	0.06	0.7	19
2820	Pineapple; Canned; Heavy Syrup	1 Slice	58	79	45	0	0	0	0	0	0	12	8	4	0.2	60	1	10	1	0.05	0.01	0.2	4
2830	Pineapple; Canned; Juice Pack	1 Cup	250	84	150	1	0	0	0	0.1	0	39	35	15	0.7	305	3	100	10	0.24	0.05	0.7	24
2840	Pineapple; Canned; Juice Pack	1 Slice	58	84	35	0	0	0	0	0	0	9	8	3	0.2	71	1	20	2	0.06	0.01	0.2	6
2850	Pineapple Juice; Canned;Unswtn	1 Cup	250	86	140	1	0	0	0	0.1	0	34	43	20	0.7	335	3	10	1	0.14	0.06	0.6	27
2860	Plantains; Raw	1 Plantn	179	65	220	2	1	0.3	0.1	0.1	0	57	5	61	1.1	893	7	2020	202	0.09	0.1	1.2	33
2870	Plantains; Cooked	1 Cup	154	67	180	1	0	0.1	0	0.1	0	48	3	43	0.9	716	8	1400	140	0.07	0.08	1.2	17
2880	Plums; Raw; 2-1/8-In Diam	1 Plum	66	85	35	1	0	0	0.3	0.1	0	9	3	7	0.1	114	0	210	21	0.03	0.06	0.3	6
2890	Plums; Raw; 1-1/2-In Diam	1 Plum	28	85	15	0	0	0	0.1	0	0	4	1	3	0	48	0	90	9	0.01	0.03	0.1	3
2900	Plums; Canned; Heavy Syrup	1 Cup	258	76	230	1	0	0	0.2	0.1	0	60	23	34	2.2	235	49	670	67	0.04	0.1	0.8	1
2910	Plums; Canned; Heavy Syrup	3 Plums	133	76	120	0	0	0	0.1	0	0	31	12	17	1.1	121	25	340	34	0.02	0.05	0.4	1
2920	Plums; Canned; Juice Pack	1 Cup	252	84	145	1	0	0	0	0	0	38	25	38	0.9	388	3	2540	254	0.06	0.15	1.2	7
2930	Plums; Canned; Juice Pack	3 Plums	95	84	55	0	0	0	0	0	0	14	10	14	0.3	146	1	960	96	0.02	0.06	0.4	3
2940	Prunes; Dried	5 Large	49	32	115	1	0	0	0.2	0.1	0	31	25	39	1.2	365	2	970	97	0.04	0.08	1	2
2950	Prunes; Dried; Cooked;Unswtned	1 Cup	212	70	225	2	0	0	0.3	0.1	0	60	49	74	2.4	708	4	650	65	0.05	0.21	1.5	6
2960	Prune Juice; Canned	1 Cup	256	81	180	2	0	0	0.1	0	0	45	31	64	3	707	10	10	1	0.04	0.18	2	10
2970	Raisins	1 Cup	145	15	435	5	1	0.2	0	0.2	0	115	71	141	3	1089	17	10	1	0.23	0.13	1.2	5
2980	Raisins	1 Packet	14	15	40	0	0	0	0	0	0	11	7	14	0.3	105	2	0	0	0.02	0.01	0.1	0

TABLE A-8 Food Composition Table *(continued)*

No.	Description of Food		Wt. (g)	Water (%)	Energy (kcal)	Protein (g)	Fat (g)	Sat (g)	Mono (g)	Poly (g)	Cholest (mg)	Carb (g)	Ca (mg)	P (mg)	Fe (mg)	K (mg)	Na (mg)	Vit A (IU)	Vit A (RE)	Thmn (mg)	Ribofl (mg)	Niacin (mg)	Vit C (mg)
Fruits and Fruit Products *(continued)*																							
2990	Raspberries; Raw	1 Cup	123	87	60	1	1	0	0.1	0.4	0	14	27	15	0.7	187	0	160	16	0.04	0.11	1.1	31
3000	Raspberries; Frozen; Sweetened	10 Oz	284	73	295	2	0	0	0	0.3	0	74	43	48	1.8	324	3	170	17	0.05	0.13	0.7	47
3010	Raspberries; Frozen; Sweetened	1 Cup	250	73	255	2	0	0	0	0.2	0	65	38	43	1.6	285	3	150	15	0.05	0.11	0.6	41
3020	Rhubarb; Cooked; Added Sugar	1 Cup	240	68	280	1	0	0	0	0.1	0	75	348	19	0.5	230	2	170	17	0.04	0.06	0.5	8
3030	Strawberries; Raw	1 Cup	149	92	45	1	1	0	0.1	0.3	0	10	21	28	0.6	247	1	40	4	0.03	0.1	0.3	84
3040	Strawberries; Frozen; Sweetend	10 Oz	284	73	275	2	0	0	0.1	0.2	0	74	31	37	1.7	278	9	70	7	0.05	0.14	1.1	118
3050	Strawberries; Frozen; Sweetend	1 Cup	255	73	245	1	0	0	0	0.2	0	66	28	33	1.5	250	8	60	6	0.04	0.13	1	106
3060	Tangerines; Raw	1 Tangrn	84	88	35	1	0	0	0	0	0	9	12	8	0.1	132	1	770	77	0.09	0.02	0.1	26
3070	Tangerines; Canned; Light Syrp	1 Cup	252	83	155	1	0	0	0	0.1	0	41	18	25	0.9	197	15	2120	212	0.13	0.11	1.1	50
3080	Tangerine Juice; Canned;Swtned	1 Cup	249	87	125	1	0	0	0	0.1	0	30	45	35	0.5	443	2	1050	105	0.15	0.05	0.2	55
3090	Watermelon; Raw	1 Piece	482	92	155	3	2	0.3	0.2	1	0	35	39	43	0.8	559	10	1760	176	0.39	0.1	1	46
3100	Watermelon; Raw; Diced	1 Cup	160	92	50	1	1	0.1	0.1	0.3	0	11	13	14	0.3	186	3	590	59	0.13	0.03	0.3	15
Breads and Cereals																							
3110	Bagels; Plain	1 Bagel	68	29	200	7	2	0.3	0.5	0.7	0	38	29	46	1.8	50	245	0	0	0.26	0.2	2.4	0
3111	Bagels; Egg	1 Bagel	68	29	200	7	2	0.3	0.5	0.7	44	38	29	46	1.8	50	245	22	7	0.26	0.2	2.4	0
3120	Barley; Pearled;Light; Uncookd	1 Cup	200	11	700	16	2	0.3	0.2	0.9	0	158	32	378	4.2	320	6	0	0	0.24	0.1	6.2	0
3130	Baking Pwdr Biscuits; Homerecpe	1 Biscuit	28	28	100	2	5	1.2	2	1.3	0	13	47	36	0.7	32	195	10	3	0.08	0.08	0.8	0
3140	Baking Pwdr Biscuits; From Mix	1 Biscuit	28	29	95	2	3	0.8	1.4	0.9	0	14	58	128	0.7	56	262	20	4	0.12	0.11	0.8	0
3150	Baking Pwdr Biscuits; Refrgdogh	1 Biscuit	20	30	65	1	2	0.6	0.9	0.6	1	10	4	79	0.5	18	249	0	0	0.08	0.05	0.7	0
3160	Breadcrumbs; Dry; Grated	1 Cup	100	7	390	13	5	1.5	1.6	1	5	73	122	141	4.1	152	736	0	0	0.35	0.35	4.8	0
3170	Boston Brown Bread; W/Whtecrnm	1 Slice	45	45	95	2	1	0.3	0.1	0.1	3	21	41	72	0.9	131	113	0	0	0.06	0.04	0.7	0
3171	Boston Brown Bread; W/Yllwcrnml	1 Slice	45	45	95	2	1	0.3	0.1	0.1	3	21	41	72	0.9	131	113	32	3	0.06	0.04	0.7	0
3180	Cracked-Wheat Bread	1 Loaf	454	35	1190	42	16	3.1	4.3	5.7	0	227	295	581	12.1	608	1966	0	0	1.73	1.73	15.3	0
3190	Cracked-Wheat Bread	1 Slice	25	35	65	2	1	0.2	0.2	0.3	0	12	16	32	0.7	34	106	0	0	0.1	0.09	0.8	0
3200	Cracked-Wheat Bread; Toasted	1 Slice	21	26	65	2	1	0.2	0.2	0.3	0	12	16	32	0.7	34	106	0	0	0.07	0.09	0.8	0
3210	French or Vienna Bread	1 Loaf	454	34	1270	43	18	3.8	5.7	5.9	0	230	499	386	14	409	2633	0	0	2.09	1.59	18.2	0
3220	French Bread	1 Slice	35	34	100	3	1	0.3	0.4	0.5	0	18	39	30	1.1	32	203	0	0	0.16	0.12	1.4	0
3230	Vienna Bread	1 Slice	25	34	70	2	1	0.2	0.3	0.3	0	13	28	21	0.8	23	145	0	0	0.12	0.09	1	0
3240	Italian Bread	1 Loaf	454	32	1255	41	4	0.6	0.3	1.6	0	256	77	350	12.7	336	2656	0	0	1.8	1.1	15	0
3250	Italian Bread	1 Slice	30	32	85	3	0	0	0	0.1	0	17	5	23	0.8	22	176	0	0	0.12	0.07	1	0
3260	Mixed Grain Bread	1 Loaf	454	37	1165	45	17	3.2	4.1	6.5	0	212	472	962	14.8	990	1870	0	0	1.77	1.73	18.9	0

TABLE A-8 Food Composition Table *(continued)*

No.	Description of Food		Wt. (g)	Water (%)	Energy (kcal)	Protein (g)	Fat (g)	Sat (g)	Mono (g)	Poly (g)	Cholest (mg)	Carb (g)	Ca (mg)	P (mg)	Fe (mg)	K (mg)	Na (mg)	Vit A (IU)	Vit A (RE)	Thmn (mg)	Ribofl (mg)	Niacin (mg)	Vit C (mg)
Breads and Cereals *(continued)*																							
3270	Mixed Grain Bread	1 Slice	25	37	65	2	1	0.2	0.2	0.4	0	12	27	55	0.8	56	106	0	0	0.1	0.1	1.1	0
3280	Mixed Grain Bread; Toasted	1 Slice	23	27	65	2	1	0.2	0.2	0.4	0	12	27	55	0.8	56	106	0	0	0.08	0.1	1.1	0
3290	Oatmeal Bread	1 Loaf	454	37	1145	38	20	3.7	7.1	8.2	0	212	267	563	12	707	2231	0	0	2.09	1.2	15.4	0
3300	Oatmeal Bread	1 Slice	25	37	65	2	1	0.2	0.4	0.5	0	12	15	31	0.7	39	124	0	0	0.12	0.07	0.9	0
3310	Oatmeal Bread; Toasted	1 Slice	23	30	65	2	1	0.2	0.4	0.5	0	12	15	31	0.7	39	124	0	0	0.09	0.07	0.9	0
3320	Pita Bread	1 Pita	60	31	165	6	1	0.1	0.1	0.4	0	33	49	60	1.4	71	339	0	0	0.27	0.12	2.2	0
3330	Pumpernickel Bread	1 Loaf	454	37	1160	42	16	2.6	3.6	6.4	0	218	322	990	12.4	1966	2461	0	0	1.54	2.36	15	0
3340	Pumpernickel Bread	1 Slice	32	37	80	3	1	0.2	0.3	0.5	0	16	23	71	0.9	141	177	0	0	0.11	0.17	1.1	0
3350	Pumpernickel Bread; Toasted	1 Slice	29	28	80	3	1	0.2	0.3	0.5	0	16	23	71	0.9	141	177	0	0	0.09	0.17	1.1	0
3360	Raisin Bread	1 Loaf	454	33	1260	37	18	4.1	6.5	6.7	0	239	463	395	14.1	1058	1657	0	0	1.5	2.81	18.6	0
3370	Raisin Bread	1 Slice	25	33	65	2	1	0.2	0.3	0.4	0	13	25	22	0.8	59	92	0	0	0.08	0.15	1	0
3380	Raisin Bread; Toasted	1 Slice	21	24	65	2	1	0.2	0.3	0.4	0	13	25	22	0.8	59	92	0	0	0.06	0.15	1	0
3390	Rye Bread; Light	1 Loaf	454	37	1190	38	17	3.3	5.2	5.5	0	218	363	658	12.3	926	3164	0	0	1.86	1.45	15	0
3400	Rye Bread; Light	1 Slice	25	37	65	2	1	0.2	0.3	0.3	0	12	20	36	0.7	51	175	0	0	0.1	0.08	0.8	0
3410	Rye Bread; Light; Toasted	1 Slice	22	28	65	2	1	0.2	0.3	0.3	0	12	20	36	0.7	51	175	0	0	0.08	0.08	0.8	0
3420	Wheat Bread	1 Loaf	454	37	1160	43	19	3.9	7.3	4.5	0	213	572	835	15.8	627	2447	0	0	2.09	1.45	20.5	0
3430	Wheat Bread	1 Slice	25	37	65	2	1	0.2	0.4	0.3	0	12	32	47	0.9	35	138	0	0	0.12	0.08	1.2	0
3440	Wheat Bread; Toasted	1 Slice	23	28	65	3	1	0.2	0.4	0.3	0	12	32	47	0.9	35	138	0	0	0.1	0.08	1.2	0
3450	White Bread	1 Loaf	454	37	1210	38	18	5.6	6.5	4.2	0	222	572	490	12.9	508	2334	0	0	2.13	1.41	17	0
3460	White Bread; Slice 18 Per Loaf	1 Slice	25	37	65	2	1	0.3	0.4	0.2	0	12	32	27	0.7	28	129	0	0	0.12	0.08	0.9	0
3470	White Bread; Toasted 18 Per Loaf	1 Slice	22	28	65	2	1	0.3	0.4	0.2	0	12	32	27	0.7	28	129	0	0	0.09	0.08	0.9	0
3480	White Bread; Slice 22 Per Loaf	1 Slice	20	37	55	2	1	0.2	0.3	0.2	0	10	25	21	0.6	22	101	0	0	0.09	0.06	0.7	0
3490	White Bread; Toasted 22 Per Loaf	1 Slice	17	28	55	2	1	0.2	0.3	0.2	0	10	25	21	0.6	22	101	0	0	0.07	0.06	0.7	0
3500	White Bread Cubes	1 Cup	30	37	80	2	1	0.4	0.4	0.3	0	15	38	32	0.9	34	154	0	0	0.14	0.09	1.1	0
3510	White Bread Crumbs; Soft	1 Cup	45	37	120	4	2	0.6	0.6	0.4	0	22	57	49	1.3	50	231	0	0	0.21	0.14	1.7	0
3520	Whole-Wheat Bread	1 Loaf	454	38	1110	44	20	5.8	6.8	5.2	0	206	327	1180	15.5	799	2887	0	0	1.59	0.95	17.4	0
3530	Whole-Wheat Bread	1 Slice	28	38	70	3	1	0.4	0.4	0.3	0	13	20	74	1	50	180	0	0	0.1	0.06	1.1	0
3540	Whole-Wheat Bread; Toasted	1 Slice	25	29	70	3	1	0.4	0.4	0.3	0	13	20	74	1	50	180	0	0	0.08	0.06	1.1	0
3550	Bread Stuffing; From Mx; Drytype	1 Cup	140	33	500	9	31	6.1	13.3	9.6	0	50	92	136	2.2	126	1254	910	273	0.17	0.2	2.5	0
3560	Bread Stuffing; From Mx; Moist	1 Cup	203	61	420	9	26	5.3	11.3	8	67	40	81	134	2	118	1023	850	256	0.1	0.18	1.6	0
3570	Corn Grits; Ckd;Reg;Whte;Nosalt	1 Cup	242	85	145	3	0	0	0.1	0.2	0	31	0	29	1.5	53	0	0	0	0.24	0.15	2	0

TABLE A-8 Food Composition Table *(continued)*

No.	Description of Food		Wt. (g)	Water (%)	Energy (kcal)	Protein (g)	Fat (g)	Sat (g)	Mono (g)	Poly (g)	Cholest (mg)	Carb (g)	Ca (mg)	P (mg)	Fe (mg)	K (mg)	Na (mg)	Vit A (IU)	Vit A (RE)	Thmn (mg)	Ribofl (mg)	Niacin (mg)	Vit C (mg)
Breads and Cereals *(continued)*																							
3571	Corn Grits; Ckd;Reg;Whte;W/Salt	1 Cup	242	85	145	3	0	0	0.1	0.2	0	31	0	29	1.5	53	540	0	0	0.24	0.15	2	0
3572	Corn Grits; Ckd;Reg;Yllw;Nosalt	1 Cup	242	85	145	3	0	0	0.1	0.2	0	31	0	29	1.5	53	0	145	14	0.24	0.15	2	0
3573	Corn Grits; Ckd;Reg;Yllw;W/Salt	1 Cup	242	85	145	3	0	0	0.1	0.2	0	31	0	29	1.5	53	540	145	14	0.24	0.15	2	0
3580	Corn Grits; Cooked; Instant	1 Pkt	137	85	80	2	0	0	0	0.1	0	18	7	16	1	29	343	0	0	0.18	0.08	1.3	0
3590	Crm Wheat; Ckd;Reg;Inst;No Salt	1 Cup	244	86	140	4	0	0.1	0	0.2	0	29	54	43	10.9	46	5	0	0	0.24	0	1.5	0
3591	Crm Wheat; Ckd;Reg;Inst;W/Salt	1 Cup	244	86	140	4	0	0.1	0	0.2	0	29	54	43	10.9	46	390	0	0	0.24	0.07	1.5	0
3592	Crm Wheat; Ckd; Quick; No Salt	1 Cup	244	86	140	4	0	0.1	0	0.2	0	29	54	102	10.9	46	142	0	0	0.24	0.07	1.5	0
3593	Crm Wheat; Ckd;Quick; W/Salt	1 Cup	244	86	140	4	0	0.1	0	0.2	0	29	54	102	10.9	46	390	0	0	0.24	0.07	1.5	0
3600	Cream of Wheat; Ckd;Mix N Eat	1 Pkt	142	82	100	3	0	0	0	0.1	0	21	20	20	8.1	38	241	1250	376	0.43	0.28	5	0
3610	Malt-O-Meal; W/O Salt	1 Cup	240	88	120	4	0	0	0	0.1	0	26	5	24	9.6	31	2	0	0	0.48	0.24	5.8	0
3611	Malt-O-Meal; With Salt	1 Cup	240	88	120	4	0	0	0	0.1	0	26	5	24	9.6	31	324	0	0	0.48	0.24	5.8	0
3620	Oatmeal; Ckd;Rg;Qck;Inst;W/O Salt	1 Cup	234	85	145	6	2	0.4	0.8	1	0	25	19	178	1.6	131	2	40	5	0.26	0.05	0.3	0
3621	Oatmeal; Ckd;Rg;Qck;Inst;W/Salt	1 Cup	234	85	145	6	2	0.4	0.8	1	0	25	19	178	1.6	131	374	40	4	0.26	0.05	0.3	0
3630	Oatmeal; Ckd;Instnt;Plain;Fortf	1 Pkt	177	86	105	4	2	0.3	0.6	0.7	0	18	163	133	6.3	99	285	1510	453	0.53	0.28	5.5	0
3640	Oatmeal; Ckd;Instnt;Flvrd;Fortf	1 Pkt	164	76	160	5	2	0.3	0.7	0.8	0	31	168	148	6.7	137	254	1530	460	0.53	0.38	5.9	0
3650	All-Bran Cereal	1 Oz	28.4	3	70	4	1	0.1	0.1	0.3	0	21	23	264	4.5	350	320	1250	375	0.37	0.43	5	15
3660	Cap'n Crunch Cereal	1 Oz	28.4	3	120	1	3	1.7	0.3	0.4	0	23	5	36	7.5	37	213	40	4	0.5	0.55	6.6	0
3670	Cheerios Cereal	1 Oz	28.4	5	110	4	2	0.3	0.6	0.7	0	20	48	134	4.5	101	307	1250	375	0.37	0.43	5	15
3680	Corn Flakes; Kellogg's	1 Oz	28.4	3	110	2	0	0	0	0	0	24	1	18	1.8	26	351	1250	375	0.37	0.43	5	15
3690	Corn Flakes; Toasties	1 Oz	28.4	3	110	2	0	0	0	0	0	24	1	12	0.7	33	297	1250	375	0.37	0.43	5	0
3700	40% Bran Flakes; Kellogg's	1 Oz	28.4	3	90	4	1	0.1	0.1	0.3	0	22	14	139	8.1	180	264	1250	375	0.37	0.43	5	0
3710	40% Bran Flakes; Post	1 Oz	28.4	3	90	3	0	0.1	0.1	0.2	0	22	12	179	4.5	151	260	1250	375	0.37	0.43	5	0
3720	Froot Loops Cereal	1 Oz	28.4	3	110	2	1	0.2	0.1	0.1	0	25	3	24	4.5	26	145	1250	375	0.37	0.43	5	15
3730	Golden Grahams Cereal	1 Oz	28.4	2	110	2	1	0.7	0.1	0.2	0	24	17	41	4.5	63	346	1250	375	0.37	0.43	5	15
3740	Grape-Nuts Cereal	1 Oz	28.4	3	100	3	0	0	0	0.1	0	23	11	71	1.2	95	197	1250	375	0.37	0.43	5	0
3750	Honey Nut Cheerios Cereal	1 Oz	28.4	3	105	3	1	0.1	0.3	0.3	0	23	20	105	4.5	99	257	1250	375	0.37	0.43	5	15
3760	Lucky Charms Cereal	1 Oz	28.4	3	110	3	1	0.2	0.4	0.4	0	23	32	79	4.5	59	201	1250	375	0.37	0.43	5	15
3770	Nature Valley Granola Cereal	1 Oz	28.4	4	125	3	5	3.3	0.7	0.7	0	19	18	89	0.9	98	58	20	2	0.1	0.05	0.2	0
3780	100% Natural Cereal	1 Oz	28.4	2	135	3	6	4.1	1.2	0.5	0	18	49	104	0.8	140	12	20	2	0.09	0.15	0.6	0
3790	Product 19 Cereal	1 Oz	28.4	3	110	3	0	0	0	0.1	0	24	3	40	18	44	325	5000	1501	1.5	1.7	20	60
3800	Raisin Bran; Kellogg's	1 Oz	28.4	8	90	3	1	0.1	0.1	0.3	0	21	10	105	3.5	147	207	960	288	0.28	0.34	3.9	0

TABLE A-8 Food Composition Table *(continued)*

No.	Description of Food		Wt. (g)	Water (%)	Energy (kcal)	Protein (g)	Fat (g)	Sat (g)	Mono (g)	Poly (g)	Cholest (mg)	Carb (g)	Ca (mg)	P (mg)	Fe (mg)	K (mg)	Na (mg)	Vit A (IU)	Vit A (RE)	Thmn (mg)	Ribofl (mg)	Niacin (mg)	Vit C (mg)
Breads and Cereals *(continued)*																							
3810	Raisin Bran; Post	1 Oz	28.4	9	85	3	1	0.1	0.1	0.3	0	21	13	119	4.5	175	185	1250	375	0.37	0.43	5	0
3820	Rice Krispies Cereal	1 Oz	28.4	2	110	2	0	0	0	0.1	0	25	4	34	1.8	29	340	1250	375	0.37	0.43	5	15
3830	Shredded Wheat Cereal	1 Oz	28.4	5	100	3	1	0.1	0.1	0.3	0	23	11	100	1.2	102	3	0	0	0.07	0.08	1.5	0
3840	Special K Cereal	1 Oz	28.4	2	110	6	0	0	0	0	0	21	8	55	4.5	49	265	1250	375	0.37	0.43	5	15
3850	Super Sugar Crisp Cereal	1 Oz	28.4	2	105	2	0	0	0	0.1	0	26	6	52	1.8	105	25	1250	375	0.37	0.43	5	0
3860	Sugar Frosted Flakes; Kellogg	1 Oz	28.4	3	110	1	0	0	0	0	0	26	1	21	1.8	18	230	1250	375	0.37	0.43	5	15
3870	Sugar Smacks Cereal	1 Oz	28.4	3	105	2	1	0.1	0.1	0.2	0	25	3	31	1.8	42	75	1250	375	0.37	0.43	5	15
3880	Total Cereal	1 Oz	28.4	4	100	3	1	0.1	0.1	0.3	0	22	48	118	18	106	352	5000	1501	1.5	1.7	20	60
3890	Trix Cereal	1 Oz	28.4	3	110	2	0	0.2	0.1	0.1	0	25	6	19	4.5	27	181	1250	375	0.37	0.43	5	15
3900	Wheaties Cereal	1 Oz	28.4	5	100	3	0	0.1	0	0.2	0	23	43	98	4.5	106	354	1250	375	0.37	0.43	5	15
3910	Buckwheat Flour; Light; Sifted	1 Cup	98	12	340	6	1	0.2	0.4	0.4	0	78	11	86	1	314	2	0	0	0.08	0.04	0.4	0
3920	Bulgur; Uncooked	1 Cup	170	10	600	19	3	1.2	0.3	1.2	0	129	49	575	9.5	389	7	0	0	0.48	0.24	7.7	0
Cakes, Cookies, Pasta																							
3930	Angelfood Cake; From Mix	1 Cake	635	38	1510	38	2	0.4	0.2	1	0	342	527	1086	2.7	845	3226	0	0	0.32	1.27	1.6	0
3940	Angelfood Cake; From Mix	1 Piece	53	38	125	3	0	0	0	0.1	0	29	44	91	0.2	71	269	0	0	0.03	0.11	0.1	0
3950	Coffeecake; Crumb; From Mix	1 Cake	430	30	1385	27	41	11.8	16.7	9.6	279	225	262	748	7.3	469	1853	690	194	0.82	0.9	7.7	1
3960	Coffeecake; Crumb; From Mix	1 Piece	72	30	230	5	7	2	2.8	1.6	47	38	44	125	1.2	78	310	120	32	0.14	0.15	1.3	0
3970	Devil's Food Cake; Chocfrst;Fmx	1 Cake	1107	24	3755	49	136	55.6	51.4	19.7	598	645	653	1162	22.1	1439	2900	1660	498	1.11	1.66	10	1
3980	Devil's Food Cake; Chocfrst;Fmx	1 Piece	69	24	235	3	8	3.5	3.2	1.2	37	40	41	72	1.4	90	181	100	31	0.07	0.1	0.6	0
3990	Devil's Food Cake; Chocfrst;Fmx	1 Cupcak	35	24	120	2	4	1.8	1.6	0.6	19	20	21	37	0.7	46	92	50	16	0.04	0.05	0.3	0
4000	Gingerbread Cake; From Mix	1 Cake	570	37	1575	18	39	9.6	16.4	10.5	6	291	513	570	10.8	1562	1733	0	0	0.86	1.03	7.4	1
4010	Gingerbread Cake; From Mix	1 Piece	63	37	175	2	4	1.1	1.8	1.2	1	32	57	63	1.2	173	192	0	0	0.09	0.11	0.8	0
4020	Yellow Cake W/Choc Frst;Frmix	1 Cake	1108	26	3735	45	125	47.8	48.8	21.8	576	638	1008	2017	15.5	1208	2515	1550	465	1.22	1.66	11.1	1
4030	Yellow Cake W/Choc Frst;Frmix	1 Piece	69	26	235	3	8	3	3	1.4	36	40	63	126	1	75	157	100	29	0.08	0.1	0.7	0
4040	Carrot Cake; Cremchese Frstg;Rec	1 Cake	1536	23	6175	63	328	66	135.2	107.5	1183	775	707	998	21	1720	4470	2240	246	1.83	1.97	14.7	23
4050	Carrot Cake; Cremchese Frstg;Rec	1 Piece	96	23	385	4	21	4.1	8.4	6.7	74	48	44	62	1.3	108	279	140	15	0.11	0.12	0.9	1
4060	Fruitcake; Dark; From Homerecip	1 Cake	1361	18	5185	74	228	47.6	113	51.7	640	783	1293	1592	37.6	6138	2123	1720	422	2.41	2.55	17	504
4070	Fruitcake; Dark; From Homerecip	1 Piece	43	18	165	2	7	1.5	3.6	1.6	20	25	41	50	1.2	194	67	50	13	0.08	0.08	0.5	16
4080	Sheetcake; W/O Frstng;Homerecip	1 Cake	777	25	2830	35	108	29.5	45.1	25.6	552	434	497	793	11.7	614	2331	1320	373	1.24	1.4	10.1	2
4090	Sheetcake; W/O Frstng;Homerecip	1 Piece	86	25	315	4	12	3.3	5	2.8	61	48	55	88	1.3	68	258	150	41	0.14	0.15	1.1	0
4100	Sheetcake; W/Whfrstng;Homercip	1 Cake	1096	21	4020	37	129	41.6	50.4	26.3	636	694	548	822	11	669	2488	2190	647	1.21	1.42	9.9	2

TABLE A-8 Food Composition Table *(continued)*

No.	Description of Food	Wt. (g)	Water (%)	Energy (kcal)	Protein (g)	Fat (g)	Sat (g)	Mono (g)	Poly (g)	Cholest (mg)	Carb (g)	Ca (mg)	P (mg)	Fe (mg)	K (mg)	Na (mg)	Vit A (IU)	Vit A (RE)	Thmn (mg)	Ribofl (mg)	Niacin (mg)	Vit C (mg)
Cakes, Cookies, Pasta *(continued)*																						
4110	Sheetcake; W/Whfrstng;Homercip 1 Piece	121	21	445	4	14	4.6	5.6	2.9	70	77	61	91	1.2	74	275	240	71	0.13	0.16	1.1	0
4120	Pound Cake; From Home Recipe 1 Loaf	514	22	2025	33	94	21.1	40.9	26.7	555	265	339	473	9.3	483	1645	3470	1033	0.93	1.08	7.8	1
4130	Pound Cake; From Home Recipe 1 Slice	30	22	120	2	5	1.2	2.4	1.6	32	15	20	28	0.5	28	96	200	60	0.05	0.06	0.5	0
4140	Pound Cake; Commercial 1 Loaf	500	24	1935	26	94	52	30	4	1100	257	146	517	8	443	1857	2820	715	0.96	1.12	8.1	0
4150	Pound Cake; Commercial 1 Slice	29	24	110	2	5	3	1.7	0.2	64	15	8	30	0.5	26	108	160	41	0.06	0.06	0.5	0
4160	Snack Cakes;Devils Food;Cremflsm Cake	28	20	105	1	4	1.7	1.5	0.6	15	17	21	26	1	34	105	20	4	0.06	0.09	0.7	0
4170	Snack Cakes;Sponge Creme Fllngsm Cake	42	19	155	1	5	2.3	2.1	0.5	7	27	14	44	0.6	37	155	30	9	0.07	0.06	0.6	0
4180	White Cake W/Wht Frstng;Comml 1 Cake	1140	24	4170	43	148	33.1	61.6	42.2	46	670	536	1585	15.5	832	2827	640	194	3.19	2.05	27.6	0
4190	White Cake W/Wht Frstng;Comml 1 Piece	71	24	260	3	9	2.1	3.8	2.6	3	42	33	99	1	52	176	40	12	0.2	0.13	1.7	0
4200	Yellowcake W/Chocfrstng;Comml 1 Cake	1108	23	3895	40	175	92	58.7	10	609	620	366	1884	19.9	1972	3080	1850	488	0.78	2.22	10	0
4210	Yellowcake W/Chocfrstng;Comml 1 Piece	69	23	245	2	11	5.7	3.7	0.6	38	39	23	117	1.2	123	192	120	30	0.05	0.14	0.6	0
4220	Cheesecake 1 Cake	1110	46	3350	60	213	119.9	65.5	14.4	2053	317	622	977	5.3	1088	2464	2820	833	0.33	1.44	5.1	56
4230	Cheesecake 1 Piece	92	46	280	5	18	9.9	5.4	1.2	170	26	52	81	0.4	90	204	230	69	0.03	0.12	0.4	5
4240	Brownies W/Nuts;Frstng;Cmmrcl 1 Browne	25	13	100	1	4	1.6	2	0.6	14	16	13	26	0.6	50	59	70	18	0.08	0.07	0.3	0
4250	Brownies W/Nuts;Frm Home Recp 1 Browne	20	10	95	1	6	1.4	2.8	1.2	18	11	9	26	0.4	35	51	20	6	0.05	0.05	0.3	0
4260	Chocolate Chip Cookies;Commrcl 4 Cookie	42	4	180	2	9	2.9	3.1	2.6	5	28	13	41	0.8	68	140	50	15	0.1	0.23	1	0
4270	Chocolate Chip Cookies;Hme Rcp 4 Cookie	40	3	185	2	11	3.9	4.3	2	18	26	13	34	1	82	82	20	5	0.06	0.06	0.6	0
4280	Chocolate Chip Cookies;Refrig 4 Cookie	48	5	225	2	11	4	4.4	2	22	32	13	34	1	62	173	30	8	0.06	0.1	0.9	0
4290	Fig Bars 4 Cookie	56	12	210	2	4	1	1.5	1	27	42	40	34	1.4	162	180	60	6	0.08	0.07	0.7	0
4300	Oatmeal W/Raisins Cookies 4 Cookie	52	4	245	3	10	2.5	4.5	2.8	2	36	18	58	1.1	90	148	40	12	0.09	0.08	1	0
4310	Peanut Butter Cookie;Home Recp 4 Cookie	48	3	245	4	14	4	5.8	2.8	22	28	21	60	1.1	110	142	20	5	0.07	0.07	1.9	0
4320	Sandwich Type Cookie 4 Cookie	40	2	195	2	8	2	3.6	2.2	0	29	12	40	1.4	66	189	0	0	0.09	0.07	0.8	0
4330	Shortbread Cookie; Commercial 4 Cookie	32	6	155	2	8	2.9	3	1.1	27	20	13	39	0.8	38	123	30	8	0.1	0.09	0.9	0
4340	Shortbread Cookie; Home Recipe 2 Cookie	28	3	145	2	8	1.3	2.7	3.4	0	17	6	31	0.6	18	125	300	89	0.08	0.06	0.7	0
4350	Sugar Cookie; From Refrig Dogh 4 Cookie	48	4	235	2	12	2.3	5	3.6	29	31	50	91	0.9	33	261	40	11	0.09	0.06	1.1	0
4360	Vanilla Wafers 10 Cooke	40	4	185	2	7	1.8	3	1.8	25	29	16	36	0.8	50	150	50	14	0.07	0.1	1	0
4370	Corn Chips 1 Oz	28.4	1	155	2	9	1.4	2.4	3.7	0	16	35	52	0.5	52	233	110	11	0.04	0.05	0.4	1
4380	Cornmeal;Whole-Grnd;Unbolt;Dry 1 Cup	122	12	435	11	5	0.5	1.1	2.5	0	90	24	312	2.2	346	1	620	62	0.46	0.13	2.4	0
4390	Cornmeal;Bolted;Dry Form 1 Cup	122	12	440	11	4	0.5	0.9	2.2	0	91	21	272	2.2	303	1	590	59	0.37	0.1	2.3	0
4400	Cornmeal;Degermed;Enriched;Dry 1 Cup	138	12	500	11	2	0.2	0.4	0.9	0	108	8	137	5.9	166	1	610	61	0.61	0.36	4.8	0
4410	Cornmeal;Degermed;Enrched;Cook1 Cup	240	88	120	3	0	0	0.1	0.2	0	26	2	34	1.4	38	0	140	14	0.14	0.1	1.2	0

TABLE A-8 Food Composition Table *(continued)*

No.	Description of Food		Wt. (g)	Water (%)	Energy (kcal)	Protein (g)	Fat (g)	Sat (g)	Mono (g)	Poly (g)	Cholest (mg)	Carb (g)	Ca (mg)	P (mg)	Fe (mg)	K (mg)	Na (mg)	Vit A (IU)	Vit A (RE)	Thmn (mg)	Ribofl (mg)	Niacin (mg)	Vit C (mg)
Cakes, Cookies, Pasta *(continued)*																							
4420	Cheese Crackers; Plain	10 Crack	10	4	50	1	3	0.9	1.2	0.3	6	6	11	17	0.3	17	112	20	5	0.05	0.04	0.4	0
4430	Cheese Crackers; Sandwch;Peant	1 Sandwh	8	3	40	1	2	0.4	0.8	0.3	1	5	7	25	0.3	17	90	0	1	0.04	0.03	0.6	0
4440	Graham Cracker; Plain	2 Crackr	14	5	60	1	1	0.4	0.6	0.4	0	11	6	20	0.4	36	86	0	0	0.02	0.03	0.6	0
4450	Melba Toast; Plain	1 Piece	5	4	20	1	0	0.1	0.1	0.1	0	4	6	10	0.1	11	44	0	0	0.01	0.01	0.1	0
4460	Rye Wafers; Whole-Grain	2 Wafers	14	5	55	1	1	0.3	0.4	0.3	0	10	7	44	0.5	65	115	0	0	0.06	0.03	0.5	0
4470	Saltines	4 Crackr	12	4	50	1	1	0.5	0.4	0.2	4	9	3	12	0.5	17	165	0	0	0.06	0.05	0.6	0
4480	Snack Type Crackers	1 Crackr	3	3	15	0	1	0.2	0.4	0.1	0	2	3	6	0.1	4	30	0	0	0.01	0.01	0.1	0
4490	Wheat; Thin Crackers	4 Crackr	8	3	35	1	1	0.5	0.5	0.4	0	5	3	15	0.3	17	69	0	1	0.04	0.03	0.4	0
4500	Whole-Wheat Wafers; Crackers	2 Cracker	8	4	35	1	2	0.5	0.6	0.4	0	5	3	22	0.2	31	59	0	0	0.02	0.03	0.4	0
4510	Croissants	1 Crosst	57	22	235	5	12	3.5	6.7	1.4	13	27	20	64	2.1	68	452	50	13	0.17	0.13	1.3	0
4520	Danish Pastry; Plain; No Nuts	1 Ring	340	27	1305	21	71	21.8	28.6	15.6	292	152	360	347	6.5	316	1302	360	99	0.95	1.02	8.5	0
4530	Danish Pastry; Plain; No Nuts	1 Pastry	57	27	220	4	12	3.6	4.8	2.6	49	26	60	58	1.1	53	218	60	17	0.16	0.17	1.4	0
4540	Danish Pastry; Plain; No Nuts	1 Oz	28.4	27	110	2	6	1.8	2.4	1.3	24	13	30	29	0.5	26	109	30	8	0.08	0.09	0.7	0
4550	Danish Pastry; Fruit	1 Pastry	65	30	235	4	13	3.9	5.2	2.9	56	28	17	80	1.3	57	233	40	11	0.16	0.14	1.4	0
4560	Doughnuts; Cake Type; Plain	1 Donut	50	21	210	3	12	2.8	5	3	20	24	22	111	1	58	192	20	5	0.12	0.12	1.1	0
4570	Doughnuts; Yeast-Leavend;Glzed	1 Donut	60	27	235	4	13	5.2	5.5	0.9	21	26	17	55	1.4	64	222	0	0	0.28	0.12	1.8	0
4580	English Muffins; Plain	1 Muffin	57	42	140	5	1	0.3	0.2	0.3	0	27	96	67	1.7	331	378	0	0	0.26	0.19	2.2	0
4590	English Muffins; Plain; Toastd	1 Muffin	50	29	140	5	1	0.3	0.2	0.3	0	27	96	67	1.7	331	378	0	0	0.23	0.19	2.2	0
4600	French Toast; Home Recipe	1 Slice	65	53	155	6	7	1.6	2	1.6	112	17	72	85	1.3	86	257	110	32	0.12	0.16	1	0
4610	Macaroni; Cooked; Firm	1 Cup	130	64	190	7	1	0.1	0.1	0.3	0	39	14	85	2.1	103	1	0	0	0.23	0.13	1.8	0
4620	Macaroni; Cooked; Tender;Cold	1 Cup	105	72	115	4	0	0.1	0.1	0.2	0	24	8	53	1.3	64	1	0	0	0.15	0.08	1.2	0
4630	Macaroni; Cooked; Tender; Hot	1 Cup	140	72	155	5	1	0.1	0.1	0.2	0	32	11	70	1.7	85	1	0	0	0.2	0.11	1.5	0
4640	Blueberry Muffins; Home Recipe	1 Muffin	45	37	135	3	5	1.5	2.1	1.2	19	20	54	46	0.9	47	198	40	9	0.1	0.11	0.9	1
4650	Bran Muffins; Home Recipe	1 Muffin	45	35	125	3	6	1.4	1.6	2.3	24	19	60	125	1.4	99	189	230	30	0.11	0.13	1.3	3
4660	Corn Muffins; Home Recipe	1 Muffin	45	33	145	3	5	1.5	2.2	1.4	23	21	66	59	0.9	57	169	80	15	0.11	0.11	0.9	0
4670	Blueberry Muffins; From Com Mix	1 Muffin	45	33	140	3	5	1.4	2	1.2	45	22	15	90	0.9	54	225	50	11	0.1	0.17	1.1	0
4680	Bran Muffins; From Commerl Mix	1 Muffin	45	28	140	3	4	1.3	1.6	1	28	24	27	182	1.7	50	385	100	14	0.08	0.12	1.9	0
4690	Corn Muffins; From Commerl Mix	1 Muffin	45	30	145	3	6	1.7	2.3	1.4	42	22	30	128	1.3	31	291	90	16	0.09	0.09	0.8	0
4700	Noodles; Egg; Cooked	1 Cup	160	70	200	7	2	0.5	0.6	0.6	50	37	16	94	2.6	70	3	110	34	0.22	0.13	1.9	0
4710	Noodles; Chow Mein; Canned	1 Cup	45	11	220	6	11	2.1	7.3	0.4	5	26	14	41	0.4	33	450	0	0	0.05	0.03	0.6	0
4720	Pancakes; Buckwheat; From Mix	1 Pancake	27	58	55	2	2	0.9	0.9	0.5	20	6	59	91	0.4	66	125	60	17	0.04	0.05	0.2	0

TABLE A-8 Food Composition Table *(continued)*

No.	Description of Food		Wt. (g)	Water (%)	Energy (kcal)	Protein (g)	Fat (g)	Sat (g)	Mono (g)	Poly (g)	Cholest (mg)	Carb (g)	Ca (mg)	P (mg)	Fe (mg)	K (mg)	Na (mg)	Vit A (IU)	Vit A (RE)	Thmn (mg)	Ribofl (mg)	Niacin (mg)	Vit C (mg)
Cakes, Cookies, Pasta *(continued)*																							
4730	Pancakes; Plain; Home Recipe	1 Pancak	27	50	60	2	2	0.5	0.8	0.5	16	9	27	38	0.5	33	115	30	10	0.06	0.07	0.5	0
4740	Pancakes; Plain; From Mix	1 Pancak	27	54	60	2	2	0.5	0.9	0.5	16	8	36	71	0.7	43	160	30	7	0.09	0.12	0.8	0
4750	Piecrust; From Home Recipe	1 Shell	180	15	900	11	60	14.8	25.9	15.7	0	79	25	90	4.5	90	1100	0	0	0.54	0.4	5	0
4760	Piecrust; From Mix	2 Crust	320	19	1485	20	93	22.7	41	25	0	141	131	272	9.3	179	2602	0	0	1.06	0.8	9.9	0
4770	Apple Pie	1 Pie	945	48	2420	21	105	27.4	44.4	26.5	0	360	76	208	9.5	756	2844	280	28	1.04	0.76	9.5	9
4780	Apple Pie	1 Piece	158	48	405	3	18	4.6	7.4	4.4	0	60	13	35	1.6	126	476	50	5	0.17	0.13	1.6	2
4790	Blueberry Pie	1 Pie	945	51	2285	23	102	25.5	44.4	27.4	0	330	104	217	12.3	945	2533	850	85	1.04	0.85	10.4	38
4800	Blueberry Pie	1 Piece	158	51	380	4	17	4.3	7.4	4.6	0	55	17	36	2.1	158	423	140	14	0.17	0.14	1.7	6
4810	Cherry Pie	1 Pie	945	47	2465	25	107	28.4	46.3	27.4	0	363	132	236	9.5	992	2873	4160	416	1.13	0.85	9.5	0
4820	Cherry Pie	1 Piece	158	47	410	4	18	4.7	7.7	4.6	0	61	22	40	1.6	166	480	700	70	0.19	0.14	1.6	0
4830	Creme Pie	1 Pie	910	43	2710	20	139	90.1	23.7	6.4	46	351	273	919	6.8	796	2207	1250	391	0.36	0.89	6.4	0
4840	Creme Pie	1 Piece	152	43	455	3	23	15	4	1.1	8	59	46	154	1.1	133	369	210	65	0.06	0.15	1.1	0
4850	Custard Pie	1 Pie	910	58	1985	56	101	33.7	40	19.1	1010	213	874	1028	9.1	1247	2612	2090	573	0.82	1.91	5.5	0
4860	Custard Pie	1 Piece	152	58	330	9	17	5.6	6.7	3.2	169	36	146	172	1.5	208	436	350	96	0.14	0.32	0.9	0
4870	Lemon Meringue Pie	1 Pie	840	47	2140	31	86	26	34.4	17.6	857	317	118	412	8.4	420	2369	1430	395	0.59	0.84	5	25
4880	Lemon Meringue Pie	1 Piece	140	47	355	5	14	4.3	5.7	2.9	143	53	20	69	1.4	70	395	240	66	0.1	0.14	0.8	4
4890	Peach Pie	1 Pie	945	48	2410	24	101	24.6	43.5	26.5	0	361	95	274	11.3	1408	2533	6900	690	1.04	0.95	14.2	28
4900	Peach Pie	1 Piece	158	48	405	4	17	4.1	7.3	4.4	0	60	16	46	1.9	235	423	1150	115	0.17	0.16	2.4	5
4910	Pecan Pie	1 Pie	825	20	3450	42	189	28.1	101.5	47	569	423	388	850	27.2	1015	1823	1320	322	1.82	0.99	6.6	0
4920	Pecan Pie	1 Piece	138	20	575	7	32	4.7	17	7.9	95	71	65	142	4.6	170	305	220	54	0.3	0.17	1.1	0
4930	Pumpkin Pie	1 Pie	910	59	1920	36	102	38.2	40	18.2	655	223	464	628	8.2	1456	1947	22480	2493	0.82	1.27	7.3	0
4940	Pumpkin Pie	1 Piece	152	59	320	6	17	6.4	6.7	3	109	37	78	105	1.4	243	325	3750	416	0.14	0.21	1.2	0
4950	Fried Pie; Apple	1 Pie	85	43	255	2	14	5.8	6.6	0.6	14	31	12	34	0.9	42	326	30	3	0.09	0.06	1	1
4960	Fried Pie; Cherry	1 Pie	85	42	250	2	14	5.8	6.7	0.6	13	32	11	41	0.7	61	371	190	19	0.06	0.06	0.6	1
4970	Popcorn; Air-Popped; Unsalted	1 Cup	8	4	30	1	0	0	0.1	0.2	0	6	1	22	0.2	20	0	10	1	0.03	0.01	0.2	0
4980	Popcorn; Popped; Veg Oil;Saltd	1 Cup	11	3	55	1	3	0.5	1.4	1.2	0	6	3	31	0.3	19	86	20	2	0.01	0.02	0.1	0
4990	Popcorn; Sugar Syrup Coated	1 Cup	35	4	135	2	1	0.1	0.3	0.6	0	30	2	47	0.5	90	0	30	3	0.13	0.02	0.4	0
5000	Pretzels; Stick	10 Pretz	3	3	10	0	0	0	0	0	0	2	1	3	0.1	3	48	0	0	0.01	0.01	0.1	0
5010	Pretzels; Twisted; Dutch	1 Pretz	16	3	65	2	1	0.1	0.2	0.2	0	13	4	15	0.3	16	258	0	0	0.05	0.04	0.7	0
5020	Pretzels; Twisted; Thin	10 Pretz	60	3	240	6	2	0.4	0.8	0.6	0	48	16	55	1.2	61	966	0	0	0.19	0.15	2.6	0
5030	Rice; Brown; Cooked	1 Cup	195	70	230	5	1	0.3	0.3	0.4	0	50	23	142	1	137	0	0	0	0.18	0.04	2.7	0

TABLE A-8 Food Composition Table *(continued)*

No.	Description of Food		Wt. (g)	Water (%)	Energy (kcal)	Protein (g)	Fat (g)	Sat (g)	Mono (g)	Poly (g)	Cholest (mg)	Carb (g)	Ca (mg)	P (mg)	Fe (mg)	K (mg)	Na (mg)	Vit A (IU)	Vit A (RE)	Thmn (mg)	Ribofl (mg)	Niacin (mg)	Vit C (mg)
Cakes, Cookies, Pasta *(continued)*																							
5040	Rice; White; Raw	1 Cup	185	12	670	12	1	0.2	0.2	0.3	0	149	44	174	5.4	170	9	0	0	0.81	0.06	6.5	0
5050	Rice; White; Cooked	1 Cup	205	73	225	4	0	0.1	0.1	0.1	0	50	21	57	1.8	57	0	0	0	0.23	0.02	2.1	0
5060	Rice; White; Instant; Cooked	1 Cup	165	73	180	4	0	0.1	0.1	0.1	0	40	5	31	1.3	0	0	0	0	0.21	0.02	1.7	0
5070	Rice; White; Parboiled; Raw	1 Cup	185	10	685	14	1	0.1	0.1	0.2	0	150	111	370	5.4	278	17	0	0	0.81	0.07	6.5	0
5080	Rice; White; Parboiled; Cooked	1 Cup	175	73	185	4	0	0	0	0.1	0	41	33	100	1.4	75	0	0	0	0.19	0.02	2.1	0
5090	Rolls; Dinner; Commercial	1 Roll	28	32	85	2	2	0.5	0.8	0.6	0	14	33	44	0.8	36	155	0	0	0.14	0.09	1.1	0
5100	Rolls; Frankfurter+Hamburger	1 Roll	40	34	115	3	2	0.5	0.8	0.6	0	20	54	44	1.2	56	241	0	0	0.2	0.13	1.6	0
5110	Rolls; Hard	1 Roll	50	25	155	5	2	0.4	0.5	0.6	0	30	24	46	1.4	49	313	0	0	0.2	0.12	1.7	0
5120	Rolls; Hoagie or Submarine	1 Roll	135	31	400	11	8	1.8	3	2.2	0	72	100	115	3.8	128	683	0	0	0.54	0.33	4.5	0
5130	Rolls; Dinner; Home Recipe	1 Roll	35	26	120	3	3	0.8	1.2	0.9	12	20	16	36	1.1	41	98	30	8	0.12	0.12	1.2	0
5140	Spaghetti; Cooked; Firm	1 Cup	130	64	190	7	1	0.1	0.1	0.3	0	39	14	85	2	103	1	0	0	0.23	0.13	1.8	0
5150	Spaghetti; Cooked; Tender	1 Cup	140	73	155	5	1	0.1	0.1	0.2	0	32	11	70	1.7	85	1	0	0	0.2	0.11	1.5	0
5160	Toaster Pastries	1 Pastry	54	13	210	2	6	1.7	3.6	0.4	0	38	104	104	2.2	91	248	520	52	0.17	0.18	2.3	4
5170	Tortillas; Corn	1 Tortla	30	45	65	2	1	0.1	0.3	0.6	0	13	42	55	0.6	43	1	80	8	0.05	0.03	0.4	0
5180	Waffles; From Home Recipe	1 Waffle	75	37	245	7	13	4	4.9	2.6	102	26	154	135	1.5	129	445	140	39	0.18	0.24	1.5	0
5190	Waffles; From Mix	1 Waffle	75	42	205	7	8	2.7	2.9	1.5	59	27	179	257	1.2	146	515	170	49	0.14	0.23	0.9	0
5200	Wheat Flour; All-Purpose; Siftd	1 Cup	115	12	420	12	1	0.2	0.1	0.5	0	88	18	100	5.1	109	2	0	0	0.73	0.46	6.1	0
5210	Wheat Flour; All-Purpose; Unsif	1 Cup	125	12	455	13	1	0.2	0.1	0.5	0	95	20	109	5.5	119	3	0	0	0.8	0.5	6.6	0
5220	Cake or Pastry Flour; Sifted	1 Cup	96	12	350	7	1	0.1	0.1	0.3	0	76	16	70	4.2	91	2	0	0	0.58	0.38	5.1	0
5230	Self-Rising Flour; Unsifted	1 Cup	125	12	440	12	1	0.2	0.1	0.5	0	93	331	583	5.5	113	1349	0	0	0.8	0.5	6.6	0
5240	Whole-Wheat Flour;Hrd Wht;Stir	1 Cup	120	12	400	16	2	0.3	0.3	1.1	0	85	49	446	5.2	444	4	0	0	0.66	0.14	5.2	0
Nuts and Seeds																							
5250	Almonds; Slivered	1 Cup	135	4	795	27	70	6.7	45.8	14.8	0	28	359	702	4.9	988	15	0	0	0.28	1.05	4.5	1
5260	Almonds; Whole	1 Oz	28.4	4	165	6	15	1.4	9.6	3.1	0	6	75	147	1	208	3	0	0	0.06	0.22	1	0
5270	Black Beans; Dry; Cooked;Drand	1 Cup	171	66	225	15	1	0.1	0.1	0.5	0	41	47	239	2.9	608	1	0	0	0.43	0.05	0.9	0
5280	Great Northn Beans;Dry;Ckd;Drn	1 Cup	180	69	210	14	1	0.1	0.1	0.6	0	38	90	266	4.9	749	13	0	0	0.25	0.13	1.3	0
5290	Lima Beans; Dry; Cooked;Draned	1 Cup	190	64	260	16	1	0.2	0.1	0.5	0	49	55	293	5.9	1163	4	0	0	0.25	0.11	1.3	0
5300	Pea Beans; Dry; Cooked;Drained	1 Cup	190	69	225	15	1	0.1	0.1	0.7	0	40	95	281	5.1	790	13	0	0	0.27	0.13	1.3	0
5310	Pinto Beans;Dry;Cooked;Drained	1 Cup	180	65	265	15	1	0.1	0.1	0.5	0	49	86	296	5.4	882	3	0	0	0.33	0.16	0.7	0
5320	Beans;Dry;Canned;W/Frankfurter	1 Cup	255	71	365	19	18	7.4	8.8	0.7	30	32	94	303	4.8	668	1374	330	33	0.18	0.15	3.3	0
5330	Beans;Dry;Canned;W/Pork+Tomsce	1 Cup	255	71	310	16	7	2.4	2.7	0.7	10	48	138	235	4.6	536	1181	330	33	0.2	0.08	1.5	5

TABLE A-8 Food Composition Table *(continued)*

No.	Description of Food		Wt. (g)	Water (%)	Energy (kcal)	Protein (g)	Fat (g)	Sat (g)	Mono (g)	Poly (g)	Cholest (mg)	Carb (g)	Ca (mg)	P (mg)	Fe (mg)	K (mg)	Na (mg)	Vit A (IU)	Vit A (RE)	Thmn (mg)	Ribofl (mg)	Niacin (mg)	Vit C (mg)
Nuts and Seeds *(continued)*																							
5340	Beans; Dry;Canned;W/Pork+Swtsce	1 Cup	255	66	385	16	12	4.3	4.9	1.2	10	54	161	291	5.9	536	969	330	33	0.15	0.1	1.3	5
5350	Red Kidney Beans; Dry; Canned	1 Cup	255	76	230	15	1	0.1	0.1	0.6	0	42	74	278	4.6	673	968	10	1	0.13	0.1	1.5	0
5360	Black-Eyed Peas; Dry; Cooked	1 Cup	250	80	190	13	1	0.2	0	0.3	0	35	43	238	3.3	573	20	30	3	0.4	0.1	1	0
5370	Brazil Nuts	1 Oz	28.4	3	185	4	19	4.6	6.5	6.8	0	4	50	170	1	170	1	0	0	0.28	0.03	0.5	0
5380	Carob Flour	1 Cup	140	3	255	6	0	0	0.1	0.1	0	126	390	102	5.7	1275	24	0	0	0.07	0.07	2.2	0
5390	Cashew Nuts; Dry Roasted; Saltd	1 Cup	137	2	785	21	63	12.5	37.4	10.7	0	45	62	671	8.2	774	877	0	0	0.27	0.27	1.9	0
5391	Cashew Nuts; Dry Roastd; Unsalt	1 Cup	137	2	785	21	63	12.5	37.4	10.7	0	45	62	671	8.2	774	21	0	0	0.27	0.27	1.9	0
5400	Cashew Nuts; Dry Roastd; Salted	1 Oz	28.4	2	165	4	13	2.6	7.7	2.2	0	9	13	139	1.7	160	181	0	0	0.06	0.06	0.4	0
5401	Cashew Nuts; Dry Roastd; Unsalt	1 Oz	28.4	2	165	4	13	2.6	7.7	2.2	0	9	13	139	1.7	160	4	0	0	0.06	0.06	0.4	0
5410	Cashew Nuts; Oil Roastd; Salted	1 Cup	130	4	750	21	63	12.4	36.9	10.6	0	37	53	554	5.3	689	814	0	0	0.55	0.23	2.3	0
5411	Cashew Nuts; Oil Roastd; Unsalt	1 Cup	130	4	750	21	63	12.4	36.9	10.6	0	37	53	554	5.3	689	22	0	0	0.55	0.23	2.3	0
5420	Cashew Nuts; Oil Roastd; Salted	1 Oz	28.4	4	165	5	14	2.7	8.1	2.3	0	8	12	121	1.2	150	177	0	0	0.12	0.05	0.5	0
5421	Cashew Nuts; Oil Roastd; Unsalt	1 Oz	28.4	4	165	5	14	2.7	8.1	2.3	0	8	12	121	1.2	150	5	0	0	0.12	0.05	0.5	0
5430	Chestnuts; European; Roasted	1 Cup	143	40	350	5	3	0.6	1.1	1.2	0	76	41	153	1.3	847	3	30	3	0.35	0.25	1.9	37
5440	Chickpeas; Cooked; Drained	1 Cup	163	60	270	15	4	0.4	0.9	1.9	0	45	80	273	4.9	475	11	0	0	0.18	0.09	0.9	0
5450	Coconut; Raw; Piece	1 Piece	45	47	160	1	15	13.4	0.6	0.2	0	7	6	51	1.1	160	9	0	0	0.03	0.01	0.2	1
5460	Coconut; Raw; Shredded	1 Cup	80	47	285	3	27	23.8	1.1	0.3	0	12	11	90	1.9	285	16	0	0	0.05	0.02	0.4	3
5470	Coconut; Dried; Sweetnd;Shredd	1 Cup	93	13	470	3	33	29.3	1.4	0.4	0	44	14	99	1.8	313	244	0	0	0.03	0.02	0.4	1
5480	Filberts; (Hazelnuts) Chopped	1 Cup	115	5	725	15	72	5.3	56.5	6.9	0	18	216	359	3.8	512	3	80	8	0.58	0.13	1.3	1
5490	Filberts; (Hazelnuts) Chopped	1 Oz	28.4	5	180	4	18	1.3	13.9	1.7	0	4	53	88	0.9	126	1	20	2	0.14	0.03	0.3	0
5500	Lentils; Dry; Cooked	1 Cup	200	72	215	16	1	0.1	0.2	0.5	0	38	50	238	4.2	498	26	40	4	0.14	0.12	1.2	0
5510	Macadamia Nuts; Oilrstd; Salted	1 Cup	134	2	960	10	103	15.4	80.9	1.8	0	17	60	268	2.4	441	348	10	1	0.29	0.15	2.7	0
5511	Macadamia Nuts; Oilrstd; Unsalt	1 Cup	134	2	960	10	103	15.4	80.9	1.8	0	17	60	268	2.4	441	9	10	1	0.29	0.15	2.7	0
5520	Macadamia Nuts; Oilrstd; Salted	1 Oz	28.4	2	205	2	22	3.2	17.1	0.4	0	4	13	57	0.5	93	74	0	0	0.06	0.03	0.6	0
5521	Macadamia Nuts; Oilrstd;Unsalt	1 Oz	28.4	2	205	2	22	3.2	17.1	0.4	0	4	13	57	0.5	93	2	0	0	0.06	0.03	0.6	0
5530	Mixed Nuts W/Peanuts; Dry;Saltd	1 Oz	28.4	2	170	5	15	2	8.9	3.1	0	7	20	123	1	169	190	0	0	0.06	0.06	1.3	0
5531	Mixed Nuts W/Peanuts; Dry;Unslt	1 Oz	28.4	2	170	5	15	2	8.9	3.1	0	7	20	123	1	169	3	0	0	0.06	0.06	1.3	0
5540	Mixed Nuts W/Peanuts; Oil;Saltd	1 Oz	28.4	2	175	5	16	2.5	9	3.8	0	6	31	131	0.9	165	185	10	1	0.14	0.06	1.4	0
5541	Mixed Nuts W/Peanuts; Oil;Unslt	1 Oz	28.4	2	175	5	16	2.5	9	3.8	0	6	31	131	0.9	165	3	10	1	0.14	0.06	1.4	0
5550	Peanuts; Oil Roasted; Salted	1 Cup	145	2	840	39	71	9.9	35.5	22.6	0	27	125	734	2.8	1019	626	0	0	0.42	0.15	21.5	0
5551	Peanuts; Oil Roasted; Unsalted	1 Cup	145	2	840	39	71	9.9	35.5	22.6	0	27	125	734	2.8	1019	22	0	0	0.42	0.15	21.5	0

TABLE A-8 Food Composition Table *(continued)*

No.	Description of Food		Wt. (g)	Water (%)	Energy (kcal)	Protein (g)	Fat (g)	Sat (g)	Mono (g)	Poly (g)	Cholest (mg)	Carb (g)	Ca (mg)	P (mg)	Fe (mg)	K (mg)	Na (mg)	Vit A (IU)	Vit A (RE)	Thmn (mg)	Ribofl (mg)	Niacin (mg)	Vit C (mg)
Nuts and Seeds *(continued)*																							
5560	Peanuts; Oil Roasted; Salted	1 Oz	28.4	2	165	8	14	1.9	6.9	4.4	0	5	24	143	0.5	199	122	0	0	0.08	0.03	4.2	0
5561	Peanuts; Oil Roasted; Unsalted	1 Oz	28.4	2	165	8	14	1.9	6.9	4.4	0	5	24	143	0.5	199	4	0	0	0.08	0.03	4.2	0
5570	Peanut Butter	1 Tbsp	16	1	95	5	8	1.4	4	2.5	0	3	5	60	0.3	110	75	0	0	0.02	0.02	2.2	0
5580	Peas; Split; Dry; Cooked	1 Cup	200	70	230	16	1	0.1	0.1	0.3	0	42	22	178	3.4	592	26	80	8	0.3	0.18	1.8	0
5590	Pecans; Halves	1 Cup	108	5	720	8	73	5.9	45.5	18.1	0	20	39	314	2.3	423	1	140	14	0.92	0.14	1	2
5600	Pecans; Halves	1 Oz	28.4	5	190	2	19	1.5	12	4.7	0	5	10	83	0.6	111	0	40	4	0.24	0.04	0.3	1
5610	Pine Nuts	1 Oz	28.4	6	160	3	17	2.7	6.5	7.3	0	5	2	10	0.9	178	20	10	1	0.35	0.06	1.2	1
5620	Pistachio Nuts	1 Oz	28.4	4	165	6	14	1.7	9.3	2.1	0	7	38	143	1.9	310	2	70	7	0.23	0.05	0.3	0
5630	Pumpkin and Squash Kernels	1 Oz	28.4	7	155	7	13	2.5	4	5.9	0	5	12	333	4.2	229	5	110	11	0.06	0.09	0.5	0
5640	Refried Beans; Canned	1 Cup	290	72	295	18	3	0.4	0.6	1.4	0	51	141	245	5.1	1141	1228	0	0	0.14	0.16	1.4	17
5650	Sesame Seeds	1 Tbsp	8	5	45	2	4	0.6	1.7	1.9	0	1	11	62	0.6	33	3	10	1	0.06	0.01	0.4	0
5660	Soybeans; Dry; Cooked; Drained	1 Cup	180	71	235	20	10	1.3	1.9	5.3	0	19	131	322	4.9	972	4	50	5	0.38	0.16	1.1	0
5670	Miso	1 Cup	276	53	470	29	13	1.8	2.6	7.3	0	65	188	853	4.7	922	8142	110	11	0.17	0.28	0.8	0
5680	Tofu	1 Piece	120	85	85	9	5	0.7	1	2.9	0	3	108	151	2.3	50	8	0	0	0.07	0.04	0.1	0
5690	Sunflower Seeds	1 Oz	28.4	5	160	6	14	1.5	2.7	9.3	0	5	33	200	1.9	195	1	10	1	0.65	0.07	1.3	0
5700	Tahini	1 Tbsp	15	3	90	3	8	1.1	3	3.5	0	3	21	119	0.7	69	5	10	1	0.24	0.02	0.8	1
5710	Walnuts; Black; Chopped	1 Cup	125	4	760	30	71	4.5	15.9	46.9	0	15	73	580	3.8	655	1	370	37	0.27	0.14	0.9	0
5720	Walnuts; Black; Chopped	1 Oz	28.4	4	170	7	16	1	3.6	10.6	0	3	16	132	0.9	149	0	80	8	0.06	0.03	0.2	0
5730	Walnuts; English; Pieces	1 Cup	120	4	770	17	74	6.7	17	47	0	22	113	380	2.9	602	12	150	15	0.46	0.18	1.3	4
5740	Walnuts; English; Pieces	1 Oz	28.4	4	180	4	18	1.6	4	11.1	0	5	27	90	0.7	142	3	40	4	0.11	0.04	0.3	1
Meats																							
5750	Beef; Ckd;Chuck Blade; Lean+Fat	3 Oz	85	43	325	22	26	10.8	11.7	0.9	87	0	11	163	2.5	163	53	0	0	0.06	0.19	2	0
5760	Beef; Ckd;Chuck Blade; Lean Only	2.2 Oz	62	53	170	19	9	3.9	4.2	0.3	66	0	8	146	2.3	163	44	0	0	0.05	0.17	1.7	0
5770	Beef; Ckd;Bttm Round; Lean+Fat	3 Oz	85	54	220	25	13	4.8	5.7	0.5	81	0	5	217	2.8	248	43	0	0	0.06	0.21	3.3	0
5780	Beef; Ckd;Bttm Round; Lean Only	2.8 Oz	78	57	175	25	8	2.7	3.4	0.3	75	0	4	212	2.7	240	40	0	0	0.06	0.2	3	0
5790	Ground Beef; Broiled; Lean	3 Oz	85	56	230	21	16	6.2	6.9	0.6	74	0	9	134	1.8	236	65	0	0	0.04	0.18	4.4	0
5800	Ground Beef; Broiled; Regular	3 Oz	85	54	245	20	18	6.9	7.7	0.7	76	0	9	144	2.1	248	70	0	0	0.03	0.16	4.9	0
5810	Beef Heart; Braised	3 Oz	85	65	150	24	5	1.2	0.8	1.6	164	0	5	213	6.4	198	54	0	0	0.12	1.31	3.4	5
5820	Beef Liver; Fried	3 Oz	85	56	185	23	7	2.5	3.6	1.3	410	7	9	392	5.3	309	90	###	9120	0.18	3.52	12.3	23
5830	Beef Roast; Rib; Lean+Fat	3 Oz	85	46	315	19	26	10.8	11.4	0.9	72	0	8	145	2	246	54	0	0	0.06	0.16	3.1	0
5840	Beef Roast; Rib; Lean Only	2.2 Oz	61	57	150	17	9	3.6	3.7	0.3	49	0	5	127	1.7	218	45	0	0	0.05	0.13	2.7	0

TABLE A-8 Food Composition Table *(continued)*

No.	Description of Food		Wt. (g)	Water (%)	Energy (kcal)	Protein (g)	Fat (g)	Sat (g)	Mono (g)	Poly (g)	Cholest (mg)	Carb (g)	Ca (mg)	P (mg)	Fe (mg)	K (mg)	Na (mg)	Vit A (IU)	Vit A (RE)	Thmn (mg)	Ribofl (mg)	Niacin (mg)	Vit C (mg)
Meats *(continued)*																							
5850	Beef Roast; Eye O Rnd; Lean+Fat	3 Oz	85	57	205	23	12	4.9	5.4	0.5	62	0	5	177	1.6	308	50	0	0	0.07	0.14	3	0
5860	Beef Roast; Eye O Rnd; Lean	2.6 Oz	75	63	135	22	5	1.9	2.1	0.2	52	0	3	170	1.5	297	46	0	0	0.07	0.13	2.8	0
5870	Beef Steak; Sirloin; Broil; Lean+Fat	3 Oz	85	53	240	23	15	6.4	6.9	0.6	77	0	9	186	2.6	306	53	0	0	0.1	0.23	3.3	0
5880	Beef Steak; Sirloin; Broil; Lean	2.5 Oz	72	59	150	22	6	2.6	2.8	0.3	64	0	8	176	2.4	290	48	0	0	0.09	0.22	3.1	0
5890	Beef; Canned; Corned	3 Oz	85	59	185	22	10	4.2	4.9	0.4	80	0	17	90	3.7	51	802	0	0	0.02	0.2	2.9	0
5900	Beef; Dried; Chipped	2.5 Oz	72	48	145	24	4	1.8	2	0.2	46	0	14	287	2.3	142	3053	0	0	0.05	0.23	2.7	0
5910	Lamb; Chops;Arm;Braised;Lean+Fat	2.2 Oz	63	44	220	20	15	6.9	6	0.9	77	0	16	132	1.5	195	46	0	0	0.04	0.16	4.4	0
5920	Lamb; Chops;Arm;Braised;Lean	1.7 Oz	48	49	135	17	7	2.9	2.6	0.4	59	0	12	111	1.3	162	36	0	0	0.03	0.13	3	0
5930	Lamb; Chops;Loin;Broil;Lean+Fat	2.8 Oz	80	54	235	22	16	7.3	6.4	1	78	0	16	162	1.4	272	62	0	0	0.09	0.21	5.5	0
5940	Lamb; Chops;Loin;Broil;Lean	2.3 Oz	64	61	140	19	6	2.6	2.4	0.4	60	0	12	145	1.3	241	54	0	0	0.08	0.18	4.4	0
5950	Lamb; Leg;Roasted; Lean+Fat	3 Oz	85	59	205	22	13	5.6	4.9	0.8	78	0	8	162	1.7	273	57	0	0	0.09	0.24	5.5	0
5960	Lamb; Leg;Roasted; Lean Only	2.6 Oz	73	64	140	20	6	2.4	2.2	0.4	65	0	6	150	1.5	247	50	0	0	0.08	0.2	4.6	0
5970	Lamb; Rib; Roasted; Lean+Fat	3 Oz	85	47	315	18	26	12.1	10.6	1.5	77	0	19	139	1.4	224	60	0	0	0.08	0.18	5.5	0
5980	Lamb; Rib; Roasted; Lean Only	2 Oz	57	60	130	15	7	3.2	3	0.5	50	0	12	111	1	179	46	0	0	0.05	0.13	3.5	0
5990	Pork; Cured; Bacon; Regul;Cked	3 Slice	19	13	110	6	9	3.3	4.5	1.1	16	0	2	64	0.3	92	303	0	0	0.13	0.05	1.4	6
6000	Pork; Cured; Bacon;Canadn;Cked	2 Slice	46	62	85	11	4	1.3	1.9	0.4	27	1	5	136	0.4	179	711	0	0	0.38	0.09	3.2	10
6010	Pork; Cured; Ham; Rosted;Ln+Fat	3 Oz	85	58	205	18	14	5.1	6.7	1.5	53	0	6	182	0.7	243	1009	0	0	0.51	0.19	3.8	0
6020	Pork; Cured; Ham; Rosted;Lean	2.4 Oz	68	66	105	17	4	1.3	1.7	0.4	37	0	5	154	0.6	215	902	0	0	0.46	0.17	3.4	0
6030	Pork; Cured; Ham; Canned;Roast	3 Oz	85	67	140	18	7	2.4	3.5	0.8	35	0	6	188	0.9	298	908	0	0	0.82	0.21	4.3	19
6040	Pork; Luncheon Meat;Canned	2 Slices	42	52	140	5	13	4.5	6	1.5	26	1	3	34	0.3	90	541	0	0	0.15	0.08	1.3	0
6050	Pork; Luncheon Meat;Choppd Ham	2 Slices	42	64	95	7	7	2.4	3.4	0.9	21	0	3	65	0.3	134	576	0	0	0.27	0.09	1.6	8
6060	Pork; Luncheon Meat;Ckd Ham;Rg	2 Slices	57	65	105	10	6	1.9	2.8	0.7	32	2	4	141	0.6	189	751	0	0	0.49	0.14	3	16
6070	Pork; Luncheon Meat;Ckd Ham;Ln	2 Slices	57	71	75	11	3	0.9	1.3	0.3	27	1	4	124	0.4	200	815	0	0	0.53	0.13	2.8	15
6080	Pork Chop; Loin; Broil; Lean+Fat	3.1 Oz	87	50	275	24	19	7	8.8	2.2	84	0	3	184	0.7	312	61	10	3	0.87	0.24	4.3	0
6090	Pork Chop; Loin; Broil; Lean	2.5 Oz	72	57	165	23	8	2.6	3.4	0.9	71	0	4	176	0.7	302	56	10	1	0.83	0.22	4	0
6100	Pork Chop; Loin;Panfry; Lean+Fat	3.1 Oz	89	45	335	21	27	9.8	12.5	3.1	92	0	4	190	0.7	323	64	10	3	0.91	0.24	4.6	0
6110	Pork Chop; Loin;Panfry; Lean	2.4 Oz	67	54	180	19	11	3.7	4.8	1.3	72	0	3	178	0.7	305	57	10	1	0.84	0.22	4	0
6120	Pork Fresh Ham; Roastd; Lean+Fat	3 Oz	85	53	250	21	18	6.4	8.1	2	79	0	5	210	0.9	280	50	10	2	0.54	0.27	3.9	0
6130	Pork Fresh Ham; Roastd; Lean	2.5 Oz	72	60	160	20	8	2.7	3.6	1	68	0	5	202	0.8	269	46	10	1	0.5	0.25	3.6	0
6140	Pork Fresh Rib; Roastd; Lean+Fat	3 Oz	85	51	270	21	20	7.2	9.2	2.3	69	0	9	190	0.8	313	37	10	3	0.5	0.24	4.2	0
6150	Pork Fresh Rib; Roastd; Lean	2.5 Oz	71	57	175	20	10	3.4	4.4	1.2	56	0	8	182	0.7	300	33	10	2	0.45	0.22	3.8	0

TABLE A-8 Food Composition Table *(continued)*

No.	Description of Food		Wt. (g)	Water (%)	Energy (kcal)	Protein (g)	Fat (g)	Sat (g)	Mono (g)	Poly (g)	Cholest (mg)	Carb (g)	Ca (mg)	P (mg)	Fe (mg)	K (mg)	Na (mg)	Vit A (IU)	Vit A (RE)	Thmn (mg)	Ribofl (mg)	Niacin (mg)	Vit C (mg)
Meats *(continued)*																							
6160	Pork Shoulder; Braisd; Lean+Fat	3 Oz	85	47	295	23	22	7.9	10	2.4	93	0	6	162	1.4	286	75	10	3	0.46	0.26	4.4	0
6170	Pork Shoulder; Braisd; Lean	2.4 Oz	67	54	165	22	8	2.8	3.7	1	76	0	5	151	1.3	271	68	10	1	0.4	0.24	4	0
6180	Bologna	2 Slices	57	54	180	7	16	6.1	7.6	1.4	31	2	7	52	0.9	103	581	0	0	0.1	0.08	1.5	12
6190	Braunschweiger	2 Slices	57	48	205	8	18	6.2	8.5	2.1	89	2	5	96	5.3	113	652	8010	2405	0.14	0.87	4.8	6
6200	Brown And Serve Sausage;Brwnd	1 Link	13	45	50	2	5	1.7	2.2	0.5	9	0	1	14	0.1	25	105	0	0	0.05	0.02	0.4	0
6210	Frankfurter; Cooked	1 Frank	45	54	145	5	13	4.8	6.2	1.2	23	1	5	39	0.5	75	504	0	0	0.09	0.05	1.2	12
6220	Pork; Link; Cooked	1 Link	13	45	50	3	4	1.4	1.8	0.5	11	0	4	24	0.2	47	168	0	0	0.1	0.03	0.6	0
6230	Salami; Cooked Type	2 Slices	57	60	145	8	11	4.6	5.2	1.2	37	1	7	66	1.5	113	607	0	0	0.14	0.21	2	7
6240	Salami; Dry Type	2 Slices	20	35	85	5	7	2.4	3.4	0.6	16	1	2	28	0.3	76	372	0	0	0.12	0.06	1	5
6250	Sandwich Spread; Pork; Beef	1 Tbsp	15	60	35	1	3	0.9	1.1	0.4	6	2	2	9	0.1	17	152	10	1	0.03	0.02	0.3	0
6260	Vienna Sausage	1 Sausag	16	60	45	2	4	1.5	2	0.3	8	0	2	8	0.1	16	152	0	0	0.01	0.02	0.3	0
6270	Veal Cutlet; Med Fat;Brsd;Brld	3 Oz	85	60	185	23	9	4.1	4.1	0.6	86	0	9	196	0.8	258	56	0	0	0.06	0.21	4.6	0
6280	Veal Rib; Med Fat; Roasted	3 Oz	85	55	230	23	14	6	6	1	109	0	10	211	0.7	259	57	0	0	0.11	0.26	6.6	0
Dinners																							
6290	Beef and Vegetable Stew; Hm Rcp	1 Cup	245	82	220	16	11	4.4	4.5	0.5	71	15	29	184	2.9	613	292	5690	568	0.15	0.17	4.7	17
6300	Beef Potpie; Home Recipe	1 Piece	210	55	515	21	30	7.9	12.9	7.4	42	39	29	149	3.8	334	596	4220	517	0.29	0.29	4.8	6
6310	Chicken A La King; Home Recipe	1 Cup	245	68	470	27	34	12.9	13.4	6.2	221	12	127	358	2.5	404	760	1130	272	0.1	0.42	5.4	12
6320	Chicken And Noodles; Home Recp	1 Cup	240	71	365	22	18	5.1	7.1	3.9	103	26	26	247	2.2	149	600	430	130	0.05	0.17	4.3	0
6330	Chicken Chow Mein; Canned	1 Cup	250	89	95	7	0	0.1	0.1	0.8	8	18	45	85	1.3	418	725	150	28	0.05	0.1	1	13
6340	Chicken Chow Mein; Home Recipe	1 Cup	250	78	255	31	10	4.1	4.9	3.5	75	10	58	293	2.5	473	718	280	50	0.08	0.23	4.3	10
6350	Chicken Potpie; Home Recipe	1 Piece	232	57	545	23	31	10.3	15.5	6.6	56	42	70	232	3	343	594	7220	735	0.32	0.32	4.9	5
6360	Chili Con Carne W/Beans; Cnnd	1 Cup	255	72	340	19	16	5.8	7.2	1	28	31	82	321	4.3	594	1354	150	15	0.08	0.18	3.3	8
6370	Chop Suey W/Beef+Pork;Hmrcp	1 Cup	250	75	300	26	17	4.3	7.4	4.2	68	13	60	248	4.8	425	1053	600	60	0.28	0.38	5	33
6380	Macaroni And Cheese; Canned	1 Cup	240	80	230	9	10	4.7	2.9	1.3	24	26	199	182	1	139	730	260	72	0.12	0.24	1	0
6390	Macaroni And Cheese; Home Recipe	1 Cup	200	58	430	17	22	9.8	7.4	3.6	44	40	362	322	1.8	240	1086	860	232	0.2	0.4	1.8	1
6400	Quiche Lorraine	1 Slice	176	47	600	13	48	23.2	17.8	4.1	285	29	211	276	1	283	653	1640	454	0.11	0.32	0	0
6410	Spaghetti; Tom Sauce Chees;Cnd	1 Cup	250	80	190	6	2	0.4	0.4	0.5	3	39	40	88	2.8	303	955	930	120	0.35	0.28	4.5	10
6420	Spaghetti; Tom Sauce Chees;Hmrp	1 Cup	250	77	260	9	9	3	3.6	1.2	8	37	80	135	2.3	408	955	1080	140	0.25	0.18	2.3	13
6430	Spaghetti; Meatballs;Tomsac;Cnd	1 Cup	250	78	260	12	10	2.4	3.9	3.1	23	29	53	113	3.3	245	1220	1000	100	0.15	0.18	2.3	5
6440	Spaghetti; Meatballs;Tomsa;Hmrp	1 Cup	248	70	330	19	12	3.9	4.4	2.2	89	39	124	236	3.7	665	1009	1590	159	0.25	0.3	4	22
6450	Cheeseburger; Regular	1 Sandwh	112	46	300	15	15	7.3	5.6	1	44	28	135	174	2.3	219	672	340	65	0.26	0.24	3.7	1

TABLE A-8 Food Composition Table *(continued)*

No.	Description of Food		Wt. (g)	Water (%)	Energy (kcal)	Protein (g)	Fat (g)	Sat (g)	Mono (g)	Poly (g)	Cholest (mg)	Carb (g)	Ca (mg)	P (mg)	Fe (mg)	K (mg)	Na (mg)	Vit A (IU)	Vit A (RE)	Thmn (mg)	Ribofl (mg)	Niacin (mg)	Vit C (mg)
Dinners *(continued)*																							
6460	Cheeseburger; 4 Oz Patty	1 Sandwh	194	46	525	30	31	15.1	12.2	1.4	104	40	236	320	4.5	407	1224	670	128	0.33	0.48	7.4	3
6470	Enchilada	1 Enchld	230	72	235	20	16	7.7	6.7	0.6	19	24	97	198	3.3	653	1332	2720	352	0.18	0.26	0	0
6480	English Muffin; Egg; Cheese; Bacon	1 Sandwh	138	49	360	18	18	8	8	0.7	213	31	197	290	3.1	201	832	650	160	0.46	0.5	3.7	1
6490	Fish Sandwich; Reg; W/Cheese	1 Sandwh	140	43	420	16	23	6.3	6.9	7.7	56	39	132	223	1.8	274	667	160	25	0.32	0.26	3.3	2
6500	Fish Sandwich; Lge; W/O Cheese	1 Sandwh	170	48	470	18	27	6.3	8.7	9.5	91	41	61	246	2.2	375	621	110	15	0.35	0.23	3.5	1
6510	Hamburger; Regular	1 Sandwh	98	46	245	12	11	4.4	5.3	0.5	32	28	56	107	2.2	202	463	80	14	0.23	0.24	3.8	1
6520	Hamburger; 4 Oz Patty	1 Sandwh	174	50	445	25	21	7.1	11.7	0.6	71	38	75	225	4.8	404	763	160	28	0.38	0.38	7.8	1
6530	Pizza; Cheese	1 Slice	120	46	290	15	9	4.1	2.6	1.3	56	39	220	216	1.6	230	699	750	106	0.34	0.29	4.2	2
6540	Roast Beef Sandwich	1 Sandwh	150	52	345	22	13	3.5	6.9	1.8	55	34	60	222	4	338	757	240	32	0.4	0.33	6	2
6550	Taco	1 Taco	81	55	195	9	11	4.1	5.5	0.8	21	15	109	134	1.2	263	456	420	57	0.09	0.07	1.4	1
Poultry																							
6560	Chicken; Fried; Batter; Breast	4.9 Oz	140	52	365	35	18	4.9	7.6	4.3	119	13	28	259	1.8	281	385	90	28	0.16	0.2	14.7	0
6570	Chicken; Fried; Batter;Drmstck	2.5 Oz	72	53	195	16	11	3	4.6	2.7	62	6	12	106	1	134	194	60	19	0.08	0.15	3.7	0
6580	Chicken; Fried; Flour; Breast	3.5 Oz	98	57	220	31	9	2.4	3.4	1.9	87	2	16	228	1.2	254	74	50	15	0.08	0.13	13.5	0
6590	Chicken; Fried; Flour; Drmstck	1.7 Oz	49	57	120	13	7	1.8	2.7	1.6	44	1	6	86	0.7	112	44	40	12	0.04	0.11	3	0
6600	Chicken; Roasted; Breast	3.0 Oz	86	65	140	27	3	0.9	1.1	0.7	73	0	13	196	0.9	220	64	20	5	0.06	0.1	11.8	0
6610	Chicken; Roasted; Drumstick	1.6 Oz	44	67	75	12	2	0.7	0.8	0.6	41	0	5	81	0.6	108	42	30	8	0.03	0.1	2.7	0
6620	Chicken; Stewed; Light+Dark	1 Cup	140	67	250	38	9	2.6	3.3	2.2	116	0	20	210	1.6	252	98	70	21	0.07	0.23	8.6	0
6630	Chicken Liver; Cooked	1 Liver	20	68	30	5	1	0.4	0.3	0.2	126	0	3	62	1.7	28	10	3270	983	0.03	0.35	0.9	3
6640	Duck; Roasted; Flesh Only	1/2 Duck	221	64	445	52	25	9.2	8.2	3.2	197	0	27	449	6	557	144	170	51	0.57	1.04	11.3	0
6650	Turkey; Roasted; Dark Meat	4 Pieces	85	63	160	24	6	2.1	1.4	1.8	72	0	27	173	2	246	67	0	0	0.05	0.21	3.1	0
6660	Turkey; Roasted; Light Meat	2 Pieces	85	66	135	25	3	0.9	0.5	0.7	59	0	16	186	1.1	259	54	0	0	0.05	0.11	5.8	0
6670	Turkey; Roasted; Light+Dark	1 Cup	140	65	240	41	7	2.3	1.4	2	106	0	35	298	2.5	417	98	0	0	0.09	0.25	7.6	0
6680	Turkey; Roasted; Light+Dark	3 Pieces	85	65	145	25	4	1.4	0.9	1.2	65	0	21	181	1.5	253	60	0	0	0.05	0.15	4.6	0
6690	Chicken; Canned; Boneless	5 Oz	142	69	235	31	11	3.1	4.5	2.5	88	0	20	158	2.2	196	714	170	48	0.02	0.18	9	3
6700	Chicken Frankfurter	1 Frank	45	58	115	6	9	2.5	3.8	1.8	45	3	43	48	0.9	38	616	60	17	0.03	0.05	1.4	0
6710	Chicken Roll; Light	2 Slices	57	69	90	11	4	1.1	1.7	0.9	28	1	24	89	0.6	129	331	50	14	0.04	0.07	3	0
6720	Gravy And Turkey; Frozen	5 Oz	142	85	95	8	4	1.2	1.4	0.7	26	7	20	115	1.3	87	787	60	18	0.03	0.18	2.6	0
6730	Turkey Ham; Cured Turkey Thigh	2 Slices	57	71	75	11	3	1	0.7	0.9	32	0	6	108	1.6	184	565	0	0	0.03	0.14	2	0
6740	Turkey Loaf; Breast Meat W/O C	2 Slices	42	72	45	10	1	0.2	0.2	0.1	17	0	3	97	0.2	118	608	0	0	0.02	0.05	3.5	0
6741	Turkey Loaf; Breast Meat; W/C	2 Slices	42	72	45	10	1	0.2	0.2	0.1	17	0	3	97	0.2	118	608	0	0	0.02	0.05	3.5	11

TABLE A-8 Food Composition Table *(continued)*

No.	Description of Food		Wt. (g)	Water (%)	Energy (kcal)	Protein (g)	Fat (g)	Sat (g)	Mono (g)	Poly (g)	Cholest (mg)	Carb (g)	Ca (mg)	P (mg)	Fe (mg)	K (mg)	Na (mg)	Vit A (IU)	Vit A (RE)	Thmn (mg)	Ribofl (mg)	Niacin (mg)	Vit C (mg)
Poultry *(continued)*																							
6750	Turkey Patties; Brd;Battd;Frid	1 Patty	64	50	180	9	12	3	4.8	3	40	10	9	173	1.4	176	512	20	7	0.06	0.12	1.5	0
6760	Turkey Roast; Frzn;Lght+Drk;Ck	3 Oz	85	68	130	18	5	1.6	1	1.4	45	3	4	207	1.4	253	578	0	0	0.04	0.14	5.3	0
Soups, Sauces, and Gravies																							
6770	Clam Chowder; New Eng; W/Milk	1 Cup	248	85	165	9	7	3	2.3	1.1	22	17	186	156	1.5	300	992	160	40	0.07	0.24	1	3
6780	Cr of Chicken Soup W/Mlk; Cnd	1 Cup	248	85	190	7	11	4.6	4.5	1.6	27	15	181	151	0.7	273	1047	710	94	0.07	0.26	0.9	1
6790	Cr of Mushrm Soup W/Mlk; Cnd	1 Cup	248	85	205	6	14	5.1	3	4.6	20	15	179	156	0.6	270	1076	150	37	0.08	0.28	0.9	2
6800	Tomato Soup With Milk; Canned	1 Cup	248	85	160	6	6	2.9	1.6	1.1	17	22	159	149	1.8	449	932	850	109	0.13	0.25	1.5	68
6810	Bean with Bacon Soup; Canned	1 Cup	253	84	170	8	6	1.5	2.2	1.8	3	23	81	132	2	402	951	890	89	0.09	0.03	0.6	2
6820	Beef Broth; Boulln; Consm;Cnnd	1 Cup	240	98	15	3	1	0.3	0.2	0	0	0	14	31	0.4	130	782	0	0	0	0.05	1.9	0
6830	Beef Noodle Soup; Canned	1 Cup	244	92	85	5	3	1.1	1.2	0.5	5	9	15	46	1.1	100	952	630	63	0.07	0.06	1.1	0
6840	Chicken Noodle Soup; Canned	1 Cup	241	92	75	4	2	0.7	1.1	0.6	7	9	17	36	0.8	55	1106	710	71	0.05	0.06	1.4	0
6850	Chicken Rice Soup; Canned	1 Cup	241	94	60	4	2	0.5	0.9	0.4	7	7	17	22	0.7	101	815	660	66	0.02	0.02	1.1	0
6860	Clam Chowder; Manhattan; Cnd	1 Cup	244	90	80	4	2	0.4	0.4	1.3	2	12	34	59	1.9	261	1808	920	92	0.06	0.05	1.3	3
6870	Cr of Chicken Soup W/H_2O; Cnd	1 Cup	244	91	115	3	7	2.1	3.3	1.5	10	9	34	37	0.6	88	986	560	56	0.03	0.06	0.8	0
6880	Cr of Mushroom Soup W/H_2O; Cnd	1 Cup	244	90	130	2	9	2.4	1.7	4.2	2	9	46	49	0.5	100	1032	0	0	0.05	0.09	0.7	1
6890	Minestrone Soup; Canned	1 Cup	241	91	80	4	3	0.6	0.7	1.1	2	11	34	55	0.9	313	911	2340	234	0.05	0.04	0.9	1
6900	Pea; Green; Soup; Canned	1 Cup	250	83	165	9	3	1.4	1	0.4	0	27	28	125	2	190	988	200	20	0.11	0.07	1.2	2
6910	Tomato Soup W/Water; Canned	1 Cup	244	90	85	2	2	0.4	0.4	1	0	17	12	34	1.8	264	871	690	69	0.09	0.05	1.4	66
6920	Vegetable Beef Soup; Canned	1 Cup	244	92	80	6	2	0.9	0.8	0.1	5	10	17	41	1.1	173	956	1890	189	0.04	0.05	1	2
6930	Vegetarian Soup; Canned	1 Cup	241	92	70	2	2	0.3	0.8	0.7	0	12	22	34	1.1	210	822	3010	301	0.05	0.05	0.9	1
6940	Bouillon; Dehydrtd; Unprepared	1 Pkt	6	3	15	1	1	0.3	0.2	0	1	1	4	19	0.1	27	1019	0	0	0	0.01	0.3	0
6950	Onion Soup; Dehydrtd; Unprpred	1 Pkt	7	4	20	1	0	0.1	0.2	0	0	4	10	23	0.1	47	627	0	0	0.02	0.04	0.4	0
6960	Chicken Noodle Soup;Dehyd;Prpd	1 Pkt	188	94	40	2	1	0.2	0.4	0.3	2	6	24	24	0.4	23	957	50	5	0.05	0.04	0.7	0
6970	Onion Soup; Dehydratd; Prepred	1 Pkt	184	96	20	1	0	0.1	0.2	0.1	0	4	9	22	0.1	48	635	0	0	0.02	0.04	0.4	0
6980	Tomato Veg Soup; Dehyd;Prepred	1 Pkt	189	94	40	1	1	0.3	0.2	0.1	0	8	6	23	0.5	78	856	140	14	0.04	0.03	0.6	5
6990	Cheese Sauce W/Milk; Frm Mix	1 Cup	279	77	305	16	17	9.3	5.3	1.6	53	23	569	438	0.3	552	1565	390	117	0.15	0.56	0.3	2
7000	Hollandaise Sce; W/H_2O; Frm Mix	1 Cup	259	84	240	5	20	11.6	5.9	0.9	52	14	124	127	0.9	124	1564	730	220	0.05	0.18	0.1	0
7010	White Sauce W/Milk From Mix	1 Cup	264	81	240	10	13	6.4	4.7	1.7	34	21	425	256	0.3	444	797	310	92	0.08	0.45	0.5	3
7020	White Sauce; Medium; Home Recp	1 Cup	250	73	395	10	30	9.1	11.9	7.2	32	24	292	238	0.9	381	888	1190	340	0.15	0.43	0.8	2
7030	Barbecue Sauce	1 Tbsp	16	81	10	0	0	0	0.1	0.1	0	2	3	3	0.1	28	130	140	14	0	0	0.1	1

TABLE A-8 Food Composition Table *(continued)*

No.	Description of Food		Wt. (g)	Water (%)	Energy (kcal)	Protein (g)	Fat (g)	Sat (g)	Mono (g)	Poly (g)	Cholest (mg)	Carb (g)	Ca (mg)	P (mg)	Fe (mg)	K (mg)	Na (mg)	Vit A (IU)	Vit A (RE)	Thmn (mg)	Ribofl (mg)	Niacin (mg)	Vit C (mg)
Soups, Sauces, and Gravies *(continued)*																							
7040	Soy Sauce	1 Tbsp	18	68	10	2	0	0	0	0	0	2	3	38	0.5	64	1029	0	0	0.01	0.02	0.6	0
7050	Beef Gravy; Canned	1 Cup	233	87	125	9	5	2.7	2.3	0.2	7	11	14	70	1.6	189	1305	0	0	0.07	0.08	1.5	0
7060	Chicken Gravy; Canned	1 Cup	238	85	190	5	14	3.4	6.1	3.6	5	13	48	69	1.1	259	1373	880	264	0.04	0.1	1.1	0
7070	Mushroom Gravy; Canned	1 Cup	238	89	120	3	6	1	2.8	2.4	0	13	17	36	1.6	252	1357	0	0	0.08	0.15	1.6	0
7080	Brown Gravy From Dry Mix	1 Cup	261	91	80	3	2	0.9	0.8	0.1	2	14	66	47	0.2	61	1147	0	0	0.04	0.09	0.9	0
7090	Chicken Gravy From Dry Mix	1 Cup	260	91	85	3	2	0.5	0.9	0.4	3	14	39	47	0.3	62	1134	0	0	0.05	0.15	0.8	3
Sugar, Sweets, and Sweeteners																							
7100	Caramels; Plain or Chocolate	1 Oz	28.4	8	115	1	3	2.2	0.3	0.1	1	22	42	35	0.4	54	64	0	0	0.01	0.05	0.1	0
7110	Milk Chocolate Candy; Plain	1 Oz	28.4	1	145	2	9	5.4	3	0.3	6	16	50	61	0.4	96	23	30	10	0.02	0.1	0.1	0
7120	Milk Chocolate Candy; W/Almond	1 Oz	28.4	2	150	3	10	4.8	4.1	0.7	5	15	65	77	0.5	125	23	30	8	0.02	0.12	0.2	0
7130	Milk Chocolate Candy; W/Peanuts	1 Oz	28.4	1	155	4	11	4.2	3.5	1.5	5	13	49	83	0.4	138	19	30	8	0.07	0.07	1.4	0
7140	Milk Chocolate Candy; W/Rice C	1 Oz	28.4	2	140	2	7	4.4	2.5	0.2	6	18	48	57	0.2	100	46	30	8	0.01	0.08	0.1	0
7150	Semisweet Chocolate	1 Cup	170	1	860	7	61	36.2	19.9	1.9	0	97	51	178	5.8	593	24	30	3	0.1	0.14	0.9	0
7160	Sweet (Dark) Chocolate	1 Oz	28.4	1	150	1	10	5.9	3.3	0.3	0	16	7	41	0.6	86	5	10	1	0.01	0.04	0.1	0
7170	Fondant; Uncoated	1 Oz	28.4	3	105	0	0	0	0	0	0	27	2	0	0.1	1	57	0	0	0	0	0	0
7180	Fudge; Chocolate; Plain	1 Oz	28.4	8	115	1	3	2.1	1	0.1	1	21	22	24	0.3	42	54	0	0	0.01	0.03	0.1	0
7190	Gum Drops	1 Oz	28.4	12	100	0	0	0	0	0.1	0	25	2	0	0.1	1	10	0	0	0	0	0	0
7200	Hard Candy	1 Oz	28.4	1	110	0	0	0	0	0	0	28	0	2	0.1	1	7	0	0	0.1	0	0	0
7210	Jelly Beans	1 Oz	28.4	6	105	0	0	0	0	0.1	0	26	1	1	0.3	11	7	0	0	0	0	0	0
7220	Marshmallows	1 Oz	28.4	17	90	1	0	0	0	0	0	23	1	2	0.5	2	25	0	0	0	0	0	0
7230	Custard; Baked	1 Cup	265	77	305	14	15	6.8	5.4	0.7	278	29	297	310	1.1	387	209	530	146	0.11	0.5	0.3	1
7240	Gelatin Dessert; Prepared	1/2 Cup	120	84	70	2	0	0	0	0	0	17	2	23	0	0	55	0	0	0	0	0	0
7250	Honey	1 Cup	339	17	1030	1	0	0	0	0	0	279	17	20	1.7	173	17	0	0	0.02	0.14	1	3
7260	Honey	1 Tbsp	21	17	65	0	0	0	0	0	0	17	1	1	0.1	11	1	0	0	0	0.01	0.1	0
7270	Jams and Preserves	1 Tbsp	20	29	55	0	0	0	0	0	0	14	4	2	0.2	18	2	0	0	0	0.01	0	0
7280	Jams and Preserves	1 Pkt	14	29	40	0	0	0	0	0	0	10	3	1	0.1	12	2	0	0	0	0	0	0
7290	Jellies	1 Tbsp	18	28	50	0	0	0	0	0	0	13	2	0	0.1	16	5	0	0	0	0.01	0	1
7300	Jellies	1 Pkt	14	28	40	0	0	0	0	0	0	10	1	0	0	13	4	0	0	0	0	0	1
7310	Popsicle	1 Popcle	95	80	70	0	0	0	0	0	0	18	0	0	0	4	11	0	0	0	0	0	0
7320	Pudding; Chocolate; Canned	5 Oz	142	68	205	3	11	9.5	0.5	0.1	1	30	74	117	1.2	254	285	100	31	0.04	0.17	0.6	0
7330	Pudding; Tapioca; Canned	5 Oz	142	74	160	3	5	4.8	0	0	0	28	119	113	0.3	212	252	0	0	0.03	0.14	0.4	0

TABLE A-8 Food Composition Table *(continued)*

No.	Description of Food		Wt. (g)	Water (%)	Energy (kcal)	Protein (g)	Fat (g)	Sat (g)	Mono (g)	Poly (g)	Cholest (mg)	Carb (g)	Ca (mg)	P (mg)	Fe (mg)	K (mg)	Na (mg)	Vit A (IU)	Vit A (RE)	Thmn (mg)	Ribofl (mg)	Niacin (mg)	Vit C (mg)
Sugar, Sweets, and Sweeteners *(continued)*																							
7340	Pudding; Vanilla; Canned	5 Oz	142	69	220	2	10	9.5	0.2	0.1	1	33	79	94	0.2	155	305	0	0	0.03	0.12	0.6	0
7350	Pudding; Choc; Instant; From Mix	1/2 Cup	130	71	155	4	4	2.3	1.1	0.2	14	27	130	329	0.3	176	440	130	33	0.04	0.18	0.1	1
7360	Pudding; Choc; Cooked From Mix	1/2 Cup	130	73	150	4	4	2.4	1.1	0.1	15	25	146	120	0.2	190	167	140	34	0.05	0.2	0.1	1
7370	Pudding; Rice; From Mix	1/2 Cup	132	73	155	4	4	2.3	1.1	0.1	15	27	133	110	0.5	165	140	140	33	0.1	0.18	0.6	1
7380	Pudding; Tapioca; From Mix	1/2 Cup	130	75	145	4	4	2.3	1.1	0.1	15	25	131	103	0.1	167	152	140	34	0.04	0.18	0.1	1
7390	Pudding; Vanilla; Instant; From Mix	1/2 Cup	130	73	150	4	4	2.2	1.1	0.2	15	27	129	273	0.1	164	375	140	33	0.04	0.17	0.1	1
7400	Pudding; Vanilla; Cooked; Frm Mix	1/2 Cup	130	74	145	4	4	2.3	1	0.1	15	25	132	102	0.1	166	178	140	34	0.04	0.18	0.1	1
7410	Sugar; Brown; Pressed Down	1 Cup	220	2	820	0	0	0	0	0	0	212	187	56	4.8	757	97	0	0	0.02	0.07	0.2	0
7420	Sugar; White; Granulated	1 Cup	200	1	770	0	0	0	0	0	0	199	3	0	0.1	7	5	0	0	0	0	0	0
7430	Sugar; White; Granulated	1 Tbsp	12	1	45	0	0	0	0	0	0	12	0	0	0	0	0	0	0	0	0	0	0
7440	Sugar; White; Granulated	1 Pkt	6	1	25	0	0	0	0	0	0	6	0	0	0	0	0	0	0	0	0	0	0
7450	Sugar; Powdered; Sifted	1 Cup	100	1	385	0	0	0	0	0	0	100	1	0	0	4	2	0	0	0	0	0	0
7460	Syrup; Chocolate Flavored Thin	2 Tbsp	38	37	85	1	0	0.2	0.1	0.1	0	22	6	49	0.8	85	36	0	0	0	0.02	0.1	0
7470	Syrup; Chocolate Flvred; Fudge	2 Tbsp	38	25	125	2	5	3.1	1.7	0.2	0	21	38	60	0.5	82	42	40	13	0.02	0.08	0.1	0
7480	Molasses; Cane; Blackstrap	2 Tbsp	40	24	85	0	0	0	0	0	0	22	274	34	10.1	1171	38	0	0	0.04	0.08	0.8	0
7490	Table Syrup (Corn And Maple)	2 Tbsp	42	25	122	0	0	0	0	0	0	32	1	4	0	7	19	0	0	0	0	0	0
Vegetables																							
7500	Alfalfa Seeds; Sprouted; Raw	1 Cup	33	91	10	1	0	0	0	0.1	0	1	11	23	0.3	26	2	50	5	0.03	0.04	0.2	3
7510	Artichokes; Globe; Cooked; Drn	1 Artchk	120	87	55	3	0	0	0	0.1	0	12	47	72	1.6	316	79	170	17	0.07	0.06	0.7	9
7520	Asparagus; Ckd Frm Raw; Dr;Cut	1 Cup	180	92	45	5	1	0.1	0	0.2	0	8	43	110	1.2	558	7	1490	149	0.18	0.22	1.9	49
7530	Asparagus; Ckd Frm Raw;Dr;Sper	4 Spears	60	92	15	2	0	0	0	0.1	0	3	14	37	0.4	186	2	500	50	0.06	0.07	0.6	16
7540	Asparagus; Ckd Frm Frz;Drn;Cut	1 Cup	180	91	50	5	1	0.2	0	0.3	0	9	41	99	1.2	392	7	1470	147	0.12	0.19	1.9	44
7550	Asparagus; Ckd Frm Frz;Dr;Sper	4 Spears	60	91	15	2	0	0.1	0	0.1	0	3	14	33	0.4	131	2	490	49	0.04	0.06	0.6	15
7560	Asparagus; Canned;Spears;W/Salt	4 Spears	80	95	10	1	0	0	0	0.1	0	2	11	30	0.5	122	278	380	38	0.04	0.07	0.7	13
7561	Asparagus; Canned;Spears;Nosalt	4 Spears	80	95	10	1	0	0	0	0.1	0	2	11	30	0.5	122	3	380	38	0.04	0.07	0.7	13
7570	Bamboo Shoots; Canned; Drained	1 Cup	131	94	25	2	1	0.1	0	0.2	0	4	10	33	0.4	105	9	10	1	0.03	0.03	0.2	1
7580	Lima Beans; Thick Seed;Frzn;Ckd	1 Cup	170	74	170	10	1	0.1	0	0.3	0	32	37	107	2.3	694	90	320	32	0.13	0.1	1.8	22
7590	Lima Beans; Baby; Frzn;Cked;Drn	1 Cup	180	72	190	12	1	0.1	0	0.3	0	35	50	202	3.5	740	52	300	30	0.13	0.1	1.4	10
7600	Snap Bean; Raw;Ckd;Drnd;Green	1 Cup	125	89	45	2	0	0.1	0	0.2	0	10	58	49	1.6	374	4	830	83	0.09	0.12	0.8	12
7601	Snap Bean; Raw;Ckd;Drnd;Yellow	1 Cup	125	89	45	2	0	0.1	0	0.2	0	10	58	49	1.6	374	4	101	10	0.09	0.12	0.8	12
7610	Snap Bean; Frz;Ckd;Drnd;Green	1 Cup	135	92	35	2	0	0	0	0.1	0	8	61	32	1.1	151	18	710	71	0.06	0.1	0.6	11

TABLE A-8 Food Composition Table *(continued)*

No.	Description of Food		Wt. (g)	Water (%)	Energy (kcal)	Protein (g)	Fat (g)	Sat (g)	Mono (g)	Poly (g)	Cholest (mg)	Carb (g)	Ca (mg)	P (mg)	Fe (mg)	K (mg)	Na (mg)	Vit A (IU)	Vit A (RE)	Thmn (mg)	Ribofl (mg)	Niacin (mg)	Vit C (mg)
Vegetables *(continued)*																							
7611	Snap Bean;Frz;Ckd;Drnd;Yellow	1 Cup	135	92	35	2	0	0	0	0.1	0	8	61	32	1.1	151	18	151	15	0.06	0.1	0.6	11
7620	Snap Bean;Cnnd;Drnd;Green;Salt	1 Cup	135	93	25	2	0	0	0	0.1	0	6	35	26	1.2	147	339	470	47	0.02	0.08	0.3	6
7621	Snap Bean;Cnnd;Drnd;Grn;Nosalt	1 Cup	135	93	25	2	0	0	0	0.1	0	6	35	26	1.2	147	3	470	47	0.02	0.08	0.3	6
7622	Snap Bean;Cnnd;Drnd;Yllw; Salt	1 Cup	135	93	25	2	0	0	0	0.1	0	6	35	26	1.2	147	339	142	14	0.02	0.08	0.3	6
7623	Snap Bean;Cnnd;Drnd;Yllw;Nosal	1 Cup	135	93	25	2	0	0	0	0.1	0	6	35	26	1.2	147	3	142	14	0.02	0.08	0.3	6
7630	Bean Sprouts; Mung; Raw	1 Cup	104	90	30	3	0	0	0	0.1	0	6	14	56	0.9	155	6	20	2	0.09	0.13	0.8	14
7640	Bean Sprouts; Mung; Cookd;Dran	1 Cup	124	93	25	3	0	0	0	0	0	5	15	35	0.8	125	12	20	2	0.06	0.13	1	14
7650	Beets; Cooked; Drained; Diced	1 Cup	170	91	55	2	0	0	0	0	0	11	19	53	1.1	530	83	20	2	0.05	0.02	0.5	9
7660	Beets; Cooked; Drained; Whole	2 Beets	100	91	30	1	0	0	0	0	0	7	11	31	0.6	312	49	10	1	0.03	0.01	0.3	6
7670	Beets; Canned; Drained; W/Salt	1 Cup	170	91	55	2	0	0	0	0.1	0	12	26	29	3.1	252	466	20	2	0.02	0.07	0.3	7
7671	Beets; Canned; Drained; No Salt	1 Cup	170	91	55	2	0	0	0	0.1	0	12	26	29	3.1	252	78	20	2	0.02	0.07	0.3	7
7680	Beet Greens; Cooked; Drained	1 Cup	144	89	40	4	0	0	0.1	0.1	0	8	164	59	2.7	1309	347	7340	734	0.17	0.42	0.7	36
7690	Blackeye Peas; Immatr;Raw;Cked	1 Cup	165	72	180	13	1	0.3	0.1	0.6	0	30	46	196	2.4	693	7	1050	105	0.11	0.18	1.8	3
7700	Blackeye Peas;Immtr;Frzn;Cked	1 Cup	170	66	225	14	1	0.3	0.1	0.5	0	40	39	207	3.6	638	9	130	13	0.44	0.11	1.2	4
7710	Broccoli; Raw	1 Spear	151	91	40	4	1	0.1	0	0.3	0	8	72	100	1.3	491	41	2330	233	0.1	0.18	1	141
7720	Broccoli; Raw; Cooked; Drained	1 Spear	180	90	50	5	1	0.1	0	0.2	0	10	82	86	2.1	293	20	2540	254	0.15	0.37	1.4	113
7730	Broccoli; Raw; Cooked; Drained	1 Cup	155	90	45	5	0	0.1	0	0.2	0	9	71	74	1.8	253	17	2180	218	0.13	0.32	1.2	97
7740	Broccoli; Frzn; Cooked; Draned	1 Piece	30	91	10	1	0	0	0	0	0	2	15	17	0.2	54	7	570	57	0.02	0.02	0.1	12
7750	Broccoli; Frzn; Cooked; Draned	1 Cup	185	91	50	6	0	0	0	0.1	0	10	94	102	1.1	333	44	3500	350	0.1	0.15	0.8	74
7760	Brussels Sprouts; Raw; Cooked	1 Cup	155	87	60	4	1	0.2	0.1	0.4	0	13	56	87	1.9	491	33	1110	111	0.17	0.12	0.9	96
7770	Brussels Sprouts; Frzn; Cooked	1 Cup	155	87	65	6	1	0.1	0	0.3	0	13	37	84	1.1	504	36	910	91	0.16	0.18	0.8	71
7780	Cabbage; Common; Raw	1 Cup	70	93	15	1	0	0	0	0.1	0	4	33	16	0.4	172	13	90	9	0.04	0.02	0.2	33
7790	Cabbage; Comn; Cooked; Drned	1 Cup	150	94	30	1	0	0	0	0.2	0	7	50	38	0.6	308	29	130	13	0.09	0.08	0.3	36
7800	Cabbage; Chinese; Pak-Choi;Ckd	1 Cup	170	96	20	3	0	0	0	0.1	0	3	158	49	1.8	631	58	4370	437	0.05	0.11	0.7	44
7810	Cabbage; Chinese;Pe-Tsai; Raw	1 Cup	76	94	10	1	0	0	0	0.1	0	2	59	22	0.2	181	7	910	91	0.03	0.04	0.3	21
7820	Cabbage; Red; Raw	1 Cup	70	92	20	1	0	0	0	0.1	0	4	36	29	0.3	144	8	30	3	0.04	0.02	0.2	40
7830	Cabbage; Savoy; Raw	1 Cup	70	91	20	1	0	0	0	0	0	4	25	29	0.3	161	20	700	70	0.05	0.02	0.2	22
7840	Carrots; Raw; Whole	1 Carrot	72	88	30	1	0	0	0	0.1	0	7	19	32	0.4	233	25	###	2025	0.07	0.04	0.7	7
7850	Carrots; Raw; Grated	1 Cup	110	88	45	1	0	0	0	0.1	0	11	30	48	0.6	355	39	###	3094	0.11	0.06	1	10
7860	Carrots; Cooked from Raw	1 Cup	156	87	70	2	0	0.1	0	0.1	0	16	48	47	1	354	103	###	3830	0.05	0.09	0.8	4
7870	Carrots; Cooked from Frozen	1 Cup	146	90	55	2	0	0	0	0.1	0	12	41	38	0.7	231	86	25850	2585	0.04	0.05	0.6	4

TABLE A-8 Food Composition Table *(continued)*

No.	Description of Food		Wt. (g)	Water (%)	Energy (kcal)	Protein (g)	Fat (g)	Sat (g)	Mono (g)	Poly (g)	Cholest (mg)	Carb (g)	Ca (mg)	P (mg)	Fe (mg)	K (mg)	Na (mg)	Vit A (IU)	Vit A (RE)	Thmn (mg)	Ribofl (mg)	Niacin (mg)	Vit C (mg)
Vegetables *(continued)*																							
7880	Carrots; Canned; Drn; W/Salt	1 Cup	146	93	35	1	0	0.1	0	0.1	0	8	37	35	0.9	261	352	20110	2011	0.03	0.04	0.8	4
7881	Carrots; Canned;Drnd; W/O Salt	1 Cup	146	93	35	1	0	0.1	0	0.1	0	8	37	35	0.9	261	61	20110	2011	0.03	0.04	0.8	4
7890	Cauliflower; Raw	1 Cup	100	92	25	2	0	0	0	0.1	0	5	29	46	0.6	355	15	20	2	0.08	0.06	0.6	72
7900	Cauliflower; Cooked from Raw	1 Cup	125	93	30	2	0	0	0	0.1	0	6	34	44	0.5	404	8	20	2	0.08	0.07	0.7	69
7910	Cauliflower; Cooked from Frozen	1 Cup	180	94	35	3	0	0.1	0	0.2	0	7	31	43	0.7	250	32	40	4	0.07	0.1	0.6	56
7920	Celery; Pascal Type; Raw; Stalk	1 Stalk	40	95	5	0	0	0	0	0	0	1	14	10	0.2	114	35	50	5	0.01	0.01	0.1	3
7930	Celery; Pascal Type; Raw; Piece	1 Cup	120	95	20	1	0	0	0	0.1	0	4	43	31	0.6	341	106	150	15	0.04	0.04	0.4	8
7940	Collards; Cooked from Raw	1 Cup	190	96	25	2	0	0.1	0	0.2	0	5	148	19	0.8	177	36	4220	422	0.03	0.08	0.4	19
7950	Collards; Cooked from Frozen	1 Cup	170	88	60	5	1	0.1	0.1	0.4	0	12	357	46	1.9	427	85	10170	1017	0.08	0.2	1.1	45
7960	Corn; Cooked from Raw; Yellow	1 Ear	77	70	85	3	1	0.2	0.3	0.5	0	19	2	79	0.5	192	13	170	17	0.17	0.06	1.2	5
7961	Corn; Cooked from Raw; White	1 Ear	77	70	85	3	1	0.2	0.3	0.5	0	19	2	79	0.5	192	13	0	0	0.17	0.06	1.2	5
7970	Corn; Cooked from Frozen; Yellow	1 Ear	63	73	60	2	0	0.1	0.1	0.2	0	14	2	47	0.4	158	3	130	13	0.11	0.04	1	3
7971	Corn; Cooked from Frozen; White	1 Ear	63	73	60	2	0	0.1	0.1	0.2	0	14	2	47	0.4	158	3	0	0	0.11	0.04	1	3
7980	Corn; Cooked from Frozen; Yellow	1 Cup	165	76	135	5	0	0	0	0.1	0	34	3	78	0.5	229	8	410	41	0.11	0.12	2.1	4
7981	Corn; Cooked from Frozen; White	1 Cup	165	76	135	5	0	0	0	0.1	0	34	3	78	0.5	229	8	0	0	0.11	0.12	2.1	4
7990	Corn; Cnd; Crm Stl;Yllw; W/Salt	1 Cup	256	79	185	4	1	0.2	0.3	0.5	0	46	8	131	1	343	730	250	25	0.06	0.14	2.5	12
7991	Corn; Cnd; Crm Stl;Yllw; No Sal	1 Cup	256	79	185	4	1	0.2	0.3	0.5	0	46	8	131	1	343	8	250	25	0.06	0.14	2.5	12
7992	Corn; Cnd; Crm Stl;Whit; W/Salt	1 Cup	256	79	185	4	1	0.2	0.3	0.5	0	46	8	131	1	343	730	0	0	0.06	0.14	2.5	12
7993	Corn; Cnd; Crm Stl;Whit; No Sal	1 Cup	256	79	185	4	1	0.2	0.3	0.5	0	46	8	131	1	343	8	0	0	0.06	0.14	2.5	12
8000	Corn;Cnd; Whl Krnl;Yllw; W/Salt	1 Cup	210	77	165	5	1	0.2	0.3	0.5	0	41	11	134	0.9	391	571	510	51	0.09	0.15	2.5	17
8001	Corn;Cnd; Whl Krnl;Yllw; No Sal	1 Cup	210	77	165	5	1	0.2	0.3	0.5	0	41	11	134	0.9	391	6	510	51	0.09	0.15	2.5	17
8002	Corn;Cnd; Whl Krnl;Whte; W/Salt	1 Cup	210	77	165	5	1	0.2	0.3	0.5	0	41	11	134	0.9	391	571	0	0	0.09	0.15	2.5	17
8003	Corn;Cnd; Whl Krnl;Whte; No Sal	1 Cup	210	77	165	5	1	0.2	0.3	0.5	0	41	11	134	0.9	391	6	0	0	0.09	0.15	2.5	17
8010	Cucumber; W/Peel	6 Slices	28	96	5	0	0	0	0	0	0	1	4	5	0.1	42	1	10	1	0.01	0.01	0.1	1
8020	Dandelion Greens; Cooked; Drnd	1 Cup	105	90	35	2	1	0.1	0	0.3	0	7	147	44	1.9	244	46	12290	1229	0.14	0.18	0.5	19
8030	Eggplant; Cooked; Steamed	1 Cup	96	92	25	1	0	0	0	0.1	0	6	6	21	0.3	238	3	60	6	0.07	0.02	0.6	1
8040	Endive; Curly; Raw	1 Cup	50	94	10	1	0	0	0	0	0	2	26	14	0.4	157	11	1030	103	0.04	0.04	0.2	3
8050	Jerusalem-Artichoke; Raw	1 Cup	150	78	115	3	0	0	0	0	0	26	21	117	5.1	644	6	30	3	0.3	0.09	2	6
8060	Kale; Cooked From Raw	1 Cup	130	91	40	2	1	0.1	0	0.3	0	7	94	36	1.2	296	30	9620	962	0.07	0.09	0.7	53
8070	Kale; Cooked From Frozen	1 Cup	130	91	40	4	1	0.1	0	0.3	0	7	179	36	1.2	417	20	8260	826	0.06	0.15	0.9	33
8080	Kohlrabi; Cooked; Drained	1 Cup	165	90	50	3	0	0	0	0.1	0	11	41	74	0.7	561	35	60	6	0.07	0.03	0.6	89

TABLE A-8 Food Composition Table *(continued)*

No.	Description of Food		Wt. (g)	Water (%)	Energy (kcal)	Protein (g)	Fat (g)	Sat (g)	Mono (g)	Poly (g)	Cholest (mg)	Carb (g)	Ca (mg)	P (mg)	Fe (mg)	K (mg)	Na (mg)	Vit A (IU)	Vit A (RE)	Thmn (mg)	Ribofl (mg)	Niacin (mg)	Vit C (mg)
Vegetables *(continued)*																							
8090	Lettuce; Butterhead; Raw; Head	1 Head	163	96	20	2	0	0	0	0.2	0	4	52	38	0.5	419	8	1580	158	0.1	0.1	0.5	13
8100	Lettuce; Butterhead; Raw; Leaf	1 Leaf	15	96	0	0	0	0	0	0	0	0	5	3	0	39	1	150	15	0.01	0.01	0	1
8110	Lettuce; Crisphead; Raw; Head	1 Head	539	96	70	5	1	0.1	0	0.5	0	11	102	108	2.7	852	49	1780	178	0.25	0.16	1	21
8120	Lettuce; Crisphead; Raw; Wedge	1 Wedge	135	96	20	1	0	0	0	0.1	0	3	26	27	0.7	213	12	450	45	0.06	0.04	0.3	5
8130	Lettuce; Crisphead; Raw; Pieces	1 Cup	55	96	5	1	0	0	0	0.1	0	1	10	11	0.3	87	5	180	18	0.03	0.02	0.1	2
8140	Lettuce; Looseleaf	1 Cup	56	94	10	1	0	0	0	0.1	0	2	38	14	0.8	148	5	1060	106	0.03	0.04	0.2	10
8150	Mushrooms; Raw	1 Cup	70	92	20	1	0	0	0	0.1	0	3	4	73	0.9	259	3	0	0	0.07	0.31	2.9	2
8160	Mushrooms; Cooked; Drained	1 Cup	156	91	40	3	1	0.1	0	0.3	0	8	9	136	2.7	555	3	0	0	0.11	0.47	7	6
8170	Mushrooms; Canned; Drnd; W/Salt	1 Cup	156	91	35	3	0	0.1	0	0.2	0	8	17	103	1.2	201	663	0	0	0.13	0.03	2.5	0
8180	Mustard Greens; Cooked; Draned	1 Cup	140	94	20	3	0	0	0.2	0.1	0	3	104	57	1	283	22	4240	424	0.06	0.09	0.6	35
8190	Okra Pods; Cooked	8 Pods	85	90	25	2	0	0	0	0	0	6	54	48	0.4	274	4	490	49	0.11	0.05	0.7	14
8200	Onions; Raw; Chopped	1 Cup	160	91	55	2	0	0.1	0.1	0.2	0	12	40	46	0.6	248	3	0	0	0.1	0.02	0.2	13
8210	Onions; Raw; Sliced	1 Cup	115	91	40	1	0	0.1	0	0.1	0	8	29	33	0.4	178	2	0	0	0.07	0.01	0.1	10
8220	Onions; Raw; Cooked; Drained	1 Cup	210	92	60	2	0	0.1	0	0.1	0	13	57	48	0.4	319	17	0	0	0.09	0.02	0.2	12
8230	Onions; Spring; Raw	6 Onion	30	92	10	1	0	0	0	0	0	2	18	10	0.6	77	1	1500	150	0.02	0.04	0.1	14
8240	Onion Rings; Breaded;Frzn;Prpd	2 Rings	20	29	80	1	5	1.7	2.2	1	0	8	6	16	0.3	26	75	50	5	0.06	0.03	0.7	0
8250	Parsley; Raw	10 Sprig	10	88	5	0	0	0	0	0	0	1	13	4	0.6	54	4	520	52	0.01	0.01	0.1	9
8260	Parsley; Freeze-Dried	1 Tbsp	0.4	2	0	0	0	0	0	0	0	0	1	2	0.2	25	2	250	25	0	0.01	0	1
8270	Parsnips; Cooked; Drained	1 Cup	156	78	125	2	0	0.1	0.2	0.1	0	30	58	108	0.9	573	16	0	0	0.13	0.08	1.1	20
8280	Peas; Edible Pod; Cooked;Drned	1 Cup	160	89	65	5	0	0.1	0	0.2	0	11	67	88	3.2	384	6	210	21	0.2	0.12	0.9	77
8290	Peas; Green;Cnnd;Drnd; W/Salt	1 Cup	170	82	115	8	1	0.1	0.1	0.3	0	21	34	114	1.6	294	372	1310	131	0.21	0.13	1.2	16
8291	Peas; Green;Cnnd;Drnd; W/O Salt	1 Cup	170	82	115	8	1	0.1	0.1	0.3	0	21	34	114	1.6	294	3	1310	131	0.21	0.13	1.2	16
8300	Peas;Grn; Frozen Cooked;Draned	1 Cup	160	80	125	8	0	0.1	0	0.2	0	23	38	144	2.5	269	139	1070	107	0.45	0.16	2.4	16
8310	Peppers; Hot Chili; Raw; Red	1 Pepper	45	88	20	1	0	0	0	0	0	4	8	21	0.5	153	3	4840	484	0.04	0.04	0.4	109
8311	Peppers; Hot Chili; Raw; Green	1 Pepper	45	88	20	1	0	0	0	0	0	4	8	21	0.5	153	3	350	35	0.04	0.04	0.4	109
8320	Peppers; Sweet; Raw; Green	1 Pepper	74	93	20	1	0	0	0	0.2	0	4	4	16	0.9	144	2	390	39	0.06	0.04	0.4	95
8321	Peppers; Sweet; Raw; Red	1 Pepper	74	93	20	1	0	0	0	0.2	0	4	4	16	0.9	144	2	4220	422	0.06	0.04	0.4	141
8330	Peppers; Sweet; Cooked; Green	1 Pepper	73	95	15	0	0	0	0	0.1	0	3	3	11	0.6	94	1	280	28	0.04	0.03	0.3	81
8331	Peppers; Sweet; Cooked; Red	1 Pepper	73	95	15	0	0	0	0	0.1	0	3	3	11	0.6	94	1	2740	274	0.04	0.03	0.3	121
8340	Potatoes; Baked with Skin	1 Potato	202	71	220	5	0	0.1	0	0.1	0	51	20	115	2.7	844	16	0	0	0.22	0.07	3.3	26
8350	Potatoes; Baked Flesh Only	1 Potato	156	75	145	3	0	0	0	0.1	0	34	8	78	0.5	610	8	0	0	0.16	0.03	2.2	20

TABLE A-8 Food Composition Table *(continued)*

No.	Description of Food		Wt. (g)	Water (%)	Energy (kcal)	Protein (g)	Fat (g)	Sat (g)	Mono (g)	Poly (g)	Cholest (mg)	Carb (g)	Ca (mg)	P (mg)	Fe (mg)	K (mg)	Na (mg)	Vit A (IU)	Vit A (RE)	Thmn (mg)	Ribofl (mg)	Niacin (mg)	Vit C (mg)
Vegetables *(continued)*																							
8360	Potatoes; Boiled; Peeled After	1 Potato	136	77	120	3	0	0	0	0.1	0	27	7	60	0.4	515	5	0	0	0.14	0.03	2	18
8370	Potatoes; Boiled; Peeled Before	1 Potato	135	77	115	2	0	0	0	0.1	0	27	11	54	0.4	443	7	0	0	0.13	0.03	1.8	10
8380	Potatoes; French-Frd; Frzn; Oven	10 Strip	50	53	110	2	4	2.1	1.8	0.3	0	17	5	43	0.7	229	16	0	0	0.06	0.02	1.2	5
8390	Potatoes; French-Frd; Frzn; Fried	10 Strip	50	38	160	2	8	2.5	1.6	3.8	0	20	10	47	0.4	366	108	0	0	0.09	0.01	1.6	5
8400	Potatoes; Au Gratin; From Mix	1 Cup	245	79	230	6	10	6.3	2.9	0.3	12	31	203	233	0.8	537	1076	520	76	0.05	0.2	2.3	8
8410	Potatoes; Au Gratin; Home Recipe	1 Cup	245	74	325	12	19	11.6	5.3	0.7	56	28	292	277	1.6	970	1061	650	93	0.16	0.28	2.4	24
8420	Potatoes; Hashed Brown; Fr Frzn	1 Cup	156	56	340	5	18	7	8	2.1	0	44	23	112	2.4	680	53	0	0	0.17	0.03	3.8	10
8430	Potatoes; Mashed; Recpe; W/Milk	1 Cup	210	78	160	4	1	0.7	0.3	0.1	4	37	55	101	0.6	628	636	40	12	0.18	0.08	2.3	14
8440	Potatoes; Mashed; Recpe; Mlk+Mar	1 Cup	210	76	225	4	9	2.2	3.7	2.5	4	35	55	97	0.5	607	620	360	42	0.18	0.08	2.3	13
8450	Potatoes; Mashed; Frm Dehydrted	1 Cup	210	76	235	4	12	7.2	3.3	0.5	29	32	103	118	0.5	489	697	380	44	0.23	0.11	1.4	20
8460	Potato Salad Made W/Mayonnaise	1 Cup	250	76	360	7	21	3.6	6.2	9.3	170	28	48	130	1.6	635	1323	520	83	0.19	0.15	2.2	25
8470	Potatoes; Scalloped; From Mix	1 Cup	245	79	230	5	11	6.5	3	0.5	27	31	88	137	0.9	497	835	360	51	0.05	0.14	2.5	8
8480	Potatoes; Scalloped; Home Recp	1 Cup	245	81	210	7	9	5.5	2.5	0.4	29	26	140	154	1.4	926	821	330	47	0.17	0.23	2.6	26
8490	Potato Chips	10 Chips	20	3	105	1	7	1.8	1.2	3.6	0	10	5	31	0.2	260	94	0	0	0.03	0	0.8	8
8500	Pumpkin; Cooked from Raw	1 Cup	245	94	50	2	0	0.1	0	0	0	12	37	74	1.4	564	2	2650	265	0.08	0.19	1	12
8510	Pumpkin; Canned	1 Cup	245	90	85	3	1	0.4	0.1	0	0	20	64	86	3.4	505	12	###	5404	0.06	0.13	0.9	10
8520	Radishes; Raw	4 Radish	18	95	5	0	0	0	0	0	0	1	4	3	0.1	42	4	0	0	0	0.01	0.1	4
8530	Sauerkraut; Canned	1 Cup	236	93	45	2	0	0.1	0	0.1	0	10	71	47	3.5	401	1560	40	4	0.05	0.05	0.3	35
8540	Seaweed; Kelp; Raw	1 Oz	28.4	82	10	0	0	0.1	0	0	0	3	48	12	0.8	25	66	30	3	0.01	0.04	0.1	0
8550	Seaweed; Spirulina; Dried	1 Oz	28.4	5	80	16	2	0.8	0.2	0.6	0	7	34	33	8.1	386	297	160	16	0.67	1.04	3.6	3
8560	Spinach; Raw	1 Cup	55	92	10	2	0	0	0	0.1	0	2	54	27	1.5	307	43	3690	369	0.04	0.1	0.4	15
8570	Spinach; Cooked From Raw; Drnd	1 Cup	180	91	40	5	0	0.1	0	0.2	0	7	245	101	6.4	839	126	14740	1474	0.17	0.42	0.9	18
8580	Spinach; Cooked Fr Frzen; Drnd	1 Cup	190	90	55	6	0	0.1	0	0.2	0	10	277	91	2.9	566	163	14790	1479	0.11	0.32	0.8	23
8590	Spinach; Canned; Drnd;W/Salt	1 Cup	214	92	50	6	1	0.2	0	0.4	0	7	272	94	4.9	740	683	18780	1878	0.03	0.3	0.8	31
8591	Spinach; Canned; Drnd;W/O Salt	1 Cup	214	92	50	6	1	0.2	0	0.4	0	7	272	94	4.9	740	58	18780	1878	0.03	0.3	0.8	31
8600	Spinach Souffle	1 Cup	136	74	220	11	18	7.1	6.8	3.1	184	3	230	231	1.3	201	763	3460	675	0.09	0.3	0.5	3
8610	Squash; Summer; Cooked; Draind	1 Cup	180	94	35	2	1	0.1	0	0.2	0	8	49	70	0.6	346	2	520	52	0.08	0.07	0.9	10
8620	Squash; Winter; Baked	1 Cup	205	89	80	2	1	0.3	0.1	0.5	0	18	29	41	0.7	896	2	7290	729	0.17	0.05	1.4	20
8630	Sweet Potatoes; Baked; Peeled	1 Potato	114	73	115	2	0	0	0	0.1	0	28	32	63	0.5	397	11	24880	2488	0.08	0.14	0.7	28
8640	Sweet Potatoes; Boiled W/O Peel	1 Potato	151	73	160	2	0	0.1	0	0.2	0	37	32	41	0.8	278	20	25750	2575	0.08	0.21	1	26
8650	Sweet Potatoes; Candied	1 Piece	105	67	145	1	3	1.4	0.7	0.2	8	29	27	27	1.2	198	74	4400	440	0.02	0.04	0.4	7

TABLE A-8 Food Composition Table *(continued)*

No.	Description of Food		Wt. (g)	Water (%)	Energy (kcal)	Protein (g)	Fat (g)	Sat (g)	Mono (g)	Poly (g)	Cholest (mg)	Carb (g)	Ca (mg)	P (mg)	Fe (mg)	K (mg)	Na (mg)	Vit A (IU)	Vit A (RE)	Thmn (mg)	Ribofl (mg)	Niacin (mg)	Vit C (mg)
Vegetables *(continued)*																							
8660	Sweet Potatoes; Canned; Mashed	1 Cup	255	74	260	5	1	0.1	0	0.2	0	59	77	133	3.4	536	191	38570	3857	0.07	0.23	2.4	13
8670	Sweet Potatoes; Canned; Vac Pack	1 Piece	40	76	35	1	0	0	0	0	0	8	9	20	0.4	125	21	3190	319	0.01	0.02	0.3	11
8680	Tomatoes; Raw	1 Tomato	123	94	25	1	0	0	0	0.1	0	5	9	28	0.6	255	10	1390	139	0.07	0.06	0.7	22
8690	Tomatoes; Canned; S+L; W/Salt	1 Cup	240	94	50	2	1	0.1	0.1	0.2	0	10	62	46	1.5	530	391	1450	145	0.11	0.07	1.8	36
8691	Tomatoes; Canned; S+L; W/O Salt	1 Cup	240	94	50	2	1	0.1	0.1	0.2	0	10	62	46	1.5	530	31	1450	145	0.11	0.07	1.8	36
8700	Tomato Juice; Canned with Salt	1 Cup	244	94	40	2	0	0	0	0.1	0	10	22	46	1.4	537	881	1360	136	0.11	0.08	1.6	45
8701	Tomato Juice; Canned W/O Salt	1 Cup	244	94	40	2	0	0	0	0.1	0	10	22	46	1.4	537	24	1360	136	0.11	0.08	1.6	45
8710	Tomato Paste; Canned W/O Salt	1 Cup	262	74	220	10	2	0.3	0.4	0.9	0	49	92	207	7.8	2442	170	6470	647	0.41	0.5	8.4	111
8711	Tomato Paste; Canned with Salt	1 Cup	262	74	220	10	2	0.3	0.4	0.9	0	49	92	207	7.8	2442	2070	6470	647	0.41	0.5	8.4	111
8720	Tomato Puree; Canned W/O Salt	1 Cup	250	87	105	4	0	0	0	0.1	0	25	38	100	2.3	1050	50	3400	340	0.18	0.14	4.3	88
8721	Tomato Puree; Canned with Salt	1 Cup	250	87	105	4	0	0	0	0.1	0	25	38	100	2.3	1050	998	3400	340	0.18	0.14	4.3	88
8730	Tomato Sauce; Canned with Salt	1 Cup	245	89	75	3	0	0.1	0.1	0.2	0	18	34	78	1.9	909	1482	2400	240	0.16	0.14	2.8	32
8740	Turnips; Cooked; Diced	1 Cup	156	94	30	1	0	0	0	0.1	0	8	34	30	0.3	211	78	0	0	0.04	0.04	0.5	18
8750	Turnip Greens; Cooked from Raw	1 Cup	144	93	30	2	0	0.1	0	0.1	0	6	197	42	1.2	292	42	7920	792	0.06	0.1	0.6	39
8760	Turnip Greens; Cooked from Frozen	1 Cup	164	90	50	5	1	0.2	0	0.3	0	8	249	56	3.2	367	25	13080	1308	0.09	0.12	0.8	36
8770	Vegetable Juice Cocktail; Canned	1 Cup	242	94	45	2	0	0	0	0.1	0	11	27	41	1	467	883	2830	283	0.1	0.07	1.8	67
8780	Vegetables; Mixed; Canned	1 Cup	163	87	75	4	0	0.1	0	0.2	0	15	44	68	1.7	474	243	18990	1899	0.08	0.08	0.9	8
8790	Vegetables; Mixed; Cked Fr Frz	1 Cup	182	83	105	5	0	0.1	0	0.1	0	24	46	93	1.5	308	64	7780	778	0.13	0.22	1.5	6
8800	Water Chestnuts; Canned	1 Cup	140	86	70	1	0	0	0	0	0	17	6	27	1.2	165	11	10	1	0.02	0.03	0.5	2
Spices and Condiments																							
8810	Baking Powder;Sas; Ca Po4	1 Tsp	3	2	5	0	0	0	0	0	0	1	58	87	0	5	329	0	0	0	0	0	0
8820	Baking Powder;Sas;Capo4+Caso4	1 Tsp	2.9	1	5	0	0	0	0	0	0	1	183	45	0	4	290	0	0	0	0	0	0
8830	Baking Powder; Strght Phosphat	1 Tsp	3.8	2	5	0	0	0	0	0	0	1	239	359	0	6	312	0	0	0	0	0	0
8840	Baking Powder; Low Sodium	1 Tsp	4.3	1	5	0	0	0	0	0	0	1	207	314	0	891	0	0	0	0	0	0	0
8850	Catsup	1 Cup	273	69	290	5	1	0.2	0.2	0.4	0	69	60	137	2.2	991	2845	3820	382	0.25	0.19	4.4	41
8860	Catsup	1 Tbsp	15	69	15	0	0	0	0	0	0	4	3	8	0.1	54	156	210	21	0.01	0.01	0.2	2
8870	Celery Seed	1 Tsp	2	6	10	0	1	0	0.3	0.1	0	1	35	11	0.9	28	3	0	0	0.01	0.01	0.1	0
8880	Chili Powder	1 Tsp	2.6	8	10	0	0	0.1	0.1	0.2	0	1	7	8	0.4	50	26	910	91	0.01	0.02	0.2	2
8890	Chocolate; Bittersweet or Baking	1 Oz	28.4	2	145	3	15	9	4.9	0.5	0	8	22	109	1.9	235	1	10	1	0.01	0.07	0.4	0
8900	Cinnamon	1 Tsp	2.3	10	5	0	0	0	0	0	0	2	28	1	0.9	12	1	10	1	0	0	0	1
8910	Curry Powder	1 Tsp	2	10	5	0	0	0	0	0	0	1	10	7	0.6	31	1	20	2	0.01	0.01	0.1	0

TABLE A-8 Food Composition Table *(concluded)*

No.	Description of Food		Wt. (g)	Water (%)	Energy (kcal)	Protein (g)	Fat (g)	Sat (g)	Mono (g)	Poly (g)	Cholest (mg)	Carb (g)	Ca (mg)	P (mg)	Fe (mg)	K (mg)	Na (mg)	Vit A (IU)	Vit A (RE)	Thmn (mg)	Ribofl (mg)	Niacin (mg)	Vit C (mg)
Spices and Condiments *(continued)*																							
8920	Garlic Powder	1 Tsp	2.8	6	10	0	0	0	0	0	0	2	2	12	0.1	31	1	0	0	0.01	0	0	0
8930	Gelatin; Dry	1 Envelp	7	13	25	6	0	0	0	0	0	0	1	0	0	2	6	0	0	0	0	0	0
8940	Mustard; Prepared; Yellow	1 Tsp	5	80	5	0	0	0	0.2	0	0	0	4	4	0.1	7	63	0	0	0	0.01	0	0
8950	Olives; Canned; Green	4 Medium	13	78	15	0	2	0.2	1.2	0.1	0	0	8	2	0.2	7	312	40	4	0	0	0	0
8960	Olives; Canned; Ripe; Mission	3 Small	9	73	15	0	2	0.3	1.3	0.2	0	0	10	2	0.2	2	68	10	1	0	0	0	0
8970	Onion Powder	1 Tsp	2.1	5	5	0	0	0	0	0	0	2	8	7	0.1	20	1	0	0	0.01	0	0	0
8980	Oregano	1 Tsp	1.5	7	5	0	0	0	0	0.1	0	1	24	3	0.7	25	0	100	10	0.01	0	0.1	1
8990	Paprika	1 Tsp	2.1	10	5	0	0	0	0	0.2	0	1	4	7	0.5	49	1	1270	127	0.01	0.04	0.3	1
9000	Pepper; Black	1 Tsp	2.1	11	5	0	0	0	0	0	0	1	9	4	0.6	26	1	0	0	0	0.01	0	0
9010	Pickles; Cucumber; Dill	1 Pickle	65	93	5	0	0	0	0	0.1	0	1	17	14	0.7	130	928	70	7	0	0.01	0	4
9020	Pickles; Cucumber; Fresh Pack	2 Slices	15	79	10	0	0	0	0	0	0	3	5	4	0.3	30	101	20	2	0	0	0	1
9030	Pickles; Cucumber; Sweet Gherkin	1 Pickle	15	61	20	0	0	0	0	0	0	5	2	2	0.2	30	107	10	1	0	0	0	1
9040	Relish; Sweet	1 Tbsp	15	63	20	0	0	0	0	0	0	5	3	2	0.1	30	107	20	2	0	0	0	1
9050	Salt	1 Tsp	5.5	0	0	0	0	0	0	0	0	0	14	3	0	0	2132	0	0	0	0	0	0
9060	Vinegar; Cider	1 Tbsp	15	94	0	0	0	0	0	0	0	1	1	1	0.1	15	0	0	0	0	0	0	0
9070	Yeast; Bakers; Dry; Active	1 Pkg	7	5	20	3	0	0	0.1	0	0	3	3	90	1.1	140	4	0	0	0.16	0.38	2.6	0
9080	Yeast; Brewers; Dry	1 Tbsp	8	5	25	3	0	0	0	0	0	3	17	140	1.4	152	10	0	0	1.25	0.34	3	07

Appendix B

Harvesting and Storing Fruits, Nuts, and Vegetables

HARVESTING AND STORING VEGETABLES

Harvesting

For the best possible quality, vegetables should be harvested at the peak of maturity. This is possible when the vegetables will be used for home consumption, sold at a local market, or processed. The best harvesting time varies with each vegetable crop. Some will hold their quality for only a few days, while others can maintain quality over a period of several weeks. Frequent and timely harvest of crops is needed if the vegetable grower wishes to supply the market or the kitchen table with high-quality produce over a period of time. Specifications for harvesting each kind of vegetable crop should be consulted for the best results.

Harvesting is done by hand and through the use of machines. Mechanical harvesting is especially useful for the commercial grower. When harvesting a crop, care must be taken to prevent injury to the crop by machinery and handling.

Storing

When storing vegetables, proper stage of maturity is essential. There are specific temperatures and humidity levels for storage of

(A) Mechanical harvesting of vegetable crops such as potatoes requires special machinery that is designed to separate the crop from leaves, stems, and soil. (B) Metal parts are often rubber-coated to avoid bruising the tubers.

vegetable crops. Most crops need 90 to 95 percent humidity. Most homeowners find it difficult to reproduce the exact temperatures and humidity levels, but commercial growers can accomplish storage through specialized facilities.

Vegetables differ in the amount of time that they can be stored. Some may be stored for several months, others for only a few days. For the vegetables that do store well, storage is essential to prolong the marketing period.

Temperatures and humidities for storing vegetables.

Crop	Temperature °F	Relative Humidity Percent
Asparagus	32	85-90
Beans, snap	45-50	85-90
Beans, lima	32	85-90
Beets	32	90-95
Broccoli	32	90-95
Brussels sprouts	32	90-95
Cabbage	32	90-95
Carrots	32	90-95
Cauliflower	32	85-90
Corn	31-32	85-90
Cucumbers	45-50	85-95
Eggplants	45-50	85-90
Lettuce	32	90-95
Cantaloupes	40-45	85-90
Onions	32	70-75
Parsnips	32	90-95
Peas, green	32	85-90
Potatoes	38-40	85-90

Temperatures and humidities for storage and storage life of vegetables.

Storage Conditions							
Vegetable	Temperature (°F)	Relative Humidity (%)	Storage Life	Vegetable	Temperature (°F)	Relative Humidity (%)	Storage Life
Pea, English	32	95-98	1-2 weeks	Squash, winter	50	50-70	___[4]
Pea, southern	40-41	95	6-8 days	Strawberry	32	90-95	5-7 days
Pepper, chili (dry)	32-50	60-70	6 months	Sweet corn	32	95-98	5-8 days
Pepper, sweet	45-55	90-95	2-3 weeks	Sweet potato	55-60[3]	85-90	4-7 months
Potato, early	___[1]	90-95	___[1]	Tamarillo	37-40	85-95	10 weeks
Potato, late	___[2]	90-95	5-10 months	Taro	45-50	85-90	4-5 months
Pumpkin	50-55	50-70	2-3 months	Tomato, mature green	55-70	90-95	1-3 weeks
Radish, spring	32	95-100	3-4 weeks	Tomato, firm ripe	46-50	90-95	4-7 days
Radish, winter	32	95-100	2-4 months	Turnip	32	95	4-5 months
Rhubarb	32	95-100	2-4 weeks	Turnip greens	32	95-100	10-14 days
Rutabaga	32	98-100	4-6 months	Water chestnut	32-36	98-100	1-2 months
Salsify	32	95-98	2-4 months	Watercress	32	95-100	2-3 weeks
Spinach	32	95-100	10-14 days	Yam	61	70-80	6-7 months
Squash, summer	41-50	95	1-2 weeks				

Adapted from R. E. Hardenburg, A. E. Watada, and C. Y. Wang, *The Commercial Storage of Fruits, Vegetables, and Florist and Nursery Stocks*, USDA Agriculture Handbook 66 (1986).

[1]Spring- or summer-harvested potatoes are usually not stored. However, they can be held 4-5 months at 40°F if cured 4 or more days at 60-70°F before storage. Potatoes for chips should be held at 70°F or conditioned for best chip quality.

[2]Fall-harvested potatoes should be cured at 50-60°F and high relative humidity for 10-14 days. Storage temperatures for table stock or seed should be lowered gradually to 38-40°F. Potatoes intended for processing should be stored at 50-55°F; those stored at lower temperatures or with a high reducing sugar content should be conditioned at 70°F for 1-4 weeks, or until cooking tests are satisfactory.

[3]Sweet potatoes should be cured immediately after harvest by holding at 85°F and 90-95% relative humidity for 4-7 days,

[4]Winter squash varieties differ in storage life.

Some vegetables can be stored for extended periods of time in refrigerated storage units such as this large, climate-controlled potato storage structure.

Fresh vegetables utilized for storage should be free of skin breaks, bruises, decay, and disease. Such damage or disease will decrease the life of the vegetable in storage.

Refrigeration storage is recommended. This type of storage reduces respiration and other metabolic activities. It slows ripening, moisture loss and wilting, spoilage from bacteria, fungus, or yeast, and undesirable growths, such as potato sprouts.

One activity that can increase the successful storage of vegetables is precooling. Precooling is the process of rapid removal of the heat from the crop before storage or shipment. One method of precooling is termed hydrocooling. This is a method wherein vegetables are immersed in cold water and stay under water long enough to lower the temperature to the desired level.

HARVEST AND STORAGE OF FRUITS AND NUTS

Harvesting

Harvesting fruits can be a daily practice. The advantage of home-grown fruit is that it can be picked at its peak of ripeness. However, fruits for commercial use must be picked several days ahead so they will withstand shipping and handling.

For the best quality of fruit, harvest apples, pears, and quince when they begin to drop, soften, and become fully colored. Some varieties will ripen over a two-week period, requiring picking every day. Other varieties will ripen all at once.

Peaches, plums, apricots, and cherries should be harvested when the green disappears from the surface skin of the fruit. A yellow undercolor should be developed by this time. The fruit should be soft when pressed lightly in a cupped hand .

The nut varieties ripen from August to November. Harvest nuts immediately after they fall from the tree. Most of the nuts that do not fall can be knocked off the tree with poles or mechanical harvesters.

The nut varieties of pecan, hickory, chestnut, and Persian walnut will lose their husks when ripe. However, the husks of black walnuts and other types of walnuts will need to be removed.

Grapes should be harvested only when fully ripe. The best way to judge when the grape is at full maturity is to sample an occasional grape.

Small bush and cane fruits, such as strawberries and raspberries, should be harvested when they are fully ripe. The best indicator is when the fruit is fully colored. When these fruits and berries reach maturity, they should be picked every day. When harvesting, pick when the fruit is dry to avoid mildew and mold damage.

Storage

Fruit storing is limited to those varieties that mature late in the fall or those that can be purchased at the market during the winter months. The length of time fruits can be stored depends on the variety, stage of maturity, and soundness of the fruit at harvest.

Fruit should be harvested for home use when it acquires the color of ripe fruit and when it begins to be soft to the touch. (Courtesy of PhotoDisc)

A fruit storage facility in which temperature and humidity are controlled.

Red Delicious apples ready for harvest in an orchard near Wenatchee, Washington. (Courtesy of USDA/ARS #K-3853-5)

For long-term storage of apples, the temperature should be as close to 32°F as possible. Apples can be stored in many ways, but they should be protected from freezing. A cellar or other area below ground level that is cooled by night air is a good place for apple storage. There should be moderate humidity in the storage area. Pears have the same storage requirements as apples.

Nuts should be air dried before they are stored. Nuts, especially pecans, keep longer if they are left in the shell and refrigerated at 35°F. They can also be frozen if they are kept in their shells. Chestnuts have special requirements. They should be stored at 35° to 40°F and high humidity, shortly after harvest.

Commodity	Freezing Point	Storage Conditions		Length of Storage Period
		Temperature	Humidity	
	°F	°F		
Fruits:				
Apples	29°	32°	Moderate Moisture	Fall/Winter
Grapefruit	29.8°	32°	Moderate Moisture	4 to 6 weeks
Grapes	28.1°	32°	Moderate Moisture	1 to 2 months
Oranges	30.5°	32°	Moderate Moisture	4 to 6 weeks
Pears	29.2°	32°	Moderate Moisture	4 to 6 weeks

Fruit storage temperatures as adapted from USDA.

Large trucks and modern highways move agricultural commodities quickly from farm to the processor or market. (Courtesy of FFA)

Small bush and cane fruits, including grapes, are not suitable for storage. These fruits are too perishable to keep very long and are best utilized as soon as they are harvested.

Appendix C

Beef, Poultry, Pork, and Lamb Retail and Wholesale Cuts

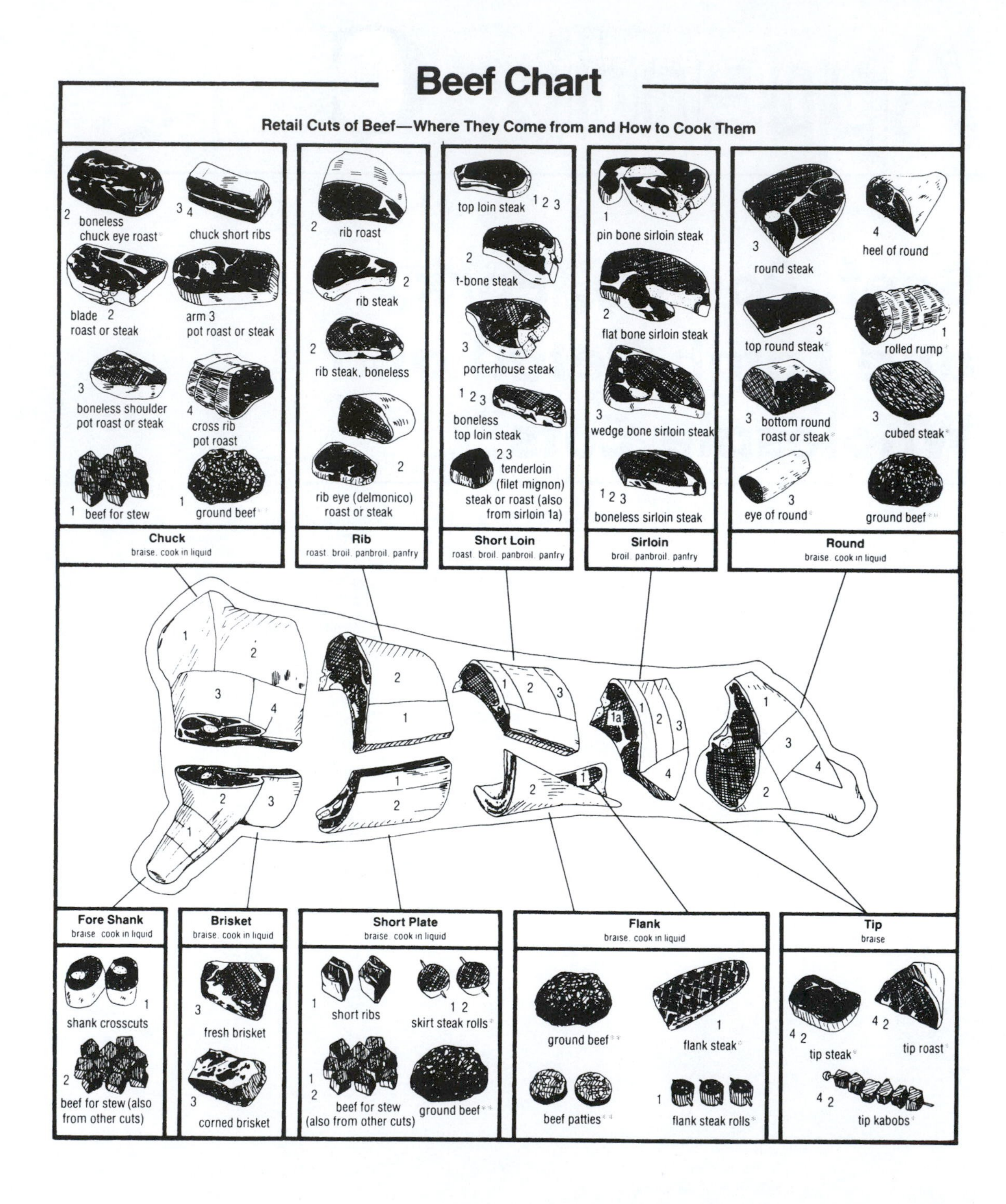
Beef Chart
Retail Cuts of Beef—Where They Come from and How to Cook Them
2 boneless chuck eye roast
3 4 chuck short ribs
blade 2 roast or steak
arm 3 pot roast or steak
3 boneless shoulder pot roast or steak
4 cross rib pot roast
1 beef for stew
1 ground beef
Chuck
braise, cook in liquid
2 rib roast
2 rib steak
2 rib steak, boneless
2 rib eye (delmonico) roast or steak
Rib
roast, broil, panbroil, panfry
top loin steak 1 2 3
2 t-bone steak
3 porterhouse steak
1 2 3 boneless top loin steak
2 3 tenderloin (filet mignon) steak or roast (also from sirloin 1a)
Short Loin
roast, broil, panbroil, panfry
1 pin bone sirloin steak
2 flat bone sirloin steak
3 wedge bone sirloin steak
1 2 3 boneless sirloin steak
Sirloin
broil, panbroil, panfry
3 round steak
4 heel of round
3 top round steak
1 rolled rump
3 bottom round roast or steak
3 cubed steak
3 eye of round
ground beef
Round
braise, cook in liquid
Fore Shank
braise, cook in liquid
1 shank crosscuts
2 beef for stew (also from other cuts)
Brisket
braise, cook in liquid
3 fresh brisket
3 corned brisket
Short Plate
braise, cook in liquid
1 short ribs
1 2 skirt steak rolls
1 2 beef for stew (also from other cuts)
ground beef
Flank
braise, cook in liquid
ground beef
1 flank steak
beef patties
1 flank steak rolls
Tip
braise
4 2 tip steak
4 2 tip roast
4 2 tip kabobs

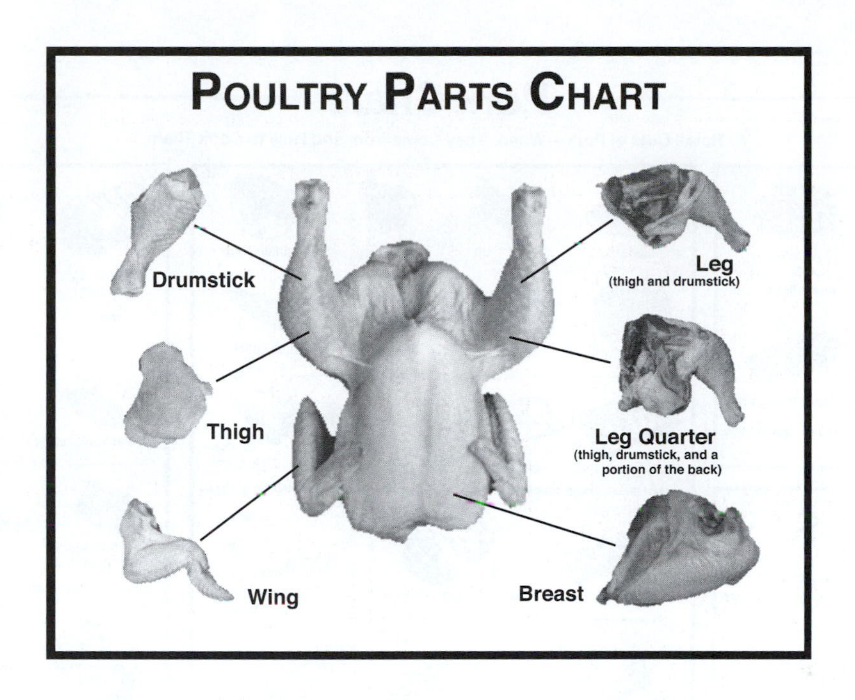
POULTRY PARTS CHART
Drumstick
Leg
(thigh and drumstick)
Thigh
Leg Quarter
(thigh, drumstick, and a portion of the back)
Wing
Breast

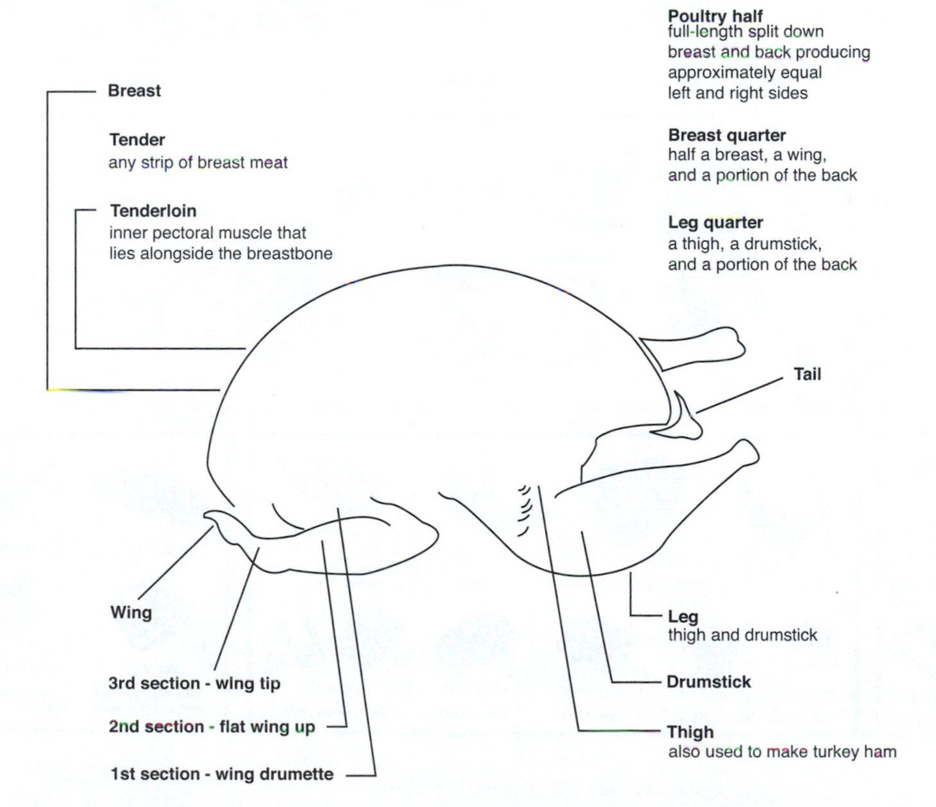
Poultry half
full-length split down breast and back producing approximately equal left and right sides
Breast
Tender
any strip of breast meat
Breast quarter
half a breast, a wing, and a portion of the back
Tenderloin
inner pectoral muscle that lies alongside the breastbone
Leg quarter
a thigh, a drumstick, and a portion of the back
Tail
Wing
Leg
thigh and drumstick
3rd section - wing tip
Drumstick
2nd section - flat wing up
Thigh
also used to make turkey ham
1st section - wing drumette

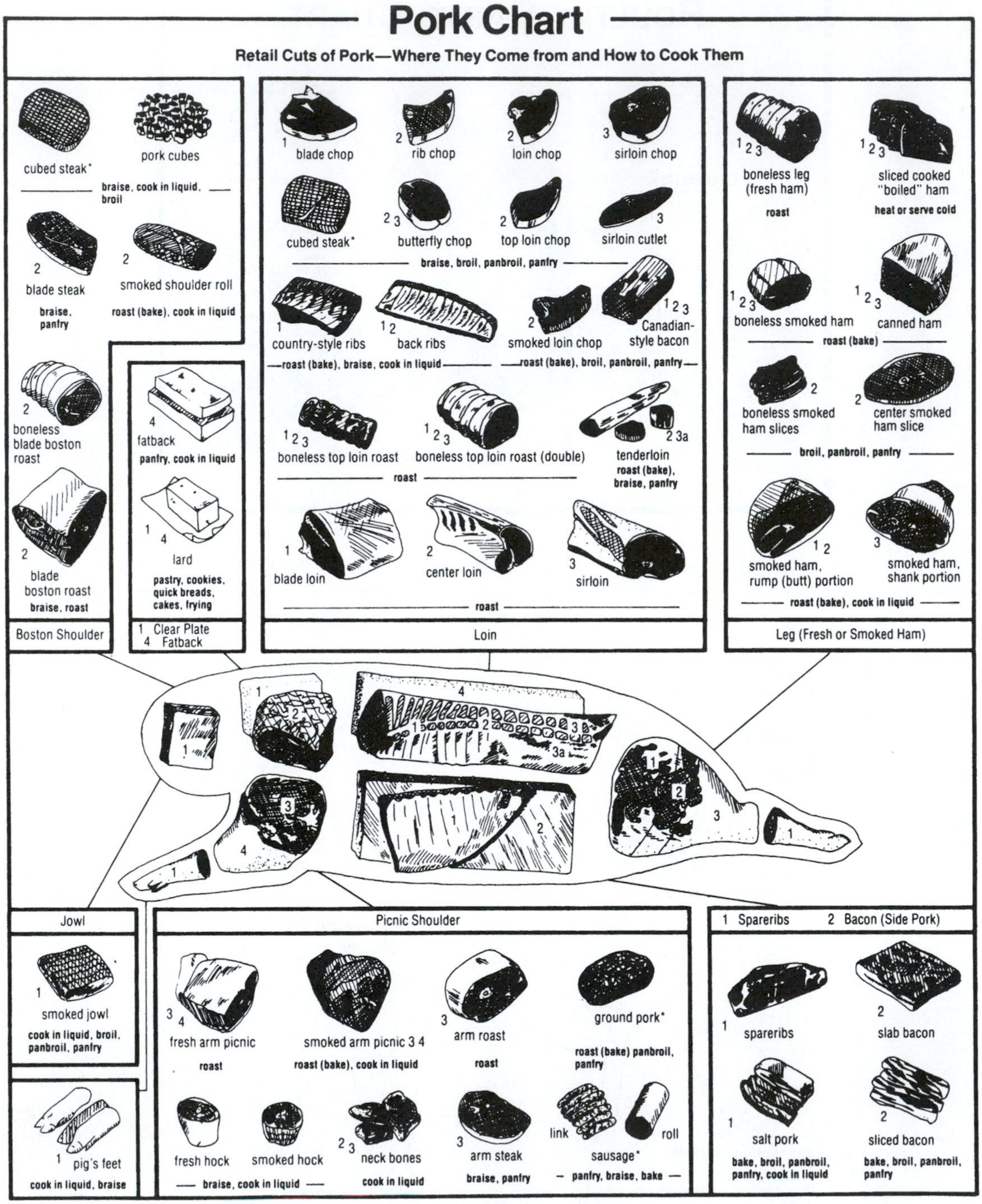

Pork Chart

Retail Cuts of Pork—Where They Come from and How to Cook Them

cubed steak*
pork cubes
braise, cook in liquid, broil
2 blade steak
braise, panfry
2 smoked shoulder roll
roast (bake), cook in liquid
2 boneless blade boston roast
2 blade boston roast
braise, roast
Boston Shoulder

4 fatback
panfry, cook in liquid
1 4 lard
pastry, cookies, quick breads, cakes, frying
1 Clear Plate
4 Fatback

1 blade chop
2 rib chop
2 loin chop
3 sirloin chop
cubed steak*
2 3 butterfly chop
2 top loin chop
3 sirloin cutlet
braise, broil, panbroil, panfry
1 country-style ribs
1 2 back ribs
roast (bake), braise, cook in liquid
2 smoked loin chop
1 2 3 Canadian-style bacon
roast (bake), broil, panbroil, panfry
1 2 3 boneless top loin roast
1 2 3 boneless top loin roast (double)
roast
2 3a tenderloin
roast (bake), braise, panfry
1 blade loin
2 center loin
3 sirloin
roast
Loin

1 2 3 boneless leg (fresh ham)
roast
1 2 3 sliced cooked "boiled" ham
heat or serve cold
1 2 3 boneless smoked ham
1 2 3 canned ham
roast (bake)
2 boneless smoked ham slices
2 center smoked ham slice
broil, panbroil, panfry
1 2 smoked ham, rump (butt) portion
3 smoked ham, shank portion
roast (bake), cook in liquid
Leg (Fresh or Smoked Ham)

Jowl
1 smoked jowl
cook in liquid, broil, panbroil, panfry
1 pig's feet
cook in liquid, braise

Picnic Shoulder
3 4 fresh arm picnic
roast
smoked arm picnic 3 4
roast (bake), cook in liquid
3 arm roast
roast
ground pork*
roast (bake) panbroil, panfry
fresh hock
smoked hock
braise, cook in liquid
2 3 neck bones
cook in liquid
3 arm steak
braise, panfry
link sausage* roll
panfry, braise, bake

1 Spareribs 2 Bacon (Side Pork)
1 spareribs
2 slab bacon
1 salt pork
bake, broil, panbroil, panfry, cook in liquid
2 sliced bacon
bake, broil, panbroil, panfry

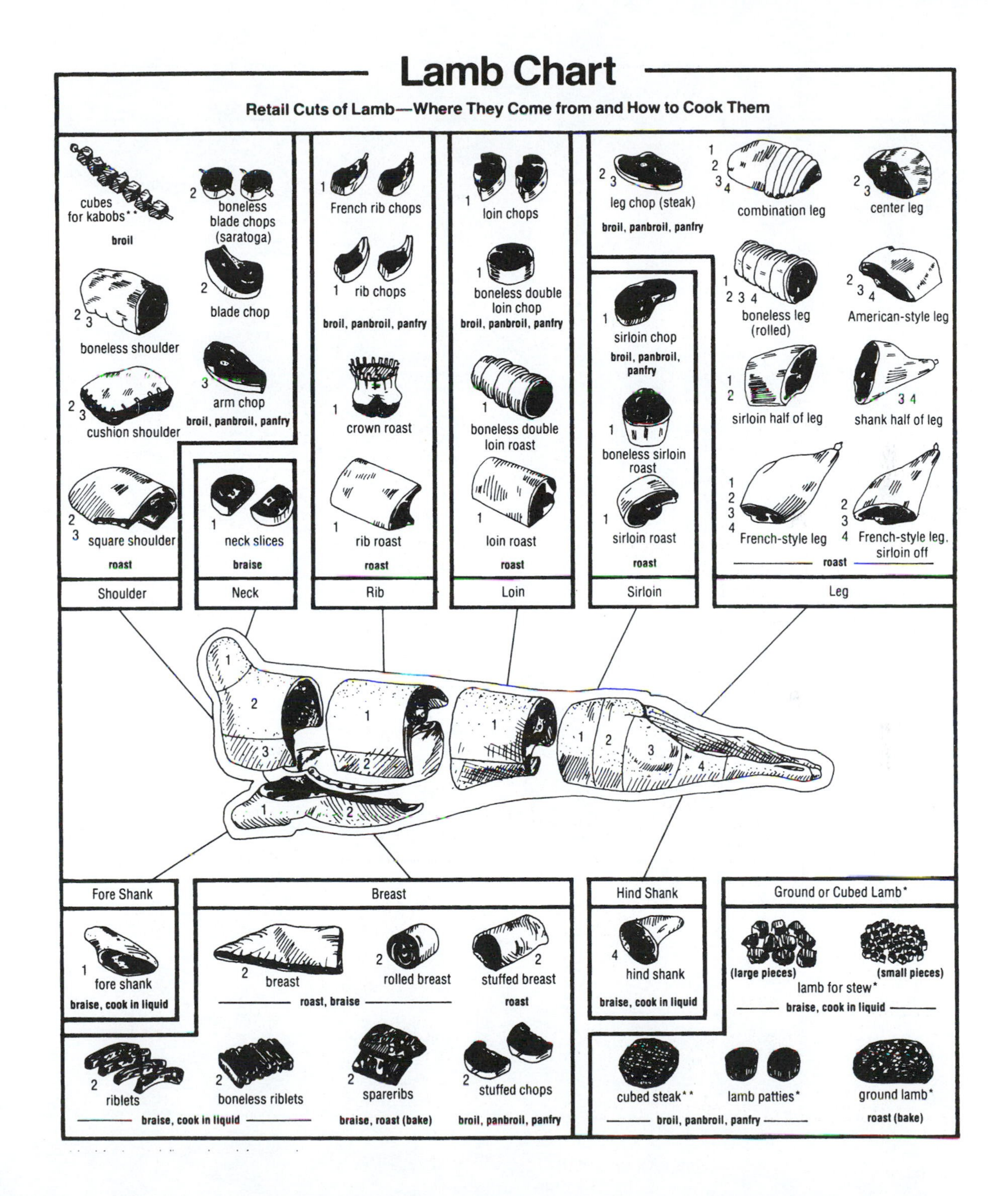
Lamb Chart
Retail Cuts of Lamb—Where They Come from and How to Cook Them
cubes for kabobs**
broil
2 boneless blade chops (saratoga)
2 blade chop
2 3 boneless shoulder
3 arm chop
broil, panbroil, panfry
2 3 cushion shoulder
2 3 square shoulder
roast
1 neck slices
braise
1 French rib chops
1 rib chops
broil, panbroil, panfry
1 crown roast
1 rib roast
roast
1 loin chops
1 boneless double loin chop
broil, panbroil, panfry
1 boneless double loin roast
1 loin roast
roast
2 3 leg chop (steak)
broil, panbroil, panfry
1 2 3 4 combination leg
2 3 center leg
1 sirloin chop
broil, panbroil, panfry
1 boneless sirloin roast
1 sirloin roast
roast
1 2 3 4 boneless leg (rolled)
2 3 4 American-style leg
1 2 sirloin half of leg
3 4 shank half of leg
1 2 3 4 French-style leg
2 3 4 French-style leg, sirloin off
roast
Shoulder
Neck
Rib
Loin
Sirloin
Leg
Fore Shank
Breast
Hind Shank
Ground or Cubed Lamb*
1 fore shank
braise, cook in liquid
2 breast
2 rolled breast
roast, braise
2 stuffed breast
roast
4 hind shank
braise, cook in liquid
(large pieces)
(small pieces)
lamb for stew*
braise, cook in liquid
2 riblets
2 boneless riblets
braise, cook in liquid
2 spareribs
braise, roast (bake)
2 stuffed chops
broil, panbroil, panfry
cubed steak**
lamb patties*
broil, panbroil, panfry
ground lamb*
roast (bake)

Glossary

Like a foreign language, terms unique to food science can be baffling to the newcomer. Individuals traveling to a foreign country to do business are expected to know the language of the country. The same is true for the individual wanting to learn about food science. Indeed, the term "glossary" means obscure or foreign words of a field. Successful individuals use the glossary and learn the language. Words not found in the glossary may be listed in the index and defined within a chapter of the book.

A

absorption (1) Penetration of liquid into a solid that contains a porous structure; (2) adherence of molecules of liquid, gas, or solid to the surface of a solid.

acidulants Make a food acid or sour; added to foods primarily to change the taste and to control microbial growth.

activated carbon An absorbent, formed by the heating of carbon materials at very high temperatures to expose large internal surface areas; frequently used to decolorize materials such as sugar liquors.

active dry yeast Tiny dehydrated granules of yeast that are in a dormant phase until they are exposed to water.

aftertaste A taste that remains in the mouth after a food has been swallowed.

agar A hydrocolloid made from marine algae; frequently used as a bacterial culture medium.

agglomeration Gathering into a cluster, mass, or ball.

agglutination Sticking together as with glue.

aggregation Clumping together.

aging Holding of beef in a cooler or beef in the refrigerator is commonly referred to as the "aging period"; process tenderizes meat.

agitating retort See **retort**.

albumen Also known as egg white; contains about 75 calories (kcal) of energy; provides humans with a high-quality protein containing all the essential amino acids.

alcohol by weight A measurement (weight per volume) of the alcohol content of a solution in terms of the percentage weight of alcohol per volume of beer.

alcohols Chemical compounds characterized by an OH group.

aldehyde Class of organic compounds characterized by the presence of the unsaturated carbonyl group (H–C=O) and a hydrogen atom attached to the carbon represented by (R–C=O).

aleurone layer Outer layer of the endosperm of certain seeds; contains protein bodies that store enzymes concerned with the breakdown of storage material in the endosperm.

algorithm A set of rules used to perform operations or calculations; can be used by food-processing equipment to perform its controller operation.

alkaline Basic pH.

alkaloid Nitrogenous heterocyclic compounds product of plant metabolism many of which are poisonous.

alkane A class of saturated hydrocarbons containing only single bonds represented by the formula C_nH_{2n}+2.

alkene A class of unsaturated aliphatic hydrocarbons containing one double bond represented by C_nH_{2n}.

all-grain beer A beer made entirely from malt as opposed to one made from malt extract, or from malt extract and malted barley.

all-purpose flour A blend of soft and hard wheat flours with a medium amount of gluten, suitable for most baking purposes including conventional handmade yeast breads.

allied industry Supporting industry associated with food.

alpha-tocopherol A chemical with vitamin E activity represented by $C_{29}H_5O_2$.

amaranth A seed that can be crushed or ground to flour and added to breads; does not have significant amounts of gluten.

ameliorate To improve.

amide A class of compound that contains an acyl group: R–C–NH_2.

amino acid A basic building block of protein containing at least one amino group (NH_2) and at least one carboxyl group (–COOH) or acid group of small molecules, each having both an organic acid group (–COOH) and an amino acid group (–NH_2), that are the building units for protein molecules.

amino group A chemical group (NH_2) characteristic of all amino acids.

amorphous Has no crystalline structure.

amphophilic Liking or being attracted to both water and fat.

amphoteric Elements or compounds that act either as an acid or base (gain or lose electrons) depending upon the medium they are in.

amylase An enzyme that hydrolyzes starch to produce dextrins, maltose, and glucose.

amylopectin The long-chain branched fraction of starch.

amylose The long-chain or linear fraction of starch.

anabolism Reactions involving the synthesis of compounds.

anaerobic Without atmospheric oxygen.

anhydrous A solid containing no water bound to the molecule as in a hydrate or not water of crystallization.

anion A negative ion.

antagonism The competitive or inhibiting effect of one substance upon another of similar molecular structure.

antemortem Before slaughter or death.

antibiotic A substance that inhibits the growth of bacteria.

anticaking An additive used to inhibit or prevent caking of dry materials.

antimicrobial agents Substances that prevent or inhibit the growth of microorganisms.

antioxidant A substance that can stop an oxidation reaction; a substance that slows down or interferes with the deterioration of fats through oxidation.

AOAC Abbreviation for Association of Official Analytical Chemists.

aquaculture The art, science, and business of cultivating plants and animals in water.

aroma An odor detected by the olfactory sense.

aromatic compounds Compounds that have an aroma or odor.

ascorbic acid Vitamin C.

aseptic packaging Filling a container previously sterilized without recontaminating either the product or the container.

aseptically Free from disease-producing microorganisms.

ash The residue remaining after total combustion of a solution or mixture; used as a measure of the inorganic (mineral) components of a food.

aspartame A high-intensity alternative sweetener with the trade name NutraSweet®; approximately 180 times sweeter than sucrose; essentially calorie free as small quantities are used. If bought at the grocery stores as Equal®, it is mixed with dextrose and maltodextrin.

astringency The puckering, drawing, or shrinking sensation produced by certain compounds in food.

atmospheric pressure Force per unit area

exerted against a surface by the weight of the air above that surface.

atomic number An experimentally determined number typical of a chemical element that represents the number of protons in the nucleus, which in a neutral atom equals the number of electrons outside the nucleus, and that determines the place of the element in the periodic table.

ATP (adenosine triphosphate) A compound containing high-energy phosphate bonds in which the body cell traps energy from the metabolism of carbohydrate, fat, or protein; the energy in ATP is then used to do mechanical or chemical work in the body.

Aw Vapor pressure of food product at a specified temperature.

B

bacteria Microorganisms usually consisting of a single cell composed of proteinaceous substances; some cause disease, others are used in food processing.

bacteriophage An organism that surrounds and gradually disintegrates the bacterial cell and thus inhibits the growth.

bagasse The crushed plant fiber remaining after the extraction of the sugar-containing juice from sugarcane; high in cellulose and used as fuel.

bagel A traditional, doughnut-shaped roll with a characteristic dense texture achieved by a short rise, followed by boiling and then baking the product.

bake To cook covered or uncovered in an oven usually by dry heat; usually done in an oven but occasionally under coals, in ashes, or on heated stones or metals.

baker's sugar A refined specialty product that has an average crystal size smaller than that of normal table sugar.

baker's yeast Yeast used for raising bread, typically from the taxonomic group *Saccharomyces cerevisiae.*

bar International unit of pressure equal to 29.531 in. of mercury at 32°F.

barley A cereal of the genus *Hordeum*, a member of the Gramineae or grass family of plants. The two varieties of barley are classed according to the number of rows and grains on each of the ears of the plant: two- and six-rowed barley.

barrel (1) A large cylindrical container of greater length than breadth and with bulging sides once made of wood coated with tar (pitch) to prevent infection, now made of aluminum or stainless steel; (2) A standard liquid measure: in the United States, 31 fi gallons (119.2369 liters).

baste To pour liquid composed of drippings, fat, and water, or sugar and water over a food while cooking.

batch The amount of material prepared or required for one operation.

beading The appearance of tiny droplets of syrup on the surface of a baked meringue as it stands.

beat To mix with an over-and-over motion to smooth a mixture and to introduce air; also accomplished by a rotary beater or electric mixer.

beer A generic name for alcoholic beverages produced by fermenting a cereal or a mixture of cereals.

beet sugar Sugar (sucrose) processed from the sugar beet plant.

beta-amylase An enzyme that hydrolyzes starch by breaking off two glucose units at a time, thus producing maltose.

BHA The antioxidant butylated hydroxyanisole.

BHT The antioxidant butylated hydroxytoluene.

bioavailability In a form that can be used by the body.

biofilms Films formed by organisms.

bioproducts Designates a wide variety of corn-refining products made from natural, renewable raw materials that replace products made from nonrenewable resources; items that are produced by chemical synthesis.

biosensors Devices sensitive to a physical or chemical stimulus, such as heat or an ion, that transmit information about a life process.

biotechnology The collection of industrial processes or tools that involve the use of biological systems, such as plants, animals, and microorganisms.

biotin A water-soluble vitamin; functions in fatty acid synthesis.

birefringence The ability of a substance to refract light in two slightly different directions to form two rays; this produces a dark cross on each starch granule when viewed with a polarizing microscope.

bitterness The quality or state of being bitter; in beer, the bitter flavor and aroma are caused by the tannins and the isohumulones of hops.

black tea After rolling, lumps of tea are broken and spread in a fermentation room to oxidize, which turns the leaves to a copper color. The leaves are finally hot-air dried in a process that stops fermentation and turns the leaves black.

blackstrap molasses A type of molasses that are generally used as animal feed or biological (fermentation) feed stock; the by-product of sugar extraction from sugar-containing liquors.

blanc mange A thickened milk-based dessert.

blanching Pretreating with steam or boiling water (a) to partially inactivate enzymes and shrink food before canning, freezing, or drying, by heating with steam or boiling water; (b) to aid in removal of skins from nuts and fruits by dipping into boiling water from 1 to 5 minutes; (c) to reduce strong flavor or set color of food by plunging into boiling water.

bleached Flour processed with a "bleaching agent." Fresh ground wheat flour does not result in consistently good products. Over time, flour ages and whitens, and within several months it produces a better product. To hasten the improvement process, modern flour mills bleach and age flour chemically through the addition of a bleaching agent.

bleaching To make whiter or lighter especially by physical or chemical removal of color.

blend To mix two or more ingredients until they are well combined.

blood spot Also called meat spots; occasionally found on an egg yolk.

bloom Refers to the way the top of bread opens up during baking along the cuts made in the top crust.

BOD Biological Oxygen Demand, a measure of water quality.

boil To cook in boiling liquid. A liquid is boiling when bubbles are breaking on the surface and steam is being released; in a slowly boiling liquid, bubbles are small; in a rapidly boiling liquid, bubbles are large; as boiling changes from slow to rapid, more steam is formed but there is no increase in temperature. Boiling temperature of water at sea level is 100°C or 212°F. It is reduced by rise above sea level, approximately 1°C for every 970 feet of elevation. It is increased by solution of solids in the water and by pressure of enclosed steam as in a pressure saucepan.

boiling point The temperature at which a liquid vaporizes.

bomb calorimeter Instrument for measuring the energy content of food.

botulinum toxin A very potent toxin produced by *Clostridium botulinum* bacteria; in a low-acid environment, the high temperatures achieved in a pressure canner are required for complete destruction of the spores of this microbe.

bound water Water that is held so tightly by another molecule (usually a large molecule such as a protein) that it no longer has the properties of free water; water that is not easily removed from the food.

braise To cook in a covered utensil with a small amount of liquid; a moist heat method of cooking.

bread flour A special flour, higher in gluten, that can be used for making yeast breads by hand; recommended for use in a bread machine.

breaded Coated with bread crumbs or similar flour product.

brew The infusion and boiling stages of tea-making or the beer-making process.

brewer's yeast An inactive yeast product that is a by-product of beer-making and is specially processed to be a nutritional supplement for humans.

brewing Process of making beer, ale, or other similar cereal beverages that are fermented but not distilled.

brine A salt solution.

brix A hydrometer used for testing the sugar concentration of syrups.

broiling A dry-heat method of cooking usually by radiation, otherwise, direct exposure to heat source; to cook by direct heat.

bromelin A proteolytic enzyme found in pineapple; used in meat tenderization.

brown sugar A finished sugar product consisting of sugar crystals and darker non-sucrose materials. Soft brown sugars are brown sugars crystallized directly by specialized crystallization processes. Brown sugars can also be made by blending white crystallized sugar and dark syrups.

buckwheat A seed of a small plant, ground into light or dark (greater fiber, stronger flavor) flour.

buffer A mixture of compounds that protects solutions from a substantial change in pH; a substance that resists change in acidity or alkalinity.

bulk density The weight per unit volume of a large mass of material; comparative for the conditions that apply; reported as pounds per cubic foot (lb./ft.^{3}) for solids and pounds per gallon (lb./gal.) for liquids.

bulking agent A substance used in relatively small amounts to affect the texture and body of some manufactured foods made without sugar or with reduced amounts.

bumping A phenomenon whereby a viscous product "bumps" or "burps" when heated. In a microwave oven this occurs due to entrapped steam pockets that will "suddenly" bump. In a thickened pudding it may occur as the starch gelatinizes and steam builds up if heated through direct heat.

buttermilk Liquid by-product of butter making. Churning breaks the fat globule membrane so the emulsion breaks, fat coalesces, and water (buttermilk) escapes.

butyric acid A saturated fatty acid with four carbon atoms that is found in relatively large amounts in butter.

BV biological value.

by-products secondary or incidental products from a manufacturing process.

C

caffeine Stimulant that is a plant alkaloid in coffee, tea, and selected carbonated beverages.

caking Formed into a crust or compact mass.

calorie A unit of heat measurement; the small calorie used in chemistry. The kilocalorie is used in nutrition. One kilocalorie is equal to 1,000 small calories.

cane sugar A sugar (sucrose) product processed from sugarcane.

canners sugar A refined, granulated specialty product that meets exacting standards for microbiological quality.

capon A male chicken desexed before 6 weeks of age less than 8 months old; a male bird castrated when young.

caramel A product formed by sugar decomposition due to heating of sucrose, or it may also be made to be a confectionery product due to the Maillard reaction.

caramelization Sucrose heated past the molten point so that it dehydrates and decomposes; the development of brown color and caramel flavor as dry sugar is heated to a high temperature and chemical decomposition occurs in the sugar.

carbohydrate A category of organic compounds with carbon, hydrogen, and oxygen; in sugars, the ratio is an approximate C:H:O [1:2:1] ratio.

carbon dioxide (CO_2) Leavening agent produced by chemical or biological means for baked products.

carbonate A salt or ester of carbonic acid.

carbonated Having carbon dioxide gas injected or dissolved in a liquid, creating an effervescence of pleasant taste and texture.

carbonator Used to add carbon dioxide to a liquid.

carbonyl group A ketone (–C=O) or an aldehyde (HC=O) group.

carcinogen A cancer-causing substance.

carotenoids Fat-soluble, yellow-orange pigments that are produced by plants; may be stored in the fatty tissues of animals.

case hardening During the drying process, food cooks on the outside before it dries on the inside.

casein A major protein found in milk.

caseinate A protein salt derived from milk.

casings Tubular intestinal membrane of sheep, cattle, or hogs, or a synthetic facsimile used for sausage, salami, and the like.

catabolism Breaking down of complex substances into simpler ones with the release of energy; the opposite of anabolism.

catalase An enzyme that breaks down hydrogen peroxide to water and oxygen.

catalyst A substance that changes the rate of a chemical reaction without being used up in the reaction; enzymes are catalysts.

catalyze To make a reaction occur at a more rapid rate by the addition of a substance called a catalyst, which itself undergoes no permanent chemical change.

catechin A specific tannin; closely related to anthocyanins and anthoxanthins.

cation A positive ion.

caustic Capable of destroying living tissue.

cellulose A plant carbohydrate of long chains of glucose; indigestible to humans.

centrifuge Equipment that uses centrifugal force to separate solids and liquids; different spinning speeds used depend upon separation needs.

certificate A document providing evidence of status of qualifications.

chelating agent Substance that binds strongly to multivalent cations, by virtue of a number of anionic groups acting like pincers, for example, EDTA (ethylene diamine tetraacetic acid).

chicory A plant whose root is roasted and ground for use as a coffee substitute.

chitin A water-insoluble polysaccharide containing amine groups; exoskeleton of insects and crustaceans.

choice One of the quality grades.

cholesterol A sterol with the formula $C_{27}H_{46}O$ abundant in animal fat, brain and nervous tissue, and eggs; functions in the body as a part of membranes, as a precursor of steroid hormones and bile acids.

choline A dietary component of many foods, is part of several major phospholipids that are critical for normal membrane structure and function; used by the kidney to maintain water balance and by the liver; used to produce the important neurotransmitter acetylcholine.

chroma Intensity or purity of a color.

chromatography A process in which a chemical mixture carried by a liquid or gas is separated into components as a result of differential distribution of the solutes as they flow around or over a stationary liquid or solid phase.

churning The process that breaks the fat globule membrane so the emulsion breaks, fat coalesces, and water (buttermilk) escapes.

cis Configuration has the hydrogen atoms on the same side of the double bond, particularly with unsaturated fatty acids.

clarifier a material or piece of equipment that will remove suspended solids or colloidal materials from a liquid.

clarify To make clear a cloudy liquid such as heated soup stock by adding raw egg white and/or egg shell; as the proteins coagulate, they trap tiny particles from the liquid that can then be strained out.

climacteric Fruits that produce ethylene gas during ripening; are ethylene sensitive.

coagulate To form a clot, a semisolid mass, or a gel, after initial denaturation of a protein; to produce a firm mass or gel by denaturation of protein molecules followed by formation of new crosslinks.

coagulation Aggregation of protein macromolecules into clumps or aggregates of semisolid material.

coalesce To grow together or to unite into a whole.

coenzyme Nonprotein compound that aids to form the active portion of an enzyme system.

cold shortening A carcass that is chilled too rapidly; causes subsequent toughness.

coliform Relating to, resembling, or being *E. coli.*

collage A fibrous type of protein molecule found in the connective tissue of animals; produces gelatin when it is partially hydrolyzed.

colloidal Dispersion state of subdivision of dispersed particles; intermediate between very small particles in true solution and large particles in suspension.

color The property of a material in which specific visual wavelengths of the electromagnetic spectrum are absorbed and/or reflected.

commercial sterility The condition where all pathogenic and toxin-forming organisms have been destroyed, as well as other organisms capable of growth and spoilage under normal handling and storage conditions.

comminute To reduce to small fine particles.

competencies Abilities or capabilities of employees.

complex carbohydrates Carbohydrates made up of many small sugar units joined together –for example, starch and cellulose.

compliance Conformity in fulfilling official requirements.

compressed yeast Fresh (not dried) yeast that is extruded and cut into a cake form. It must be refrigerated at all times and has a relatively short shelf life of 4 to 6 weeks.

concentration Reducing the weight and volume of a product.

conching A flavor development process that puts the chocolate through a "kneading" action and takes its name from the shell-like shape of the containers originally employed.

conduction Heating transfers heat by direct contact of the heated molecules to those at a lower energy level.

conductivity The measurement of the electrical conductance or the heat energy in a substance. Conductance and conductivity ash refers to solutions and is influenced by inorganic salts in a number of food-processing industries.

confectioners sugar A refined sugar product whose granule sizes range from coarse to powdered.

congealing A liquid oil becomes solid at a certain temperature.

conglomerate Multiple crystals or particles cemented together.

consumer Person or organization that purchases or uses a service or commodity.

controlled atmosphere storage The monitoring and controlling of content of gases in the storage warehouse atmosphere; a low oxygen content slows down plant respiration and delays senescence (aging).

convection Air currents aid in distributing heat throughout. As liquids and gases are heated, they become lighter (less dense) and rise, whereas cooler molecules of the liquid or gas move to the bottom of a container or closed compartment.

convection oven Oven has a fan built into it that circulates the air and cooks the food more evenly than conventional ovens.

cool storage Is considered any temperature from 68° to 28°F (16° to –2°C).

copolymer Product of chemical reaction in which two molecules combine to form larger molecules that contain repeating structural units.

COP Clean-out-of-place.

corn sugar Processed sugar products from acid or enzyme hydrolyzed cornstarch.

corn syrup (1) A syrup made by partial hydrolysis of cornstarch to dextrose, maltose, and dextrins; (2) the purified concentrated aqueous solution of nutritive saccharides obtained from edible starch.

cotyledon Part of the embryo in seeds, acting either as a storage organ or in absorbing food reserves from the endosperm. Dicotyledonous plants have two cotyledons, and monocotyledonous plants have one.

couche A large piece of linen or canvas used to wrap dough for rising; seasoned by dusting it with flour.

covalent bond A strong chemical bond that joins two atoms together.

cracklings The adipose tissue residue left from rendered pork fat that has had the lard extracted.

cream of tartar Potassium acid tartrate, the partial salt of tartaric acid, an organic acid; a weak acid substance commonly added to fondant to produce variable amounts of invert sugar from the hydrolysis of sucrose.

cream (1) Butterfat of milk; (2) To work one or more foods until soft and creamy. The hands, a spoon, electric mixer, or other implements may be used.

creative thinking Ability to generate new ideas by making nonlinear or unusual connections or by changing or reshaping goals to imagine new possibilities; using imagination freely, combining ideas and information in new ways.

crepe A thin pancake.
critical control point Any point in the process where loss of control may result in a health risk.
critical temperature The temperature above which a gas can exist only as a gas, regardless of the pressure, because the motion of the molecules is so violent.
croissant A French classic roll, crescent shaped and made from buttered layers of yeast dough much like a puff pastry.
cross-contamination Contamination of one substance by another; for example, cooked chicken is contaminated with salmonella organisms when it is cut on the same board used for cutting the raw chicken.
crustaceans Shellfish with a segmented, crust shell and jointed appendages.
cryogenic Being or relating to very low temperatures.
cryoprotectants Substances that offer protection to such sensitive molecules as proteins during freezing and frozen storage.
crystalline The aggregation of molecules of a substance in a set, ordered pattern, forming individual crystals.
crystallization The formation of crystals from the solidification of dispersed elements in a precise orderly structure.
crystallize To form crystals from the solidification of dispersed elements in a precise orderly structure.
cubeb An essential oil used in flavorings.
cuisine A style of cooking or manner of preparing food.
cultural diversity Term to describe the American workplace representing people from different backgrounds.
curd Substance consisting mainly of casein, obtained by coagulation of milk and used as food or made into cheese.
curing A preservative method; more often used for flavor and color enhancement.
cut (1) To divide food material with knife or scissors; (2) to incorporate fat into dry ingredients with a pastry blender or two knives, with the least possible amount of blending.
cut and fold A combination of two motions–to cut vertically through mixture and to turn over by sliding tool across bottom of mixing bowl at each turn; proper folding prevents loss of air.
cyclamate Any group of nonnutritive sweeteners with general formula $C_6H_{11}NHSO_4$; an artificial sweetener.
cytoplasm Pertaining to the protoplasm of a cell, exclusive of the nucleus.

D

D value The time in minutes at a specified temperature to reduce the number of microorganisms by one log cycle.
D.E. A measure of the total reducing sugars, expressed as dextrose (glucose).
dark rye flour A coarse rye flour ground from the whole rye grain.
data sheet Similar to a résumé; contains pertinent information about potential employees.
deboning Removal of bone from meat.
decolorization The process of removing colored impurities from sugar solutions by absorption with activated carbon, bone char, and/or ion-exchange resin.
degumming The first step in the oil refining process, oils are mixed with water, which removes valuable emulsifiers such as lecithin; enhanced by adding phosphoric or citric acid or silica gel.
dehydrated food A food dried by artificial means to less than 5 percent moisture.
dehydration The almost complete removal of water from a product.
dehydrofrozen A product held in frozen form during dehydration.
dehydrogenase An enzyme that catalyzes a chemical reaction in which hydrogen is removed; similar to an oxidation reaction.
dehydrogenation Removal of hydrogen from a compound.
Delaney clause Government action enforced by the FDA that basically says the food industry cannot add any substance to food if it induces cancer when ingested by man or animal.
demographic Having to do with vital and social statistics.
denature A change in the molecular structure from the native structure of a protein.

density/mass Weight per unit volume; potatoes with higher density or specific gravity are heavier for their size; the weight (in vacuum) per unit volume at a specific temperature.

deodorization Removal of odor or smell.

dermal A layer of protective tissue.

developing country Not yet highly industrialized.

dextrinization Addition of various water-soluble gummy polysaccharides obtained from starch by the action of heat, acids, or enzymes and used as adhesives, as sizes for paper and textiles, as thickening agents, and in beer.

dextrins Polysaccharides composed of many glucose units; produced at the beginning stages of starch hydrolysis (breakdown); somewhat smaller than starch molecules.

dextrose An alternate name for glucose, a monosaccharide having the chemical formula $C_6H_{12}O_6$.

diacetyl Flavoring agent in butter; chemical formula: $CH_3COCOCH_3$.

diffusion The movement of a substance from an area of higher concentration to an area of lower concentration.

digestion The process that beaks down food into molecules small enough to absorb.

diglyceride Glycerol combined with two fatty acids.

disaccharide A sugar composed of two simple sugars or monosaccharides; two monosaccharides linked together; simple sugars with two basic units.

dissolve To break into parts or to pass into solution.

distribution Division and classification; deals with those aspects favorable to product sales, including product form, weight and bulk, storage requirements, and storage stability.

disulfide A bond between two sulfur atoms (–S–S–), each of which is also joined to another chemical group; these bonds often tie protein chains together; sometime called disulfide linkages.

double bonds Two bonds between atoms; in food science typically carbon atoms (–C=C–).

draft beer Beer drawn from casks or kegs rather than canned or bottled.

DRI Dietary Reference Intakes.

dried A food product from which most of the water has been removed.

dried corn syrup Corn syrup from which the water has been partially removed.

Dutch-processed A mild alkali treatment of chocolate to change and darken color and improve flavor.

E

effervescence The bubbling-up or fizz in drinks caused by dissolved carbon dioxide gas.

eggs The ova or female reproductive cell of chickens or other birds.

electrical stimulation Brief exposure to high voltage electrical current to improve tenderness of many cuts of the beef carcass; used before slaughter of animals to render unconscious.

electrode An electronic conductor, often a metal plate, used to collect or emit electrons; often used in pH meters and batteries.

electromagnetic energy Energy that has an electric and a magnetic component; for example, microwaves are electromagnetic energy.

electron Chemical properties of an element are determined by the number of electrons in the outermost energy level of an atom; in its elemental state, the number of electrons of an atom equals the number of protons and the atom is electrically neutral. Electrons travel around the nucleus at very high speed.

electron transfer Also called oxidation-reduction reactions; chemical reactions such as rusting and photography.

electronegativity Assumed negative potential when in contact with a dissimilar substance.

element One of a limited number of substances, such as hydrogen and carbon; composed of atoms; listed in the Periodic Table of Elements.

emulsifier A substance that acts as a bridge at the interface between two immiscible liquids and allows the formation of an emulsion; a substance that aids in producing a fine division of fat globules; in ice cream, it also stabilizes the dispersion of air in the foam structure. Eggs contain the natural emulsifier lecithin.

emulsion A system consisting of a liquid dispersed in an immiscible liquid usually in droplets of larger than colloidal size–for example, fat in milk.

encapsulate To enclose in a capsule. Flavoring materials may be combined with substances such as gum acacia or modified starch to provide an encapsulation matrix and then spray-dried. Salt, sodium chloride, is encapsulated with partially hydrogenated vegetable oil in order to keep the salt out of solution so it will not interfere with the swelling of gums such as carrageenan.

endoplasm Inner portion of the cytoplasm of a cell.

endosperm Seed tissue surrounding the embryo, containing food reserves.

energy Ability to do work; in foods measured in term of calories.

enhancers Substances that supplement, increase, or modify the original flavor or nutrient content of a food.

enriched flours Flour with added niacin, thiamin, riboflavin, folic acid, and iron to compensate for some of the nutrients lost during the milling process.

enrober Covers and surrounds each candy center (nuts, nougats, fruit, and so on) with a blanket of chocolate.

entrepreneur One who starts and conducts a business assuming full control and risk.

enzymatic browning Coloring of food caused by enzymes and prevented by blanching a food before drying.

enzymatic reactions Those catalyzed by enzymes, special proteins produced by living cells.

enzyme Organic catalyst produced by living cells that changes the rate of a reaction without being used up in the reaction.

epidemiology The study of causes and control of diseases prevalent in human population group.

essential amino acid One that is required in the diet.

essential oils Concentrated flavoring oils extracted from food substances, such as oil of orange or oil of peppermint.

ester A type of chemical compound that results from combination of an organic acid (–COOH) with an alcohol (–OH) with the removal of one molecule of water.

ethylene A small gaseous molecule (C_2H_4) produced by fruits and vegetables as an initiator of the ripening process.

evaporation The removal of water. Generally, it is removed as a vapor either due to boiling temperatures or at lower temperatures in a vacuum chamber.

eviscerated Removal of the internal organs.

expenditures Expenses of money, time, or energy.

extraction Drawing out, pulling out, or removing.

extratries Chemical interaction of packaging materials with foods.

extrusion Shaping through force.

F

facultative Microorganisms that are both aerobic and anaerobic.

Fahrenheit A thermometer scale in which the freezing point of water is 32°F and the boiling point is 212°F.

famine A great shortage of food.

fats An ester of glycerol and three fatty acids. Fats add richness, tenderness, calories, and flavor to many products.

fatty acid A chemical molecule consisting of carbon and hydrogen atoms bonded in a chainlike structure; combined through its acid group (–COOH) with the alcohol glycerol to form triglycerides.

fermentation Enzymatic decomposition of carbohydrates under anaerobic conditions.

ferrous Iron-containing.

fiber Indigestible substances including cellulose, hemicelluloses, and pectin (all polysaccharides), and also lignin, which is a noncarbohydrate material found particularly in woody parts of a vegetable.

ficin Used as a meat tenderizer.

filtration The process of separating a solid from a liquid by applying a force to move the liquid through a barrier while retaining the solid. The force may be gravity.

fine sugar (fine granulated) Refined sugar product of sugar crystals whose average size and distribution are within the range of normal table sugar. Some producers call this extra-fine granulated sugar.

flavonoid pigments Phenolic compounds related to flavones. They have hydroxyl, methoxyl, or sugar groups substituted for some of the hydrogen atoms of the flavone; a group of plant pigments with similar chemical structures; they include both anthoxanthins, which are white, and anthocyanins, which are red-blue.

flavor A blend of taste, smell, and general touch sensations evoked by the presence of a substance in the mouth.

floc A precipitate that remains suspended in a solution.

flocculation Aggregation into a mass.

foam The dispersion of a gas in a liquid, such as a beaten egg-white mixture.

follow-up letter Letter written immediately after an interview.

fondant A creamy preparation of fine sugar crystals used in after-dinner mints and other candies; usually prepared from a supersaturated solution; also a type of powdered sugar.

food additive A substance, other than usual ingredients, that is added to a food product for a specific purpose, for example, flavoring, preserving, stabilizing, thickening.

Food and Agricultural Organization (FAO) An agency of the United Nations that conducts research, provides technical assistance, conducts education programs, maintains statistics on world food, and publishes reports with the World Health Organization.

Food, Drug, and Cosmetic Act Regulates the labeling for all foods other than meat and poultry.

food infection Illness produced by the presence and growth of pathogenic microorganisms in the gastrointestinal tract; they are often, but not necessarily, present in large numbers.

food intoxication Illness produced by microbial toxin production in a food product that is consumed; the toxin produces the illness.

food labeling Labels have the product name, the manufacturer's name and address, the amount of the product in the package, and the product ingredients.

food safety A judgment of the acceptability of the risk involved in eating a food; if risk is relatively low, a food substance may be considered.

food security When all people, at all times, have physical and economic access to sufficient, safe, and nutritious food to meet their dietary needs and food preferences for an active and healthy life.

food soil Unwanted matter on food-contact surfaces.

foreign aid Large quantities of food, supplies, and money sent to people in need in many foreign countries.

FPC Fish protein concentrate.

freeze-drying A drying process that involves first freezing the product and then placing it in a vacuum chamber, the ice sublimes (goes from solid to vapor phase without going through the liquid phase). The dried food is more flavorful and fresher in appearance because it does not become hot in the drying process.

freezer burn Drying out while stored in a freezer.

freezing mixtures Mixtures of crushed ice and salt that become very cold, below the freezing point of plain water, because of the rapid melting of the ice by the salt and the attempt of the system to reach equilibrium; freezing mixtures are used to freeze ice creams in ice cream freezers.

fructose A sugar sometimes called levulose or fruit sugar; a monosaccharide with the chemical formula $C_6H_{12}O_6$.

fry To cook in fat deep enough to float the food; also called "deep-fat fry" or "French-fry"; to cook in small amount of hot fat or drippings; also called "pan-fry" or "sauté".

fumigant An organic compound that is a gas at fairly low temperatures; it is used for insect and disease control.

G

gastroenteritis Inflammation of the gastrointestinal tract.

gel A colloidal dispersion that shows some rigidity and will, when unmolded, keep the shape of the container in which it had been placed; a semirigid structure at room temperature.

gelatinization Changes that occur in the first stages of heating starch granules in a moist environment; includes swelling of granules as water is absorbed and disruption of the organized granule structure.

gelation The process of gelling.

generation time Time it takes for microorganisms to reproduce.

germ The small structure at the lower end of the kernel; rich in fat, protein, and mineral and contains most of the riboflavin content of the kernel.

germination The sprouting of a seed.

glazing Dipping a fish in cold water and then freezing a layer before dipping the fish again.

gliadin One of the wheat proteins that makes up one portion of gluten, the primary structural component.

globulins Simple proteins that are soluble in dilute salt solutions.

glucoamylase An enzyme that hydrolyzes starch by breaking off one glucose unit at a time, thus producing glucose immediately.

glucose A sugar sometimes called dextrose or blood sugar; a monosaccharide with the chemical formula $C_6H_{12}O_6$. This is the basic building block of starch.

glucose isomerase An enzyme that changes glucose to fructose.

gluten A protein in wheat and a limited number of other cereals that is formed when water is added to flour and, with kneading, gives structure to baked products.

glycerol Colorless liquid with chemical formula $C_3H_8O_3$; used as sweetener and preservative.

glycogen A complex carbohydrate–a polysaccharide–used for carbohydrate storage in the liver and muscles of the body; sometimes called animal starch.

GMP Good Manufacturing Practices guidelines that a company uses to evaluate the design and construction of food processing plants and equipment.

grades Positions in a scale of ranks or qualities.

grain mills Machine designed to grind wheat and other grains to make flour.

granulated sugar White crystalline sugar or sucrose; the sugar referred to in most recipes that call for "sugar."

GRAS The list of food additives that are "Generally Recognized As Safe" by a panel of experts; list is maintained and periodically reevaluated by the FDA.

gravity flow Movement of a liquid pulled by gravity.

green tea Leaf is heated before rolling in order to destroy the enzymes, and the leaf then remains green throughout processing; unfermented tea.

grill To cook by direct heat.

grind To put through a food chopper.

gum Any of several colloidal polysaccharide substances of plant origin that are gelatinous when moist but harden on drying and are salts of complex organic acids.

H

HACCP Hazard Analysis and Critical Control Point; a preventative food safety system.

hard water Water that contains calcium, magnesium, and iron bicarbonates or sulfates.

hard wheat Generally grown in northern climates; especially suited to bread making because of a high level of the gluten-forming wheat protein; a specific genus of wheat.

hazard A source of danger, long- or short-term, such as microbial food poisoning, cancer, birth defects, and so on.

headspace The volume above a liquid or solid in a container.

heat capacity The amount of energy required to raise a unit mass of a material one degree.

heat transfer The process by which energy in the form of heat is exchanged between two bodies.

hepatitis Inflammation of the liver.

hermetically Made airtight by fusion or sealing.

hexose A simple sugar or monosaccharide with six carbon atoms.

HFCS High-Fructose Corn Syrup.

homebrewing The art of making beer at home. In the United States, homebrewing was legalized by President Carter on February 1, 1979, by an act of Congress introduced by Senator Alan Cranston. The Cranston Bill allows a single person to brew up to 100 gallons of beer annually for personal enjoyment and up to 200 gallons in a household of two persons or more aged 18 and older.

homeostasis Tendency of a system or organism to maintain internal stability.

homogenization A process in which whole milk is forced, under pressure, through very small openings, dividing the fat globules into very tiny particles.

homogenize To subdivide particles, usually fat globules, into very small uniform-sized pieces.

hops The dried ripe cones of the female flowers of this plant used in brewing and medicine.

hot-fill *See* **hot-pack**.

hot-pack Filling unsterilized containers with sterilized food that is still hot enough to render the package commercially sterile.

HPLC High-performance liquid chromatography; form of column chromatography in which a liquid medium passes over a solid phase, usually at high pressure.

HTST High temperature, short time; a method of pasteurization.

hue Property of light by which an object is classified in reference to the color spectrum—red, blue, green, or yellow.

humectant A substance that can absorb moisture easily; a substance that retains moisture.

hunger Lack of food; desire or need for food.

HVAC Heating, ventilation, and air-conditioning systems.

hydration capacity The ability of a substance, such as flour, to absorb water.

hydrocolloid A substance with particles of colloidal size that is greatly attracted to water and absorbs it readily; colloidal materials such as vegetable gums, that bind water and have thickening and/or gelling properties; large molecules, such as those that make up vegetable gums, that form colloidal dispersions, hold water, and often serve as thickeners and stabilizers in processed foods.

hydrocooling Cooling with water.

hydrogen bond The relatively weak chemical bond that forms between a hydrogen atom and another atom with a slight negative charge, such as an oxygen or a nitrogen atom; each atom in this case is already covalently bonded to other atoms in the molecule of which it is part.

hydrogenation Addition of hydrogen to oil; a selective process that can be controlled to produce various levels of hardening, from very slight to almost solids.

hydrolysates Products of hydrolysis, the chemical process of decomposition involving the splitting of a bond and the addition of the hydrogen cation and the hydroxide anion of water.

hydrolysis A chemical reaction in which a linkage between subunits of a large molecule is broken; a molecule of water enters the reaction and becomes part of the end products.

hydrolytes Products of hydrolysis.

hydrolyze To break a molecular linkage utilizing a molecule of water; to break chemical linkages, by the addition of water, to yield smaller molecules.

hydrometer A device for measuring specific gravity. The hydrometer is placed in a solution and the relative displacement within the solution is compared to that within pure water.

hydrophobic force or bond Force arising from the tendency of hydrophobic molecules to aggregate in a hydrophilic environment, thus maximizing the number of hydrogen bonds between water molecules, and hence giving the lowest energy.

hydrostatic retort Cans flow continuously; *see* **retort**.

hygroscopic Absorbing or attracting moisture from the air.

hypobaric The pressure is reduced along with temperatures and humidity for cold storage.

I

immersion freezing Intimate contact occurs between the food or package and the refrigerant.

immiscible Describing substances that cannot be mixed or blended.

impeller A rotor for transmitting motion.

Infant Health Formula Act Provides that manufactured formulas contain the known essential nutrients at the correct levels.

inoculation Introducing a microorganism into surroundings suited to its growth.

insoluble Does not readily dissolve in water.

inspection The examining of food products or processes carefully and critically in order to assure proper sanitary practices, labeling, and/or safety for the consumer. Organizing of food products and classifying them according to quality, such as grade A, B, or C; based on defined standards.

instant yeast Instant yeast is a specially processed form of active dry yeast that can be mixed into a dough dry (rather than dissolved) and reduces rising time up to 50 percent.

integrated A system in which the components are interconnected to perform a function.

Integrated pest management (IPM) The control of one or more pests by a broad spectrum of techniques ranging from biological means to pesticides.

interfering agent Hinders formation or reaction.

international unit A standard unit of potency for a vitamin.

inversion The reaction of the hydrolysis of sucrose to yield an equal mixture of glucose and fructose.

invert sugar The result of the hydrolysis of sucrose to yield an equal mixture of glucose and fructose.

invertase The enzyme that catalyzes reaction of the hydrolysis of sucrose to yield an equal mixture of glucose and fructose.

invisible fat Fat that occurs naturally in food products such as meats, dairy products, nuts, and seeds.

iodine value Chemical tests to determine the degree of unsaturation of the fatty acids in a fat; the test is based on the amount of iodine absorbed by a fat on a per 100 grams basis. The higher the iodine value, the greater the degree of unsaturation.

ion An electrically charged (+ or –) atom or group of atoms.

ion-exchange A process that uses specially fabricated porous beads that are chemically modified to exchange one ion for another as a solution is passed through them. Different types of ion-exchange columns exist.

ionic bond A chemical bond formed when a complete transfer of electrons from one atom to another occurs.

ionization Converted into ions.

ionomer Type of plastic material formed by ionic bonds.

irradiation Energy moving through space in invisible waves.

ISM Electromagnetic spectrum frequency bands allocated for industrial, scientific, and medical purposes.

isomerization A molecular change resulting in a molecule containing the same elements in the same proportions but having a slightly different structure and, hence, different properties. For example, in carotenoids, heat causes a change in the position of the double bonds between carbon atoms.

J

juice The fluid expressed from a food product. In sugar it is the juice with dissolved sucrose pressed out of sugar beet root or sugarcane stalk.

Julian date A number–1 through 365–indicating day of the year.

K

ketones Chemical compounds characterized by a –C=0 group.

kilocalorie One kilocalorie is equal to 1,000 small calories; the small calorie is used in chemistry, whereas the kilocalorie is used in nutrition.

kinetic motion The very rapid vibration and movement of tiny molecules or ions dispersed in true solution.

knead The action used to manipulate bread dough that forms the gluten network in dough.

L

labile Unstable.

lagering Brewing by slow fermentation and maturing under refrigeration

lag time The amount of time required for an organism to reach the log growth phase.

lakes Any of various usually bright clear organic pigments composed basically of a soluble dye absorbed on or combined with an inorganic carrier.

laminar Particles of fluid move in parallel or adjacent layers; each layer has a constant velocity but relative to neighboring layers.

landfilling Disposing food-processing wastes in a landfill.

latent heat The heat or energy required to change the state of a substance, that is, from liquid to gas, without changing the temperature of the substance.

leavening Making baked products lighter by helping them rise; yeast, baking powder, and baking soda are common leavening agents.

legume Any of a large family of plants characterized by true pods enclosing seeds; for example, dried beans and peas.

letter of application Sent with a résumé or data sheet when applying for a job.

letter of inquiry Sent to a potential employee requesting possibility of employment.

levulose Is also called fructose or fruit sugar; a monosaccharide with the chemical formula $C_6H_{12}O_6$.

lignin A woody, fibrous, noncarbohydrate material produced in mature plants; component of the fiber complex.

limiting amino acid Required in the diet; the amino acid that is present in the lowest quantity compared to need.

linoleic acid A polyunsaturated fatty acid with 18 carbon atoms and 2 double bonds.

linolenic acid A polyunsaturated fatty acid with 18 carbon atoms and 3 double bonds between carbon atoms; omega-3 fatty acid.

lipase An enzyme that catalyzes the hydrolysis of triglycerides to yield glycerol and fatty acids.

lipids A broad group of fatlike substances with similar properties.

lipolysis Breakdown of lipids.

lipoproteins Proteins combined with lipid or fatty material such as phospholipids.

liquid The ingredient that has flow or viscosity; may be milk, water, juices or other liquids; will dissolve and disperse ingredients.

liquid sugars A finished sugar product sold in the liquid state; a concentrated solution of sugar products and water.

liquor A liquid off the top of a mixture of liquid and solids. In sugar-containing solutions, a high-Brix, solution recovered at various stages of the refining process.

logarithmic Expression of numbers using exponents.

LTLT–Low-Temperature Longer Time pasteurization; heating to 145°F for at least 30 minutes.

lycopene A reddish, fat-soluble pigment of the carotenoid type.

M

macromineral Minerals in the diet that are used in larger amounts; includes calcium, phosphorus, potassium, sodium, chloride, magnesium, and sulfur.

magma A mixture of sugar crystals and liquid that forms a thick slurry.

Maillard reaction A special type of browning reaction involving a combination of proteins and sugars as a first step; may occur in relatively dry foods on long storage as well as in foods heated to high temperatures; the reaction between the amino group of an amino acid and protein and the reducing sugar to cause a brown color.

malnutrition A person eats but does not receive the amounts of nutrients needed to keep the body healthy.

malt Processed barley steeped in water, germinated on malting floors or in germination boxes or drums and later dried in kilns for the purpose of converting the insoluble starch in barley to the soluble substances and sugars in malt.

maltase An enzyme that hydrolyzes maltose to glucose.

maltodextrin A mixture of small molecules resulting from starch hydrolysis.

maltose A double sugar or disaccharide made up of two glucose units.

manufacturing Converts raw agricultural products to more refined or finished products.

marbling The distribution of fat throughout the muscles of meat animals.

marinading Soaking in a prepared liquid (oil and acid mixture) for a time; tenderizes and seasons.

marketing The selling of foods, which involves wholesale, retail, institutions, restaurants, and the consumer.

mashing One of the steps of brewing; the infusion of malt, water, and crushed cereal grains at temperatures that encourage the complete conversion of the cereal starch into sugars.

maturing agent A substance that brings about some oxidative changes in white flour and improves its baking properties.

Meat Inspection Act Federal act authorized in 1906 and administered by the Food Safety and Inspection Service (FSIS) of the Department of Agriculture (USDA).

mechanically separated Separation of bone from meat using automated equipment.

melt To liquefy by heat.

melting point The temperature at which a solid fat becomes a liquid oil.

mesophilic bacteria Bacteria that grow best at moderate temperatures.

metabolic Having to do with any of the chemical changes that occur in living cell.

metabolism A general term used to refer to all of the chemical reactions that occur in a living system. Metabolism can be divided into two processes: (1) anabolism, or reactions involving the synthesis of compounds, and (2) catabolism, or reactions involving the breakdown of compounds.

microfiltration A membrane process that filters out or separates particles of very small size (0.02 to 2.00 microns), including starch, emulsified oils, and bacteria.

microminerals Minerals in the diet that are used in smaller amounts; includes chromium, cobalt, copper, fluorine, iodine, iron, manganese, molybdenum, nickel, selenium, silicon, tin, vanadium, and zinc.

micron One millionth of a meter.

microprocessor or chip Small semiconductor circuit containing numerous electronic devices, such as transistors, that perform complex calculations or operations. Used mainly in computers or/and automated instruments. The entire algorithm for a microwave oven controller can be incorporated into one single chip.

microsensors Miniaturized, and possibly biological, indicators capable of accurately measuring the physiological state of plants, indicating temperature-abuse for refrigerated foods, or monitoring shelf-life of food.

microwaves Electromagnetic energy that has an electric and a magnetic component.

middlings The inner portion of the kernel.

mill (1) To beat; best done with a rotary beater; for example, milk dishes, such as cocoa, are milled to remove scum formed during heating; (2) refers to grinding of grain to produce flour.

mill starch A starch-gluten suspension in the process of germ separation.

millet A small yellow seed that lends texture and flavor to breads.

milling The separating of the bran covering, germ, and endosperm to the extent desired.

mince To cut or chop into very small pieces.

mitochondria Microscopic sausage-shaped bodies in the cell cytoplasm that contain the enzymes necessary for energy metabolism.

modified starches Natural starches that have been treated chemically to create some specific change in chemical structure, such as linking parts of the molecules together or adding some new chemical groups to the molecules; the chemical charges create new physical properties that improve starches in food preparation.

molasses A sugar product extracted directly from sorghum or a by-product of the sugar extraction; a relative impure sugar syrup.

molecule Molecules are the smallest identifiable unit into which a pure substance can be divided and still retain the composition and chemical properties of that substance.

mollusks An invertebrate with a calcareous shell that is one or more pieces enclosing a soft, unsegmented body.

monoglyceride Glycerol combined with only one fatty acid.

monosaccharide A simple sugar unit, such as glucose, fructose, and galactose; a simple sugar with a single basic unit.

monounsaturated fat Lacking a hydrogen bond on the carbon chain.

mucoprotein A complex or conjugated protein containing a carbohydrate substance combined with a protein.

mycotoxins Toxins produced by molds.

myoglobin The name of the protein that is the primary color pigment of meat.

N

naan An East Indian flat bread, baked in an oven and leavened with wild yeast; also nan.

neutron Particle found in atomic nuclei having no charge.

niacin One of the B vitamins.

nib The roasted ground kernel of the cacao bean.

nitrogen base A molecule with a nitrogen-containing chemical group that makes the molecule alkaline.

nitrogen packed Packed in a nitrogen gas filled bag to avoid the effect that oxygen has on the product.

nonclimacteric Fruits not producing ethylene gas during ripening.

noncrystalline Types of sugar candy that includes hard candies, brittles, chewy candies, and gummy candies.

nonnutritive Used to describe a sweet product that contains little, if any, nutritive value; also called artificial sweetener.

nonreducing end In a polysaccharide or oligosaccharide chain, the end(s) that does not act as a reducing agent. Branched chains have more than one nonreducing end.

nonvolatile Lacking the ability to readily change to a vapor or to evaporate; not able to vaporize or form a gas at ordinary temperatures.

NPU Net protein utilization.

nutrition labeling Expression of the nutrient and caloric content of food products.

Nutrition Labeling and Education Act Protects consumers against partial truths, mixed messages, and fraud regarding nutrition information.

nutritionally enhanced Processed foods with added nutrient or nutrients such as vitamin C, B vitamins, iodine, iron, and so forth.

nutritive sweetener A sweetener that has a caloric value.

O

objective Evaluation having to do with a known object as distinguished from existing in the mind; in food science, measurement of the characteristics of food with a laboratory instrument such as pH meter to indicate acidity or a viscometer to measure viscosity or consistency.

obligative Restricted to a particular condition such as organisms that can only survive in the absence of oxygen.

offal All parts of a carcass that are considered by-products, such as feet, tongue, hide, and so on.

ohmic heating Heating of a food product by using an alternating current flowing between two electrodes.

oil Fats that are liquid at room temperature.

olfactory Having to do with the sense of smell.

oligosaccharide The general term for sugars composed of a few–often between three and ten–simple sugars or monosaccharides; a carbohydrate containing 2 to 20 sugar residues (the upper limit is not well-defined); intermediate-size molecules containing approximately ten or fewer basic units.

oolong Type of tea that begins like black tea; when the leaf is fired or dried, a coppery color forms around the edge of the leaf while the center remains green; the oolong flavor is fruity and pungent.

organic acid An acid containing carbon atoms; for example, citric acid and acetic acid, generally weak acids characterized by a carboxyl (–COOH) group.

organic compound A compound that has carbon included into the chemical formula; can be natural or human-made material.

organic foods Foods grown and/or produced under conditions that supposedly replenish and maintain soil fertility, use only nationally approved materials in their production, and have verifiable records of the production system.

organoleptic Perceived by any sense organ.

osmosis Movement of water through a semipermeable membrane into a solution where the solvent concentration is higher thus equalizing the concentration on either side of the membrane.

osmotic pressure The force that a dissolved substance exerts on a semipermeable membrane.

ovalbumin A major protein found in egg white.

ovary Part of the seed-bearing organ of a flower; an enlarged hollow part containing ovules that develop into seeds.

overlapping operations Food processing that includes a combination of unit operations to achieve the total process.

oxalic acid An organic acid that forms an insoluble salt with calcium.

oxidase Enzymes that catalyze oxidation reactions.

oxidation A chemical change that involves the addition of oxygen; for example, polyphenols are oxidized to produce different flavor and color compounds; a chemical reaction in which oxygen is added; addition of oxygen to carotenoid pigments lightens the color; chemical reactions in which oxygen is added or hydrogen is removed or electrons are lost; gain in oxygen or loss of electrons.

oxidation-reduction reactions Loss of electrons from an atom produces a positive oxidation state, while the gain of electrons results in negative oxidation or reduction.

P

pan broil To cook uncovered on hot metal such as a grill or frying pan; the utensil may be oiled just enough to prevent sticking.

pan fry To cook in a small amount of fat.

papain A vegetable enzyme used to tenderize meat.

parboil To boil until partially cooked. Foods with strong or salt flavor are often parboiled, as are tough foods that are to be roasted or cooked in hot fat.

pare Peel to remove outer cover.

parenchyma Thin-walled, highly vacuolated cells, forming the ground tissue of plants; often photosynthetic; relatively unspecialized, sometimes capable of cell division.

pasteurization The process of heating a food to a specified temperature for a specified time to destroy pathogenic organisms.

pasteurize To treat with mold heat to destroy pathogens–but not all microorganisms–present in a food product.

pathogenic Capable of causing disease.

pectic enzymes Enzymes such as pectinase that hydrolyze the large pectin molecules.

pectin A gel-forming polysaccharide (polygalacturonic acid) found in plant tissue; a complex carbohydrate (polysaccharide) composed of galacturonic acid subunits, partially esterified with methyl alcohol and capable of forming a gel.

pectin esterase An enzyme that catalyzes the hydrolysis of a methyl ester group from the large pectin molecule, producing pectic acid. Pectic acid tends to form insoluble salts with such ions as calcium (Ca^{2+}). These insoluble salts cause the cloud in orange juice to become destablized and settle.

pectinase An enzyme that hydrolyzes the linkages that hold the small building blocks of galacturonic acid together in the pectic substances, producing smaller molecules.

penetration (1) Refers to distance heat and/or electromagnetic waves will "penetrate" a product; (2) a marketing term indicating the number or percentage of a product sold in a given market.

pentose A simple sugar or monosaccharide with five carbon atoms.

peptide A variable number of amino acids joined together.

peptide bond Formed by the condensation of the amino group ($-NH_2$) of one amino acid with the acid group ($-COOH$) of another amino acid resulting in the loss of water.

peptide linkage Linkage between two amino acids that connects the amino group of one and the acid (carboxyl) group of the other.

PER Protein efficiency ratio.

per capita Per person.

permanently hard water Contains calcium,

magnesium, and iron sulfates that do not precipitate when boiled.

permeate To penetrate through the pores.

peroxide value Indicates the degree of oxidation that has taken place in a fat or oil. The test is based on the amount of peroxides that form at the site of double bonds. These peroxides release iodine from potassium iodide when it is added to the system.

PET Polyethylene Terephthalate; a type of plastic that is easy to recycle.

pH A scale of 1 to 14, indicating the degree of acidity or alkalinity; 1 being most acid, 7 neutral, and 14 most alkaline.

phenolic compound An organic compound that includes in its chemical structure an unsaturated ring with –OH groups on it; polyphenols have more than one –OH group; organic compounds that include in their chemical structure an unsaturated ring with –OH groups on it; these compounds are easily oxidized, producing a brownish discoloration.

phenols Organic compounds that include, as part of their chemical structures, an unsaturated ring with an –OH group on it.

phospholipid A type of lipid characterized chemically by glycerol combined with two fatty acids; phosphoric acid, and a nitrogen-containing base, for example, lecithin.

phytochemical Chemical from plants.

pigment Any biological substances that produce color in tissues.

pizza A round savory tart made with a crisp yeast dough, covered with tomato sauce, mozzarella cheese, and a variety of other ingredients such as meat, seafood, vegetables, fruit, and other condiments.

plant exudates Materials that ooze out of certain plants; some that ooze from certain tree trunks and branches are gums.

plasma (1) A component of blood; (2) a gas whose atoms or molecules have been ionized and is characterized by a colored glow. "Plasma" is often seen in a neon bulb or under low pressure in microwave freeze-drying experiments.

plasticity The ability to be molded or shaped; in plastic fats, both solid crystals and liquid oil are present.

pneumatic trailer A type of bulk carrier that allows unloading of the product by air pressure.

poach To cook in a hot liquid, carefully handling the food to retain its form.

polar Chemical molecules that have electric charges (positive or negative) and tend to be soluble in water; having two opposite natures, such as both positive and negative charges.

polarimetry The amount of rotation experienced by a beam of polarized light as it passes through a solution. The rotation reflects the type and amount of solution. The amount of sucrose can be measured by a polarimeter and is called saccharimetry.

polarized light Light that vibrates in one plane.

polarized molecules Molecules that have both positive (+) and negative (–) charges on them, creating two poles.

polymer A giant molecule formed from smaller molecules that are chemically linked together; a large molecule formed by linking together many smaller molecules of a similar kind; molecules of relatively high molecular weight that are composed of many small molecules acting as building blocks.

polymerize To form large molecules by combining smaller chemical units.

polyphenol An organic compound with more than one –OH group attached to the unsaturated ring of carbon atoms; some produce bitterness in coffee and tea.

polyphenoloxidase A mixed function oxidase that catalyzes enzymatic browning.

polysaccharide A complex carbohydrate made up of many simple sugar (monosaccharide) units linked together; in the case of starch, the simple sugars are all glucose; carbohydrate polymer consisting of at least 20 monosaccharides or monosaccharide derivatives; complex carbohydrates with many basic units (up to thousands).

polyunsaturated fatty acid Fatty acid with two or more double bonds between carbon atoms; for example, linoleic acid with two double bonds.

post-consumer recycle The collection, separation, and purification of the consumer's disposed food packages.

postmortem After death.
potable Drinkable.
potassium bromate An oxidizing substance often added to bread dough to strengthen the gluten of high-protein flour.
powdered sugar A sugar product produced by grinding a mixture of granulated sugar and corn starch.
PPO Polyphenoloxidase.
precipitate To become insoluble and separate out of a solution or dispersion.
precursor Substance that "comes before"; precursor of vitamin A is a substance out of which the body cells can make vitamin A; in flavor study, it is a compound that is nonflavorful but can be changed, usually by heat or enzymes, into a flavorful substance.
pressed sugar A sugar product formed by molding and pressing damp granulated sugar into shapes such as cubes and drying the product.
pretzel A yeast dough that is typically rolled into a long rope and often knotted. They can be crisp or soft and chewy.
primary Containers that come in direct contact with the food.
printability Ability to take on ink for printing.
processed meats Combined with other spices and additives to form a new product such as hot dogs, sausage, bologna, and jerky.
production In the food industry, production includes such industries as farming, ranching, orchard management, fishing, and aquaculture.
proofing The last rising of bread dough after it is molded into a loaf and placed in the baking pan.
proofing yeast Dissolving yeast in warm liquid about / cup water and about 1 teaspoon sugar and setting it aside for 5 to 10 minutes until it develops foam on top.
propellant Compressed inert gas that dispenses the contents of a container.
protease An enzyme that breaks down or digests proteins.
protein Large molecules of long chains of amino acids.
protein efficiency ratio (PER) A measure of protein quality assessed by determining the extent of weight gain in experimental animals when fed the test item.
protein hydrolysate The resulting mixture when a protein is broken down or hydrolyzed by an enzyme or other means to smaller units called peptides and amino acids.
protein quality Refers to amino acid content of a protein.
proteinase An enzyme that hydrolyzes protein to smaller fragments, eventually producing amino acids.
proteins Long chains of amino acids.
proton Positively charged particle found in all atomic nuclei.
protoplast That part of the plant cell lying within the cell wall; the plasma membrane and all that lies within it. This term is often used to describe plant cells from which the cell wall has been removed.
protozoa One-celled animals.
proximate analysis Approximate composition of food products consisting of its proportions of water, carbohydrate, protein, fat, and ash.
PSI (pounds per square inch) A measure of pressure referenced to atmospheric pressure.
psychrophilic Thriving at a relatively low temperature.
psychrotrophic Bacteria that grow best at cold temperatures (cold-loving bacteria).
pungency A sharp, biting quality.
purity The degree of singularity of a constituent in a product, generally expressed as a percentage. For example, in sugar it indicates the degree of sucrose and other extraneous products.

Q

quality assurance (QA) Continual monitoring of incoming raw and finished products to ensure compliance with compositional standards, microbiological standards, and various government regulations; requires many diverse technical and analytical skills.
queso The Spanish word for cheese.
quick bread Any bread product leavened with a chemical leavener (baking soda and an acid, such as buttermilk, or baking powder) rather than yeast; includes muffins, biscuits, popovers, pancakes, and the like.

quinones Cyclic, conjugated di-ketones that are highly reactive.

R

radiation The transfer from electromagnetic radiation of a body due to the vibration of its molecules.

radioisotope A chemical element that spontaneously emits radiation (such as electrons or alpha particles) and that exhibits the same atomic number and nearly identical chemical behavior but with differing atomic mass and different physical properties.

rancidity A special type of spoilage in fats that involves oxidation of unsaturated fatty acids; the deterioration of fats, usually by an oxidation process, resulting in objectionable flavors and odors.

raw sugar The intermediate crystalline product of cane sugar factories resulting from the evaporation of water from sugarcane stalk juice. True raw sugar cannot be sold in the United States. It contains too many impurities. These are washed off, and it is sometimes called "turbindado sugar" and possibly labeled raw sugar.

RDA Recommended Daily Allowances.

reciprocating Moving forward and backward alternately.

recycle To obtain, treat, or process used materials for reuse.

reducing end The end of a polysaccharide or oligosaccharide chain that contains a free glycosidic hydroxyl, and thus acts as a reducing agent; only one reducing end per molecule.

reducing sugar A sugar with a free-aldehyde or ketone group that has the ability to chemically "reduce" other chemical compounds and thus become oxidized itself; glucose, fructose, maltose, and lactose, but not sucrose, are reducing sugars.

reduction reactions Chemical reactions in which there is a gain in hydrogen or in electrons.

refining The separating of corn seed and soybeans, respectively, into component parts and converting these to high-value products.

reflectance color Observed color.

refractive index The bending, refraction, and diffraction of a beam of light of defined wavelength when shown through a solution. The index is the bending of the solution compared to the bending through pure water.

refractometer An instrument for measuring the refractive index of a solution.

refrigeration Holding a food product at temperatures that range from 40° to 45°F (4.5° to 7°C).

rehydrated Having water added to replace that lost during drying.

rendering Freeing fat from connective tissue by means of heat.

rennet Used for enzyme coagulation of milk.

respiration A metabolic process by which cells consume oxygen and give off carbon dioxide; continues after harvest.

résumé Written information for a prospective employer that may include any of the following: career objectives, work experience, education background, accomplishments, awards, or skills; also called a data sheet.

retentate A concentrated fluid.

retinol equivalent Vitamin A activity.

retort Container in which a product is heated.

retorted Canned or sterilized by processing in a pressurized vessel in order to raise the temperature to 250°F (121°C) to kill microorganisms.

retrogradation The process in which starch molecules, particularly the amylose fraction, reassociate or bond together in an ordered structure after disruption by gelatinization.

reverse osmosis A process of "dewatering" whereby ions and small molecules do not pass through a membrane but water does pass through.

rheology The study of the science of deformation of matter.

riboflavin One of the B vitamins.

rigor mortis One of the changes after slaughter, the contraction and stiffening of muscle.

ripening To age or cure to develop characteristic flavor, odor, body, texture, and color.

rise A stage in the process of making yeast breads where the dough is set in a warm, draft-free place for a period of time (usually an hour or so) while the yeast ferments some of the sugars in the dough, forming carbon dioxide. This causes the bread to increase in size. A rising period usually lasts until the dough doubles in size.

risk A measure of the probability and severity of harm to human health.

roast To bake, applied to certain foods such as meats.

roe Fish eggs.

ropey Capable of being drawn into a thread or tending to adhere in stringy masses.

roux Thickening agent made by heating a blend of fat and flour.

S

saccharimeter A polarimeter to measure the optical rotation of sucrose solutions. This is calibrated to read the rotation in degrees S or degrees Z.

salmonella A bacterium that may cause food poisoning.

salmonellosis Illness produced by ingestion of salmonella organisms.

salt A chemical compound derived from an acid by replacement of the hydrogen (H+), wholly or in part, with a metal or an electrically positive ion, for example, sodium citrate; sodium chloride crystals used as a flavoring.

sanitization Removal of or neutralization of elements that may be injurious to health.

saponification value Indicates the average molecular weight of the fatty acids in a fat. This value represents the number of milligrams of potassium hydroxide needed to saponify (convert to soap) one gram of fat. The saponification value increases and decreases inversely (opposite of) the average molecular weight.

saturated fatty acid A fatty acid with no double bonds between carbon atoms; it holds all of the hydrogen that can be attached to the carbon atoms.

saturated solution A solution containing all the solute that it can dissolve at that temperature.

saturation Indicates the level at which a solute is dissolved into a solution at a specific temperature.

sauté To cook in a small amount of fat.

scald To heat a liquid, usually milk, until bubbles appear around the edge, approximately 198° to 203°F (92° to 95°C); to blanch, as when preparing vegetables for freezing.

sear To brown the surface quickly by intense heat in order to develop color and flavor and to improve appearance; usually applied to meat.

secondary A container that holds several primary containers together.

secondary amines Derivatives of ammonia (NH_3) in which two of the hydrogen atoms are replaced by other carbon-containing chemical groups (R–NH–R).

sediment The solid material that settles out from a liquidlike dispersion.

semipermeable Membrane allowing passage of water and solute selectively to equalize solute concentration on either side of the membrane.

senescence The state of growing old or aging.

separator A machine that separates the cream and skim portions of the milk.

septicemia The presence of pathogenic microorganisms in the blood.

sequestrant A substance that binds or isolates other substances; for example, some molecules can tie up trace amounts of minerals that may have unwanted effects in a food product.

shelf life The time required for a food product to reach an unacceptable quality.

shorts A by-product of flour milling consisting of small particles of bran, germ, aleurone layer, and coarse flour.

shucked Removed from the shell.

simmer To cook in liquid below the boiling point, a liquid is simmering when bubbles form slowly and break just below the surface, about 185°F (85°C).

single-celled protein (SCP) Refers to protein obtained from single-celled organisms such as yeast, bacteria, and algae grown on specifically prepared media.

slurry A thin mixture of water and a fine insoluble material such as flour.

smoke point The temperature at which smoke comes continuously from the surface of a fat heated under standardized conditions.

smoking A preservation method for meat that inhibits microbial growth, protects fat from rancidity, contributes to the characteristic color, and creates unique flavors in processed meats.

soft wheat A general term for varieties of wheat that contain relatively small amounts of gluten.

solids-not-fat Includes the carbohydrates, lactose, protein, and minerals of milk.

soluble Able to be dissolved or liquefied.

solute Dissolved or dispersed substance.

solution (1) A mixture resulting from the dispersion of small molecules or ions (called the solute) in a liquid such as water (called the solvent); (2) the resulting mixture of a solute dissolved in a solvent.

solvent A liquid in which other substances may be dissolved.

sorbitan monostearate An emulsifier used in yeast manufacturing to aid in the drying process. Sorbitan monostearate protects the yeast from excess drying and also aids in the rehydration of the yeast cells.

sorbitol A sugar alcohol similar to glucose in chemical structure but with an alcohol group (–C–OH) replacing the aldehyde group (H–C=O) of glucose. It occurs naturally in fruit and berries. It is slowly absorbed by the body, and it is a caloric sweetener.

sous vide Literally means "under vacuum."

SPC Standard plate count.

specialty sugars Varied and unique from granulated sugar; may be "engineered" for a special need.

species A group of taxonomic classification consisting of organisms that can breed together.

specific gravity The weight of a volume of material divided by the weight of an equal volume of water.

specific heat The number of calories needed to raise the temperature of 1 gram of a given substance 1 degree C; the specific heat of water has been set at 1.0; fats and sugars have lower specific heats, thus requiring less heat than does water to raise their temperature an equal number of degrees.

spectrophotometry A method of chemical analysis based on the transmission or absorption of light.

spores A microorganism in a dormant state or a one-celled reproductive organ of a fungus. Spores may be activated by appropriate environmental conditions.

stability Resistance to chemical change, disintegration, or degradation.

stabilizer A water-holding substance, such as a vegetable gum, that interferes with ice crystal formation and contributes to a smooth texture in frozen desserts.

standard plate count (SPC) A test that determines the presence of microbiological organisms in a food.

standardized To have a rule or principle used for the basis of judgment applied.

standards Set up and established by authority as a rule for the measure of quantity, weight, extent, value, or quality. Set by the USDA to specifically describe a food; to be labeled as such, a food must meet these specifications.

starch A carbohydrate polymer made up of many units of glucose.

starch granule Composed of millions of starch molecules laid down in a very organized manner; the shape of the granule is typical for each plant species. Starch molecules are organized into tight little bundles, called granules. As they are stored in the seeds or roots of plants, the granules with characteristic shapes and sizes can be seen under the microscope.

starter culture A concentrated number of the organisms desired to start the fermentation process.

steam To cook in steam, with or without pressure.

steeping Extracting flavor, color, or other qualities from a substance by allowing it to stand in liquid below the boiling point; soaking.

sterilization The process of heating a material sufficiently to destroy essentially all microorganisms.

sterilize To destroy essentially all microorganisms.

sterol A type of fat or lipid molecule with a complex chemical structure; for example, cholesterol.

stew To simmer in a small to moderate quantity of liquid.

still retort See **retort**.

stir To mix with circular motion, to blend food materials, or to obtain a uniform consistency as in sauces.

stoma (plural **stomata**) Small opening in the epidermis of leaves and some stems that opens to permit gas exchange and closes in conditions of water stress; flanked by stomatal guard cells, which regulate the opening of the stoma.

storage tissue Located in the cytoplasm in leucoplasts; dominant in roots, tubers, bulbs, and seeds.

straight grade Flours that should contain all the flour streams resulting from the milling process; most patent flours on the market include about 85 percent of the straight flour.

strain A subgroup of a species in taxonomic classification that has a common ancestor with distinctive characteristics but is not different enough from other organisms to be a separate species.

stunting Reduced growth of children who do not have adequate nutrition.

subcutaneous Under the skin.

sublimation Water goes from a solid to a gas without passing through the liquid phase.

substrate Substance acted upon by an enzyme.

succulent Having juicy tissues; not dry.

sugar Sucrose, the disaccharide with the chemical formula $C_{12}H_{22}O_{11}$.

sugar refinery The factory where sugar is extracted from either sugar beet or sugarcane.

sulfhydryl compound A chemical substance that contains an –SH group.

surface active agents An emulsifier; these improve the uniformity of a food–the fineness of grain–the smoothness and body of foods such as bakery goods, ice creams, and confectionery products.

surface area Measure of exposed surface; measured in square inches, meters, centimeters, millimeters and so on.

surface tension The tension or force at the surface of a liquid that produces a resistance to spreading or dispersing; due to the attraction of the liquid molecules for each other.

surimi A minced fish flesh washed to remove solubles including pigments (color) and flavors leaving an odorless, flavorless, high-protein product.

sustainable Enduring without giving away or yielding.

sweeteners Any food that adds a sweet flavor to foods.

symbiotic An interdependent or mutually beneficially relationship.

synchrometer A metering device that measures syrup and water in fixed proportion to the carbonator.

syneresis The oozing of liquid from a rigid gel; sometimes called weeping.

synergism An interaction in which the effect of the mixture is greater than the effect of the sum of component part; two or more factors acting cooperatively, so that their combined effects when acting together exceed the sum of their effects when each acts alone.

synthetic Compound produced by chemically combining two or more simple compounds or elements in the laboratory.

syrup A viscous, concentrated sugar solution that occurs due to evaporation of liquid.

T

tactile Having to do with the sense of touch.

tannins Term applied to phenolic compounds that contribute to both astringency and enzymatic browning.

tariffs A schedule of duties imposed by a government on imports or exports.

taste bud A group of cells, including taste cells, supporting cells, and nerve fibers.

taste receptor Tiny ends of the taste cells that come in contact with the substance being tasted.

taste sensations Perceived through stimulation of taste buds on the tongue; primary tastes are sweet, salty, sour, and bitter.

teff flour The smallest of grains and therefore has a high ratio of bran and germ.

tempering Holding a substance at a specified temperature to give it the desired consistency; frozen foods may be tempered by holding them just below 32°F (0°C).

temporarily hard water Contains calcium, magnesium, and iron bicarbonates that precipitate when boiled.
tensile strength The ability to resist stretching; the strength when under tension.
tertiary A container that groups several secondary holders together into shipping units.
texture (1) Arrangement of the parts of a material showing the structure; for example, the texture of baked flour products such as a slice of bread may be fine and even or coarse and open; or the texture of a cream sauce may be smooth or lumpy; (2) the arrangement of the particles or constituent parts of a material that gives it its characteristic structure.
textured protein Products that are usually at least 50 percent protein and contain the eight essential amino acids and the vitamins and minerals found in meats; soybean protein is most commonly used; other plant proteins–wheat gluten, yeast protein, and most other edible proteins–can be used singly or in combination.
thermization To heat.
thermophilic Prefering hot temperatures.
thermotrophic Can tolerate high temperature.
thiamin One of the B vitamins.
total sugars A measurement of the total amount of saccharides.
toxins Poisonous to humans.
TQM Total quality management.
traditional active dry yeast Active dry yeast should be dissolved before using for best results.
tramp material Soil and extraneous material.
trans Configuration in which the hydrogen atoms are on opposite sides of the double bond, particularly with unsaturated fatty acids.
translucency Partially transparent.
trends General direction of a market.
Trichinella spiralis A tiny parasite that may be present in some fresh pork and, if not destroyed by cooking, causes a disease called trichinosis.
triglycerides Neutral fat molecule made up of three fatty acids joined to one glycerol molecule through a special chemical linkage called an ester. A type of lipid consisting chemically of one molecule of glycerol combined with three fatty acids.
tuber A short, thickened, fleshy part of an underground stem, such as a potato; new plants develop from the buds or eyes; an enlarged underground stem (for example, the potato).
turbidity Cloudiness in a fluid; the opposite of translucence.
turbinado A semirefined light brown crystalline sugar. It is a steam-cleaned, partially refined sugar. In the United States it is sold as raw sugar.

U

UHT Ultra high temperature pasteurizing.
ultrafiltration A membrane process that filters out or separates particles of extremely small size (0.02 to 0.2 microns), including proteins, gums, glucose, and pigments; filtration through an extremely fine filter.
ultrapasteurization Heating milk to 138°F or higher for 2 seconds, followed by rapid cooling to 45°F or lower.
unbleached flour White flour without bleaching or aging agents added to hasten the aging process.
undernutrition A person does not get enough food to have a healthy life.
underproof Under risen or fermented.
underweight Means of defining malnutrition; below weight for age in children.
unleavened Bread or dough product containing no yeast or chemical (baking soda, baking powder) leavener.
unsaturated fatty acid A general term used to refer to any fatty acid with one or more double bonds between carbon atoms; capable of binding more hydrogen at these points of unsaturation.
U.S. Fancy Grade designations for vegetables; more uniform in shape and have fewer defects.
U.S. Grade A, B, C Quality grades for fruits and vegetables.
U.S. No. 1, 2, 3 Quality grades for fruits, vegetables, and pigs.
USP Abbreviation for United States Pharma-

copoeia. Standards for various products including sugar are available from this organization.

V

vacuum–A space partially or nearly completely exhausted of air.

vacuum cooling Cooling under conditions of no air (atmosphere).

vacuum drying Drying a product in a vacuum chamber in which water vaporizes at a lower temperature than at atmospheric pressure.

vacuum evaporation Removal of water under conditions of no air (atmosphere).

vacuum packed Air-free and airtight in a heavy foil package to minimize exposure to oxygen.

value Magnitude, relative worth; estimated or assigned worth.

van der Waals bond Very weak bonds formed between nonpolar molecules or nonpolar parts of a molecule.

vapor pressure The pressure produced over the surface of a liquid as a result of the change of some of the liquid molecules into a gaseous state and their escape from the body of the liquid.

vascular Includes the xylem and phloem of plants.

veal Describes a young calf or meat from a calf 1 to 4 months old.

vegan A person who eats only food of plant origin.

vegetable fruit Botanically, a fruit is the ovary and surrounding tissues, including the seeds, of a plant; fruit is the fruit part of a plant that is not sweet and is usually served with the main course of a meal.

vegetable gums Polysaccharide substances that are derived from plants, including seaweed and various shrubs or trees; have the ability to hold water, and often act as thickeners, stabilizers, or gelling agents in various food products; for example, algin, carrageenan, and gum arabic.

vinegar A sour liquid consisting of dilute acetic acid obtained from the fermentation of wine, beer, cider, or similar products.

vinification Conversion of fruit juices, like grape juice, into wine by fermentation.

viscosity Resistance to flow; increase in thickness or consistency.

visible fat (1) Refined fats and oils used in food preparation, including edible oils, margarine, butter, lard, and shortenings; (2) fats seen and easily trimmed from meat.

vitamin fortified Vitamins added to processed foods.

vitamins Vitamins are chemical compounds in our food that are needed in very small amounts (in milligrams and micrograms) to regulate the chemical reactions in our bodies. They are fat-soluble and water-soluble.

vitelline membrane The covering of the yolk of an egg.

volatile Evaporates readily.

W

warmed-over flavor Describes the rapid onset of lipid oxidation that occurs in cooked meats during refrigerated storage; oxidized flavors are detectable after only 48 hours.

wastewater Water left over after being used in food processing.

wasting Gradually reducing the fullness and strength of the body.

water activity See **Aw**.

waxy Starch granules devoid of amylose.

weeping Starch gives up water while cooling and during storage.

wet scrubber Process to remove pollutants.

whey Liquid by-product of cheese making.

whip To beat rapidly to produce expansion due to incorporation of air; applied to cream, eggs, and gelatin dishes.

white sauce A starch-thickened sauce made from fat, flour, liquid, and seasonings.

whole wheat Wheat flour milled using the entire wheat berry (germ, endosperm, bran).

winterization Oils go through a process so that they will not become cloudy when chilled.

wort The clear filtrate from enzymatic action on the malt during the mashing process of beer making.

X

xanthan gum A microbial produced from the fermentation of corn sugar; used as a thickener, emulsifier and stabilizer in foods such as dairy products and salad dressings.

Y

yeast A single-celled fungus in the species *Saccharomyces cerevisiae* that ferments sugars; by-products of fermentation are principally carbon dioxide and alcohol. Carbon dioxide raises breads.

yeast bread Any bread whose primary leavening action results from the fermentation of sugar by yeast.

yeast fermentation A process in which enzymes produced by the yeast break down sugars to carbon dioxide and alcohol, and also produce some flavor substances.

yield grade Classifies carcasses on the basis of the proportion of usable meat to bone and fat, and are used in conjunction with quality grades to determine the monetary value of a carcass.

Z

Z value The temperature required to decrease the time necessary to obtain a one log reduction in cell numbers to one tenth of the original value.

zest (1) The perfumy outermost skin layer of citrus fruit (usually oranges or lemons); removed with the aid of a paring knife or vegetable peeler. Only the colored portion of the skin (and not the white pith) is considered the zest. (2) Also a term used to mean "spice."

zinfandel A red wine grape.

zuppa The Italian word for "soup."

zwieback Bread baked, cut into slices, and then returned to the oven until very crisp and dry.

zymurgy The science or study of fermentation.

Index

Please note: Page numbers in *italics* refer to figures and page numbers in **boldface** refer to tables.

C

T